Urban Agroecology

Advances in Agroecology

Series Editors:
Clive A. Edwards, The Ohio State University, Columbus, Ohio
Stephen R. Gliessman, University of California, Santa Cruz, California

Landscape Ecology in Agroecosystems Management, *edited by Lech Ryszkowski*

Tropical Agroecosystems, *edited by John H. Vandermeer*

Soil Tillage in Agroecosystems, *edited by Adel El Titi*

Multi-Scale Integrated Analysis of Agroecosystems, *by Mario Giampietro*

Soil Organic Matter in Sustainable Agriculture, *edited by Fred Magdoff and Ray R. Weil*

Agroecosystems in a Changing Climate, *edited by Paul C.D. Newton, R. Andrew Carran, Grant R. Edwards, and Pascal A. Niklaus*

Integrated Assessment of Health and Sustainability of Agroecosystems, *by Thomas Gitau, Margaret W. Gitau, and David Waltner-Toews*

Sustainable Agroecosystem Management: Integrating Ecology, Economics and Society, *edited by Patrick J. Bohlen and Gar House*

The Conversion to Sustainable Agriculture: *Principles, Processes, and Practices, edited by Stephen R. Gliessman and Martha Rosemeyer*

Sustainable Agriculture and New Biotechnologies, *edited by Noureddine Benkeblia*

Global Economic and Environmental Aspects of Biofuels, *edited by David Pimentel*

Microbial Ecology in Sustainable Agroecosystems, *edited by Tanya Cheeke, David C. Coleman, and Diana H. Wall*

Land Use Intensification: *Effects on Agriculture, Biodiversity, and Ecological Processes, edited by David Lindenmayer, Saul Cunningham, and Andrew Young*

Agroecology, Ecosystems, and Sustainability, *edited by Noureddine Benkeblia*

Agroecology: *A Transdisciplinary, Participatory and Action-oriented Approach, edited by V. Ernesto Méndez, Christopher M. Bacon, Roseann Cohen, Stephen R. Gliessman*

Energy in Agroecosystems: *A Tool for Assessing Sustainability, by Gloria I. Guzmán Casado and Manuel González de Molina*

Agroecology in China: *Science, Practice, and Sustainable Management, edited by Luo Shiming and Stephen R. Gliessman*

Climate Change and Crop Production: *Foundations for Agroecosystem Resilience, edited by Noureddine Benkeblia*

Environmental Resilience and Food Law: *Agrobiodiversity and Agroecology, edited by Gabriela Steier, Alberto Giulio Cianci*

Political Agroecology: *Advancing the Transition to Sustainable Food Systems, authored by Manuel González de Molina, Paulo F. Petersen, Francisco Garrido Peña, Francisco R. Caporal*

Urban Agroecology: *Interdisciplinary Research and Future Directions, edited by Monika Egerer, Hamutahl Cohen*

For more information about this series, please visit: https://www.crcpress.com/Advances-in-Agroecology/book-series/CRCADVAGROECO

Urban Agroecology

Interdisciplinary Research and Future Directions

Edited by
Monika Egerer and Hamutahl Cohen

CRC Press is an imprint of the
Taylor & Francis Group, an **informa** business

First edition published 2021
by CRC Press
6000 Broken Sound Parkway NW, Suite 300, Boca Raton, FL 33487-2742
and by CRC Press
2 Park Square, Milton Park, Abingdon, Oxon, OX14 4RN

Library of Congress Cataloging-in-Publication Data

A catalog record for this title has been requested

ISBN: 978-0-367-63664-7 (hbk)
ISBN: 978-0-367-26001-9 (pbk)
ISBN: 978-0-429-29099-2 (ebk)

Typeset in Times LT Std
by KnowledgeWorks Global Ltd.

Contents

Preface

In the summer of 2015, during the 17th International Agroecology Shortcourse at the Sustainable Living Center on the University of California campus at Santa Cruz (UCSC), I remember having extensive conversations with a participant in the course about the importance of food produced in urban settings. Back then, Monika Egerer was a graduate student in the PhD program in Environmental Studies at UCSC, and had a special interest in using agroecology as a framework for enabling cities to grow more of the food that they needed. Over the two weeks, she and I grappled with the question "What is the role of agroecology in cities?" Since then, Dr. Egerer has continued in this field, developing urban agroecology as a field centered on science, practice, and social justice. With her co-editor Dr. Hamutahl Cohen, they call for the creation of spaces for food to be grown in cities—an Urban Agriculture (UA) movement. This book charts the course for accelerating this movement as *Urban Agroecology*.

The past several decades have seen a dramatic acceleration of human populations relocating from rural regions to urban centers. In the process, people have moved away from the places where food is grown, and into places dependent on food being transported from where it is produced to places where it is consumed. This places consumers at risk, especially those with limited access or less economic capacity, or in times when production and distribution systems breakdown, as we are seeing today with the COVID-19 pandemic. The many food and farming system weaknesses, vulnerabilities, structural problems, and political challenges are confronting us all at once as a result of the coronavirus crisis: panic buying, breakdown of supply chains, farm and food workers at greater risk, more hunger and malnutrition, and other problems due to the simultaneous breakdown of so many parts of the food system. This is especially a problem in many inner-city areas that suffer from being what are called "food deserts." There is a growing civic movement in many urban and suburban areas around the world to create spaces for urban food production. But the public investment and policy support remains inadequate and inequitable, and urban agriculture too often is seen as a marginal and insignificant part of the food system.

As the chapters in this book demonstrate, growing food in cities is not just about convenience, lifestyle, or part of the "foodie" culture. It is about building resilience and flexibility back into our food systems. It is about people being more in charge of their own food, and in the process creating food sovereignty, choice, and fairness in the food system. It is about paying attention to *how* food is produced as well as *where* it is produced, and *by* and *for whom*. Food is a basic human right, not just a commodity. We are talking about change, and it doesn't need to happen just on farms; it also needs to happen in and around urban areas.

Monika and Hamutahl have done a remarkable job of outlining the foundations of how urban agroecology can promote this change. In fact, the way the book is structured is what I feel is an excellent example of how to "do" agroecology in general! Agroecology should link science, practice, and action for social change. It should also link the ecological and social elements of sustainability in a transdisciplinary process of knowledge generation. This book does all of this, and more. It begins by providing an introduction to urban food systems, how agroecology can be applied to them, and why there is a need to move from urban food production to urban agroecology. Early chapters use ecological science to show how biodiversity conservation, ecosystem processes, and complex ecological interactions are some of the beneficial add-ons of food production systems in urban settings. Soil science explores the practice of soil management for long-term soil health. Cultural studies then explore how biological diversity and cultural diversity are intertwined in the socio-ecological complexities of urban food systems.

The agroecological approach evolves further in chapters that touch upon human health and well-being that benefit from urban food system development, as well as how access to land is critical in allowing urban dwellers to firmly establish the roots of the movement. Besides the need for land, chapters delve deeply into how urban agroecology can be a transformative social movement, how the movement can be organized, and how it can impact local policy and urban planning agendas in effective and permanent ways. It is exciting to see both the comparison and integration of natural science and social science methodologies for doing urban agroecology so that it is both action-oriented as well as interdisciplinary. The last several chapters lay out the challenges that such interdisciplinary research faces, as well as the kinds of education and extension programs needed to train and guide the agents of change in urban agroecology for the future. The book ends with a concluding view towards the future of what the next steps and needs are for urban agroecology to become the truly transformative process that we need.

As an agroecologist, I am always interested to know what draws people to work in agroecology.

Monika impressed me with her willingness to work at the interface between ecological and social complexity, which is necessary in the highly dynamic food spaces of urban systems. These are systems where high biological diversity and high socio-cultural diversity collide in amazing spaces of nature and cultural conservation. Hamutahl was drawn to urban agroecology by the potential of urban agroecosystems as places to promote wildlife, but also as places for food system change and human well-being. In her early career, she has worked to develop molecular methodologies that reveal the invisible biodiversity of microbial life in urban agroecosystems, enriching the field by advancing novel techniques. Both of them want to position themselves to promote urban food systems at a time when green spaces in cities are shrinking and the distance of people from food production is growing. They are both examples, in my mind, of the new generation of agroecologists who are rising to the challenge of applying their science to changing practice, changing policy, and changing minds.

Returning, then, to the initial question, "What is the role of agroecology in cities?". You'll find many of the answers in this book.

Steve Gliessman
May 2020

Acknowledgments

It is an exciting time for urban agroecology. New questions, movements, challenges, and opportunities are taking hold in cities worldwide, promoting research, innovation, and social justice. These developments have motivated this edited volume. We foremost recognize and thank urban gardeners and farmers, the agents of change that are working on the ground to change urban food systems. Gardeners have been cultivating, processing, and distributing food in and around urban areas for centuries. Today, urban gardening involves a diverse range of individuals, communities, and organizations. All of which are restructuring industrial supply chains, providing food for their communities, and supporting wildlife in their backyards. As you read each chapter, you will see how practitioners have inspired much of the research presented here. Thank you for your work, your motivation, and your passion to re-envision and re-design urban agri-food systems.

We acknowledge that urban agroecology stems from the field of agroecology. We thank the agroecology researchers who leverage community ecology, agronomy, plant pathology, entomology, rural sociology, anthropology, and many other fields to understand agroecological processes. As urban agroecologists, we draw from the rural agroecology literature to understand ecological dynamics and their social-political dimensions in cities. Much of rural agroecology was developed in close association with peasant-based movements connected to food sovereignty. In particular, smallholder and traditional agriculture largely in the Global South. We would like to acknowledge the traditional owners of the lands on which this work has taken place, and forms of indigenous Knowledge that play a fundamental role in agroecological principles and practices.

We graciously thank the contributing authors of each chapter that have dedicated an immeasurable amount of time and intellectual energy to the book. We thank them for proposing new ideas, drafting and revising their work, and pushing forward what it means to engage in urban agroecology. It has been enlightening to watch the development of each chapter.

A sincere thank you to Alice Oven and CRC Press for the invitation and opportunity to put together this edited volume as part of the Advances in Agroecology section. We are honored to be included in this exciting body of work. In particular, thank you to Steve Gliessman in his role in motivating this work, and for kindly writing the foreword to this volume. We acknowledge and thank the publisher staff and team for their work, especially Adwiti Pradhan. And many thanks and recognition to Charlotte Grenier for illustrating this book's cover, which captures the lively and dynamic nature of urban agroecosystems.

Finally, we thank our readers for their interest in the book, whether just a couple of chapters or the whole volume. We hope that you enjoy the book as much as we have enjoyed putting it together.

Editors

Monika Egerer is a tenure-track professor in Urban Productive Ecosystems in the School of Life Sciences at the Technical University of Munich. Her research investigates relationships between biodiversity conservation, ecosystem service provision, and human wellbeing in urban agroecosystems. Her work aims to bridge theory and practice to create productive systems in cities that offer food, habitat and community.

Hamutahl Cohen is a postdoctoral researcher in Entomology at the University of California, Riverside. Her research investigates drivers of insect declines in changing urban landscapes and the subsequent impacts to ecosystem services in farms and gardens. She leverages natural ecology and molecular and microbial methodologies to address questions of applied significance for promoting food security.

List of Contributors

Antonio Roman-Alcalá is an educator, researcher, and organizer based in California. He has been involved in US food movements primarily through urban farming and participatory democratic network organizing, and his research focuses on collaborative efforts to improve movement effectiveness. His recent projects include co-founding the Agroecology Research-Action Collective and linking local mutual aid and food systems efforts in the Bay Area.

Joshua Earl Arnold is a Ph.D. candidate at the University of California Berkeley, and the Professor of Sustainable Agriculture at Warren Wilson College. His current research focuses on parasitic Hymenoptera and habitat manipulations to increase biological control in urban agroecosystems. Joshua is committed to food systems research that addresses food security and environmental justice issues through participatory research.

Leah Atwood is a partnership builder, land steward, and goat shepherdess on occupied Chochenyo and Karkin Ohlone land also known as El Sobrante, CA. Her current work focuses on building capacity for agroecological learning networks, connecting resources for land justice and rematriation, and strengthening cooperative partnerships for community food sovereignty. She holds a BS from UC Berkeley and is a founding member of Wild and Radish Community and Collective Agroecology.

Renata Blumberg is an Assistant Professor in the Department of Nutrition and Food Studies at Montclair State University. She conducts research on alternative food networks in the United States and Eastern Europe, embodied geographies of food, critical management studies on institutional food systems, and feminist agroecologies and pedagogies.

Timothy Bowles is an Assistant Professor of Agroecology and Sustainable Agricultural Systems in the Department of Environmental Science, Policy and Management at the University of California Berkeley. His research focuses on agricultural diversification practices at multiple scales, their socio-economic context, and how they affect plant-soil-microbe interactions that underpin key ecosystem processes.

Evan Bowness is a PhD candidate and public scholar at the University of British Columbia's Centre for Sustainable Food Systems. His dissertation research in Brazil and Canada considers the relationship between growing food in the city and social mobilization for food sovereignty. As a community-engaged, visual, and public sociologist, his work focuses on topics related to food movements, the corporate food regime and industrial extractivism.

Kirsten Valentine Cadieux is the Director of Environmental Studies and Associate Professor in the Anthropology Department at Hamline University. She builds public processes for learning together about the political ecology and moral economy of community food and agriculture systems. Understanding and valuing environments and food systems helps us nourish each other better.

Martha Caswell is the Co-director of the Agroecology and Livelihoods Collaborative (ALC) at the University of Vermont. She is also a PhD student in the Rubenstein School of Environment and Natural Resources, and a Gund Institute for the Environment Graduate Fellow. Her research and teaching focus on agroecology, food sovereignty and critical participatory action research (PAR).

Lucy Diekmann is an Urban Agriculture and Food Systems advisor with University of California Cooperative Extension. Through research and extension work, she engages with urban farms and garden networks, food access organizations, food system collaboratives, and local government partners.

Monika Egerer is a tenure-track professor in Urban Productive Ecosystems in the School of Life Sciences at the Technical University of Munich. Her research investigates relationships between biodiversity conservation, ecosystem service provision, and human wellbeing in urban agroecosystems. Her work aims to bridge theory and practice to create productive systems in cities that offer food, habitat and community.

Leonie K. Fischer is a trained gardener, landscape planner, and received her PhD in the natural sciences. Before joining the Universität Stuttgart in October 2019 as Professor of Landscape Planning and Ecology she worked as a postdoc researcher at the Technische Universität Berlin. Main focuses of her work are ecological relationships in human influenced habitats, the human-nature intersection in cities and urban restoration.

Gordon Fitch is a Ph.D. student in Ecology and Evolutionary Biology at the University of Michigan. His research focuses on the eco-evolutionary effects of anthropogenic land use—especially urbanization and agriculture—on plant-bee interactions.

Nina S. Fogel is a PhD student in Dr. Gerardo Camilo's lab at Saint Louis University studying the patterns of bee diversity in residential gardens. She is also conducting a citizen science project to determine if photo surveys can be a surrogate to lethal sampling for measuring bee populations.

Ana García is an independent researcher in Alicante, Spain. Her research focuses on the social and governance related dimensions of urban agroecological movements.

Daniel López García (PhD in Agroecology) coordinates the Agroecology Area of the Entretantos Foundation, and the Technical Secretariat of the Spanish Network of Cities for Agroecology. Working mainly through Participatory Action-Research, he is specialized in Local Agroecological Dynamization, Sustainable Food Systems, Local Food Policies and Alternative Food Networks.

María A. Juncos-Gautier is a doctoral candidate of the Faculty of Environmental and Urban Change at York University in Toronto, Canada. She is also a research associate of the Agroecology and Livelihoods Collaborative (ALC) at the University of Vermont and a past recipient of the Organization of American States' Scholarship for Doctoral Studies.

Michelle Glowa, PhD is an assistant professor in the Anthropology and Social Change department. Her research interests include critical political ecology, land-based social movements, and agri-food studies.

Rachelle K. Gould is an Assistant Professor of Sustainability and Global Justice at the University of Vermont. She is affiliated with the Rubenstein School of Environment and Natural Resources, the Environmental Program, and the Gund Institute for the Environment. Her research focuses on cultural ecosystem services, relational values, and environmental education, all undergirded by issues of justice and representation.

Sacha K. Heath is a Biodiversity Postdoctoral Fellow with the Living Earth Collaborative in Saint Louis, Missouri, USA. She is most interested in ecological research that informs and evaluates biodiversity conservation. She has studied birds in agricultural, restored, and protected systems, and is currently using bioacoustics to study bird, bat, and anuran occupancy in residential gardens located along an urbanization gradient and with different degrees of native habitat modification.

Dustin Herrmann is an Assistant Project Scientist in the Department of Botany and Plant Sciences at the University of California, Riverside. His work furthers our integrated social and environmental knowledge to inform just and ecological city making.

Trey Hull is a Ph.D. student at Saint Louis University in the Department of Integrated and Applied Sciences, co-located in Dr. Gerardo Camilo's Bee lab and Dr. Vasit Sagan's Remote Sensing lab. Trey's dissertation is focused on using remote sensing with traditional surveillance techniques to study urban disease ecology.

Stephanie E. Hurley is an Associate Professor of Ecological Landscape Design in the Department of Plant and Soil Science and a Fellow of the Gund Institute for Environment at the University of Vermont. Her teaching and research span landscape design, ecological restoration, water quality, food systems, and urban agroecology.

Shalene Jha is an Associate Professor in Department of Integrative Biology at the University of Texas, Austin. She investigates ecological and evolutionary processes from genes to landscapes, to quantify global change impacts on plant-animal interactions, movement ecology, and the provisioning of ecosystem services.

Natalia Pinzón Jiménez is a Doctoral Candidate in Geography at the University of California, Davis. She specializes in distance education and risk evaluation for small-scale farmers. Natalia also runs her own consulting business helping organizations create online courses informed by critical pedagogy and agroecology. Natalia immigrated as a child to the United States from Colombia and her multicultural background greatly informs her work and research.

Ingo Kowarik is Professor of Ecosystem Science/Plant Ecology at the Technische Universität Berlin. His main research areas include urban ecology, biodiversity conservation in cities and human-nature relationships. He also serves as State Commissioner for Nature Conservation in Berlin.

Bárbara Lazcano is a researcher in agroecology at the Solidaridad Internacional Kanda AC (SiKanda) in Oaxaca de Juárez, México

Storm Lewis is a Mellon Mays Research Fellow and a Doris Duke Scholar at Smith College. She is a senior majoring in Environmental Science and Policy with a concentration in Sustainable Food. She worked in the Agroecology and Livelihoods Collaborative (ALC) at the University of Vermont in the summer of 2019. Her research focuses on agroecology principles in urban areas.

Alex Liebman is a PhD Student in Geography at Rutgers University. He is interested in Big Data science in agricultural research and its implications for climate change, peasant livelihoods, and agroecological landscapes. He explores how agricultural data science reproduces forms of standardization that constitute racialized and exclusionary forms of international development and environmental management across the global South, focused on agrarian regions.

Heidi Liere is an Assistant Professor of Environmental Studies at Seattle University. She is a specialist in insect community ecology and agroecology and her research investigates the factors that affect biodiversity in human-managed systems and how, in turn, biodiversity helps sustain vital ecosystem services in those systems.

Azucena Lucatero is a PhD student at the University of California Santa Cruz. She studies the insect ecology of urban gardens with special interests in biological pest control, community and population ecology, and landscape ecology.

Pauline Marsh is a researcher and teacher with the Centre for Rural Health, University of Tasmania, Australia. Applying qualitative social research methods, she explores various aspects of the reciprocal relationships between human health and natural environments. Pauline has expertise in the application of concept of therapeutic landscapes to real world settings.

Ana C. Galvis-Martinez is a woman, a single mother of a young man, a devoted educator and a Latina immigrant in the United States She works in agroecology, food justice and sustainable development. Ana holds a bachellor degree in Biology from the University of Antioquia, in Colombia, a Master's degree in sustainable agriculture from ECOSUR, in Mexico, and a Master's degree in Latin American studies from the University of California, Berkeley. Currently she works as consultant for farmercampus.com and she is the founder of holisticsustainabilities.org and cafepanamericano.org

V. Ernesto Méndez is the Co-director of the Agroecology and Livelihoods Collaborative (ALC) at the University of Vermont (UVM). He is also Professor of Agroecology and Interim Chair of the Plant and Soil Science Department at UVM. His research and teaching focus on an agroecology that aspires to be inclusive, transdisciplinary and transformative.

Manuel González de Molina is full professor of Modern History at Pablo de Olavide University (Sevilla, Spain), and coordinator of the Agro-Ecosystems History Laboratory. His work focuses on the study of the contemporary rural world, the biophysical analysis of agroecosystems and food systems and the design of public policies from an agroecological point of view. Direct a Master Degree Program on Agroecology (International University of Andalusia and Pablo de Olavide University) since 1996.

Helda Morales is a Professor at El Colegio de la Frontera Sur. Her research centers on agroecolology, sustainable food systems, women in agroecology, and education.

Jennifer C. Mullikin is a PhD student in Dr. Gerardo Camilo's lab at Saint Louis University studying specialist bee pollination in cities. She previously worked as an ecological consultant conducting wetland, plant, and bat surveys in urban and agricultural areas.

Jennifer A. Nicklay, a PhD student in Land and Atmospheric Science at the University of Minnesota - Twin Cities, conducts research across physical and social science boundaries with urban growers using science as a tool to create food systems grounded in reconciliation and justice. She is also committed to developing robust training and support systems for emerging community-engaged scholars and facilitating democratic learning spaces that embody multiple knowledge systems.

Theresa W. Ong is Assistant Professor of Environmental Studies at Dartmouth College. Her research focuses on the application of ecological theory to understand and motivate agroecological transitions across urban, peri-urban and rural settings.

Marcia R. Ostrom is associate professor and extension specialist in Sustainable Food and Farming Systems at the School of the Environment at Washington State University. She teaches courses on agroecology and sustainable food systems, helps to lead a statewide Extension Food System Team, and organizes participatory research and extension projects with immigrants in agriculture.

Stacy Philpott is a Professor of Environmental Studies at the University of California Santa Cruz and Directs the Center for Agroecology and Sustainable Food Systems. She studies intersections between biodiversity, ecosystem services, and human well-being in urban and tropical agroecosystems.

Brooke Porter is a dedicated seed sower, educator, agroecologist, and award-winning photographer. She holds a BA from Mills College and is presently finishing her MSc in Agroecology, at the Norwegian University of Life Sciences in conjunction with Spain's Universidad de Córdoba. Her current research looks at popular education initiatives devoted to teaching agroecology in Brazil and the United States.

Coleman Rainey is a Ph.D. student in Environmental Science, Policy, and Management at University of California Berkeley.

Paul Rogé is an activist scholar and agroecology educator in the San Francisco Bay Area. His research focuses on participatory and community-based research, urban agriculture, agroecological transitions, and strategies for addressing climate challenges. In education, he currently coordinates Merritt College's Certificate of Achievement in Urban Agroecology. Dr. Rogé is also a founding member of the Cooperative New School for Urban Studies and Environmental Justice.

Denyse Márquez Sánchez is a biology major at Carleton College in Northfield, Minnesota with an expected graduation date in 2021. She is a UC Santa Cruz Doris Duke Scholar and worked in the Agroecology and Livelihoods Collaborative (ALC) at the University of Vermont in the summer of 2019. Her research interests are in ecology, environmental anthropology, and food systems.

Alana Bowen Siegner is an agroecological food systems researcher and climate change educator. She recently completed her PhD at UC Berkeley, where her research focused specifically on local food system transformations and integrating food systems into experiential, farm-based climate change education. She worked as part of an interdisciplinary research team on urban agroecology in the East Bay region of the San Francisco Bay Area for two years and has farmed in the San Juan Islands of Washington state, on urban farms in the East Bay, and on small farms in Vermont, Rhode Island, and Alaska. Alana is passionate about growing food collectively, healthfully, equitably and sustainably. She aspires to open a "climate farm school" where young adults can come to learn about climate change solutions through agroecological farming systems.

Jonah Landor-Yamagata received his MS in Urban Ecosystem Sciences from the Technische Universität Berlin, where his research focused on urban foraging in Berlin. He has a BA in Environmental Studies from the University of California, Santa Cruz, where he was an early participant in the Program for Community and Agroecology (PICA). Jonah currently works doing community-based urban habitat restoration in Richmond, California.

Introduction: The Role of Agroecology in Cities

Monika Egerer and Hamutahl Cohen

CONTENTS

I. Introduction

This book is a collection of essays illustrating how agricultural landscapes and urban landscapes can be managed and designed for ecologically and socially sustainable food production. Because agricultural intensification and urbanization degrade environmental quality and social wellbeing, they also motivate visions of alternative food systems and urban life. As a result, diverse forms of urban agriculture centered on ecologically-based and local knowledge-intensive practices have flourished in recent decades across the world. Urban agricultural systems include home gardens, community gardens, market farms, urban orchards, "guerilla" gardens, and many other forms of urban food production systems. In this book, we explore how the principles of agroecology, though historically developed in rural systems, have evolved to encompass the management and interdisciplinary study of urban agroecosystems.

Agroecology refers to science, movement, and practice. But in urban areas, the implementation of agroecology takes new and interesting forms. This book presents our current understanding of biodiversity, ecosystem services, and social benefits associated with urban agroecosystems. The contributing authors illustrate how people working within these systems have expanded the scope of agroecology to the urban context through interdisciplinary and transdisciplinary research. Here, interdisciplinarity refers to integrating and synthesizing knowledge, methods, and approaches from various disciplines. There are multiple definitions of transdisciplinarity, but we use it to refer to research that goes beyond (or "transcends") academic disciplinary boundaries by unifying intellectual frameworks, for example, by bridging interdisciplinary research with participatory approaches involving the community outside of academia. This collection of literature leverages interdisciplinary approaches to address both the social-ecological complexity and the land use history of cities. The chapters illustrate that people are integral to the study of urban ecosystems because their behaviors, values, and attitudes shape social and ecological features in cities to influence the broader urban regional environment. We asked researchers across disciplinary boundaries to examine the ecological, practical and social dimensions of urban agroecosystems. Their contributions reflect an urban agroecology that incorporates social-ecological dimensions and the complexity of urban food production systems.

In this introduction, we first briefly describe the history of the field of agroecology and its main ideas to contextualize the development of urban agroecology. We then highlight how the field of urban agroecology is extending its reach outward toward new horizons. In doing so, we *(1)* describe a typology of urban agroecosystems, *(2)* explore the importance of interdisciplinarity in urban agroecological research, and *(3)* provide a roadmap for this book.

II. Agroecology: A Brief History, Key Terms, and Principles

Broadly, agroecology encapsulates the complex interactions among farm scale and field scale agroecosystem practices and processes to landscape scale surrounding conditions; an agroecological perspective considers the ecological principles and practices that underlie the ecology of the agroecosystem, and considers the social systems that manage and benefit from the agroecosystem. It considers the structure of the system, the function of the system, and the social processes that affect these properties. Ultimately, the aim of agroecology as a field is to guide a more just and sustainable food production system. In the first half of the 20th century, the term "agroecology" largely referenced agronomic studies of crop cultivar production and development. The 1960s Green Revolution, with its focus on maximum yields, spurred a renewed interest in agroecology (see Wezel et al. 2009 for a detailed history). As scientists grappled with the effects of industrialized agriculture, a key question in agroecology became: *what practices make agriculture more sustainable?* The field of agroecology has evolved since the 1980s–90s to address this question, often drawing from both natural and social science fields to directly respond to the intensification, expansion, and industrialization of agriculture.

The conventional form of agriculture is characterized by monoculture (single species) cropping, high external inputs (e.g. of fertilizer, fossil fuels, biotechnology), altered resource flows (e.g. N, P, water), and overall low ecological complexity (Vandermeer and Perfecto 2017). This has consequences for ecosystem function (Tscharntke et al. 2005), human well-being (Rasmussen et al. 2018) and agricultural livelihoods (van der Ploeg 2010, Labao and Meyer 2001, Graeub et al. 2016). Agricultural industrialization and intensification have transformed the world's terrestrial landscapes (Geist and Lambin 2001, Foley et al. 2005), biophysical processes (Vitousek et al. 2009, Galloway et al. 2008), and climate (Smith et al. 2014), while simultaneously fundamentally changing the sociopolitical structure of the global food system (Labao and Meyer 2001). This has resulted in uneven access to both land (e.g. Jarosz 2008, Kenney-Lazar et al. 2018) and fresh and healthy food (Walker et al. 2010). In response, agroecology has put forth an alternative form of agricultural ***practice***, a rigorous ***science*** guiding the design of these alternative systems, and a social and political ***movement*** (Wezel et al. 2009, Gliessman 2016). These intertwined dimensions of agroecology have been the subject of much discussion, debate, and examination over the last few decades (e.g. Altieri 1989, Gliessman 1990, Dalgaard et al. 2003, Francis et al. 2003, Wezel and Soldat 2009, Wezel et al. 2014, Gliessman 2011, Holtz-Giménez and Altieri 2013, Méndez et al. 2013, Parmentier 2014, Reynolds et al. 2014, Gliessman 2016, D'Annolfo et al. 2017, Altieri 2018).

The Science

As a science, agroecology largely employs concepts and principles in ecology to the development of sustainable agricultural systems (Gliessman 1998). Agroecosystems have been conceptualized as systems existing in the intermediate between natural and fabricated/human constructed systems (Gliessman 2001). To move towards sustainability, some practitioners have suggested designing agroecosystems to mimic 'natural' systems. This means using practices that promote species diversity, nutrient cycling, and production, and referencing nature as a checkpoint in sustainable agriculture transitions. While most practitioners of agroecology recognize the philosophical and practical challenges of identifying the "natural" in human-modified systems, scientific practices working towards sustainability often also recognize the importance of complex relationships between species, biophysical factors, and scale and production within the context of society, politics, and environment (Francis et al. 2003). Researchers work to build these understandings by learning from traditional and indigenous knowledge systems (Francis et al. 2003, González Jácome 2011, Vandermeer and Perfecto 2013). It is widely acknowledged (though not always practiced) that the science around agroecology should integrate perspectives outside of academia. This means engaging local communities. Méndez et al. (2015) describes how agroecology has evolved into distinct "agroecologies" under a framework to both: *(1)* reinforce, expand, or develop scientific research towards advancing sustainability at the agroecosystem scale; and *(2)* address social science and food system issues through *transdisciplinary research* – which includes engaging with local, indigenous, or other forms of knowledge – and through *participatory research* – research involving diverse actors and voices that have historically been excluded from or underrepresented in academia. The development of this model for interdisciplinary and transdisciplinary research in urban agroecology is discussed by Caswell et al., who utilize a participatory framework in their local community to examine community notions of urban agroecology (Chapter 14). Bowness et al. offer practical suggestions for conducting scientific research and point to future directions and challenges for the field (Chapter 13).

The Practice

In its applied form, agroecology considers the technical methods and practices required to diversify agriculture. These methods include cover cropping, agroforestry, biological control, livestock integration and more (Arrignon 1987). Agroecological practices promote agricultural production that is environmentally friendly, socially just, and economically viable (Gliessman 1990, Pretty 2008, Wezel et al. 2014). Agroecological practices can be classified according to whether they increase efficiency (i.e. reduce nutrient inputs used), substitute an input or practice, or redesign a cropping or farm system (Wezel et al. 2014). Input substitution practices include, for example, replacing synthetic nitrogen fertilizer with biofertilizers and compost. Practices that redesign farming systems might include cultivating an agroforestry system with fruit or nut trees. Increasing resource use efficiency might include selecting relevant crop rotations with low water needs during drought conditions (Pala et al. 2007), or combining different crops with different root systems to increase nutrient cycling (Rigueiro-Rodríguez et al. 2009). There are several basic principles of agroecology that tie theory to practice. They are: diversifying crop structure and function, soil management that improves biological activity, nutrient cycling and organic matter, and practices that enhance the cycling of systems. Practices can be focused at (1) the local plot or field scale (such as tillage management), (2) the cropping system scale (such as on crop choice and spatial distribution), and (3) the landscape scale (including the management of landscape elements such as forest fragments). Agroecological practices also vary in their potential for system change and broader implementation across agricultural systems (Wezel et al. 2014). The ways in which people mobilize and partner to advance the adoption of sustainable practices, food and land sovereignty, and rural development and livelihoods is integral to the practice of agroecology. Much work focusing on agroecological practices, both at farm-level and system-level approaches, directly consider that our world is rapidly changing (Tomich et al. 2011).

The Social Movement

Agroecology explicitly recognizes the socio-political dimensions of agriculture. Agroecology works towards agricultural management that is grounded in local knowledge, and context-dependent strategies of farmers can be combined with modern ecological understandings (Vandermeer and Perfecto 2012). Agroecology is practiced and amplified by strong ties to social movements, through which people mobilize to achieve common goals of, for example, land and food sovereignty (Rosset and Martinez-Torres 2012). Together this means that agroecosystems are integrated ecological and social systems in which people have different motivations for their use. Subsequently, understanding agroecology is often best accomplished with a participatory action-oriented approach, which engages deeply with local communities and peoples to promote food sovereignty, sustainability, and the principles of agroecology

(Méndez et al. 2013). It is important to acknowledge that agroecology can be considered a stand-alone social movement, and in many regions of the world, the movement and practice may function independently from researchers who focus on agroecology as a scientific discipline. An excellent example of an agroecology-based social movement is La Via Campesina, discussed by Glowa and Roman-Alcala in Chapter 8 as a source of context for urban social movements. Furthermore, "agroecologist" identities are very diverse – and include farmers, agronomists, sociologists, social workers, and policy makers. These identities and their associated histories are all important in agroecology. In this book we aim to integrate and emphasize agroecology as a science, movement and practice and to be cognizant of all self-identified agroecologists specifically in the urban realm.

III. Agroecology for Sustainable Urban Food Systems: Developing an Urban Agroecology

Although the literature in agroecology is primarily focused on the development and employment of agroecology in rural areas, the industrialization of large-scale agriculture impacts also shapes food systems in urban areas. Today, we see agroecology taking new and exciting forms in cities to address issues associated with both agricultural industrialization and urbanization. In this section, we discuss how agroecology—and specifically the core ideas and associated practices of this field— has expanded and changed in urban contexts, what we henceforth consider *urban agroecology*. We discuss how urban agroecology, though encapsulating similar principles and practices of agroecology in the rural context, is not identical. **First**, we discuss the processes behind urbanization to contextualize the development of urban agroecology. **Second**, we provide a brief history of urban agroecology as a field related to, but distinct from, agroecology and urban ecology fields. **Third**, we discuss how agroecological principles and practices are applied to the study of urban agroecosystems, including the role of socio-political movements in urban agroecology.

Urbanization and Urban Agroecosystems

Urbanization— the growth of urban areas in population, area, and building density (Browning 1958, Chen et al. 1998, Zlotnik 2017)— is a key agent of land-use change. Nearly two-thirds of the world's population now lives in cities, according to estimates of the United Nations (UN 2019). From an environmental perspective, urbanization is associated with: changes to ecological communities and native biodiversity loss (McDonald et al. 2013); alteration of microclimate and macroclimate due to urban warming and building topography (Zeng et al. 2010); and changes to a suite of biophysical properties of ecosystems such as impervious cover, topography, and soil quality (Bajocco et al. 2016). Many of the changes associated with urbanization parallel changes associated with agricultural intensification. These include the rise of social inequality and isolation, declines in public health, and in environmental injustice (UN Healthy Cities Report). These environmental and social trends certainly depend on the urban regional context, but broadly around the globe we see that urbanization deeply transforms socio-ecological processes (Grimm et al. 2008, Boone and Fragkias 2012).

The impacts of urbanization have engendered debate around contemporary agro-food systems, land use, and our connection to food production. This is especially true in cities, where many people are physically disconnected from where food is grown, and may have less land access due to patterns of urban density (Seto and Ramankutty 2016). Within urban areas, urban agriculture has grown in popularity as a global solution to ameliorate social disparities, and to provide an alternative local agri-food system to the global industrialized food system (Allen 2003, Holloway and Kneafsey 2004). Urban agriculture ranges in form and function, from home kitchen gardens to highly productive for-market urban farms (Table 1). Urban agriculture can nevertheless provide food access and security to those most food insecure (Alaimo et al. 2008), specifically those with the least access to fresh and cheap fruits and vegetables that promote healthy diets. Urban agricultural spaces also are important for connecting people to the environment and for their socio-cultural benefits. These include socialization opportunities (Saldivar-Tanaka and Krasny

TABLE 1

Different types and descriptions of some common urban agricultural systems (Urban Agroecosystems). Provided references are examples of studies that focus on this type or scale of agricultural production.

Type	Description of System and Scale of Food Production	Example Studies
Community and **allotment gardens**	Small-scale, vegetatively complex and species rich agroecosystems mostly dedicated to food production that is managed by either a group of people or by individuals/households who lease small "allotment" plots within the garden. Usually located in urban or semi-urban areas.	Colding et al. 2006, Drescher et al. 2006, Speak et al. 2015, Bell et al. 2016
Private or **home gardens**	Small-scale agroecosystems managed by individuals and households on private land, where households reside. Can be the most prevalent form of urban agriculture in some cities, and forms and management vary widely. For example, privately owned gardens cover ~22–27% of the total urban area in the UK, 36% in New Zealand, and 19.5% in Dayton, Ohio, USA.	Loram et al. 2007, Mathieu et al. 2007
Easement gardens or **verge gardens**	Gardens often regulated by the local government but located within private or community properties. Urban easements are often established to improve water quality and erosion control, but they can include a wide array of biodiversity, including food plants, depending on management type. Gardening on easements or 'verges' in Australia may also be done as a form of '*guerrilla gardening*' where local communities garden on small patches of soil when few unpaved spaces are available.	Hunter and Hunter 2008, Hunter and Brown 2012, Marshall et al. 2020
Rooftop gardens or **green roofs**	People creatively use above-ground built surfaces to grow food. Here the vegetation on the roof of a building may be used to improve insulation, create local habitat, provide decorative amenity, and cultivate food plants. May be common in very dense urban areas with little ground-level green space.	Whittinghill and Rowe 2012
Urban orchards or **food forests**	Tree-based food production systems that are often of fruits (stone, apple, citrus, etc.) and nuts that contribute to longer term food production and provide a suite of other ecosystem services. Orchards may be integrated into other urban agroecosystems, such as community gardens. Orchards can be owned and run privately or by the community. For example, schools and hospitals are establishing fruit trees that provide crops, erosion control, shade and wildlife habitat, and produce food for the local community.	Drescher et al. 2006
Market farms, commercial and non-commercial farms	Focus on commercial sale of crops grown in outdoor fields, in greenhouses, or on rooftops. Generally represents the largest scale of production (total area and size) of urban agriculture, utilizing several lots.	Lowenstein and Minor 2018
Peri-urban agriculture	Primarily located at the outskirts or periphery of cities. Typically, these are multifunctional agricultural systems that include a large variety of activities and diversification approaches and contribute to environmental, social and economic functions of and for the surrounding community.	Zasada 2011

2004), political empowerment (White 2011), cultural heritage preservation (Corlett 2005), education (Tidball and Krasny 2009), and many others, described in this book by Pauline Marsh in Chapter 6.

To understand urban agroecology in cities, we highlight the diversity of urban agroecosystems. Table 1 describes six non-mutually exclusive types of urban agroecosystems, and summarizes their diverse forms, structures and biophysical and social organizations. Urban agroecosystems are further characterized by management structure and organization, types of food and service provision, motivation or purpose, the scale of production, and more. These systems are components of broader urban food systems, and often the people involved in urban agroecology engage in issues around food justice and food sovereignty in cities. Broadly, urban food systems may also encapsulate related activities in food production, including food processing, marketing and transformations (Ericksen et al. 2009).

A Brief History of the Development of Urban Agroecology

An *urban agroecological approach* takes form at the critical juncture between agriculture, ecology, and urbanization. It addresses the relationships among urban agriculture, land-use change, and food movements, and their relationships with urban biodiversity, ecosystems, and people. Here we provide a brief, historical look at the fields of agroecology and urban ecology to understand how they intersect with and differ from urban agroecology.

Urban ecology, as a discipline, is broadly concerned with biodiversity within cities, the structure and functioning of urban ecosystems in relation to city features and non-urban counterparts (Grimm et al. 2000, Niemelä et al. 2011), and the environmental sustainability of urbanization and population growth (Rebele 1994, Niemelä 1999). Early work in this field questioned the predominant view of cities as separate from nature. Urban ecologists described cities as ecosystems (Rees 1997) in which humans and animals interact and shape the world around them (Wolch et al. 1995), and explored how human perceptions influence urban environments and life (Golany 1995). Urban ecology applies ecological theories including island biogeography theory, metapopulation theory, and biocomplexity theory (among others) to spatially situate urban ecosystems within complex fragmented landscapes and to understand patterns in urban biodiversity and local species interactions (Niemelä 2000). Today, urban ecology also recognizes urban ecosystems as socio-ecological systems (Pickett et al. 1997, Grimm et al. 2008). Under this framework, species diversity in urban ecosystems are a function of social and economic variation in urban landscapes that shape management decisions at local (Andersson et al. 2007) and landscape scales (Lowenstein et al. 2014). Modern urban ecology is moving away from ecology "in" the city (focusing on e.g. impacts of urbanization on native biodiversity), to an ecology "for" the city that explicitly integrates decision makers in the co-design and production of research and knowledge that can be applied on-the-ground (Grove et al. 2016, McPhearson et al. 2016). This framework works at multiple scales of decision-making and integrates ecology, stakeholder experience, and policy for an "action-based" research approach (Tanner et al. 2016). These urban ecology principles have helped shape urban agroecology approaches to understanding urban farms and gardens as socio-ecological systems.

Neither urban ecology nor agroecology have given the same attention to urban agricultural systems as they have other ecosystems. Urban ecologists have generally conducted little to no research on urban agriculture, focusing often on the ecology of city parks or urban forest fragments. Agroecologists have tended to, historically, address the social and ecological importance of urbanization only in relation to rural agriculture. For example, in China, there was great interest in the 1990s for the potential of urban waste and refuse as fertilizer for crops in rural areas (Yan 1990, Dehui 1991, Tao 1993), but not in urban agricultural systems. Rural agroecologists began to incorporate some mentions of urban systems, but largely considered urbanization as an external force that influenced agriculture in the rural sphere (Altieri and Francis 1992, Rosset and Altieri 1997). In one case, Torres-Limas (1994) examined the impact of urbanization on the famous chinampas system, a form of agriculture invented by the Aztecs.

In an early agroecological study of urban agriculture in Cuba, Miguel Altieri described the application of agroecological principles to sustainable practices in biodiverse urban gardens and urban farms that provided food security for the island population (Altieri et al. 1999). This work importantly connected agroecological principles, traditional knowledge, and a social movement to agriculture in the urban context. There was growing interest in urban agriculture during this period, including interest in urban-rural linkages in food systems (WinklerPrins 2002, Francis et al. 2005), community-based urban agricultural projects (Hanna and Oh 2000, Asato et al. 2002, Fraser 2002, Peña 2005), environmental education in urban agricultural spaces (Doyle and Krasny 2003), land-use planning (Schmelzkopf 1995, Irvine et al. 1999), and the community benefits of urban agriculture (Dunnett and Qasim 2000, Brown and Jameton 2000, Kurtz 2001). Ecological research in urban agriculture also expanded during this time with studies on: soil management and functioning in urban systems (De Kimpe and Morel 2000); biodiversity of small mammals (Baker et al. 2003), birds (Melles 2003, Chamberlain et al. 2004), insects (Hostetler and McIntyre 2001, Tommasi et al. 2004, Gaston et al. 2005, Zanette et al. 2005), and plants (Loram et al. 2008). This work was complemented by studies on the potential of urban agriculture vegetation for mitigating urban heat (Shashua-Bar and Hoffman 2000, Wong et al. 2003).

Urban agriculture research has proliferated in the last 15 years (Lin et al. 2015), with an early mention of the term "urban agroecology" in a 2007 study by Terrile et al. on the local development of urban agriculture in the city of Rosario, Argentina. The research conducted in urban agroecosystems today is generally divided between natural and social sciences approaches, with research focusing on biodiversity, ecosystem services, social aspects, and applications for urban policy and planning (Clucas et al. 2018). Researchers explore socio-ecological interactions at the agroecosystem scale (Corlett et al. 2003, Taylor and Lovell 2014), as well as how landscape patterns and properties affect biodiversity and ecosystem functioning (Clucas et al. 2018, Philpott and Bichier 2017). Today, urban agroecology sits at the intersection of urban ecology and rural agroecology, with more recent literature addressing the intersecting socio-ecological dimensions of urban agriculture. Yet, despite much work in urban agriculture, at the time of this writing in 2020, only 156 papers use the term "urban agroecology" in comparison to 119,000 papers using "agroecology" and 89,900 using "urban ecology" (Google Scholar). Many researchers working in cities use different terms to describe their work, even though they address principles developed in the agroecology literature. While some urban ecologists have explicitly addressed how ecological principles take form in the urban sphere (Niemelä 1999), many questions remain for urban agroecology: *how do the principles of agroecology and urban ecology inform the field of urban agroecology?* Furthermore, *is urban agroecology a subset of agroecology? Or is it a separate discipline entirely?* The contributions in this book provide clues for how our field is evolving, but these are still open questions for those of us working in this space.

We argue that understanding how agroecological principles might be applied to urban agroecosystems can help guide the recent explosion of urban agricultural growth as the popularity and social-environmental importance of urban agricultural increases globally. Agroecology may offer future directions for understanding the drivers and impacts of urban cultivation in cites. While the urban ecology field continues to push forward meaningful and exciting research around cities that informs urban agroecology, the history of agroecology research can serve as a reference for directing future research in urban agrology. Looking to the well-established agroecological literature helps pinpoint research gaps. For example, although urban agroecosystems have the potential to mitigate climate change, we still need to know how urban gardeners and farmers can adapt practices towards this goal. Furthermore, although vegetation management is correlated to several ecosystem services like pollination or pest control (e.g. Philpott and Bichier 2017), the exact mechanisms through which this occurs are largely unclear. And although social benefits to human-wellbeing are clearly perceived across social groups (Corlett 2003; Saldivar-Tanaka and Krasny 2007), we still need to know how values translate to sustainable agroecological management. Understanding how agroecology extends to urban agroecosystems may assist in answering these questions. We explore the space between agroecology and urban agroecology in the next section, reviewing how agroecological work has paved the way for urban agroecology as science, practice and movement.

Agroecological Principles and Practices Applied to Urban Agroecosystems

Agroecological principles underlie the design, structure, functioning and management of urban agroecosystems. Indeed, the core principles of agroecology (biological diversification, soil management, closed loop cycling, etc.) are often purposefully practiced by urban farmers and gardeners (Altieri 1999, Altieri 2019). For example, crop diversification is a core principle in agroecology. Home gardens and community gardens within cities can be incredibly diverse in terms of crop, ornamental and weed plant species and varieties. One case study by Colding et al. documented over 400 species in a single allotment garden in Sweden (2006), and Loram et al. recorded over 1000 species across home gardens across five UK cities (2008). Home and community gardens may also harbor rare as well as novel plant species and varieties. Within a home vegetable garden in Chicago USA, Taylor and Mione documented the cultivation of *Jaltomata darcyana* (Solanaceae) as a food plant, a species that had previously not been recorded for cultivation in neither a rural or urban context (2019). The biological diversity of planned agrobiodiversity in urban agroecosystems can be explained by intentional crop diversification, but also by the multicultural character of cities due to the confluence of human migration pathways. Important to note here are the linkages among food and agricultural traditions, cultivated urban plants, and multicultural background (Glowa et al. 2018). Human diversity may beget biological diversity, and this is mirrored in planned (plant) diversity within an agroecosystem (Clark and Jenerette 2015; Philpott et al. in press). Consequently urban agricultural systems can exemplify agroecological principles in practice through plant diversity and composition. These concepts are further discussed by Philpott et al. in Chapter 2.

Soil management in urban agroecosystems provides another excellent example of how agroecological principles and practices are applied to the urban context. Urban agricultural practitioners often use compost and mulching amendments to boost soil organic matter, nutrient cycling, soil water conservation, and soil biological activity (Gregory et al. 2016, Tresch et al. 2018). Such ecological soil management practices might be especially important to remediate land use legacies of soil contamination from heavy metals or soil compaction due to urbanization and industrialization (Egerer et al. 2018). Rainey et al. further describe how soil management grounded in principles of agroecology connects promotes transformative social movements in cities (Chapter 6).

One way urban agroecology has borrowed from agroecology is in its treatment of scale. In agroecology, ecological processes are important not only in the plot scale, but at the field and landscape-level. The importance of scale is considered at both temporal and spatial axes (e.g. Kremen and Miles 2012, Miles et al. 2012). In urban agroecology, researchers have examined how habitat across local and landscape scales is important for both biodiversity and ecosystem services, such as pest control (e.g. Philpott and Bichier 2017, Egerer et al. 2020). In Chapters 1 and 2, Heath et al. and Philpott et al., respectively, synthesize the literature around the importance of habitat features across scales for biodiversity and complex organismal interactions. Scale is also important as a concept in socio- and political ecological studies, where researchers ask how we can scale up and scale out agroecology. For example, Ferguson et al. (2019) demonstrate how to leverage school gardens to improve educational outcomes and the organizational fabric of local food systems, and Mier y Terán Giménez Cacho (2018) identifies nonlinear, multidimensional drivers that bring agroecology to scale in Central America, India, and Brazil.

Several challenges exist when applying agroecology principles and practices developed in rural contexts to the urban context. One issue is that agroecology often uses nature as a reference point for designing ecological management strategies. Yet the complexity of agroecosystems in biotic and abiotic factors, properties and structure, dynamic spatial patterns, chaos, and stochasticity, make this approach challenging and ungeneralizable across systems (Vandermeer and Perfecto 2017). In urban areas, looking towards nature to manage a high functioning production system may not be the best approach, because these systems are heavily modified by urban land use history (e.g. soil contamination) and also reflect social-political features unique to the city and that inform agricultural practices (e.g. social diversity, land tenure). The question of what is historically "natural" in an urban system is complicated because urban agroecosystems, as their core, rely on day-to-day management and cultivation by humans in landscapes that have transformed over centuries. Another challenge with applying agroecological principles is that urban agroecology has come to encompass practices that are not grounded in land-based systems. For example, urban foraging, in which people harvest and collect food and medicinal plants from 'wild' vegetation, is a practice that provides urban residents an opportunity to engage directly with their food systems without land ownership (discussed by Fischer et al. in Chapter 5).

The study and application of urban agroecology therefore requires unique considerations that differ from agroecology. Urban agroecology *must* account for and adapt to the unique socio-ecological complexities of cities. We return to these challenges in the conclusion of the book, drawing insights from how each chapter associates agroecology in relation to urban agroecology.

Socio-Political Dimensions of Urban Agroecological Systems

The socio-political dimensions of urban agroecology have an important influence on management, sustainability, and political and economic transformations. Urban agroecology (as with agroecology broadly) therefore involves fields of urban sociology, anthropology, environmental sciences, ethics, and economics to provide diverse insights into urban food systems (Francis et al. 2003). The social, political, economic conditions that shape urban farming have been studied in cities across the world. For example, White illustrates how urban farming at D Town Farms in Detroit, Michigan has been a powerful form of political resistance to racial injustice and a means to political autonomy and land sovereignty (2011). In Rome, Italy, grass roots urban farming initiatives have spread widely as a movement by Romans to bypass conventional food systems and gain more control of the local food system and the urban commons (Ledant 2017). Such social movements are often inspired or linked to the prominent agroecology movement of rural peasant farmers known as La Via Campesina (Rosset and Martinez-Torres 2010). But they also uniquely respond to the rapid urbanization and the economic pressures and neoliberal conditions that come with urban densification and growth (McClintock 2017).

One of the risks facing the development of urban agroecology as a transformative movement is the romanticization of cultivation, wherein the act of working the ground (alone, without social and political action) is a cure-all for societal ills (Pudup 2008). Instead, urban agroecology must continue to recognize how urban food systems embody culture, politics, economy, conflict, and collaboration (Tornaghi and Hoeskstra 2017). Urban agroecology must "understand cities as territories of dispute between social movements engaged in the promotion of life, and the capitalist industrial food system" (Biazoti and Almeida 2017). Addressing how urban agroecology promotes or conflicts with hegemonic food systems will allow us to re-envision modernity, to challenge urban-rural dichotomies, and to improve the multifunctionality and just provisioning of urban space (Tornaghi 2014, Tornaghi and Dehaene 2019). Perhaps the first step towards this goal is to ensure scientific endeavors are inclusive to the diverse actors, dialogues, and knowledge systems found in urban spaces. This book highlights how we extend urban agroecology into local communities: in Chapters 10 and 11, we hear how to incorporate education frameworks (Roge et al.) and participatory approaches (Diekmann and Ostrom) to develop urban knowledge networks that promote human well-being in the city.

We asked book contributors to largely focus on describing advances in urban agroecology research. While this book does not explicitly focus on agroecology as a movement or practice, the authors recognize that the political and social aspects of agroecology are integral for advancing the field as a science. For example, Caswell et al. discuss both how urban farms in Vermont connect with the regional food movement and their agroecological practices (Chapter 14). Marsh illustrates how urban agroecological practices create therapeutic landscapes that promote public health and well-being (Chapter 6). Morales et al. address the concept of leveraging urban agriculture as a form of political resistance (Chapter 7). Finally, Bowness et al. discuss how to study movements, social practices, and social conditions that shape the agroecological features of urban agriculture (Chapter 13).

IV. Looking Forward: Elucidating Social-Ecological Complexity in Urban Agroecological Systems

Our aim with this book is to primarily focus on the scientific advancements that reflect research conducted in and related to urban agroecology. Because urban agroecology is both an action-oriented science and social movement, this book will highlight our understanding of urban agroecology as a highly interdisciplinary and transdisciplinary field. The authors herein contribute work that celebrates interdisciplinary and collaborative approaches in urban agroecology through case studies, syntheses, and forward-thinking perspectives. The case studies illustrate the diverse and varied ways that scientists, stakeholders, and the public work across traditional academic divides and social sectors to collectively study the complexity of agroecosystems in the city. Several authors, such as Caswell et al. (Chapter 14), illustrate how to leverage Participatory Action Research approaches to conduct urban agroecology research that defies traditional researcher-subject relations. The chapters in this collection also illustrate some different urban regions of the world in which urban agroecology is gaining traction as a science, movement, and practice. Though many chapters are in the US, we encourage a broader application and outlook from this work to other urban areas in the world. In response to the complex social-ecological features of cities, this work pushes beyond disciplinary boundaries, beyond single-species interactions, beyond simple biodiversity metrics, and integrates ecological complexity, human values and behaviors into an interdisciplinary science of urban agroecology.

REFERENCES

Alaimo, K., Packnett, E., Miles, R.A., and Kruger, D.J., 2008. Fruit and vegetable intake among urban community gardeners. *Journal of Nutrition Education and Behavior* 40, 94–101. Available from http://www.ncbi.nlm.nih.gov/pubmed/18314085.

Allen, P., FitzSimmons, M., Goodman, M., and Warner, K., 2003. Shifting plates in the agrifood landscape: the tectonics of alternative agrifood initiatives in California. *Journal of Rural Studies* 19, 61–75. Available from http://linkinghub.elsevier.com/retrieve/pii/S0743016702000475.

Biazoti, A., Almeida, N., and Tavares, P., 2017. *Caderno de metodologias: inspirações e experimentações na construção do conhecimento agroecológico.* Editora Universidade Federal de Viçosa, Viçosa.

Altieri, M.A., 1989. Agroecology: A new research and development paradigm for world agriculture. *Agriculture, Ecosystems and Environment* 27, 37–46.

Altieri, M.A., 1999. The ecological role of biodiversity in agroecosystems. In M.G. Paoletti, ed., *Invertebrate Biodiversity as Bioindicators of Sustainable Landscapes* (pp. 19–31). Elsevier, Amsterdam.

Altieri, M. A. (2018). *Agroecology: the science of sustainable agriculture.* CRC Press.

Altieri, M.A. and Nicholls, C.I., 2018. Urban Agroecology: designing biodiverse, productive and resilient city farms. *Agro. Sur.* 46, 49–60.

Altieri, M.A. and Francis, C.A., 1992. Incorporating agroecology into the conventional agricultural curriculum. *American Journal of Alternative Agriculture* 7(1–2), 89–93.

Andersson, E., Barthel, S., and Ahrné, K., 2007. Measuring social-ecological dynamics behind the generation of ecosystem services. *Ecological Applications* 17(5), 1267–1278.

Arrignon, J., 1987. *Agro-ecologie des zones arides et sub-humides.* Editions G.-P. Masonneuve and Larose et ACCT, Paris.

Asato, E., Lattuca, A., Lemos, C., Marani, S., Ottmann, G., Terrile, R., … and Rótolo, G., 2002. Urban family agriculture, new social movements and agroecology: ecological vegetables production to mitigate the hunger and to generate a local development process in Rosario city. Argentina. In 5. Congreso de la Sociedad Española de Agricultura Ecológica, Gijón (España), 16-21 Sep 2002. SERIDA.

Bajocco, S., Ceccarelli, T., Smiraglia, D., Salvati, L., and Ricotta, C., 2016. Modeling the ecological niche of long-term land use changes: The role of biophysical factors. *Ecological Indicators* 60, 231–236.

Baker, L.E., 2004. Tending cultural landscapes and food citizenship in Toronto's community gardens. *The Geographical Review* 94(July), 305–325.

Bell, S., Fox-Kämper, R., Keshavarz, N., Benson, M., Caputo, S., Noori, S., and Voigt, A., 2016. *Urban Allotment Gardens in Europe.* Routledge.

Boone, C. G., and Fragkias, M. (Eds.)., 2012. *Urbanization and sustainability: linking urban ecology, environmental justice and global environmental change* (Vol. 3). Springer Science and Business Media.

Brown, K.H., and Jameton, A.L., 2000. Public health implications of urban agriculture. *Journal of Public Health Policy* 21(1), 20–39.

Browning, H.L., 1958. Recent trends in Latin American urbanization. *The Annals of the American Academy of Political and Social Science* 316(1), 111–120.

Chamberlain, D.E., Cannon, A.R., and Toms, M.P., 2004. Associations of garden birds with gradients in garden habitat and local habitat. *Ecography* 27(5), 589–600.

Chen, N., Valente, P., and Zlotnik, H., 1998. What do we know about recent trends in urbanization. *Migration, Urbanization, and Development: New Directions and Issues*, ed. R. Bilsborrow, 59–88. United Nations Population Fund and Kluwer Academic Publishers: New York

Clucas, B., Parker, I.D., and Feldpausch-Parker, A.M., 2018. A systematic review of the relationship between urban agriculture and biodiversity. *Urban Ecosystems* 21, 635–643.

Colding, J., Lundberg, J., and Folke, C., 2006. Incorporating Green-area User Groups in Urban Ecosystem Management. *AMBIO: A Journal of the Human Environment* 35, 237–244.

Corlett, J.L., Dean, E.A., and Grivetti, L.E., 2003. Hmong Gardens: Botanical Diversity in an Urban Setting. *Economic Botany.*

D'Annolfo, R., Gemmill-Herren, B., Graeub, B., and Garibaldi, L.A., 2017. A review of social and economic performance of agroecology. *International Journal of Agricultural Sustainability* 0(0), 1–13.

Dalgaard, T., Hutchings, N.J. and Porter, J.R., (2003) Agroecology, scaling and interdisciplinarity. *Agriculture, Ecosystems and Environment* 100 (1), 39–51.

De Kimpe, C.R. and Morel, J.-L., 2000. Urban soil management: a growing concern. *Soil Science* 165(1).

Dehui, F.T.Z.Y.G., 1991. The effects of refuse compost from urban sources on the agricultural chemical properties of soil. *Journal of Hubei University* 4.

Doyle, R., and Krasny, M., 2003. Participatory rural appraisal as an approach to environmental education in urban community gardens. *Environmental Education Research* 9(1), 91–115.

Drescher, A.W., Holmer, R.J., and Iaquinta, D.L., 2006. Urban homegardens and allotment gardens for sustainable livelihoods: Management strategies and institutional environments. In B.M. Kumar, and P.K.R. Nair, eds., *Tropical Homegardens: A Time-Tested Example of Sustainable Agroforestry* (pp. 317–338). Springer Netherlands, Dordrecht.

Dunnett, N., and Qasim, M., 2000. Perceived benefits to human well-being of urban gardens. HortTechnology 10, 40–45.

Egerer, M.H., Philpott, S.M., Liere, H., Jha, S., Bichier, P., and Lin, B.B., 2018. People or place? Neighborhood opportunity influences community garden soil properties and soil-based ecosystem services. *International Journal of Biodiversity Science, Ecosystem Services and Management 14*(1), 32–44.

Egerer, M., Liere, H., Lucatero, A., and Philpott, S.M., 2020. Plant damage in urban agroecosystems varies with local and landscape factors. *Ecosphere* 11(3).

Ericksen, P.J., Ingram, J.S., Liverman, D.M., 2009. *Food Security and Global Environmental Change: Emerging Challenges.* Elsevier.

Ferguson, B.G., Morales, H., Chung, K., and Nigh, R., 2019. Scaling out agroecology from the school garden: the importance of culture, food, and place. *Agroecology and Sustainable Food Systems* 43(7–8), 724–743.

Foley, J.A., 2005. Global Consequences of Land Use. *Science* 309(5734), 570–574.

Francis, C., Lieblein, G., Gliessman, S., Breland, T.A., Creamer, N., Harwood, R., Salomonsson, L., Helenius, J., Rickerl, D., Salvador, R., Wiedenhoeft, M., Simmons, S., Allen, P., Altieri, M., Flora, C., and Poincelot, R., 2003. Agroecology: The ecology of food systems. *Journal of Sustainable Agriculture* 22(3), 99–118.

Francis, C., Lieblein, G., Steinsholt, H., Breland, T.A., Helenius, J., Sriskandarajah, N., and Salomonsson, L., 2005. Food systems and environment: Building positive rural-urban linkages. *Human Ecology Review* 60–71.

Fraser, E.D., 2002. Urban ecology in Bangkok, Thailand: Community participation, urban agriculture and forestry. *Environments* 30(1), 37–50.

Galloway, J.N., Townsend, A.R., Erisman, J.W., Bekunda, M., Cai, Z., Freney, J.R., Martinelli, L.A., Seitzinger, S.P., and Sutton, M.A., 2008. Transformation of the nitrogen cycle: recent trends, questions, and potential solutions. *Science* 320(5878), 889–892.

Gaston, K.J., Smith, R.M., Thompson, K., and Warren, P.H., 2005. Urban domestic gardens (II): experimental tests of methods for increasing biodiversity. *Biodiversity and Conservation* 14(2), 395–413.

Geist, H.J., and Lambin, E.F., 2001. What drives tropical deforestation. *LUCC Report series* 4, 116.

Gliessman, S.R., 1990. Agroecology: researching the ecological basis for sustainable agriculture. In *Agroecology* (pp. 3–10). Springer, New York, NY.

Gliessman, S.R., Engles, E. and Krieger, R., 1998. *Agroecology: ecological processes in sustainable agriculture.* CRC Press, Boca Raton, FL

Gliessman, S.R., 2001. *Agroecology: ecological processes in sustainable agriculture* . Ed. Of Univ. Federal University of Rio Grande do Sul, UFRGS.

Gliessman, S., 2011. *Transforming food systems to sustainability with agroecology.* CRC Press. Boca Raton, FL

Gliessman, S.R., 2016. Agroecology: Roots of resistance to industrialized food systems. *Agroecology: A transdisciplinary, participatory and action-oriented approach* 23–35.

Golany, G.S., 1995. *Ethics and urban design: Culture, form, and environment.* John Wiley and Sons. New York, NY.

Jácome, A.G., 2011. *Various stories: a journey back in time with Mexican farmers* . Iberoamerican University, Mexico.

Graeub, B.E., Chappell, M.J., Wittman, H., Ledermann, S., Kerr, R.B., and Gemmill-Herren, B., 2016. The state of family farms in the world. *World Development* 87, 1–15.

Gregory, M.M., Leslie, T.W., and Drinkwater, L.E., 2016. Agroecological and social characteristics of New York city community gardens: contributions to urban food security, ecosystem services, and environmental education. *Urban Ecosystems* 19, 763–794.

Grimm, N.B., Grove, J.M., Pickett, S.T.A., and Redman, C.L., 2000. Integrated Approaches to Long-Term Studies of Urban Ecological Systems.., *BioScience* 50(7), 571–584.
Grimm, N.B., Faeth, S.H., Golubiewski, N.E., Redman, C.L., Wu, J., Bai, X., and Briggs,J.M., 2008. Global change and the ecology of cities. *Science* 319, 756–760.
Grove, J.M., Childers, D.L., Galvin, M., Hines, S., Muñoz-erickson, T., Svendsen, E.S., and Al, G.E.T., 2016. Linking science and decision making to promote an ecology for the city: practices and opportunities. *Ecosystem Health and Sustainability* 2(9), 1–10.
Hanna, A.K., and Oh, P., 2000. Rethinking urban poverty: A look at community gardens. *Bulletin of Science, Technology and Society* 20(3), 207–216.
Holloway, L., Kneafsey, M., 2004. Producing–consuming food: closeness, connectedness and rurality. In L. Holloway, M. Kneafsey, Eds., *Geographies of Rural Cultures and Societies*. Ashgate, London.
Holt-Giménez, E., and Altieri, M.A., 2013. Agroecology, food sovereignty, and the new green revolution. *Agroecology and Sustainable Food systems* 37(1), 90–102.
McIntyre, N.E., and Hostetler, M.E., 2001. Effects of urban land use on pollinator (Hymenoptera: Apoidea) communities in a desert metropolis. *Basic and Applied Ecology* 2, 209–218.
Hunter, M.C.R., and Brown, D.G., 2012. Spatial contagion: Gardening along the street in residential neighborhoods. *Landscape and Urban Planning* 105, 407–416.
Hunter, M.R., and Hunter, M.D., 2008. Designing for conservation of insects in the built environment. *Insect Conservation and Diversity* 1, 189–196.
Irvine, S., Johnson, L., and Peters, K., 1999. Community gardens and sustainable land use planning: A case-study of the Alex Wilson community garden. *Local Environment* 4(1), 33–46.
Jarosz, L., (2008). The city in the country: Growing alternative food networks in Metropolitan areas. *Journal of Rural Studies* 24(3), 231–244.
Kenney-Lazar, M., 2018. Governing dispossession: Relational land grabbing in Laos. *Annals of the American Association of Geographers* 108(3), 679–694.
Kenney-Lazar, M., Suhardiman, D., & Dwyer, M. B. (2018). State spaces of resistance: industrial tree plantations and the struggle for land in Laos. *Antipode*, 50(5), 1290–1310.
Kremen, C., and Miles, A., 2012. Ecosystem services in biologically diversified versus conventional farming systems: Benefits, externalities, and trade-offs. *Ecology and Society* 17(4), 1–25.
Kurtz, H., 2001. Differentiating multiple meanings of garden and community. *Urban Geography* 22(7), 656–670.
Labao, L., Meyer, K., 2001. The great agricultural transition: crisis, change, and social consequences of twentieth century US farming. *Annu. Rev. Sociol.* 27, 103–124.
Ledant, C., 2017. Urban Agroecology in Rome. *Urban Agriculture Magazine* 33(November).
Lin, B.B., Philpott, S.M., and Jha, S., 2015. The future of urban agriculture and biodiversity-ecosystem services: challenges and next steps. *Basic and Applied Ecology* 16(3), 189–201.
Loram, A., Thompson, K., Warren, P.H., and Gaston, K.J., 2008. Urban domestic gardens (XII): The richness and composition of the flora in five UK cities. *Journal of Vegetation Science* 19(January), 321–330.
Lowenstein, D.M., Matteson, K.C., Xiao, I., Silva, A.M., and Minor, E.S.,. 2014. Humans, bees, and pollination services in the city: the case of Chicago, IL (USA). *Biodiversity and Conservation* 23,2857–2874.
Lowenstein, D. M., and Minor, E.S., 2018. Herbivores and natural enemies of brassica crops in urban agriculture. *Urban Ecosystems* 21,519–529.
Mathieu, R., Freeman, C., and Aryal, J.,. 2007. Mapping private gardens in urban areas using object-oriented techniques and very high-resolution satellite imagery. *Landscape and Urban Planning* 81,179–192.
McClintock, N., 2017. Cultiving (a) sustainability capital: urban agriculture, eco-gentrification, and the uneven valorization of social reproduction. *Annals of the American Association of Geographers* (February 2017).
McDonald, R.I., Marcotullio, P.J., and Güneralp, B., 2013. Urbanization and trends in biodiversity and ecosystem services. In *Urbanization, biodiversity and ecosystem services: Challenges and opportunities.* T. Elmqvist, M. Fragkias, J. Goodness, B. Güneralp, P.J. Marcotullio, R.I. McDonald, S. Parnell, M. Schewenius, M.S. Seto, K.C. Wilkinson (Eds.) (pp. 31–52). Springer, New.
McPhearson, T., Pickett, S.T.A., Grimm, N.B., Niemelä, J., Alberti, M., Elmqvist, T., Weber, C., Haase, D., Breuste, J., and Qureshi, S., 2016. Advancing urban ecology toward a science of cities. *BioScience* 66(3), 198–212.

Melles, S., Glenn, S., Martin, C., 2003. Urban bird diversity and landscape complexity: species–environment associations along a multiscale habitat gradient. *Conservation Ecology* 7(5) [online] URL: http://www.consecol.org/vol7/iss1/art5/.

Méndez, V.E., Bacon, C.M., Cohen, R., and Gliessman, S.R., 2015. *Agroecology: transdisciplinary, participatory and action-oriented approach.* CRC Press, Boca Raton.

Méndez, V.E., Bacon, C.M., and Cohen, R., 2013. Agroecology as a transdisciplinary, participatory, and action-oriented approach. *Agroecology and Sustainable Food Systems* 37(1), 3–18.

Mier y Terán Giménez Cacho, M., Giraldo, O.F., Aldasoro, M., Morales, H., Ferguson, B.G., Rosset, P., Khadse, A., and Campos, C., 2018. Bringing agroecology to scale: key drivers and emblematic cases. *Agroecology and Sustainable Food Systems* 42(6), 637–665.

Miles, A., Wilson, H., Altieri, M., and Nicholls, C., 2012. Habitat diversity at the field and landscape level: Conservation biological control research in California viticulture. In *Arthropod Management in Vineyards* (pp. 159–189). Springer, Dordrecht.

Niemelä, J., 1999. Ecology and urban planning. *Biodiversity and Conservation* 8(1), 119–131.

Niemelä, J., 2000. Is there a need for a theory of urban ecology? *Urban Ecosystems* 3(1996), 57–65.

Niemelä, J., Breuste, J., Elmqvist, T., Guntenspergen, G., James, P., and McIntyre, N., 2011. Introduction. In J. Niemela, J. H. Breuste, T. Elmqvist, G. Guntenspergen, P. James, & N. E. McIntyre (Eds.) *Urban Ecology* (pp. 1–4). Oxford University Press.

Pala, M., Ryan, J., Zhang, H., Singh, M., and Harris, H.C., 2007. Water-use efficiency of wheat-based rotation systems in a Mediterranean environment. *Agricultural Water Management* 93(3), 136–144.

Parmentier, S., 2014. Scaling-up agroecological approaches: What, why and how. *Oxfam-Solidarity, Brussels* 472–480.

Peña, D.G., 2005, October. Farmers Feeding Families: Agroecology in South Central Los Angeles. *In Lecture presented to the Environmental Science, Policy, and Management Colloquium.* University of California, Berkeley. October (Vol. 10).

Philpott, S.M., and Bichier, P., 2017. Local and landscape drivers of predation services in urban gardens. *Ecological Applications* 27(3), 966–976.

Pickett, S.T., Burch, W.R., Dalton, S.E., Foresman, T.W., Grove, J.M., and Rowntree, R., 1997. A conceptual framework for the study of human ecosystems in urban areas. *Urban Ecosystems* 1(4), 185–199.

Pretty, J., 2008. Agricultural sustainability: concepts, principles and evidence. *Philosophical Transactions of the Royal Society B: Biological Sciences* 363(1491), 447–465.

Pudup, M.B., 2008. It takes a garden: Cultivating citizen-subjects in organized garden projects. *Geoforum* 39(3), 1228–1240.

Rasmussen, L.V., Coolsaet, B., Martin, A., Mertz, O., Pascual, U., Corbera, E., Dawson, N., Fisher, J.A., Franks, P., and Ryan, C.M., 2018. Social-ecological outcomes of agricultural intensification. *Nature Sustainability* 1(6), 275–282.

Rebele, F., 1994. Urban ecology and special features of urban ecosystems. *Global Ecology and Biogeography Letters.* 4(6), 173–187.

Rees, W.E., 1997. Urban ecosystems: the human dimension. *Urban Eecosystems* 1(1), 63–75.

Reynolds, K., 2015. Disparity despite diversity: Social injustice in New York City's urban agriculture system. *Antipode* 47(1), 240–259.

Reynolds, H.L., Smith, A. & Farmer, J.R. (2014) Think globally, research locally: Paradigms and place in agroecological research. *American Journal of Botany*, 101, 1631–1639.

Rigueiro-Rodríguez, A., Fernández-Núñez, E., González-Hernández, P., McAdam, J.H., and Mosquera-Losada, M.R., 2009. Agroforestry systems in Europe: productive, ecological and social perspectives. In *Agroforestry in Europe* (pp. 43–65). Springer, Dordrecht.

Rosset, P.M., and Altieri, M.A., 1997. Agroecology versus input substitution: A fundamental contradiction of sustainable agriculture. *Society and Natural Resources* 10(3), 283–295.

Martínez-Torres, M.E., and Rosset, P.M., 2010. La Vía Campesina: the birth and evolution of a transnational social movement. *The Journal of Peasant Studies* 37(1), 149–175.

Rosset, P.M., and Martínez-Torres, M.E., 2012. Rural social movements and agroecology: Context, theory, and process. *Ecology and Society* 17(3), 1–12

Saldivar-Tanaka, L., and Krasny, M.E., 2004. Culturing community development, neighborhood open space, and civic agriculture: The case of Latino community gardens in New York City. *Agriculture and Human Values* 21(4), 399–412.

Schmelzkopf, K., 1995. Urban community gardens as contested space. *Geographical Review* 85(3), 364–381.

Seto, K.C., and Ramankutty, N., 2016. Hidden linkages between urbanization and food systems. *Science* 352(6288), 943–945.

Shashua-Bar, L., and Hoffman, M.E., 2000. Vegetation as a climatic component in the design of an urban street: An empirical model for predicting the cooling effect of urban green areas with trees. *Energy and Buildings* 31(3), 221–235.

Smith, R.M., Thompson, K., Hodgson, J. G., Warren, P.H., and Gaston, K.J., 2006. Urban domestic gardens (IX): Composition and richness of the vascular plant flora, and implications for native biodiversity. *Biol. Cons.* 129, 312–322.

Speak, A.F., Mizgajski, A., and Borysiak, J., 2015. Allotment gardens and parks: Provision of ecosystem services with an emphasis on biodiversity. *Urban Forestry and Urban Greening* 14, 772–781.

Smith, P., Bustamante, M., Ahammad, H., Clark, H., Dong, H., Elsiddig, E.A., Haberl, H., Harper, R., House, J., Jafari, M., and Masera, O., 2014. Agriculture, forestry and other land use (AFOLU). In *Climate* change 2014: mitigation of climate change. Contribution of Working Group III to the Fifth Assessment Report of the Intergovernmental Panel on Climate Change (pp. 811–922). Cambridge University Press.

Tanner, C.J., Adler, F.R., Grimm, N.B., Groffman, P.M., Levin, S.A, Munshi-South, J., Pataki, D.E., Pavao-Zuckerman, M., and Wilson, W.G., 2014. Urban ecology: advancing science and society. *Frontiers in Ecology and the Environment* 12(10), 574–581.

Tao, Z.Y.J., Ting, G.D.F., and Xiaoju, Y., 1993. The effects of refuse compost from urban sources on soybean plants resistance to diseases and insects pests. *Journal of Hubei University (Natural Science Edition)* (1), 18.

Taylor, J.R., and Lovell, S.T., 2014. Urban home gardens in the Global North: A mixed methods study of ethnic and migrant home gardens in Chicago, IL. *Renewable Agriculture and Food Systems* 30(01), 22–32.

Taylor, J., and Mione, T., 2019. Collection of Jaltomata darcyana (Solanaceae), previously unrecorded in cultivation, from a home garden in Chicago, IL. *Renewable Agriculture and Food Systems* 1–3.

Terrile, R.H., Ottmann, G., Guzman, E.S., Lattuca, A., Mariani, S., Timoni, R., Lemos, C., and Asato, E., 2007. Una aproximacion al proceso de agroecologizacion de la agricultural urbana en Rosario, Argentina. *Revista Brasileira de Agroecologia* 2(2).

Krasny, M.E., and Tidball, K.G., 2009. Community gardens as contexts for science, stewardship, and civic action learning. *Cities and the Environment* 2(1), 1–18.

Tidball, K.G., and Krasny, M.E., 2007. From risk to resilience: What role for community greening and civic ecology in cities? *Social learning towards a more sustainable world* ed. A. Wals 149–164. Wageningen: Wageningen Academic Publishers.

Tomich, T.P., Brodt, S., Ferris, H., Galt, R., Horwath, W.R., Kebreab, E., Leveau, J.H.J., Liptzin, D., Lubell, M., Merel, P., Michelmore, R., and Rosenstock, T., 2011. Agroecology: a review from a global-change perspective. *Annual Review of Environment and Resources* 36, 193–222.

Tommasi, D., Miro, A., Higo, H.A., and Winston, M.L., 2004. Bee diversity and abundance in an urban setting. *The Canadian Entomologist* 136(6), 851–869.

Tornaghi, C., 2014. Critical geography of urban agriculture. *Progress in Human Geography* 38(4), 551–567.

Tornaghi, C., and Dehaene, M., 2019. The prefigurative power of urban political agroecology: rethinking the urbanisms of agroecological transitions for food system transformation. *Agroecology and Sustainable Food Systems* 00(00), 1–17.

Tornaghi and Hoeskstra 2017, RUAF, 2017. Urban agroecology. Urban Agriculture Magazine, (November).

Torres-Lima, P., Canabal-Cristiani, B., and Burela-Rueda, G., 1994. Urban sustainable agriculture: The paradox of the chinampa system in Mexico City. *Agriculture and human values* 11(1), 37–46.

Tresch, S., Moretti, M., Bayon, R., Le, Mäder, P., Zanetta, A., Frey, D., and Fliessbach, A., 2018. A gardener's influence on urban soil quality. *Frontiers in Environmental Science* 6, 1–18.

Tscharntke, T., Klein, A.M., Kruess, A., Steffan-Dewenter, I., and Thies, C., 2005. Landscape perspectives on agricultural intensification and biodiversity–ecosystem service management. *Ecology letters* 8(8), 857–874.

Van Der Ploeg, J.D., 2010. The food crisis, industrialized farming and the imperial regime. *Journal of Agrarian Change* 10(1), 98–106.

Vandermeer, J.H., and Perfecto, I., 2012. Syndromes of production in agriculture: Prospects for social-ecological regime change. *Ecology and Society* 17(4).

Vandermeer, J., and Perfecto, I., 2013. Complex traditions: Intersecting theoretical frameworks in agroecological research. *Agroecology and Sustainable Food Systems* 37(1), 76–89.

Vandermeer, J., and Perfecto, I., 2017. Ecological complexity and agroecosystems: Seven themes from theory. *Agroecology and Sustainable Food Systems* 41(7), 697–722.

Vitousek, P.M., Naylor, R., Crews, T., David, M.B., Drinkwater, L.E., Holland, E., Johnes, P.J., Katzenberger, J., Martinelli, L.A., Matson, P.A., and Nziguheba, G., 2009. Nutrient imbalances in agricultural development. *Science* 324(5934), 1519–1520.

Walker, R.E., Keane, C.R., and Burke, J.G., 2010. Disparities and access to healthy food in the United States: A review of food deserts literature. *Health and Place* 16(5), 876–884.

Wezel, A., and Soldat, V., 2009. A quantitative and qualitative historical analysis of the scientific discipline of agroecology. *International Journal of Agricultural Sustainability* 7(1), 3–18.

Wezel, A., Bellon, S., Doré, T., Francis, C., Vallod, D., and David, C., 2009. Agroecology as a science, a movement and a practice. A review. *Agronomy for Sustainable Development* 29(4), 503–515.

Wezel, A., and Peeters, A., 2014. Agroecology and herbivore farming systems–principles and practices. *Options Méditerranéennes* 109(109), 753–768.

White, M.M., 2011. Sisters of the Soil: urban gardening as resistance in Detroit. *Race/Ethnicity: Multidisciplinary Global Contexts* 5(1), 13–28.

Whittinghill, L.J., and Rowe, D.B., 2012. The role of green roof technology in urban agriculture. *Renewable Agric. and Food Systems* 27, 314–322.

WinklerPrins, A.M.G.A., 2002. House-Lot gardens in santarem, para, Brazil: Linking the urban with the rural. *Urban Ecosystems* 6(1–2), 43–65.

Wolch, J.R., West, K., and Gaines, T.E., 1995. Transspecies urban theory. *Environment and Planning D: Society and Space* 13(6), 735–760.

Tan, P.Y., Wong, N.H., Chen, Y., Ong, C.L., and Sia, A., 2003, May. Thermal benefits of rooftop gardens in Singapore. In Proceedings Greening Rooftops for Sustainable Communities Conference, Chicago, Illinois.

Yan, S., 1990. Ecological engineering treatment of urban solid waste. *Research of Environmental Sciences* 1.

Zanette, L.R.S., Martins, R.P., and Ribeiro, S.P., 2005. Effects of urbanization on Neotropical wasp and bee assemblages in a Brazilian metropolis. *Landscape and Urban Planning* 71(2–4), 105–121.

Zasada, I., 2011. Multifunctional peri-urban agriculture—A review of societal demands and the provision of goods and services by farming. *Land Use Policy* 28, 639–648.

Zeng, Y., Huang, W., Zhan, F., Zhang, H., and Liu, H., 2010. Study on the urban heat island effects and its relationship with surface biophysical characteristics using MODIS imageries. *Geo-spatial Information Science* 13(1), 1–7.

Zlotnik, H., 2017. World Urbanization: Trends and Prospects. In G. Hugo, ed. *New Forms of Urbanization* (pp. 43–64). Routledge, London.

1

An Expanded Scope of Biodiversity in Urban Agriculture, with Implications for Conservation

Sacha K. Heath,[1,§] Nina S. Fogel,[2] Jennifer C. Mullikin,[2] and Trey Hull[3]
[1] *Living Earth Collaborative*
[2] *Department of Biology, Saint Louis University*
[3] *Department of Integrated and Applied Sciences, Saint Louis University*
[§] *Corresponding author: Email- skheath@wustl.edu, Living Earth Collaborative, Washington University, One Brookings Drive, St. Louis, MO 63130, USA*

CONTENTS

KEY WORDS: *biodiversity, agriculture, urban, conservation*

1.1 Introduction

Urbanization expanded dramatically across the globe during the twentieth century (Angel et al. 2011). While the urban footprint currently constitutes less than 3% of global land area, it is projected to nearly triple in area between 2000 and 2030 (Schneider et al. 2010, United Nations 2015). Compared to other land uses, urbanization has a disproportionate impact on biodiversity because human concentrations tend to spatially coincide with regions of high biodiversity (Luck 2007). At the same time, semi-natural habitats in urban areas (i.e., urban green infrastructure) can function as biodiversity refuges (Lundholm and Richardson 2010, Goddard et al. 2010, Aronson et al. 2017, Hall et al. 2017). Urban agriculture is a major component of urban green infrastructure, which includes private residential gardens, community (in North America) or allotment (in Europe) gardens, green roofs and vertical walls, and urban orchards (Lin and Egerer 2017). Given the space urban agriculture occupies in cities, and agriculture's dependence on beneficial services conferred by animals (i.e., pollination and predation), urban agroecosystems are positioned to play a key role in harboring and potentially sustaining biodiversity in cities (Lin et al. 2015, Lin and Egerer 2017, Artmann and Sartison 2018).

We set out to explore what is known by the scientific community about the role of urban agroecosystems and urban agriculture broadly in biodiversity conservation. We performed a systematic review and explicitly used broadly defined terms in an attempt to capture studies from urban agricultural settings around the world with a diversity of perspectives about what is meant by biodiversity and conservation. Several reviews within the last decade have assessed the current state of knowledge of biodiversity in cultivated urban environments. These efforts have provided much needed and succinct reviews of the topic of biodiversity and urban agriculture by focusing explicitly on residential or community gardens (Goddard et al. 2010, Guitart et al. 2012, Knapp 2014), urban food (versus ornamental) production (Clucas et al. 2018), in specific geographic regions (i.e., the "global north"; Artmann and Sartison 2018), or on specific taxa (e.g., Rodewald 2016). Thus, we aimed to complement, rather than reproduce, the efforts of previous reviews in several ways. First, we extended our definition of urban agriculture to include the production of ornamental and food plants. Second, we expanded our biodiversity-related search terms to include individual taxa, biodiversity and wildlife indices, and agrodiversity—within all global regions. Finally, we sought to expand the pool of studies on individual taxa by reviewing articles that included urban agricultural study sites as one of many site types—including those that did not explicitly study specific aspects of urban agriculture.

1.2 Methods

1.2.1 Systematic Literature Review

We performed a systematic literature review following guidelines and suggestions of Collaboration for Environmental Evidence (2013) and Livoreil et al. (2017). We used the following search terms and Boolean operators:

> *(biodiversity* OR *wildlife* OR *arthropod** OR *invertebrate** OR *insect** OR *spider** OR *bee** OR *butterfly** OR *pollinator** OR *natural enemy** OR *beneficial insect** OR *pest** OR *mammal** OR *bat** OR *bird** OR *amphibian** OR *frog** OR *salamander** OR *reptile** OR *lizard** OR *snake** OR *plant** OR *flora* OR *vegetation* OR *conservation)*

AND

> *(urban** OR *city* OR *neighborhood*)*

AND

> *(agriculture** OR *garden** OR *farm*)*

We searched the Web of Science and Agricola databases within 'Topics' and 'Keywords Anywhere', and within all years available (1900–2019 and 1970–2019, respectively). We searched all years in Google Scholar using the 'allintitle:' function and restricted the search to not include citations or patents. As of June 21, 2019, we compiled a total of 9,066 citations from these separate searches; deleting duplicates resulted in 6,688 citations. A single author manually read through titles twice, removing theses, conference abstracts, magazines, and non-refereed and predatory journals (identified as such by Ulrich's Periodicals Directory or the Directory of Open Access Journals), resulting in 1,311 titles. We sorted these titles into loose taxa- and subject- subsections among the four authors who read and retained 968 abstracts. After skimming the full texts, we retained and reported on a final set of articles.

We retained original research articles (i.e., reviews were discarded) only if they comprised all three components—*urban*, *agriculture*, and *biodiversity*—defined as follows:

1. *Urban*. Human dominated landscapes such as cities, suburbs, or peri-urban areas.
2. *Agriculture*. Human cultivated plants or animals for food, medicinal, or ornamental purposes.
3. *Biodiversity*. Any wild living organism. Inclusive in this are individual taxa, indices such as species, functional, or phylogenetic diversity and richness, colloquial terms such as "wildlife" and agro-biodiversity.

While evaluating titles, abstracts, and full papers, we made explicit attempts to reduce western hemisphere and global north biases in our review by seeking out culturally and regionally dependent terms with which we were not initially familiar.

1.2.2 Data Extraction and Analyses

We obtained and categorized data from the final article set in four ways. First, we extracted basic information such as article publication year and journal title from the citation records. Second, we used a Latent Dirichlet Allocation (LDA) topic model to assign articles to four general topic groups based on clusters of words extracted from the abstracts (R package 'revtools'; Westgate 2019, R Core Team 2019). The LDA performed 20,000 iterations of a Gibbs sampler algorithm to find an optimal set of 4 distinct topical clusters (details of the modeling procedure, data, and code can be accessed in (Appendix 1; Heath et al. 2020). Third, we extracted and categorized more detailed information from the articles manually with a set of eight variables and multiple-choice or choose-all-that-apply options for each (Table 1.1). Finally, we reviewed articles in written accounts, organized by general types of biodiversity.

TABLE 1.1

Multiple choice variables used to manually categorize the final set of articles

Variable	Description
Countries	Countries in which the study took place.
Municipalities	Municipalities in which the study took place.
Biodiversity type	General types of biodiversity studied.
Urban type	The type of urban region in which the study took place (choose all that apply[a]): urban; peri-urban; suburban; exurban; other
Agriculture type	The type of agriculture (choose all that apply): community/allotment garden or farm; residential garden; commercial garden or farm; botanical garden; park garden or farm[b]; other.
Production type	Type of agricultural production (choose all that apply): food; fiber; ornamental; livestock; other; unknown.
Conservation type	Type of conservation application of study (choose all that apply): management effects = effects of a conservation or management practice; listed = a threatened, endangered, or other category of sensitive species; distribution = the documentation of habitat use; information = conservation not explicit reason for study, but ties made to conservation application in introduction or discussion; none = no applicability to conservation mentioned; other.
Gradient paper	Was this a rural-to-urban gradient study? yes; no.

[a] If "other" was selected, the reviewer provided the category.

[b] "Park garden or farm" articles included only if a distinct garden or farm is in a public park (i.e., parks in general not included).

1.3 General Findings

We retained a total of 431 articles reporting studies that characterized biodiversity in urban agricultural settings (Appendix 2). The number of publications per year increased through time, with the highest proportion of articles published in 2017 (Figure 1.1A). Studies were published in 184 different journals; with 15 of the journals publishing at least five articles each (Figure 1.1B). Studies took place in 61 countries, with over 60% of them in the United States United Kingdom (UK; n = 109), United States (USA; n = 102), Australia (n = 35), and South Africa (n = 22); Figure 1.1C).

Our topic model, based on words from abstracts, sufficiently clustered the articles into four topic groupings (4 groups; Figure 1.2A–B). Topic groups shared highly weighted words (e.g., *gardens*), but were also differentiated along both axes by unique terms. For example, the two most distinct topic groups contained articles with abstracts characterized by the highest weighted terms *landscape*, *habitat*, *bee*, and *abundance* (Topic Group 1, Figure 1.2C) and *bird, populations, natural,* and *activity* (Topic Group 2; Figure 1.2D). The remaining topics were characterized by the words *species*, *plant*, *native*, and *trees* (Topic Group 3, Figure 1.2E) and *biodiversity*, *green*, *vegetation*, *agricultural*, and *city* (Topic Group 4, Figure 1.2F). These data-driven clusters of articles somewhat aligned with our manual groupings of studies by general biodiversity type.

The most frequent types of biodiversity reported in these studies were plants (67% of articles), invertebrates (39%), and birds (23%; Figure 1.3A). By definition, we retained studies that took place at least partially in urban areas, so the high percentage of studies in urban or suburban areas (94%, and 14%, respectively) was expected. In some cases, agriculture-related biodiversity was compared between these areas and peri-urban (10%) or exurban areas (6%; Figure 1.3B). Eighteen percent of studies took place along an urban-to-rural gradient. The types of agriculture in which most of the studies took place included residential (71%) and community/allotment (26%) gardens (Figure 1.3C). Yet, we were interested to discover other rarer types of urban agriculture characterized, including gardens on buildings and other urban green infrastructure (23 articles), and urban foraging of ornamental and wild plants (4 articles; Figure 1.3C). Ornamental plants (60%) and food (48%) were the dominant production types among articles, however 22% of articles did not clarify which type of production took place in the agricultural setting (Figure 1.3D). Thirty percent of articles reported on studies that explicitly measured conservation or management effects on biodiversity in an urban agricultural setting, while most articles fell under the category of basic research to inform conservation (71%; Figure 1.3E). Twelve percent of articles made no reference to biodiversity conservation.

1.4 Biodiversity Accounts

In the following accounts, we summarize the research on plants, invertebrates, birds, mammals, reptiles, soil microbes, amphibians and fish in urban agriculture (Figure 1.3A) and highlight the opportunities and challenges for conservation of these groups in urban agricultural systems. Researchers employed methods that spanned multiple fields of inquiry including the life and social sciences and landscape and urban planning. Biodiversity was quantified and compared using empirical measurement of specific taxa or community indices (e.g., Daniels and Kirkpatrick 2006, Mattah et al. 2017, Bożek et al. 2015, Lowenstein et al. 2017), remote sensing and image assessment (e.g., Cabral et al. 2017, da Rocha et al. 2018), interviews and surveys with residents (e.g., Kurz and Baudains 2012, Rodriguez et al. 2017, Prevot et al. 2018), and case studies (e.g., Acar et al. 2007, Fischer et al. 2019, Dennis and James 2017, Barau et al. 2013). Several research teams studied biodiversity across multiple residential gardens enrolled in wildlife-friendly gardening programs (e.g., Beumers and Martens 2015, Widows and Drake 2014, Gaston et al. 2005a), while others did so in networks of urban community gardens (e.g., Gardiner et al. 2014, Quistberg et al. 2016, Borysiak et al. 2017, Philpott et al. 2019). Despite the breadth of research covered here, information gaps remain. We discuss these and suggest directions for future research in subsequent sections of the chapter (see Section 1.7. **Information Gaps and Future Directions**).

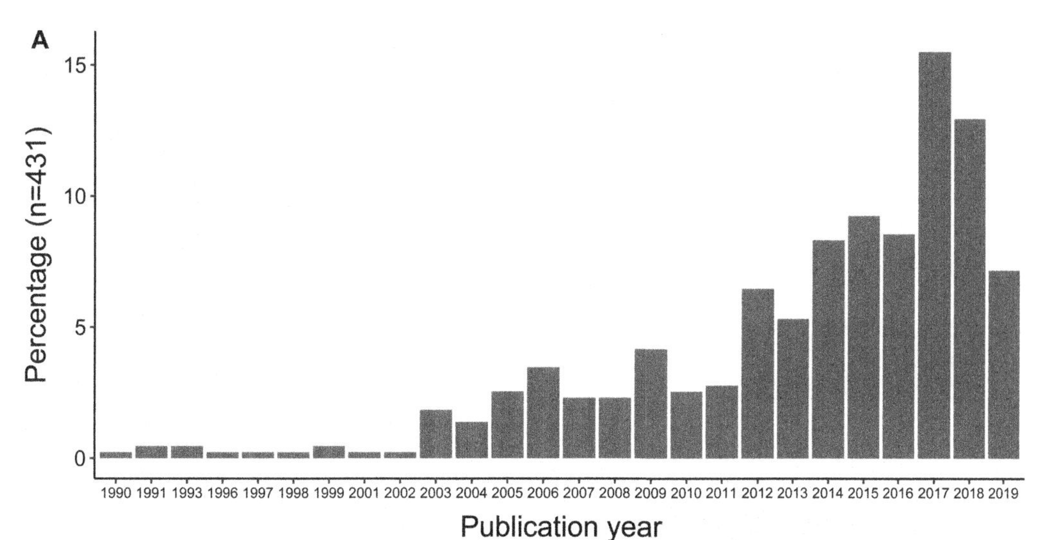

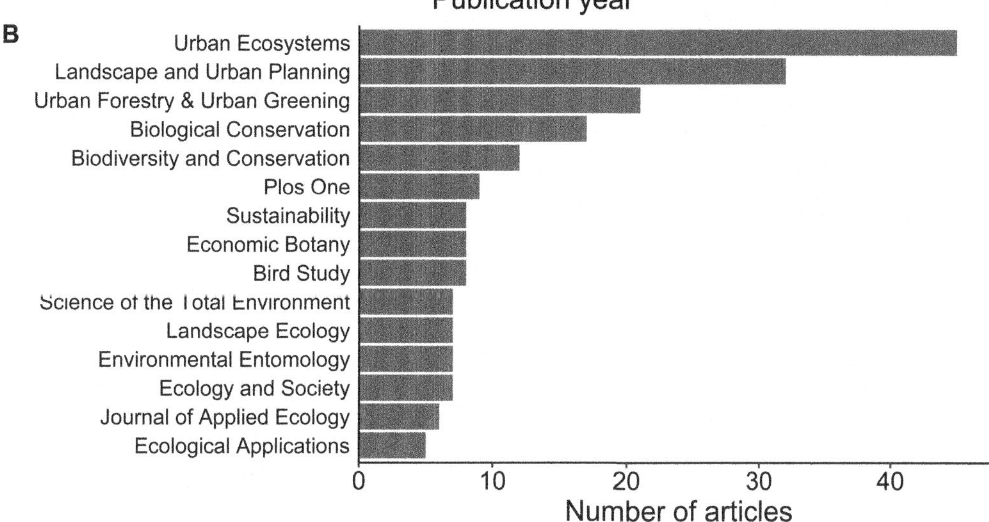

FIGURE 1.1 Percentage of publications per year, as of June 21, 2019 (A), number of articles in the top 15 journals in which at least 5 articles were published (B), and the number of studies performed in each country (C), out of the 431 publications in the review's final set.

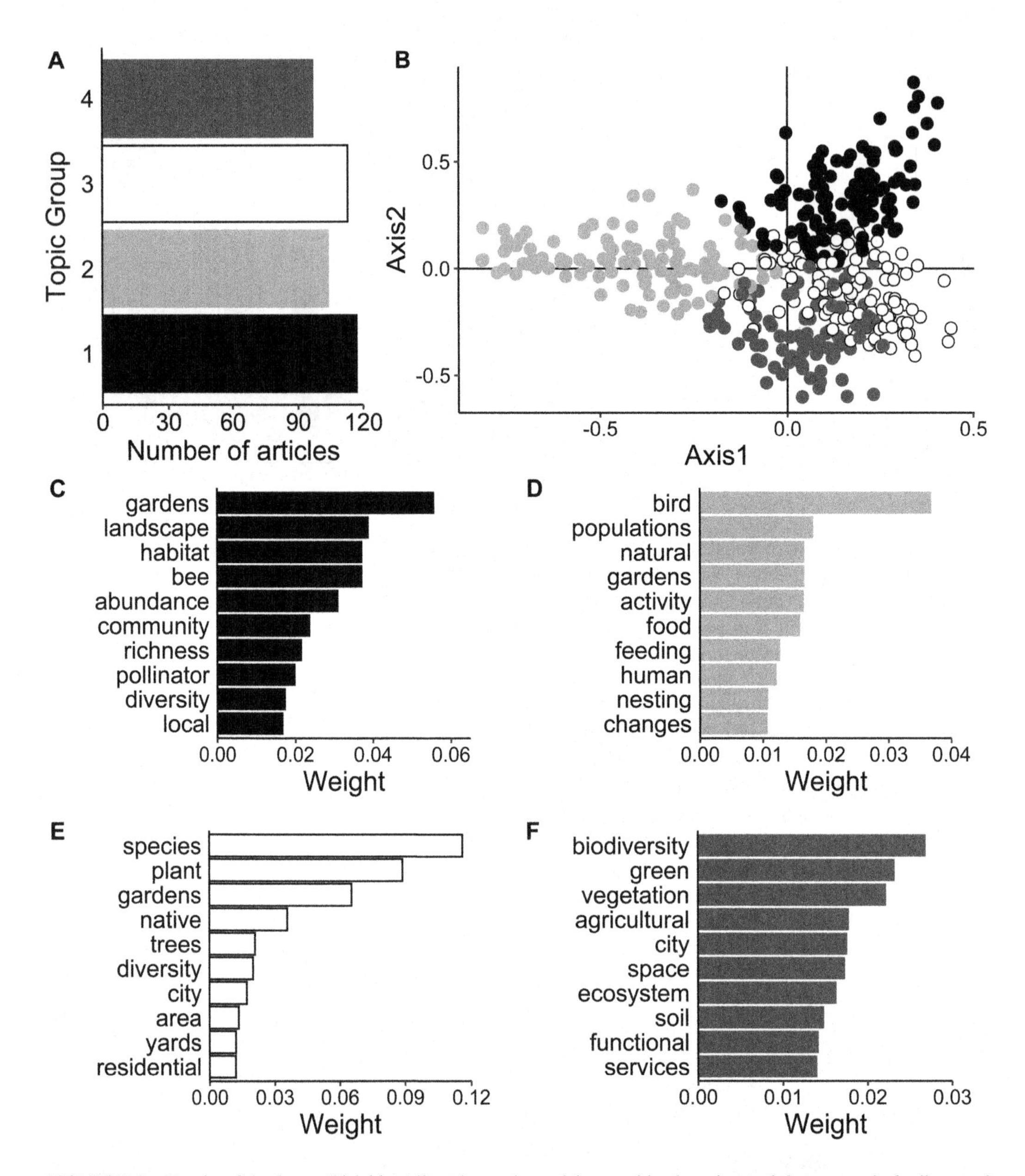

FIGURE 1.2 Results of the Latent Dirichlet Allocation topic model, a machine learning tool that generatively discerned four unique topic groups among the set of 431 articles reviewed (based on the highest-weighted words from abstracts). Each topic group (A) contains articles represented by a point plotted along two-dimensional axes (B), described by highly weighted words extracted from article abstracts (C, D, E, F; note different scales on x-axes). Only the top ten highest weighted words are presented. Colors link the same topic groups in A, B, C, D, E, and F.

1.4.1 Plants

We sub-divided the literature on plants in urban agriculture (n = 290) into five ecological and functional groupings: 1) general plant communities, 2) native plants, 3) invasive and non-native plants, 4) crop plants, and 5) medicinal plants. Interactions between plants and pollinators, and their use as habitat by invertebrate and vertebrate animals are covered in accounts for those taxa. The overarching theme across

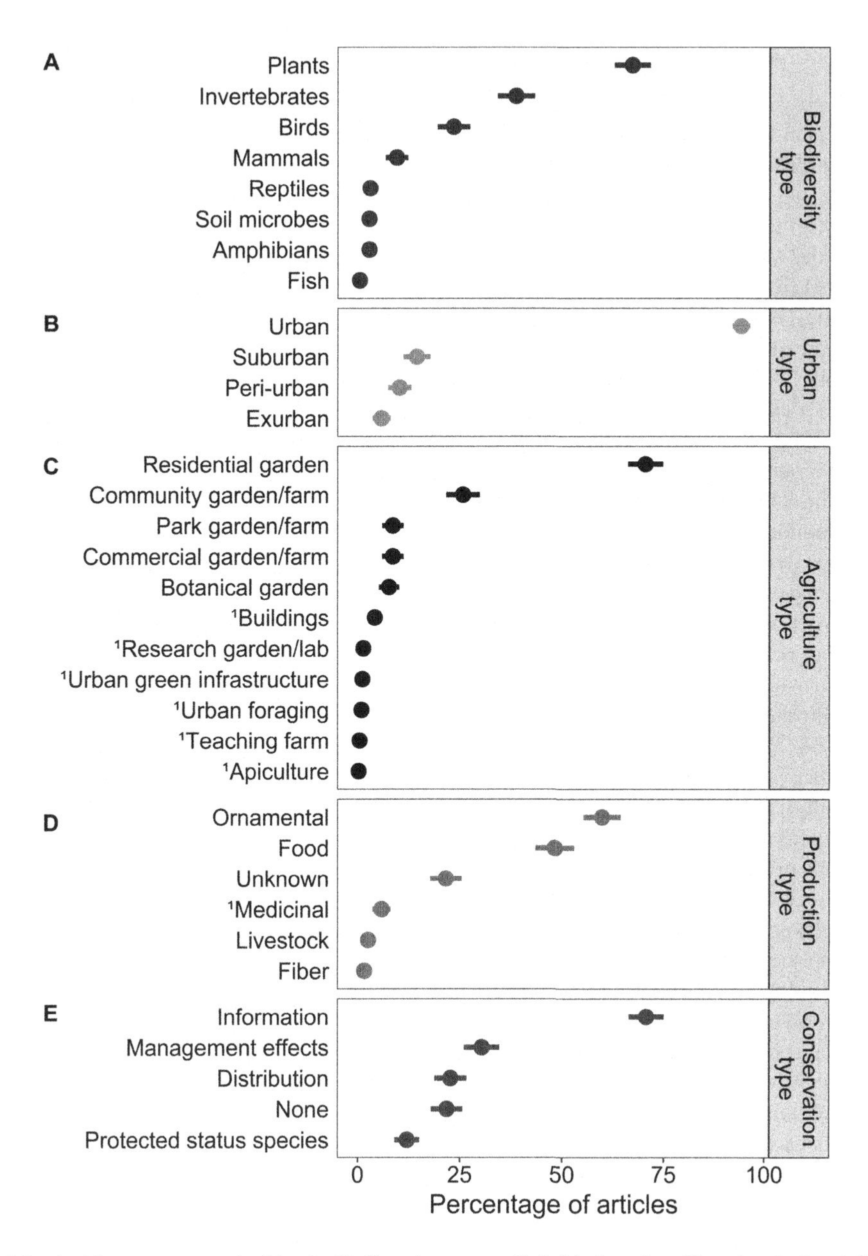

FIGURE 1.3 Articles were categorized by the biodiversity type studied (A), the urban (B) and agricultural (C) type in which the studies took place, the type of agricultural production (D), and the type of conservation application addressed by the study (E), if any. "Other" options are indicated by 1. The percentage of articles within each type category (points; lines are 95% confidence intervals) sum to greater than 100% because choose-all-that-apply answers were permitted. Variable descriptions are provided in Table 1.1.

most of the plant articles was the use of plants for human benefit, with potential side benefits for plant conservation. The literature highlighted that urban agriculture is a largely untapped resource for human communities. Indeed, plants grown in urban areas are a potential source of income if social, economic, and political structures are in place to help establish commercial outlets for growers. As with studies of urban ecology in general, many conclusions of the research were scale-dependent, and direct comparisons across cities or urban green spaces remains a challenge.

1.4.1.1 General Plant Communities

Urban agroecosystems are important reservoirs for high plant species richness. A large body of the research on plants in urban agroecosystems was dedicated to quantifying the species composition of residential gardens, dominant plant use (food, ornamental, or medicinal plantings), and the percentages of non-native and ornamental species planted over natives. Gardens have been interpreted as their own ecosystem, with gardeners driving changes in the system (Egerer et al. 2019a). Depending on what is planted and the watering scheme, gardens can support a different microclimate from the surrounding environment (Lin et al. 2017a). This is important to note because of the substantial acreage that gardens occupy in cities. For example, in Poland, allotment gardeners collectively manage the largest percentage of city land area (Klepacki and Kujawska 2018), and in Sheffield, England, domestic gardens covered approximately 23% of the city (Gaston et al. 2005b). Greater garden area has been associated with higher plant species richness (Sierra-Guerrero and Amarillo-Suarez 2017, Ward and Amatangelo 2018, Smith et al. 2006). Residential gardens with greater plant diversity provided more ecosystem services (Caballero-Serrano et al. 2016) and were important for species conservation (Šantrůčková et al. 2017, but see Gbedomon et al. 2017). Urban gardens are suitable habitat for endemic, endangered plant species, but are mostly dominated by non-native cultivated crops or invasive species (Lubbe et al. 2011).

Seventy-two percent of plant research took place in residential gardens, with studies on plant species composition in gardens conducted primarily in Europe (n = 21), North America (n = 16), and Asia (n = 15), followed by Africa (n = 10) and Australia and South America (n = 5). Most studies examined factors relating to plant species richness and diversity in residential and community gardens within a single city or region. At the continental scale, extreme temperatures, precipitation, and gardner education level had significant effects on cultivated plant species composition (Cubino et al. 2019a). Looking at gardens in seven cities across the USA, weedy, spontaneously grown plants (mainly non-native grasses and short, small-seeded plants) in urban yards homogenized the plant species pool between cities (Cubino et al. 2019b). However, the intentionally cultivated plant species differentiated yard flora in USA cities, indicating that human-driven planting decisions and the local climate determine differences in species composition. Alternatively, a study in the Republic of Benin showed that home garden plant diversity was a subset of regional plant diversity, with 60% of regional plant species growing in urban gardens (Gbedomon et al. 2017).

Interestingly, proximity to urban areas has had varying effects on plant species richness of residential gardens. In species rich communities in the Amazon, the draw of commerce from nearby urban areas influences planting decisions of the gardeners (Wezel and Ohl 2005). Similarly, along an urban-to-rural gradient in Peru, proximity to an urban area was more important than differences in cultural background between villagers in explaining residential garden plant species diversity in each village. Commercial markets for subsistence crops in urban areas influenced garden plant composition in villages closer to the city, and rural markets for tourism handicrafts or plant-based dyes for hammocks influenced gardens in villages further away (Lamont et al. 1999). In Campeche, Mexico, plant species composition, but not plant diversity, differed along the urban-to-rural gradient, with urban gardens containing more herbaceous and ornamental plants compared to the subsistence and native plants grown in rural gardens (Poot-Pool et al. 2015).

When using socioeconomic factors to predict garden plant diversity or richness, results varied by city, likely due to the different cultural norms of residents. For example, in Boston, Massachusetts, USA, garden vegetation structure was not affected by neighborhood socio-economic characteristics (Ossola et al. 2019), while in Raleigh, North Carolina, USA, neighborhood practices, ethnicity, and income all influenced landscaping practices (Peterson et al. 2012). Socioeconomic factors influenced gardens in Poznan, Poland (Borysiak et al. 2017), multiple cities in the UK (Loram et al. 2008), the Catalonia region of Spain (Cubino et al. 2016), France (Marco et al. 2010), Dunedin, New Zealand (van Heezik et al. 2010), and multiple cities in the USA (Cubino et al. 2019b, Lowenstein and Minor 2016).

1.4.1.2 Native Plants

Several studies explicitly examined the role of native plants in urban gardens (e.g., Head and Muir 2006a, Whelan et al. 2006, Reid and Oki 2008, Goddard et al. 2013), while many other studies incorporated native plants as part of overall garden plant composition analyses. Native plants—those that are endemic to a particular region and have coevolved with the flora and fauna of that region—are proposed to play a key role in food security, conservation, economic livelihoods, and overall biodiversity and ecosystem functioning in urban areas (Cousins and Witkowski 2015). There is a social and historical component to gardeners choosing native versus ornamental or non-native plants (Head and Muir 2006a). A study in Sydney, Australia and surrounding cities, for example, found that gardeners viewed growing native plants as a moral imperative and because the plants "belonged" to the area, while those who grew ornamental or non-native plants mostly did so for aesthetic reasons (Head and Muir 2006a). Survey respondents across several cities in the UK, Australia, and the Netherlands indicated that they included native plants as a form of wildlife-friendly gardening in their yards (Gaston et al. 2007, Goddard et al. 2013, Mumaw and Bekessy 2017, Beumer 2018). Because native plants come into contact with horticultural and ornamental varieties in urban and suburban areas, there is a potential for gene flow from cultivars to native plants in human dominated areas (Whelan et al. 2006). It is expected that opportunities for hybridization between cultivars and native species will increase with climate change, and research focusing on native plant resilience to stressors associated with climate change will likely become more common (Reid and Oki 2008).

1.4.1.3 Invasive and Non-Native Plants

Plants are often imported for their aesthetics, escape propagation in residential gardens, and become invasive. In Riebeek Kasteel, South Africa, one third of garden plants were or had the potential to become invasive (McLean et al. 2018). In Switzerland, people discarded their yard waste in adjacent forests, which spread non-native seeds (Gaggini et al. 2017). Furthermore, in a Spanish study it was found that invasive plants were sold in nurseries, indicating the need of appropriate labeling and regulations concerning the selling of invasive plants (Cubino et al. 2015). Many trees planted in urban areas are non-native (Pincetl et al. 2013). This phenomenon may be partially explained by the observation that people do not consider invasiveness potential when deciding which species to plant, and instead focus on trees that provide resources such as food, shade and firewood (Moro et al. 2014). As a conservation measure, and to avoid further spread of invasive plants, researchers are studying the development of sterile versions of horticultural invasive plants for commercial use (Wilson et al. 2012).

1.4.1.4 Crop Plants

The literature on urban agrodiversity (i.e., crop diversity) was unique in that, unlike research on other plant group types, agrodiversity studies were cross-disciplinary and intertwined aspects of policy with agriculture and human communities (Taylor and Lovell 2014, Avila et al. 2017, da Silva et al. 2018). Urban agrodiversity researchers sought to establish community relations and connections to local culture (Woods et al. 2016, Taylor and Lovell 2014), and worked with local governments to help establish commercial outlets for growers (Avila et al. 2017, da Silva et al. 2018). Of primary concern in crop diversity studies was the loss of genetic diversity of wild and cultivated crops through modern and intensified agricultural systems. Several studies focused on the genetic diversity of crops in urban gardens, how urban gardens could be reservoirs of diversity, and how human migrants brought cultivated varieties to new locations (da Silva et al. 2018, Avila et al. 2017, Ricogray et al. 1990). For example, pineapple landraces (i.e., local cultivars) were adequately preserved in residential gardens, creating an economic market for the residents growing pineapples in Cabaceiras do Paraguaçú, Bahia, Brazil (da Silva et al. 2018), and Chinese immigrants to the USA preserve seeds and specific landraces in community gardens in Chicago, Illinois (Taylor and Lovell 2014).

Researchers used questionnaires and site surveys of gardens to assess gardener demographics (mostly women and older residents), people's perceptions of gardening, and how these were associated with plant composition in gardens (da Silva et al. 2018, Siviero et al. 2011, Avila et al. 2017). Gardens were used to reconnect with familial culture, add culture-specific food plants to diets, and sometimes gain income through the sale of food (Taylor and Lovell 2014, Gopal and Nagendra 2014). In Bangalore, Karnataka, India, planting area and finances were extremely limited, so residents used pots to grow the fruit trees, crop plants, and medicinal herbs that were essential to social activities and diet (Gopal and Nagendra 2014). The presence of locally important medicinal plants and food crops in residential gardens were indicative of the importance of culture and socioeconomic status in the composition of the residential gardens (Avila et al. 2017). In two US studies, there were differences in the composition of plants in community gardens based on the cultural background of the gardener (Taylor and Lovell 2014, Woods et al. 2016). Similarly, socioeconomic trends influenced the type and prevalence of fruit trees grown for sustenance in home gardens. In Kinshasa, Democratic Republic of Congo, most residents wanted trees but only homeowners and those who could afford the initial investment had fruit trees (Etshekape et al. 2018). Conversely, in Kigali, Rwanda, fruit trees were favored by residents with little money or wealth while affluent residents preferred palms (Seburanga et al. 2014). In Pará, Brazil, families in poverty grew more fruit trees than other socioeconomic groups as a means of food security (Rayol et al. 2019), whereas poorer South Africans cultivated trees for firewood while wealthier people cultivated fruit trees (Kaoma et al. 2014). In Rio Branco, Acre, Brazil, however, socioeconomic status did not influence the species of food crops grown in residential gardens, while it did have an effect on yard size and therefore garden size and associated crop diversity (Siviero et al. 2011).

1.4.1.5 Medicinal Plants

Medicinal plants grown in residential gardens provided a source of primary care for residents, supplementary income, and a root to traditional knowledge (Mosina and Maroyi 2016, Mosina et al. 2015, Taylor et al. 2017, Zuin et al. 2010). In poorer communities, medicinal plants were overwhelmingly used as a supplement to income and diet (Castro et al. 2018, Dasylva et al. 2018, Karhagomba et al. 2013, Kujawska et al. 2018, Maroyi and Mosina 2014, Merlin-Uribe et al. 2013, Zuin et al. 2010). Gardens were limited in size, and market forces influenced whether gardeners planted crops for food or medicinal purposes. In the Buhozi region of the Democratic Republic of the Congo, medicinal plants were more profitable than traditional food crops (Karhagomba et al. 2013), whereas in homegardens in cities of three Brazilian states, there was a higher occurrence of food plants than medicinal plants and Brazilians living further from cities had a higher tendency to garden than city residents (Botelho et al. 2014). In Mexico City, Mexico, people who tended Chinampas gardens (i.e., a type of agricultural practice unique to the region) shifted from growing traditional crops and medicinal plants to growing horticultural plants to keep their gardens economically viable (Merlin-Uribe et al. 2013). Proximity to Mexican cities was equated with better access to roads, higher school attendance, and less time spent gardening medicinal plants or gaining traditional cultivation knowledge (Castro et al. 2018). Interestingly, those who lived closer to natural areas with access to wild medicinal plants planted fewer of them in their gardens, because wild sources were available nearby (Kujawska et al. 2018). In a survey of communities in Abaetetuba, Para, Brazil, women were the main family caregivers and tenders of their gardens, where up to 56% reported that their first line of medical treatment comes from the medicinal plants in their gardens (Palheta et al. 2017). Women also kept culture and traditional practices alive through the cultivation of medicinal plants in their gardens (Botelho et al. 2014, Kujawska et al. 2018, Mosina and Maroyi 2016 Palheta et al. 2017, Pradeiczuk et al. 2017, Zuin et al. 2010). Several studies highlight that, for these benefits to continue, local communities need greater economic outlets, gardening education, policy and conservation cooperation with local governments e.g., (Castro et al. 2018, Mosina et al. 2015).

1.4.2 Invertebrates

We reviewed the literature on invertebrate biodiversity in urban agriculture (n = 167) within five categories: 1) pollinators and pollination, 2) crop pests and natural enemies, 3) soil macrofauna, 4) disease vectors, and 5) additional invertebrate biodiversity.

1.4.2.1 Pollinators and Pollination

Fifty percent of invertebrate studies examined pollinators and the ecological function of invertebrate pollination in urban agriculture settings. Of note, is that most work on pollinators focused only on bees (n = 54), with less attention given to butterflies (n = 15) and other pollinators such as flies (n = 2). The research primarily addressed pollinator conservation indirectly, by reporting research findings that can inform conservation activities—for example, by describing pollination networks, species abundance and diversity, and how pollinators respond to specific habitat features or urbanization. Many studies examined whether ornamental flowers prevalent in home and community gardens could serve as adequate floral resources for insect pollinators (e.g., Bożek et al. 2015, Masierowska et al. 2018). In general, some flowers—whether they were horticultural varieties or native species—were very attractive to many pollinator species, some flowers were moderately attractive, whereas a vast majority of flower species were not attractive or only had a few specialist visitors (Bennet et al. 2018, Garbuzov et al. 2017, Hanley et al. 2014, Lowenstein et al. 2019, Norfolk et al. 2014, Frankie et al. 2013). Studies that examined phenology noted that having non-native ornamental flowers could have the added benefit of blooming before or after the native plant season, increasing the days where floral resources are present (Koyama et al. 2018, Mach and Potter 2018). In general, increasing biodiversity of pollinators in urban agroecosystems was correlated with the diversity of attractive floral resources, regardless of the plant's native status (Davis et al. 2017, Di Mauro et al. 2007). Conservation efforts should ensure that suggested plant lists for pollinator gardens and plants sold at gardening centers are species that pollinators utilize as floral resources (Garbuzov et al. 2017).

Community and residential gardens were often hotspots for pollinators, and even relatively small patches supported enough pollinators to effectively pollinate fruiting crops (Matteson 2009). In a survey of one small residential garden in Ukiah, California, USA, two-thirds of the city-wide bee species were found (Frankie et al. 2013), and a newly established garden in Berkeley, California, USA attracted 32 bee species in its first growing season (Wojcik et al. 2008). Another common theme was an assessment of functional diversity. Many studies note a dearth of ground nesting bees in urban agroecosystems compared to less urban areas, most likely due to the lack of uncovered ground (Mazzeo and Torretta 2015, Threlfall et al. 2015, Matteson and Langellotto 2010). Additionally, floral specialist butterflies were more restricted from urban areas as compared to generalists (Bergerot et al. 2010). Others found that pollinator diversity was driven by the interaction between local habitat and landscape level land cover patterns (Ballare et al. 2019, Di Mauro et al. 2007). Overall, pollinators benefitted from urban agroecosystems and studies remarked on residents' desire to learn more about urban pollinators (e.g., Pawelek et al. 2009).

A few studies, via interviews and surveys, documented the types of practices used by gardeners to attract pollinators to their gardens including "bee homes" (Mumaw and Bekessy 2017), "insect hotels" (Goddard et al. 2013, Beumer 2018), and nectar flower plantings (Beumer 2018; Shwartz et al. 2014). Even fewer studies explicitly measured the effects of these types of resource provisioning on pollinators, and the results were mixed. For example, pollinators responded positively to "biodiversity-friendly" certification programs in Parisian public gardens (Shwartz et al. 2013a), while bumblebees (*Bombus*) failed to occupy experimental nesting sites and very few butterfly larvae occupied experimentally grown host plants in residential backyards of the UK (Gaston et al. 2005a). Solitary bees (*Hylaeus communis, Osmia rufa*), on the other hand, did occupy artificial nests in 60% of UK residential gardens where they were placed (Gaston et al. 2005a). Given the prevalence of honeybees (*Apis mellifera*) in urban areas globally, our search captured surprisingly few papers that focused on the direct impact of managed honeybee hives on bee diversity, especially in areas where honeybees are non-native. Research from Australia and the USA, however, suggested that honeybees out-competed native bees for resources (Threlfall et al. 2015) and higher honeybee abundance was correlated with higher bee parasite abundances (Cohen et al. 2017).

1.4.2.2 Invertebrate Crop Pests and Natural Enemies

Twenty-six of our invertebrate studies examined biocontrol in urban agriculture (see Chapter 2 for more on complex ecological interactions and ecosystem services in urban agroecosystems). Nine studies were performed in food production systems, eight in ornamental gardens, one in both food and ornamental gardens, and eight studies did not differentiate the production type. While most studies (n = 15) did not explicitly address conservation of the study species, their conclusions are uniformly applicable to concrete improvements in ecological functioning and invertebrate biodiversity. Overall, these studies suggested that maintaining a heterogeneity of urban green space types could be important for conserving beneficial arthropod groups and for supporting ecological services for urban agriculture (Gardiner et al. 2014, Mata et al. 2017). Few studies examined if or how private gardeners provision for beneficial insects. A Dutch survey found that 8% of survey participants either attracted or released beneficial insects in their gardens (Beumer 2018), while Gaston et al. (2005a) recorded nine species of parasitoid wasps occupying artificial nests placed for them in gardens in the UK.

An important reason for examining ornamental cultivation in the context of urban agriculture is the mediation of pest and natural enemy interactions in food crops by ornamental plants. For example, in residential gardens of suburban South Bohemia, Czechoslovakia, a common garden plant (hydrangea; *Philadelphus coronarious*) harbored a non-pest subspecies of aphid (*Aphis fabae cirsiiacanthoidis*) which hosted a type of parasitoid that also attacked pest aphids in nearby food crops (Stary 1991). Providing floral resources for ichneumonid parasitoids (*Pimpla disparis*) increased parasitism rates of the bagworm (*Thyridopteryx ephemeraeformis*)—a common herbivore of ornamental shrubs and trees in urban areas of the eastern USA (Ellis et al. 2005). Urban cultivation of ornamental plants procured through global trade also plays a large role in transmitting invasive insects to locales outside of their natural ranges. This unrestricted trade is a key driver of alien insect spread in Europe with drastic implications for the historic vegetation of very old cities (Ciceoi et al. 2017). The harlequin lady beetle (*Harmonia axyridis*) which was introduced as a biological control agent in agricultural areas worldwide has spread to natural areas in South Africa where it is influencing the composition of lady beetle and insect herbivore communities. The areas of highest adult and larval harlequin lady beetle abundance were in urban areas and appeared associated with a non-native aphid that feeds on cultivated oak trees (Mukwevho et al. 2017).

Following a similar trajectory of research in rural agricultural systems, researchers in urban agroecosystems have begun disentangling the effects of natural vegetation on natural enemy diversity at local and landscape scales. In California, USA urban community gardens embedded in the most urbanized landscapes supported the highest abundance and species richness of an important crop pest predator (Coccinellidae; Egerer et al. 2017), and the highest predation rates of crop pests were in relatively smaller community food gardens (Philpott and Bichier 2017). Conversely, parasitoid species richness was lower in more highly urbanized landscapes, and parasitoid abundance was higher in larger community food gardens (Burks and Philpott 2017). Counter to both results, neither species richness nor functional diversity of ground beetles (Caribidae) in urban gardens responded to any local or landscape factors measured; instead variation was found among measures of specific taxonomic and morphological traits (Philpott et al. 2019). In areas of high to moderate levels of urbanization in Madison, Wisconsin, USA, parasitoid abundance increased with local flower diversity, while at larger scales parasitoid diversity decreased with increasing impervious surfaces (Bennet and Gratton 2012). In Cincinnati, Ohio, USA, lower levels of leaf damage were found in non-native versus native plant species but no differences in herbivory were found in plants within residential gardens versus more natural settings (Matter et al. 2012).

There were a few features unique to urban agroecosystems (e.g., anthropogenic structures and development patterns) that influenced pest and natural enemy communities. For example, Peralta et al. (2011) found that garden walls reduced the colonization of ornamental host plants by leaf miners (*Liriomyza commelinae*), while leaf miner parasitism decreased with increasing distances between the host plant and ground pavement (Peralta et al. 2011). Vacant lots and community gardens in Akron and Cleveland, Ohio, USA supported similar abundances of several taxonomic families of predatory arthropods, but each urban land use type also supported higher abundances of two distinct family groups (Gardiner et al. 2014). Urban residential yards with gardens sustained a higher abundance of spiders than yards with no garden (Rodriguez-Rodriguez et al. 2015) and higher species richness (but lower abundance) than vacant lots (Burkman and Gardiner 2015). Within gardens, spiders responded positively to increased mulching and flowering plants (Otoshi et al. 2015). Overall, urban gardens supported a unique spider assemblage compared to other land types (Moorhead and Philpott 2013, Burkman and Gardiner 2015). In private gardens of Chiapas, Mexico, Ichneumonoidea parasitoid abundance was positively associated with the percentage of mulch cover and bare ground, and experimental pest insect egg and larvae removal increased with increasing flower abundance and herbaceous cover (Morales et al. 2018). Conversely, Chalcoidea parasitoid abundance decreased with the number of herbaceous species in these gardens, demonstrating complex responses of multiple organisms to gardening practices.

1.4.2.3 Soil Macrofauna

Soil macrofaunal communities also play an important functional role in urban agroecosystems, though fewer studies have been conducted on these invertebrates. Increases in plant species richness across urban garden sites has been associated with greater macrofaunal species composition, and in turn, increased litter decomposition (Tresch et al. 2019). The species and functional composition of soil macrofaunal communities can vary between different types of urban green spaces within cities (Bao-Ming et al. 2012, Joimel et al. 2017), but can also be more similar within cities than between cities (Joimel et al. 2019). Conversely, within a community garden setting, Egerer et al. (2018a) argued that soil quality might be detached from the influences of geographic location or site history because of the homogenization of soil components that can take place across gardens. Given the isolation of gardens and farms in urban areas one might expect barriers to colonization by soil invertebrates, yet even rooftop gardens were occupied by species-rich springtail (Collembolan) communities, presumably by either wind dispersal or compost inputs (Joimel et al. 2018). Similarly, several types of soil invertebrates (e.g., Chilopoda, Oligochaetes, Mollusca, Isopoda, and Collembola) colonized experimental woodpiles within 23 months of placement in UK residential gardens (Gaston et al. 2005a).

1.4.2.4 Disease Vectors and Hosts

The proximity of large human populations to agricultural practices poses a unique health risk for urban agriculture. Mosquitoes and snails were the primary disease vectors and hosts associated with urban agriculture in our set of literature, with public health risk analysis being the primary research focus. Irrigation and sanitation infrastructure in urban agriculture was found to significantly influence the diversity and prevalence of mosquito species (Dongus et al. 2009; Mattah et al. 2017; Matthys et al. 2006; Ortiz-Perea et al. 2018; Robert et al. 1998). More than half of the irrigation trenches sampled from urban farms in Man, Côte d'Ivoire, and Accra, Ghana, contained *Culex* ssp. and *Anopheles* ssp. mosquito larvae (Klinkenberg et al, 2008; Matthys et al, 2006), and garden wells in Dakar, Senegal provided permanent sites for *A. arabiensis*, a major vector of malaria (Robert et al. 1998). In addition to increased larval habitat, urban agroecosystems may contribute to a city's mosquito population by providing adult mosquitoes more resting areas (Klinkenberg et al. 2008). Furthermore, extensive use of pesticides to reduce vegetable pests in areas of expanding urban agriculture—such as in the larger cities of the Republic of Benin—have been linked to insecticide resistance in malaria-carrying mosquitoes (Yadouleton et al. 2009).

Invasive snails that occupy urban gardens and farms can also pose a threat to humans as known hosts of zoonotic disease. For example, in several cities of Brazil, dense populations of the giant African snail (*Achatina fulica*) occupy ornamental and vegetable gardens, and can act as an intermediate host for two species of nematode that cause meningoencephalitis and abdominal angiostrongylosis in humans, respectively (Albuquerque et al. 2008, Thiengo et al. 2007). In urban watercress gardens of Rio de Janeiro, Brazil, snails (*Biomphalaria tenegophila*) are hosts for organisms that cause intestinal schistosomiasis in humans. These snails establish in the watercress gardens because of domestic sewage contamination of irrigation water and fertilization techniques which employ organic matter from adjacent pigsties (Baptista and Jurberg 1993). Conversely, some species of non-native snails that occupy urban gardens, such as *Hygromia cinctella* in Brooklyn, New Zealand, appear to pose low risk to native species, agriculture, or humans (Walton 2017).

1.4.2.5 Additional Invertebrate Biodiversity

There is little question that effects of urban agriculture on pollinators, natural enemies, soil macrofauna, and disease can be significant (Pinilla-Gallego et al., 2016). However, less research has been conducted on invertebrate biodiversity more generally, without the condition of direct application to human welfare. Although ants in agriculture and urbanized areas have been studied separately, little research has investigated the role of urban agriculture and how it contributes to ant diversity. In Perth, Australia, ant species richness was low in urban gardens and lawns compared to native plant regrowth adjacent to a large suburban freeway (Heterick et al. 2013). A basic taxonomic survey conducted by Taheri et al. (2017) found that urban gardens and parks of Tangier, Morocco supported a total of 38 ant species including eight of the nine non-native ant species in Morocco. This corroborated what is known elsewhere—that non-native ants (i.e., tramp ants) have been transported around the world and often thrive in urban port cities. Carabid beetle species diversity appeared unaffected by different turf grass management regimes and pesticide use, but seed-feeding species were positively associated with weed cover, suggesting that even in almost entirely uniform cultivated environments, heterogeneity can improve biodiversity (Blubaugh et al. 2011). In private residential gardens of Dunedin, New Zealand, the number of beetle species and families exceeded beetle diversity reported for less modified landscapes, likely due to the heterogeneous nature of the geography of the city, which provided a diversity of microclimates for beetles (van Heezik et al. 2016). In Cleveland, Ohio, USA, unique beetle communities were found in three different types of urban greenspaces (urban prairies, vacant lots, urban farms), while most species were influenced similarly by the amount of grass present locally and by the amount of building cover within the landscape (de la Flor et al. 2017). Ladybird beetle (Coccinellidae) abundance in urban gardens of California, USA and Michigan, USA responded to urbanization differently; diversity increased with urbanization in the former while the opposite was found in the latter (Egerer et al. 2019b). This demonstrates the importance of considering the effects of factors such as urban climate regime, and patterns and rates of urbanization on biodiversity within urban farming systems.

1.4.3 Birds

Most bird studies we retained were performed in residential gardens (86% of 101 bird studies). Fifteen of these were identified as food-producing gardens, 30 produced ornamental plants, and 56 studies did not explicitly define the production type. Only a few studies across urban agricultural types involved some form of food production in relation to bird measures. One found no strong association with either vegetables or orchards as predictors of avian abundance in backyards (Chamberlain et al. 2004). Another found that community gardens (compared to other types of urban green spaces) were key predictors of house sparrow (*Passer domesticus*) density (in the UK, where they are native), suggesting that the diversity of cultivated and uncultivated microhabitats in allotments might provide good foraging for these types of birds (Chamberlain et al. 2007a). Two additional studies analyzed decision-making around the business of swiftlet farming (Collocaliini, of which the nests are edible) in urban buildings of George Town, Malaysia (Connolly 2016, 2017).

Several investigators quantified bird responses to the amount or type of native vegetation in residential gardens. Daniels and Kirkpatrick (2006) found that while native birds of Hobart, Australia, utilized many non-native plants, native plants were preferred; conversely, non-native bird species largely used non-native plants. Urbanization around South African suburban gardens reduced the functional diversity of nectar-consuming birds (Pauw and Louw 2012), and in several cities in Scotland, native plant species that produced more nectar, at higher concentrations, and with greater sugar reward than non-native plants were preferred by native bird species (French et al. 2005). Conversely, aggressive, native, nectar-eating bird species visiting bird baths in urban gardens of several Australian municipalities resulted in shifts in bird species composition, with small native birds occurring less frequently in urban gardens than in rural ones (Cleary et al. 2016). Noisy Miners (Manorina melanocephala, considered a nuisance bird in Sydney, Australia) were associated with native *Eucalyptus* but not native *Grevillea* in residential gardens (Ashley et al. 2009). The most thorough study to date demonstrated that residential yards dominated by non-native plants in Washington, D.C., USA were associated with lower arthropod abundance, forcing diet switching to less preferred prey by birds (Narango et al. 2018). This mechanism explained a dynamic in which birds produced fewer young and population growth was unsustainable until native plant biomass reached > 70%.

Other types of resource provisioning for birds in gardens included nest boxes (Gaston et al. 2007, Goddard et al. 2013, Shwartz et al. 2014, Beumer 2018) and bird baths (Gaston et al. 2007, Goddard et al. 2013). About half of our articles on birds investigated the multibillion-dollar global industry of supplemental bird feeding in residential gardens (e.g., Gaston et al. 2007, Fuller et al. 2008, Goddard et al. 2013, Galbraith et al. 2014, Cox and Gaston 2016, Clark et al. 2019, Plummer et al. 2019). Bird feeding in urban gardens was associated with increases in bird abundance (but not species richness;

Fuller et al. 2008), shifts in winter distributions (Plummer et al. 2015), temporal shifts in feeding behavior (Ockendon et al. 2009), and a complete transformation of bird community composition in England over a 40-year period (Plummer et al. 2019). Bird feeding helped establish feral populations of rose-ringed parakeets (Psittacula krameri) in Paris, France (Clergeau and Vergnes 2011), and the presence of this species in London, England reduced the feeding rates and increased vigilance among native bird species (Peck et al. 2014). Experimental placement of bird feeders in New Zealand gardens led to dramatic increases in the non-native introduced house sparrows and spotted doves (*Spilopelia chinensis*) and decreases in a native insectivore (grey warblers (*Gerygone igata*); Galbraith et al. 2015, Galbraith et al. 2017a). Notably, however, these differences in abundance did not remain once experimental feeding was removed (Galbraith et al. 2015), and Galbraith et al. (2017b) found no observable effects on avian health for the disease agents and host species sampled. Other characteristics of the urban environment, such as greater presence of artificial lights, appeared to influence the behavior of birds at feeders (Clewley et al. 2016). While European gardeners mostly deploy seeds for birds (e.g., Tryjanowski et al. 2018), residents of a South African suburb provided a multitude of different food items to African woolly-necked storks (*Ciconia episcopus*), raising concern among conservationists (Thabethe and Downs 2018).

Birds and supplemental bird feeding can have direct and indirect effects on agroecosystems and other types of biodiversity in urban settings. For example, non-native house sparrows in Chicago, Illinois, USA were likely important consumers of bagworm (*Thyridopteryx ephemeraeformis*), an important herbivore of ornamental trees and shrubs (Ellis et al. 2005). When supplemental food was offered to garden birds in the UK, they significantly reduced the abundance and colony survival time of the pea aphid (*Acyrthosiphon pisum*), a pest of the broad bean (Orros and Fellowes 2012). On the other hand, fewer ground beetles (Carabidae, predators of common garden pests) were found under bird feeding stations than in areas of habitat away from the feeders (Orros et al. 2015), suggesting that bird feeding might also encourage intraguild predation of beneficial invertebrates. A study in Zurich, Switzerland found that top-down predation by birds was highest in heterogeneous gardens embedded in urban landscapes without much other available habitat (Frey et al. 2018).

Another issue that appears directly and indirectly tied to bird conservation and species interactions in residential garden spaces in urban environments, is the prevalence of domestic cats (van Heezik et al. 2010, Belaire et al. 2014, Flux 2007, Plummer et al. 2019, Patterson et al. 2016, Parsons et al. 2006, Gaston et al. 2005b, Baker et al. 2003). In Sheffield, England, 14% of domestic gardens were estimated to be home to at least one cat (Gaston et al. 2005b); this was extrapolated to an estimated population of 52,000 domestic cats. The number of outdoor cats had a negative association with native bird species richness and was associated with more non-native birds around Chicago, Illinois, USA (Belaire et al. 2014). Birds were the most common prey item returned to residential gardens in Dunedin, New Zealand, and cats preferred garden habitat over native vegetation fragments (van Heezik et al. 2010). In Lower Hutt, New Zealand, 223 of 558 prey items brought home by one domestic cat over a 17-year lifetime were birds; the cat caught prey both inside and outside the garden, and up to 600 m from the house (Flux 2007). Resident pairs that were killed were quickly replaced, suggesting that these types of gardens could be population sinks for birds (Flux 2007). Domestic cats were also responsible for predation of 3% of experimental nests in urban gardens of the KwaZulu-Natal Province of South Africa (Patterson et al. 2016). The presence of dogs and cats, however, was not related to the total abundance of birds overall or small birds in gardens of Sydney, Australia (Parsons et al. 2006).

1.4.4 Mammals

Forty-one studies examined mammals in urban agroecosystems. Several studies, primarily originating from North America and Europe, focused on habitats and diets of common urban mammals and did not make more than tenuous ties to conservation. For example, foxes (*Vulpes vulpes*) used gardens with hedges and fruiting plants for hiding and food (König et al. 2012, Contesse et al. 2004), martens (*Martes foina*) consumed more fruit in urban areas than natural ones due to increase availability (Hisano et al. 2016), vervet monkeys (*Chlorocebus pygerythrus*) preferred gardens with fruit trees and bird feeders (Patterson et al. 2018), and badgers (*Meles meles*) reduced their foraging distance due to food provisions in gardens (Davison et al. 2009). Deer (*Odocoileus hemionus*) and rabbits

(*Lepus townsendii*) foraged from ornamental and non-native species planted in residential gardens, especially in winter and early spring when there were few other available food sources (Conover et al. 2018, Beaudoin and Beaudoin 2012). Hedgehog (*Erinaceus europaeus*) prevalence, on the other hand, could not be predicted by habitat factors such as flowers, pesticide use, or food sources in residential gardens (Williams et al. 2015).

The literature on bats was mainly concerned with flight activity and their response to landscape microhabitats. Bat communities in urban areas can have substantial direct and indirect effects on agriculture because of the ecosystem services they provide in the form of seed dispersal (Gulraiz et al. 2016; Lim et al. 2018; McDonald-Madden et al. 2005) and insect consumption (Bartoniĉka and Zukal 2003). Bats were able to exploit resources in urban areas due to switching habitats throughout the night (Bartoniĉka and Zukal 2003), and frugivorous bats were flexible in their diet, using native and non-native fruits when foraging (Lim et al. 2018). Humans can also make urban environments more suitable for bats, as was the case of the Melbourne Botanical Garden, Australia, where management regimes, warmer temperatures from climate change and the urban heat island effect, and high tree species diversity in nearby public and private areas created suitable conditions for a colony of bats to stay year-round (Parris and Hazell 2005).

Several studies employed the use of citizen scientists (hereafter participatory scientists) but given that many of the mammal species were nocturnal, the results were biased toward when and where people were looking for them, indicating nighttime surveys as an avenue of future research (Walter et al. 2018, Williams et al. 2015, Kauhala et al. 2016). Two studies in Australia and New Zealand focused on determining what types of residential gardens non-native foxes (*V. vulpes*) and possums (*Trichosurus vulpecula*) utilized to inform invasive management plans (Marks and Bloomfield 2006, Adams et al. 2013). Several mammal species in a New Zealand residential garden fell prey to a single domestic cat, with 58% of prey items identified as mice, weasels, rats, rabbits, and hares (Flux 2007). The lone marine vertebrate study was performed in an urbanized agri- and aquaculture complex in the inner-gulf region of Thailand (Kamjing et al. 2017). The study found that smooth-coated otter (*Lutrogale perspicillata*) occupancy was negatively associated with urban and agricultural cover, but positively associated with both natural vegetation and traditional aquaculture cover, much to the chagrin of fish farmers (Kamjing et al. 2017).

Certain mammal species readily used urban agricultural areas for shelter and food, however given that many species consumed food intended for humans, conservation programs should account for the possibility that humans might have unfavorable opinions of this taxon within cities. On the other hand, survey responses from several UK urban areas indicated that gardeners purposefully fed mammals monthly (1%, n = 4,403), weekly (3%), or even daily (3%; Gaston et al. 2007), suggesting that in some regions, residents might be amenable to urban conservation programs aimed at mammals. In addition to food and shelter, mammals in urban areas require space and free passage, which private or community gardeners can address to some extent by providing gaps in fences or removing fences altogether (Widows and Drake 2014, Beumer 2018). Issues of connectivity and free passage for mammals across property lines, however, are likely best addressed at the city planning level. Lastly, our combination of search terms did not discover many studies on urban livestock, such as goats or cattle, which could have implications for native plant diversity (e.g. Shackleton et al. 2017), disease transmission, and soil health.

1.4.5 Reptiles

There were few studies of reptiles in urban agroecosystems (n = 13). When home gardeners in Dunedin, New Zealand were surveyed about their biodiversity knowledge, respondents were less familiar with reptiles than they were with other taxa such as plants and birds (van Heezik et al. 2012), while respondents in Melbourne, Australia indicated that they provided shelter for lizards in their gardens (Mumaw and Bekessy 2017). Though lizards were present in residential gardens, they were more prevalent in natural areas with tall grasses and trees, depending on the features of the climate and the species studied (van Heezik and Ludwig 2012, Gonzalez-Garcia et al. 2009). Nonetheless, there was some evidence that lizards were adapting their behavior in urban agroecosystems as compared to natural areas (Prosser et al. 2006). Residential gardens are not without threats for reptiles, however. In a residential garden of Lower

Hutt, New Zealand, cats were reported to prey on skinks (Sincidae; Flux 2007). A study on iguanas in Leon, Nicaragua was emblematic of the complexities of studying and conserving biodiversity in urban agroecosystems (Gonzalez-Garcia et al. 2009). Traditionally, black spiny-tailed iguana (*Ctenosaura similis*) have been hunted because they were considered an urban agricultural pest and a good protein source. This iguana species was found to be declining in traditional agricultural systems, but the species was able to survive in urban gardens, despite hunting pressure, due to its ability to hide in trees (Gonzalez-Garcia et al. 2009). Biodiversity studies such as this one—that are chiefly focused on urban agroecosystems as means of habitat but that are also centered on the actions and motivations of the people who use these spaces—are needed for determining the best conservation practices in urban systems.

1.4.6 Amphibians

Twelve articles addressed amphibians in urban agroecosystems. The most detailed study examined the prevalence and distribution of a non-native lesser Antillean frog (*Eleutherodactylus johnstonei*) in Western French Guiana (Ernst et al. 2011). The study found that the frog was restricted to residential areas and was significantly more likely to occur in gardens that contained potted ornamental plants as compared to gardens without pots, suggesting that the movement and transfer of potted plants could potentially facilitate spread of invasive amphibian species (Ernst et al. 2011). Other studies included amphibians in overall diversity scores used to assess residential gardens (Goddard et al. 2013, Beumer and Martens 2016, Beumer 2018). A few studies from the UK and Australia quantified the presence of amphibian habitat (e.g., ponds, streams, or fountains) in residential gardens (Gaston et al. 2005a, Gaston et al. 2007, Mumaw and Bekessy 2017). Across the UK, it was estimated that between 2.5 and 3.5 million domestic garden ponds were available as potential habitat for amphibians (Davies et al. 2009). While garden structures such as fences could impinge on the ability of amphibians to move between gardens (Widows and Drake 2014), 7 out of 19 experimental garden ponds were colonized by juvenile and adult frogs (*Rana temporaria*) within two seasons of placement, though no breeding was observed (Gaston et al. 2005a). Decaying wood piles were also offered as habitat, and frogs occupied them in 21% of gardens (Gaston et al. 2005a).

1.4.7 Soil Microbes

If we had included soil microbes in our biodiversity search terms, it is possible that we would have retained more studies on these organisms which are so functionally crucial for agroecosystems. The twelve studies our search did discover examined how different land use types can affect either soil microbial communities or the function of nutrient cycling. The type of land use within green spaces can have a significant effect on soil microbial community composition (Yadav et al. 2012, Cabral et al. 2017, Cousins et al. 2003) and nutrient quantities (Edmondson et al. 2014, Tresch et al. 2018, Pierart et al. 2018). For example, urban gardens had higher soil quality than traditional agricultural areas (Cabral et al. 2017, Cousins et al. 2003, Edmondson et al. 2014). However, the underlying cause for greater soil mycorrhizal fungal diversity can vary across sites (Cousins et al. 2003), and garden management can be a driving factor that influences soil quality and soil functions (Tresch et al. 2018). For example, slime molds and fungi that feed on, facilitate, or compete with soil microorganisms, (e.g., species of *Hypoxylon*, *Coriolus*, *Stereum*, *Bjerkandera*, and *Enteridion*), can be encouraged by allowing dead wood to decay in garden plots (Gaston et al. 2005a). Even in heavily managed urban spaces, greater soil microbial diversity was generally beneficial for above ground plant growth, as it led to an increase in the uptake of soil nutrients (Pierart et al. 2018). In turn, plant species richness has been associated with microbial activity, which can lead to increased litter decomposition (Tresch et al. 2019) and capture of soil organic carbon with increasing urban density (Tresch et al. 2018). In Mahikeng, South Africa, the subsoil horizons (20–40 cm deep) of urban agricultural plots were compacted via human tillage practices, which constrained root growth, and had further implications for crop growth and water holding capacity of soil (Materechera 2018). Heavy metals were also a concern in urban soils due to the proximity to industrial practices and past land use of a site (Pierart et al. 2018). In carrots, arbuscular mycorrhizal fungi associations led to an increase in the root uptake of harmful metals and metalloids (Pierart et al. 2018).

1.4.8 Fish

Fish were considered in only two studies of biodiversity in urban agroecosystems. The first included fish as one factor to consider in an ecological assessment of biodiversity in residential gardens (Beumer and Martens 2015). The second study used virtual gardens to explore what types of biodiversity residents of Paris, France wished to see in their city; goldfish (*Carassius auratus auratus*) and European chub (*Squalius cephalus*) were selected for 25% and 13% of gardens, respectively (Shwartz et al. 2013b). It did not occur to us to include fish as a biodiversity term in our literature search until we read these two studies. We predict that urban aquaculture will have very different implications for the conservation of urban aquatic species, depending on the size, scale, and purpose of the operation, and the socio-economic context of the urban environment.

1.5 Consideration of Space and Scale

The benefits of urban agriculture for biodiversity conservation will depend on activities and features both within individual gardens and at spatial scales beyond the control of individual gardeners. While garden size can influence garden composition and individual species responses (Loram et al. 2008a), at larger scales a higher number of smaller gardens can provide more garden cover than fewer larger gardens (Gaston et al. 2005b). Additionally, an individual garden's biodiversity potential can be influenced by city-wide patterns (Tratalos et al. 2007), while a city's biodiversity potential can be influenced by the configuration of individual gardens among other types of green space (Orsini et al. 2014). Habitat connectivity within urban areas has implications for animal and plant populations with spatial dynamics that extend beyond property lines (Cox et al. 2016, Coppola et al. 2019) or even city boundaries (Rolf et al. 2018, 2019). In cities with high human densities and little room for residential or community gardens at the ground level, vertical gardens and green roofs are suggested as a way to improve habitat connectivity (Chen et al. 2015, Walters and Midden 2018, Orsini et al. 2014), with some noted habitat benefits for plants and animals (Tonietto et al. 2011, Wang et al. 2017, Joimel et al. 2017, Joimel et al. 2018, Oh et al. 2018). Some researchers, however, question the effectiveness of managing green spaces within urban areas for conservation. For example, using a framing commonly employed in the wider field of agroecology, Collas et al. (2017) asked whether tree populations in urban areas were greater under land-sharing arrangements —with low-density housing and larger gardens— or land-sparing land management conditions in which undeveloped green spaces were managed separately from high-density housing. They found that tree densities were highest in areas of respectively lower human density, but models predicted that restoring separate woodlands and increasing human density would result in greater tree densities, at the risk of disconnecting city dwellers from nature.

1.6 Human Engagement in Biodiversity Conservation through Urban Agriculture

Thus far, our review suggests that the success of biodiversity conservation in urban agroecosystems will be dependent on the life histories and ecology of individual taxa, the types of practices implemented in urban agriculture, and the spatial configuration of gardens, farms, and other green spaces in cities. Another determinant of conservation success will be the willingness of human communities to participate in conservation activities in and outside of the agricultural setting, whether it be through personal behavior, volunteerism, support of professional conservation organizations, or civic engagement through taxation and representation. The following pool of studies reveals that urban agriculture appears particularly well suited to engage urban residents in biodiversity conservation, with room for improvement.

The act of gardening or spending time in gardens often engages people with biodiversity which can motivate them to conserve it. Observations of human interaction with biodiversity in urban gardens revealed that 12% of interactions were with individual plant species and that about 17% of these were derived from local wild or cultivated species pools (Palliwoda et al. 2017), suggesting that even highly cultivated spaces can engage people with the local flora. Explicitly gardening with iconic native species can contribute to a gardener's sense of place (Standish et al. 2013), and the simple practice of gardening has been associated with place attachment (Raymond et al. 2019). Time spent gardening can lead to generally positive attitudes about biodiversity conservation (Head and Muir 2006b).

Communities also approached gardening with pre-determined motivations for conserving biodiversity. These motivations included personal well-being and a moral responsibility to nature (Goddard et al. 2013), a desire to improve environmental sustainability, improve habitat for beneficial insects, and to maintain healthy soils and diverse plant species (Guitart et al. 2013), or to explicitly provision wildlife with natural native vegetation (Fischer et al. 2019). Gardener's preferences for these higher quality habitats for biodiversity was linked with increased environmental awareness (Kurz and Baudains 2012, Rodriguez et al. 2017, Head and Muir 2006b), which was often gained outside of gardening. For example, when compared to other activities, a gardener's engagement in participatory science had the greatest association with changes in attitudes toward biodiversity (van Heezik et al. 2012). Activities with explicit attentiveness to biodiversity (e.g., participatory science, nature watching, or environmental groups), was associated with more biodiversity and conservation knowledge, compared to activities with only implicit attention to biodiversity (e.g., participation in community gardens, community supported agriculture, personal garden use; Prevot et al. 2018, Coldwell and Evans 2017). Gardeners were also more likely to implement biodiversity conservation practices if they experienced biodiversity daily, or by routine (Prevot et al. 2018). Specific interventions with knowledge, such as onsite garden assessments, indigenous or community sources of plants, communication hubs and linkages, and experiential learning activities were also motivating factors for gardeners to conserve native species (Mumaw and Bekessy 2017).

Also noted were some potential conflicts and tradeoffs for urban agricultural practices and biodiversity conservation. The amount of area required by many animals and plant species, for example, often far exceeds what is available in individual gardens (Dennis and James 2017). Specific gardening practices such as pesticide and fertilizer use can also be detrimental to plants and animals (Shwartz et al. 2013a). Different pressures from within-communities can dictate the implementation of biodiversity-friendly practices. For example, in San Juan, Puerto Rico, older gardeners with more years of education were more likely to apply pesticides than younger gardeners (Melendez-Ackerman et al. 2016). On the other hand, the reduction of impervious surface area was more likely done by women with fewer years of education and in specific neighborhoods with larger yards (Melendez-Ackerman et al. 2016). Neighbors often have different expectations for yard cleanliness and tidiness which can conflict with biodiversity goals (Raymond et al. 2019). Indeed, while human perceptions of biodiversity best explained biodiversity features in backyards, neighbor attitudes were more likely to dictate the presence of these features in front yards (Belaire et al. 2016). Finally, some practices can benefit some species while harming others. For example, yards governed by homeowners associations (HOAs) harbored more bird and plant species, but fewer arthropods, than those without HOAs (Lerman et al. 2012). Quistberg et al. (2016) found that mulch negatively impacted bee species diversity while —in the same system—Otoshi et al. (2015) found that mulch benefitted spider activity and species richness.

1.7 Information Gaps and Future Directions

Although our search retained 431 studies, there remain information gaps that hinder our ability to generalize about biodiversity conservation through agroecosystems globally and across taxa. We organized these research gaps and our suggestions for future directions into four general themes: 1) standardize and expand scope, 2) increase specificity and novelty, 3) consider harm reduction, and 4) increase cultural perspectives.

First, there is a lack of standardization in methodology and terminology in biodiversity measurements; reporting on size, age, and quality of urban sites; and defining how to characterize sites along a gradient of urbanization (e.g., what defines urban vs. suburban). In addition, as is true throughout the field of urban ecology, many conclusions were scale-dependent, making it difficult to draw explicit comparisons between studies. Agreeing on common terminologies and methodologies while understanding the very different ways cities and urban greenspaces manifest across the globe will be a difficult but worthwhile endeavor. Additionally, long term studies are needed to elucidate the complexities of biodiversity in urban agroecosystems and to make more robust predictions about future trends and responses to management (e.g., Warren et al. 2019). Finally, the most prevalent type of biodiversity data collected was counts. There is a much greater need for population level studies that focus on rates of fecundity and mortality, the practices and factors that influence these vital rates, and whether organisms inhabiting urban agroecosystems represent sustainable populations.

Most urban agriculture research has focused on plants, insects, and birds, representing 67%, 39%, and 23% of all studies, respectively. Conversely, there were fewer studies on reptiles and amphibians (and our search terms likely insufficiently captured studies on soil microbes, and fish; Figure 1.3A). Despite the many studies involving plants generally (i.e., plant cover as a part of the landscape), more research is needed on the ecology and conservation of individual plant species in urban agricultural spaces. This includes examining urban agroecosystems as reservoirs of crop diversity and native plant conservation. Lastly, most research took place in only a few types of urban agriculture including community/allotment gardens and home gardens (Figure 1.3C). As technology and culture change and urban areas expand globally, there will be an increase of novel sites such as green roofs and vertical walls (including ones producing food), urban commercial farming, and urban aquaculture such as fish and oyster farms. Research will be needed to explore the biodiversity implications of these types of urban production sites.

Additional research is needed to determine how urban agricultural practices affect the prevalence and diversity of pests and non-native species, pesticides, and disease vectors in urban settings, so that biodiversity and agriculture goals can be met without causing additional health risks to human populations. An understudied health concern is the interaction between soil microorganisms and the uptake of beneficial and harmful metals by edible plants in urban gardens. Future research studying mosquito insecticide resistance and urban pesticide use should consider alternative methods of controlling pest prevalence such as integrated pest management and habitat modification. Given the apparent importance of gardens in urban areas for birds, and the beneficial services that birds can provide, we suggest that pest reduction by birds in urban agroecosystems, and what specific plants and pests attract them are avenues of research in need of further exploration (Davies et al. 2009, Cerra and Crain 2016). Additionally, due to the prevalence of non-native pollinators (e.g., *Apis mellifera* outside of Europe) in agricultural systems, few studies address the impacts of non-native pollinators on native biodiversity. Likewise, few studies ask how to actively combat the spread of invasive plants through labeling, removal from nurseries, and other top-down methods.

Local culture, demography, and socioeconomic factors influence planting decisions and human interactions with biodiversity in urban agriculture (e.g., Avila et al. 2017, Gopal and Nagendra 2014). Yet, the research was dominated by residential or community gardens in Europe, North America, and Australia (Figure 1.1C). Thus, more research is needed in Asia, Africa, and Central and South America to better understand the conservation potential of urban agriculture for biodiversity on a global scale. Since human-driven changes to urban plant composition and market-driven pressures on cultivated crops are major influences on plant species in urban agricultural areas, a greater research focus is needed on ways humans can implement planting decisions that benefit conservation of plants and the animals they support. Additionally, there is an overall need to examine biodiversity in response to specific land management actions people take in urban ecosystems (Verderame and Scudiero 2019; Parris and Hazell 2005; Cane 2015). This includes, but is not limited to, investigating how taxa respond to the legacy of land use (such as lead contamination); the impacts of tilling, fertilizer and pesticide applications; altered watering regimes; the spatial dynamics of area and connectivity among habitat patches; resource availability; and the intentional introduction of pest, predator, and pollinating animals.

1.8 Conclusions

This chapter has attempted to represent the state of global research on biodiversity in urban agroecosystems. We presented general trends in the literature, including the per year increase in the number of publications on the topic (Figure 1.1B). We highlight both regional (Figure 1.1C) and taxonomic (Figure 1.3A) biases, and the dominance of residential garden research (Figure 1.3C). Our topic model results corroborated these general taxonomic and topical patterns (Figure 1.2). We provided detailed accounts spotlighting specific research findings for several different types of biodiversity, highlighted major themes across taxa, and provided four key areas where information gaps exist and future work is recommended. Throughout the chapter, we have attempted to identify challenges and opportunities for biodiversity conservation in urban agroecosystems. We hope this effort has been helpful in informing the reader and guiding them toward research topics of interest.

REFERENCES

Acar, Cengiz, Habibe Acar, and Engin Eroğlu. "Evaluation of Ornamental Plant Resources to Urban Biodiversity and Cultural Changing: A Case Study of Residential Landscapes in Trabzon City (Turkey)." *Building and Environment* 42, no. 1 (2007): 218–29. https://doi.org/10/brp9nn.

Adams, Amy L., Katharine J. M. Dickinson, Bruce C. Robertson, and Yolanda van Heezik. "Predicting Summer Site Occupancy for an Invasive Species, the Common Brushtail Possum (Trichosurus Vulpecula), in an Urban Environment." *Plos One* 8, no. 3 (2013): e58422. https://doi.org/10/f4pp8x.

Albuquerque, Fabio Suzart, Marelen Campos Peso-Aguiar, and Maria Jose Teixeira Assuncao-Albuquerque. "Distribution, Feeding Behavior and Control Strategies of the Exotic Land Snail Achatina Fulica (Gastropoda Pulmonata) in the Northeast of Brazil." *Brazilian Journal of Biology* 68 (2008): 837–42. https://doi.org/10/dq3td4.

Angel, Shlomo, Jason Parent, Daniel L. Civco, Alexander Blei, and David Potere. "The Dimensions of Global Urban Expansion: Estimates and Projections for all Countries, 2000–2050." *Progress in Planning* 75, no. 2 (2011): 53–107. https://doi.org/10/fw5xhh.

Aronson, Myla F. J., Christopher A. Lepczyk, Karl L. Evans, Mark A. Goddard, Susannah B. Lerman, J. Scott MacIvor, Charles H. Nilon, and Timothy Vargo. "Biodiversity in the City: Key Challenges for Urban Green Space Management." *Frontiers in Ecology and the Environment* 15, no. 4 (2017): 189–96. https://doi.org/10/f96prt.

Artmann, Martina, and Katharina Sartison. "The Role of Urban Agriculture as a Nature-Based Solution: A Review for Developing a Systemic Assessment Framework." *Sustainability* 10 (2018). https://doi.org/10/gd74vr.

Ashley, Lisa C., Richard E. Major, and Charlotte E. Taylor. "Does the Presence of Grevilleas and Eucalypts in Urban Gardens Influence the Distribution and Foraging Ecology of Noisy Miners?" *Emu* 109 (2009): 135–42. https://doi.org/10/fqbmsd.

Avila, Julia Vieira da Cunha, Anderson Santos de Mello, Mariane Elis Beretta, Rafael Trevisan, Pedro Fiaschi, and Natalia Hanazaki. "Agrobiodiversity and in Situ Conservation in Quilombola Home Gardens with Different Intensities of Urbanization." *Acta Botanica Brasilica* 31 (2017): 1–10. https://doi.org/10/gf4z4s.

Baker, Philip J., Rachel J. Ansell, Phillippa A. A. Dodds, Claire E. Webber, and Stephen Harris. "Factors Affecting the Distribution of Small Mammals in an Urban Area." *Mammal Review* 33 (2003): 100–195. https://doi.org/10/dns8vk.

Ballare, Kimberly M., John L. Neff, Rebecca Ruppel, and Shalene Jha. "Multi-Scalar Drivers of Biodiversity: Local Management Mediates Wild Bee Community Response to Regional Urbanization." *Ecological Applications* 29 (2019): 1–15. https://doi.org/10/gf4z8d.

Baptista, Darcilio F., and Pedro Jurberg. "Factors Conditioning the Habitat and the Density of Biomphalaria Tenagophila (Orbigny, 1835) in an Isolated Schistosomiasis Focus in Rio de Janeiro City." *Memorias do Instituto Oswaldo Cruz* 88 (1993): 457–64. https://doi.org/10/cs5s57.

Barau, Aliyu Salisu, A. N. M. Ludin, and Ismail Said. "Socio-Ecological Systems and Biodiversity Conservation in African City Insights from Kano Emir's Palace Gardens." *Urban Ecosystems* 16 (2013): 783–800. https://doi.org/10/f5kkwt.

Bartoniĉka, Tomáŝ, and Jan Zukal. "Flight Activity and Habitat Use of Four Bat Species in a Small Town Revealed by Bat Detectors." *Folia Zoologica* 52 (2003): 155–66.

Beaudoin, Alwynne B., and Yves Beaudoin. "Urban White-Tailed Jackrabbits (Lepus Townsendii) Eat Spike Plants (Cordyline Australis) in Winter." *Canadian Field-Naturalist* 126 (2012): 157–59. https://doi.org/10/f4hcs4.

Belaire, J. Amy, Lynne M. Westphal, and Emily S. Minor. "Different Social Drivers, Including Perceptions of Urban Wildlife, Explain the Ecological Resources in Residential Landscapes." *Landscape Ecology* 31 (2016): 401–13. https://doi.org/10/f8ftt7.

Belaire, J. Amy, Christopher J. Whelan, and Emily S. Minor. "Having Our Yards and Sharing them too: The Collective Effects of Yards on Native Bird Species in an Urban Landscape." *Ecological Applications* 24 (2014): 2132–43. https://doi.org/10/f6r5d5.

Bennet, Della G., Dave Kelly, and John Clemens. "Food Plants and Foraging Distances for the Native Bee Lasioglossum Sordidum in Christchurch Botanic Gardens." *New Zealand Journal of Ecology* 42 (2018): 40–47. https://doi.org/10/gf42g5.

Bennett, Ashley B., and Claudio Gratton. "Local and Landscape Scale Variables Impact Parasitoid Assemblages Across an Urbanization Gradient." *Landscape and Urban Planning* 104 (2012): 26–33. https://doi.org/10/dwptgx.

Bergerot, Benjamin, Benoit Fontaine, Mathilde Renard, Antoine Cadi, and Romain Julliard. "Preferences for Exotic Flowers do not Promote Urban Life in Butterflies." *Landscape and Urban Planning* 96 (2010): 107–98. https://doi.org/10/dkzkzq.

Beumer, Carijn. "Show me your Garden and I will Tell you how Sustainable you are: Dutch Citizens' Perspectives on Conserving Biodiversity and Promoting a Sustainable Urban Living Environment Through Domestic Gardening." *Urban Forestry & Urban Greening* 30 (2018): 260–79. https://doi.org/10/gdg2jc.

Beumer, Carijn, and Pim Martens. "BIMBY's First Steps: A Pilot Study on the Contribution of Residential Front-Yards in Phoenix and Maastricht to Biodiversity, Ecosystem Services and Urban Sustainability." *Urban Ecosystems* 19 (2016): 45–76. https://doi.org/10/f8f3fd.

———. "Biodiversity in my (Back)Yard Towards a Framework for Citizen Engagement in Exploring Biodiversity and Ecosystem Services in Residential Gardens." *Sustainability Science* 10 (2015): 100–187. https://doi.org/10/f6tjjv.

Blubaugh, Carmen K., Victoria A. Caceres, Ian Kaplan, Jonathan Larson, Clifford S. Sadof, and Douglas S. Richmond. "Ground Beetle (Coleoptera: Carabidae) Phenology, Diversity, and Response to Weed Cover in a Turfgrass Ecosystem." *Environmental Entomology* 40, no. 5 (2011): 1093–1101. https://doi.org/10/d6x66k.

Borysiak, Janina, Andrzej Mizgajski, and Andrew Speak. "Floral Biodiversity of Allotment Gardens and its Contribution to Urban Green Infrastructure." *Urban Ecosystems* 20 (2017): 323–35. https://doi.org/10/f95tjt.

Botelho, Juliana de Mello, A. P. do N. Lamano-Ferreira, and Mauricio L. Ferreire. "Cultivation and use of Domestic Plants in Different Brazilian Cities." *Ciencia Rural* 44 (2014): 1810–15. https://doi.org/10/gf4z7z.

Bożek, Małgorzata, Monika Strzałkowska-Abramek, and Bożena Denisow. "Nectar and Pollen Production and Insect Visitation on Ornamentals from the Genus Hosta Tratt. (Asparagaceae)." *Journal of Apicultural Science* 59 (2015): 115–25. https://doi.org/10/gf42c9.

Burkman, Caitlin E., and Mary M. Gardiner. "Spider Assemblages within Greenspaces of a Deindustrialized Urban Landscape." *Urban Ecosystems* 18 (2015): 793–818. https://doi.org/10/f7pd6v.

Burks, Julia M., and Stacy M. Philpott. "Local and Landscape Drivers of Parasitoid Abundance, Richness, and Composition in Urban Gardens." *Environmental Entomology* 46 (2017): 201–9. https://doi.org/10/f9897j.

Caballero-Serrano, Veronica, Miren Onaindia, Josu G. Alday, David Caballero, Juan Carlos Carrasco, Brian McLaren, and Javier Amigo. "Plant Diversity and Ecosystem Services in Amazonian Homegardens of Ecuador." *Agriculture Ecosystems & Environment* 225 (2016): 116–25. https://doi.org/10/f8qq85.

Cabral, Ines, Jessica Keim, Rolf Engelmann, Roland Kraemer, Julia Siebert, and Aletta Bonn. "Ecosystem Services of Allotment and Community Gardens a Leipzig, Germany Case Study." *Urban Forestry & Urban Greening* 23 (2017): 44–53. https://doi.org/10/gf42nc.

Cane, James H. "Landscaping Pebbles Attract Nesting by the Native Ground-Nesting Bee Halictus Rubicundus (Hymenoptera Halictidae)." *Apidologie* 46 (2015): 728–34. https://doi.org/10/f7znk8.

Castro, Andy, Maite Lascurain-Rangel, Jorge A. Gomez-Diaz, and Victoria Sosa. "Mayan Homegardens in Decline: The Case of the Pitahaya (Hylocereus Undatus), a Vine Cactus with Edible Fruit." *Tropical Conservation Science* 11 (2018): 1–10. https://doi.org/10/gf42k5.

Cerra, Joshua F., and Rhiannon Crain. "Urban Birds and Planting Design Strategies for Incorporating Ecological Goals into Residential Landscapes." *Urban Ecosystems* 19 (2016): 1823–46. https://doi.org/10/f9fghk.

Chamberlain, Dan E., Andrew R. Cannon, and Mike P. Toms. "Associations of Garden Birds with Gradients in Garden Habitat and Local Habitat." *Ecography* 27 (2004): 589–600. https://doi.org/10/chz2jr.

Chamberlain, Dan E., Mike P. Toms, Rosie Cleary-McHarg, and Alex N. Banks. "House Sparrow (Passer Domesticus) Habitat use in Urbanized Landscapes." *Journal of Ornithology* 148 (2007a): 453–62. https://doi.org/10/dnsw5t.

Chen, Yu Chi, Lian Pei, and Yan Chyuan Shiau. "Application of coastal vegetation to green roofs of residential buildings in Taiwan." *Artificial Life and Robotics* 20 (2015): 86–91. https://doi.org/10/gf4z57

Ciceoi, Roxana, Cătălin Gutue, Minodora Gutue, and Ioan Roşca. "Current Status of Pests Associated with Urban Vegetation in Bucharest Area." *Acta Zoologica Bulgarica Supplement* 9 (2017): 181–90.

Clark, David N., Darryl N. Jones, and S. James Reynolds. "Exploring the Motivations for Garden Bird Feeding in South-East England." *Ecology and Society* 24 (2019): 1–15. https://doi.org/10/gf4z8g.

Cleary, Gráinne P., Holly Parsons, Adrian Davis, Bill R. Coleman, Darryl N. Jones, Kelly K. Miller, and Michael A. Weston. "Avian Assemblages at Bird Baths: A Comparison of Urban and Rural Bird Baths in Australia." *Plos One* 11 (2016): 1–12. https://doi.org/10/gf42hz.

Clergeau, Philippe, and Alan Vergnes. "Bird Feeders may Sustain Feral Rose-Ringed Parakeets Psittacula Krameri in Temperate Europe." *Wildlife Biology* 17 (2011): 248–52. https://doi.org/10/cvzkh3.

Clewley, Gary D., Kate E. Plummer, Robert A. Robinson, Clare H. Simm, and Mike P. Toms. "The Effect of Artificial Lighting on the Arrival Time of Birds using Garden Feeding Stations in Winter: A Missed Opportunity?" *Urban Ecosystems* 19 (2016): 535–46. https://doi.org/10/gf42mj.

Clucas, Barbara, Israel D. Parker, and Andrea M. Feldpausch-Parker. "A Systematic Review of the Relationship Between Urban Agriculture and Biodiversity." *Urban Ecosystems* 21 (2018): 635–43. https://doi.org/10/gdxbtj.

Cohen, Hamutahl, Robyn D. Quistberg, and Stacy M. Philpott. "Vegetation Management and Host Density Influence Bee-Parasite Interactions in Urban Gardens." *Environmental Entomology* 46 (2017): 1313–21. https://doi.org/10/gcrcp4.

Coldwell, Deborah F., and Karl L. Evans. "Contrasting Effects of Visiting Urban Green-Space and the Countryside on Biodiversity Knowledge and Conservation Support." *Plos One* 12 (2017): 1–18. https://doi.org/10/f9v356.

Collaboration for Environmental Evidence. "Guidelines for Systematic Review and Evidence Synthesis in Environmental Management. Version 4.2." *Environmental Evidence* (2013). www.environmentalevidence.org/Documents/Guidelines/Guidelines4.2.pdf.

Collas, Lydia, Rhys E. Green, Alexander Ross, Josie H. Wastell, and Andrew Balmford. "Urban Development, Land Sharing and Land Sparing the Importance of Considering Restoration." *Journal of Applied Ecology* 54 (2017): 1865–73. https://doi.org/10/gf42dd.

Connolly, Creighton. "'A Place for Everything' Moral Landscapes of 'Swiftlet Farming' in George Town, Malaysia." *Geoforum* 77 (2016): 182–91. https://doi.org/10/f9gq3q.

———. "'Bird Cages and Boiling Pots for Potential Diseases' Contested Ecologies of Urban 'Swiftlet Farming' in George Town, Malaysia." *Journal of Political Ecology* 24 (2017): 24–43. https://doi.org/10/gf42d9.

Conover, Michael R., Kristin B. Hulvey, Megan L. Monroe, and Collin B. Fitzgerald. "Susceptibility of Spring-Flowering Garden Plants to Herbivory by Mule Deer." *Wildlife Society Bulletin* 42 (2018): 131–35. https://doi.org/10/gf5pc4.

Contesse, Pascale, Daniel Hegglin, Sandra Gloor, Fabio Bontadina, and Peter Deplazes. "The Diet of Urban Foxes (Vulpes Vulpes) and the Availability of Anthropogenic Food in the City of Zurich, Switzerland." *Mammalian Biology* 69 (2004): 81–95. https://doi.org/10/df9pvh.

Coppola, Emanuela, Youssef Rouphael, Stefania De Pascale, Francesco D. Moccia, and Chiara Cirillo. "Ameliorating a Complex Urban Ecosystem Through Instrumental use of Softscape Buffers: Proposal for a Green Infrastructure Network in the Metropolitan Area of Naples." *Frontiers in Plant Science* 10 (2019): 1–11. https://doi.org/10/gf42bm.

Cousins, Jamaica R., Diane Hope, Corinna Gries, and Jean C. Stutz. "Preliminary Assessment of Arbuscular Mycorrhizal Fungal Diversity and Community Structure in an Urban Ecosystem." *Mycorrhiza* 13 (2003): 319–26. https://doi.org/10/fgkdnj.

Cousins, Stephen R., and E. T. F. Witkowski. "Indigenous Plants: Key Role Players in Community Horticulture Initiatives." *Human Ecology Review* 21, no. 1 (2015): 59–85. https://doi.org/10/gf42ch.

Cox, Daniel T. C., and Kevin J. Gaston. "Urban Bird Feeding Connecting People with Nature." *Plos One* 11 (2016): 1–13. https://doi.org/10/gbpmch.

Cox, Daniel T. C., Richard Inger, Steven Hancock, Karen Anderson, and Kevin J. Gaston. "Movement of Feeder-Using Songbirds the Influence of Urban Features." *Scientific Reports* 6 (2016): 1–9. https://doi.org/10/f9cs6t.

Cubino, Josep P., Jeannine Cavender-Bares, Sarah E. Hobbie, Diane E. Pataki, Meghan L. Avolio, Lindsay E. Darling, and Kelli L. Larson, et al. "Drivers of Plant Species Richness and Phylogenetic Composition in Urban Yards at the Continental Scale." *Landscape Ecology* 34 (2019a): 63–77. https://doi.org/10/gfxf4n.

Cubino, Josep P., Jeannine Cavender-Bares, Sarah E. Hobbie, Sharon J. Hall, Tara L. E. Trammell, Christopher Neill, Meghan L. Avolio, Lindsay E. Darling, and Peter M. Groffman. "Contribution of non-native plants to the phylogenetic homogenization of US yard floras." *Ecosphere* 10 (2019b): 1–18. https://doi.org/10/gf4z8w.

Cubino, Josep P., Josep V. Subirós, and Carles B. Lozano. "Floristic and Structural Differentiation Between Gardens of Primary and Secondary Residences in the Costa Brava (Catalonia, Spain)." *Urban Ecosystems* 19 (2016): 505–21. https://doi.org/10/f8gq5z.

———. "Propagule Pressure from Invasive Plant Species in Gardens in Low-Density Suburban Areas of the Costa Brava (Spain)." *Urban Forestry & Urban Greening* 14 (2015): 941–51. https://doi.org/10/f76nh6.

da Rocha, Léo C., Maria J. Ferreira-Caliman, Carlos A. Garofalo, and Solange C. Augusto. "A Specialist in an Urban Area are Cities Suitable to Harbour Populations of the Oligolectic Bee Centris (Melacentris) Collaris (Apidae Centridini)?" *Annales Zoologici Fennici* 55 (2018): 135–49. https://doi.org/10/gdhkpf.

da Silva, Ronilze Leite, Everton Hilo de Souza, Carlos Alberto da Silva Ledo, Regina Pelacani Claudinéia, and Fernanda V. Duarte Souza. "Urban backyards as a new model of pineapple germplasm conservation." *Plant Genetic Resources* 16, no. 6 (2018): 524–532.

Daniels, Grant D., and Jamie B. Kirkpatrick. "Does Variation in Garden Characteristics Influence the Conservation of Birds in Suburbia?" *Biological Conservation* 133 (2006): 326–35. https://doi.org/10/b85jfc.

Dasylva, Maurice, Ngor Ndour, Bienvenu Sambou, and Christophe T. Soulard. "Micro-Farms Producing Aromatic and Medicinal Plants a Figure of the Urban Agriculture in Ziguinchor, Senegal." *Cahiers Agricultures* 27 (2018): 1–9. https://doi.org/10/gf4z7q.

Davies, Zoe G., Richard A. Fuller, Alison Loram, Katherine N. Irvine, Victoria Sims, and Kevin J. Gaston. "A National Scale Inventory of Resource Provision for Biodiversity within Domestic Gardens." *Biological Conservation* 142 (2009): 761–71. https://doi.org/10/drzfh7.

Davis, Amélie Y., Eric V. Lonsdorf, Cliff R. Shierk, Kevin C. Matteson, John R. Taylor, Sarah T. Lovell, and Emily S. Minor. "Enhancing Pollination Supply in an Urban Ecosystem Through Landscape Modifications." *Landscape and Urban Planning* 162 (2017): 157–66. https://doi.org/10/f95mzc.

Davison, John, Maren Huck, Richard J. Delahay, and T. J. Roper. "Restricted Ranging Behaviour in a High-Density Population of Urban Badgers." *Journal of Zoology* 277 (2009): 45–53. https://doi.org/10/bp9v4s.

Dennis, Matthew, and Philip James. "Ecosystem Services of Collectively Managed Urban Gardens: Exploring Factors Affecting Synergies and Trade-Offs at the Site Level." *Ecosystem Services* 26 (2017): 17–26. https://doi.org/10/gckcsg.

Di Mauro, Desiree, Thomas Dietz, and Larry Rockwood. "Determining the Effect of Urbanization on Generalist Butterfly Species Diversity in Butterfly Gardens." *Urban Ecosystems* 10 (2007): 427–39. https://doi.org/10/bh7khc.

Dongus, Stefan, Dickson Nyika, Khadija Kannady, Deo Mtasiwa, Hassan Mshinda, Laura Gosoniu, and Axel W. Drescher, et al. "Urban Agriculture and Anopheles Habitats in Dar es Salaam, Tanzania." *Geospatial Health* 3 (2009): 189–210. https://doi.org/10/gf42bz.

Edmondson, Jill L., Zoe G. Davies, Kevin J. Gaston, and Jonathan R. Leake. "Urban Cultivation in Allotments Maintains Soil Qualities Adversely Affected by Conventional Agriculture." *Journal of Applied Ecology* 51 (2014): 880–89. https://doi.org/10/f6kmb5.

Egerer, Monika H., Peter Bichier, and Stacy M. Philpott. "Landscape and Local Habitat Correlates of Lady Beetle Abundance and Species Richness in Urban Agriculture." *Annals of the Entomological Society of America* 110 (2017): 103–97. https://doi.org/10/f9v54r.

Egerer, Monika H., Kevin Li, and Theresa W. Y. Ong. "Context Matters: Contrasting Ladybird Beetle Responses to Urban Environments Across Two US Regions." *Sustainability* 10 (2018b): 1–17. https://doi.org/10/gd72m2.

Egerer, Monika H., Alessandro Ossola, and Brenda B. Lin. "Creating Socioecological Novelty in Urban Agroecosystems from the Ground up." *Bioscience* 68 (2018a): 25–34. https://doi.org/10/gcwhhg.

Egerer, Monika H., Brenda B. Lin, and Dave Kendal. "Temperature Variability Differs in Urban Agroecosystems Across Two Metropolitan Regions." *Climate* 7 (2019b): 1–18. https://doi.org/10/gf4z73.

Egerer, Monika H., Brenda B. Lin, Caragh G. Threlfall, and Dave Kendal. "Temperature Variability Influences Urban Garden Plant Richness and Gardener Water use Behavior, but not Planting Decisions." *Science of the Total Environment* 646 (2019a): 111–20. https://doi.org/10/gf42kb.

Ellis, Jodie A., A D Walter, John F. Tooker, Matthew D. Ginzel, Peter F. Reagel, Emerson S. Lacey, Ashley B. Bennett, Erik M. Grossman, and Lawrence M. Hanks. "Conservation Biological Control in Urban Landscapes: Manipulating Parasitoids of Bagworm (Lepidoptera Psychidae) with Flowering Forbs." *Biological Control* 34 (2005): 107–99. https://doi.org/10/d3wzq5.

Ernst, Raffael, David Massemin, and Ingo Kowarik. "Non-Invasive Invaders from the Caribbean the Status of Johnstone's Whistling Frog (Eleutherodactylus Johnstonei) Ten Years after its Introduction to Western French Guiana." *Biological Invasions* 13 (2011): 1767–77. https://doi.org/10/br32m3.

Etshekape, P. Gabriel, Alain R. Atangana, and Damase P. Khasa. "Tree Planting in Urban and Peri-Urban of Kinshasa Survey of Factors Facilitating Agroforestry Adoption." *Urban Forestry & Urban Greening* 30 (2018): 12–23. https://doi.org/10.1016/j.ufug.2017.12.015.

Fischer, Leonie K., Daniel Brinkmeyer, Stefanie J. Karle, Kathrine Cremer, E Huttner, Martin Seebauee, and Ulrich Nowikow, et al. "Biodiverse Edible Schools: Linking Healthy Food, School Gardens and Local Urban Biodiversity." *Urban Forestry & Urban Greening* 40 (2019): 35–43. https://doi.org/10/gf42nj.

Flor, Yvan A. D. de la, Caitlin E. Burkman, Taro K. Eldredge, and Mary M. Gardiner. "Patch and Landscape-Scale Variables Influence the Taxonomic and Functional Composition of Beetles in Urban Greenspaces." *Ecosphere* 8 (2017): 1–17. https://doi.org/10/gcrcbz.

Flux, John E. C. "Seventeen Years of Predation by One Suburban Cat in New Zealand." *New Zealand Journal of Zoology* 34 (2007): 289–96. https://doi.org/10/cm7bhw.

Frankie, Gordon W., S. Bradleigh Vinson, Mark A. Rizzardi, Terry L. Griswold, Rollin E. Coville, Michael H. Grayum, L E S Martinez, Jennifer L. Foltz-Sweat, and Jaime C. Pawelek. "Relationships of Bees to Host Ornamental and Weedy Flowers in Urban Northwest Guanacaste Province, Costa Rica." *Journal of the Kansas Entomological Society* 86 (2013): 325–51. https://doi.org/10/f5qnw9.

French, Kristine, Richard E. Major, and Katherine Hely. "Use of Native and Exotic Garden Plants by Suburban Nectarivorous Birds." *Biological Conservation* 121 (2005): 545–59. https://doi.org/10/fms8kj.

Frey, David, Kevin Vega, Florian Zellweger, Jaboury Ghazoul, Dennis Hansen, and Marco Moretti. "Predation Risk Shaped by Habitat and Landscape Complexity in Urban Environments." *Journal of Applied Ecology* 55 (2018): 2343–53. https://doi.org/10/gd5wn4.

Fuller, Richard A., Philip H. Warren, Paul R. Armsworth, Olga Barbosa, and Kevin J. Gaston. "Garden Bird Feeding Predicts the Structure of Urban Avian Assemblages." *Diversity and Distributions* 14 (2008): 131–37. https://doi.org/10/d99h29.

Gaggini, Luca, Hans-Peter Rusterholz, and Bruno Baur. "Settlements as a Source for the Spread of Non-Native Plants into Central European Suburban Forests." *Acta Oecologica-International Journal of Ecology* 79 (2017): 18–25. https://doi.org/10/f9zp5g.

Galbraith, Josie A., Jacqueline R. Beggs, Darryl N. Jones, Ellery J. McNaughton, Cheryl R. Krull, and Margaret C. Stanley. "Risks and Drivers of Wild Bird Feeding in Urban Areas of New Zealand." *Biological Conservation* 180 (2014): 64–74. https://doi.org/10/f6s8gt.

Galbraith, Josie A., Jacqueline R. Beggs, Darryl N. Jones, and Margaret C. Stanley. "Supplementary Feeding Restructures Urban Bird Communities." *Proceedings of the National Academy of Sciences* 112 (2015): E2648–57. https://doi.org/10/f7cz94.

Galbraith, Josie A., Darryl N. Jones, Jacqueline R. Beggs, Katharina Parry, and Margaret C. Stanley. "Urban Bird Feeders Dominated by a Few Species and Individuals." *Frontiers in Ecology and Evolution* 5 (2017a): 1–15. https://doi.org/10/gf42bf.

Galbraith, Josie A., Margaret C. Stanley, Darryl N. Jones, and Jacqueline R. Beggs. "Experimental Feeding Regime Influences Urban Bird Disease Dynamics." *Journal of Avian Biology* 48 (2017b): 700–713. https://doi.org/10/gbjzkq.

Galluzzi, Gea, P Eyzaguirre, and Valeria Negri. "Home Gardens: Neglected Hotspots of Agro-Biodiversity and Cultural Diversity." *Biodiversity and Conservation* 19 (December 1, 2010): 3635–54. https://doi.org/10/c9jgxk.

Garbuzov, Mihail, Karin Alton, and Francis L. W. Ratnieks. "Most Ornamental Plants on Sale in Garden Centres are Unattractive to Flower-Visiting Insects." *PeerJ* 5 (2017): 1–17. https://doi.org/10/gf42hr.

Gardiner, Mary M., Scott P. Prajzner, Caitlin E. Burkman, Sandra Albro, and Parwinder S. Grewal. "Vacant Land Conversion to Community Gardens: Influences on Generalist Arthropod Predators and Biocontrol Services in Urban Greenspaces." *Urban Ecosystems* 17 (2014): 101–22. https://doi.org/10/f5v7vb.

Gaston, Kevin J., Richard A. Fuller, Alison Loram, Charlotte MacDonald, Sinead Power, and Nicola Dempsey. "Urban Domestic Gardens (XI) Variation in Urban Wildlife Gardening in the United Kingdom." *Biodiversity and Conservation* 16 (2007): 3227–38. https://doi.org/10/dkr648.

Gaston, Kevin J., Richard M. Smith, Ken Thompson, and Philip H. Warren. "Urban Domestic Gardens (II) Experimental Tests of Methods for Increasing Biodiversity." *Biodiversity and Conservation* 14 (2005a): 395–413. https://doi.org/10/bjgxxs.

Gaston, Kevin J., Philip H. Warren, Ken Thompson, and Richard M. Smith. "Urban Domestic Gardens (IV) The Extent of the Resource and its Associated Features." *Biodiversity and Conservation* 14 (2005b): 3327–49. https://doi.org/10/dgtgnc.

Gbedomon, Rodrigue C., Valère K. Salako, Aristide C. Adomou, Romain G. Kakaï, and Achille E. Assogbadjo. "Plants in Traditional Home Gardens Richness, Composition, Conservation and Implications for Native Biodiversity in Benin." *Biodiversity and Conservation* 26 (2017): 3307–27. https://doi.org/10/gck9rj.

Bao-Ming, Ge, Li Zhen-Xing, Zhang Dai-Zhen, Zhang Hua-Bin, Liu Zong-Tang, Zhou Chun-Lin, and Tang Bo-Ping. "Communities of Soil Macrofauna in Green Spaces of an Urbanizing City at East China." *Revista Chilena de Historia Natural* 85 (2012): 219–26. https://doi.org/10/f358sh.

Goddard, Mark A., Andrew J. Dougill, and Tim G. Benton. "Scaling up from Gardens: Biodiversity Conservation in Urban Environments." *Trends in Ecology and Evolution* 25, no. 2 (2010): 90–98. https://doi.org/10/crgjp5.

Goddard, Mark A., Andrew J. Dougill, and Tim G. Benton. "Why Garden for Wildlife? Social and Ecological Drivers Motivations and Barriers for Biodiversity Management in Residential Landscapes." *Ecological Economics* 86 (2013): 258–73. https://doi.org/10/f4vrgj.

Gonzalez-Garcia, Alberto, Josabel Belliure, Antonio Gomez-Sal, and Pedrarias Davilla. "The Role of Urban Greenspaces in Fauna Conservation: The Case of the Iguana Ctenosaura Similis in the 'Patios'' of Leon City, Nicaragua.'" *Biodiversity and Conservation* 18 (2009): 1909–20. https://doi.org/10/dn8gjz.

Gopal, Divya, and Harini Nagendra. "Vegetation in Bangalore's Slums: Boosting Livelihoods, Well-Being and Social Capital." *Sustainability* 6 (2014): 2459–73. https://doi.org/10/gcfngg.

Guitart, Daniela A., Catherine Pickering, and Jason Byrne. "Past Results and Future Directions in Urban Community Gardens Research." *Urban Forestry & Urban Greening* 11, no. 4 (2012): 364–73. https://doi.org/10/gf42m6.

Guitart, Daniela A., Jason A. Byrne, and Catherine M. Pickering. "Greener Growing Assessing the Influence of Gardening Practices on the Ecological Viability of Community Gardens in South East Queensland, Australia." *Journal of Environmental Planning and Management* 58 (2013): 189–212. https://doi.org/10/gf42dr.

Gulraiz, Tayiba, Arshad Javid, Muhammad Mahmood-Ul-Hassan, Syed Hussain, Hamda Azmat, and Sharoon Daud. "Role of Indian Flying Fox Pteropus Giganteus Brünnich, 1782 (Chiroptera: Pteropodidae) as a Seed Disperser in Urban Areas of Lahore, Pakistan." *Turkish Journal of Zoology* 40, no. 3 (2016): 417–22. https://doi.org/10.3906/zoo-1407-42.

Hall, Damon M., Gerardo R. Camilo, Rebecca K. Tonietto, Jeff Ollerton, Karin Ahrné, Mike Arduser, and John S. Ascher et al. "The City as a Refuge for Insect Pollinators." *Conservation Biology* 31, no. 1 (2017): 24–29.

Hanley, Mick E., Amanda J. Awbi, and Miguel Franco. "Going Native? Flower Use by Bumblebees in English Urban Gardens." *Annals of Botany* 113 (2014): 799–806. https://doi.org/10/f5vzf7.

Head, Lesley, and Pat Muir. "Edges of Connection Reconceptualising the Human Role in Urban Biogeography." *Australian Geographer* 37 (2006b): 101–87. https://doi.org/10/cjg77w.

———. "Suburban Life and the Boundaries of Nature Resilience and Rupture in Australian Backyard Gardens." *Transactions of the Institute of British Geographers* 31 (2006a): 505–24. https://doi.org/10/bdqvqx.

Heath, Sacha K., Fogel, Nina S., Mullikin, Jennifer C., and Hull, Trey. (2020). Data and code for Chapter 1: An expanded scope of biodiversity in urban agriculture, with implications for conservation. (Version 1.0.0) [Data set]. Urban Agroecology: Past, Present, and Future Directions in Interdisciplinary Research. Abingdon, UK: Taylor & Francis. http://doi.org/10.5281/zenodo.3989949

Heterick, Brian, Morgan J. Lythe, and C Smithyman. "Urbanisation Factors Impacting on Ant (Hymenoptera Formicidae) Biodiversity in the Perth Metropolitan Area, Western Australia Two Case Studies." *Urban Ecosystems* 16 (2013): 145–73. https://doi.org/10/f4wv9s.

Hisano, Masumi, Evgeniy G. Raichev, Stanislava Peeva, Hiroshi Tsunoda, Chris Newman, Ryuichi Masuda, Dian M Georgiev, and Yayoi K. Kaneko. "Comparing the Summer Diet of Stone Martens (Martes Foina) in Urban and Natural Habitats in Central Bulgaria." *Ethology Ecology & Evolution* 28 (2016): 295–311. https://doi.org/10/gf4z9z.

Joimel, Sophie, Baptiste Grard, Apolline Auclerc, Mickaël Hedde, Nolwenn Le Doaré, Sandrine Salmon, and Claire Chenu. "Are Collembola 'Flying' onto Green Roofs?" *Ecological Engineering* 111 (2018): 117–24. https://doi.org/10/gcvfhs.

Joimel, Sophie, Christophe Schwartz, Mickaël Hedde, Sayuri Kiyota, Paul H. Krogh, Johanne Nahmani, Guénola Pérès, Alan Vergnes, and Jérôme Cortet. "Urban and Industrial Land Uses have a Higher Soil Biological Quality than Expected from Physicochemical Quality." *Science of the Total Environment* 584 (2017): 614–21. https://doi.org/10/f95mcd.

Joimel, Sophie, Christophe Schwartz, Noëlie Maurel, Benjamin Magnus, Nathalie Machon, Jérémie Bel, and Jérôme Cortet. "Contrasting Homogenization Patterns of Plant and Collembolan Communities in Urban Vegetable Gardens." *Urban Ecosystems* 22 (2019): 553–66. https://doi.org/10/gfxfq3.

Kamjing, Anucha, Dusit Ngoprasert, Robert Steinmetz, Wanlop Chutipong, Tommaso Savini, and George A. Gale. "Determinants of Smooth-Coated Otter Occupancy in a Rapidly Urbanizing Coastal Landscape in Southeast Asia." *Mammalian Biology* 87 (2017): 168–75. https://doi.org/10/gcswjt.

Kaoma, Humphrey, and Charlie M. Shackleton. "Collection of urban tree products by households in poorer residential areas of three South African towns." *Urban Forestry & Urban Greening* 13 (2014): 244–52. https://doi.org/10/f56cng.

Karhagomba, Innocent B., Adhama Mirindi, Timothée B. Mushagalusa, Victor B. Nabino, Kwangoh Koh, and Hee S. Kim. "The Cultivation of Wild Food and Medicinal Plants for Improving Community Livelihood: The Case of the Buhozi Site, DR Congo." *Nutrition Research and Practice* 7 (2013): 510–18. https://doi.org/10/gf42hc.

Kauhala, Kaarina, Kati Talvitie, and Timo Vuorisalo. "Encounters between Medium-Sized Carnivores and Humans in the City of Turku, SW Finland, with Special Reference to the Red Fox." *Mammal Research* 61 (2016): 25–33. https://doi.org/10/gf42gt.

Klepacki, Piotr, and Monika Kujawska. "Urban Allotment Gardens in Poland: Implications for Botanical and Landscape Diversity." *Journal of Ethnobiology* 38 (2018): 123–37. https://doi.org/10/gdf3h4.

Klinkenberg, Eveline, Michael D. Wilson, P. J. McCall, Felix P. Amerasinghe, and Martin J. Donnelly. "Impact of Urban Agriculture on Malaria Vectors in Accra, Ghana." *Malaria Journal* 7 (2008): 1–9. https://doi.org/10/bhhths.

Knapp, Sonja. "The urban garden: a centre of interactions between people and biodiversity." In *Human-Environmental Interactions in Cities Challenges and Opportunities of Urban Land Use Planning and Green Infrastructure*, 2014.

König, Andreas, Christof Janko, Bence Barla-Szabo, Diana Fahrenhold, Claudius Heibl, Eva Perret, and Stefanie Wermuth. "Habitat Model for Baiting Foxes in Suburban Areas to Counteract Echinococcus Multilocularis." *Wildlife Research* 39 (2012): 488–95. https://doi.org/10/gf42ns.

Koyama, Asuka, Chika Egawa, Hisatomo Taki, Mika Yasuda, Natsumi Kanzaki, Tatsuya Ide, and Kimiko Okabe. "Non-Native Plants are a Seasonal Pollen Source for Native Honeybees in Suburban Ecosystems." *Urban Ecosystems* 21 (2018): 1113–22. https://doi.org/10/gf42mh.

Kujawska, Monika, Fernando Zamudio, Lía Montti, and Veronica P. Carrillo. "Effects of Landscape Structure on Medicinal Plant Richness in Home Gardens: Evidence for the Environmental Scarcity Compensation Hypothesis." *Economic Botany* 72 (2018): 150–65. https://doi.org/10/gfdk75.

Kurz, Tim, and Catherine Baudains. "Biodiversity in the Front Yard: An Investigation of Landscape Preference in a Domestic Urban Context." *Environment and Behavior* 44 (2012): 166–96. https://doi.org/10/d9cbwb.

Lamont, Susan R., W. Hardy Eshbaugh, and Adolph M. Greenberg. "Species Composition, Diversity, and Use of Homegardens among Three Amazonian Villages." *Economic Botany* 53 (1999): 312–26. https://doi.org/10/drghwd.

Lerman, Susannah B., Victoria K. Turner, and Christofer Bang. "Homeowner Associations as a Vehicle for Promoting Native Urban Biodiversity." *Ecology and Society* 17 (2012): 1–13. https://doi.org/10/gf4z8m.

Lim, Voon-Ching, Elizabeth L. Clare, Joanne E. Littlefair, Rosli Ramli, Subha Bhassu, and John-James Wilson. "Impact of Urbanisation and Agriculture on the Diet of Fruit Bats." *Urban Ecosystems* 21 (2018): 61–70. https://doi.org/10/gczbbp.

Lin, Brenda B., and Monika H. Egerer. "Urban agriculture an opportunity for biodiversity and food provision in urban landscapes." In *Urban Biodiversity: From Research to Practice*, 100–185. NA: Routledge, 2017.

Lin, Brenda B, Monika H. Egerer, Heidi Liere, Shalene Jha, Peter Bichier, and Stacy M Philpott. "Local- and Landscape-Scale Land Cover Affects Microclimate and Water Use in Urban Gardens." *Science of the Total Environment* 610–611 (2017a): 570–75. https://doi.org/10/gf42j3.

Lin, Brenda B., Stacey M. Philpott, and Shalene Jha. "The Future of Urban Agriculture and Biodiversity-Ecosystem Services: Challenges and Next Steps." *Basic and Applied Ecology* 16 (2015): 189–201. https://doi.org/10/f673hb.

Livoreil, Barbara, Julie Glanville, Neal R. Haddaway, Helen Bayliss, Alison Bethel, Frédérique Flamerie de Lachapelle, and Shannon Robalino, et al. "Systematic Searching for Environmental Evidence Using Multiple Tools and Sources." *Environmental Evidence* 6, no. 1 (2017). https://doi.org/10/gf5p8q.

Loram, Alison, Philip H. Warren, and Kevin J. Gaston. "Urban Domestic Gardens (XIV): The Characteristics of Gardens in Five Cities." *Environmental Management* 42 (2008): 361–76. https://doi.org/10/b32t9f.

Lowenstein, David M., Maryam Gharehaghaji, and David H. Wise. "Substantial Mortality of Cabbage Looper (Lepidoptera Noctuidae) from Predators in Urban Agriculture is not Influenced by Scale of Production or Variation in Local and Landscape-Level Factors." *Environmental Entomology* 46 (2017): 30–37. https://doi.org/10/gf4z9s.

Lowenstein, David M., Kevin C. Matteson, and Emily S. Minor. "Evaluating the Dependence of Urban Pollinators on Ornamental, Non-Native, and 'Weedy' Floral Resources." *Urban Ecosystems* 22 (2019): 293–302. https://doi.org/10/gf42mn.

Lowenstein, David M., and Emily S. Minor. "Diversity in Flowering Plants and their Characteristics Integrating Humans as a Driver of Urban Floral Resources." *Urban Ecosystems* 19 (2016): 1735–48. https://doi.org/10/f9gc4n.

Luck, Gary. W. "A review of the relationships between human population density and biodiversity." *Biological Reviews* 82 (2007): 607–645.

Lubbe, Catharina S., Stefan J. Siebert, and Sarel S. Cilliers. "Floristic Analysis of Domestic Gardens in the Tlokwe City Municipality, South Africa." *Bothalia* 41 (2011): 351–61. https://doi.org/10/gf4z7j.

Lundholm, Jeremy. T. and Paul J. Richardson. "Habitat Analogues for Reconciliation Ecology in Urban and Industrial Environments." *Journal of Applied Ecology*. 47 (2010): 966–975.

Mach, Bernadette M., and Daniel A. Potter. "Quantifying Bee Assemblages and Attractiveness of Flowering Woody Landscape Plants for Urban Pollinator Conservation." *Plos One* 13 (2018): 1–18. https://doi.org/10/gfz7zj.

Marco, Audrey, Carole Barthelemy, Thierry Dutoit, and Valérie Bertaudiere-Montes. "Bridging Human and Natural Sciences for a Better Understanding of Urban Floral Patterns: The Role of Planting Practices in Mediterranean Gardens." *Ecology and Society* 15 (2010): 1–18. https://doi.org/10/gf4z8n.

Marks, Clive A., and Tim E. Bloomfield. "Home-Range Size and Selection of Natal Den and Diurnal Shelter Sites by Urban Red Foxes (Vulpes Vulpes) in Melbourne." *Wildlife Research* 33 (2006): 339–47. https://doi.org/10/fqmfw7.

Maroyi, Alfred, and Gabolwelwe K. E. Mosina. "Medicinal Plants and Traditional Practices in Peri-Urban Domestic Gardens of the Limpopo Province, South Africa." *Indian Journal of Traditional Knowledge* 13 (2014): 665–72.

Masierowska, Marzena, Ernest Stawiarz, and Robert Rozwalka. "Perennial Ground Cover Plants as Floral Resources for Urban Pollinators: A Case of Geranium Species." *Urban Forestry & Urban Greening* 32 (2018): 185–94. https://doi.org/10/gdn8vb.

Mata, Luis, Caragh G. Threlfall, Nicholas S. G. Williams, Amy K. Hahs, Mallik M. Malipatil, Nigel E. Stork, and Stephen J. Livesley. "Conserving Herbivorous and Predatory Insects in Urban Green Spaces." *Scientific Reports* 7 (2017): 1–12. https://doi.org/10/f9ng47.

Materechera, Simeon A. "Soil Properties and Subsoil Constraints of Urban and Peri-Urban Agriculture within Mahikeng City in the North West Province (South Africa)." *Journal of Soils and Sediments* 18 (2018): 494–505. https://doi.org/10/gczsqm.

Mattah, Precious A. D., G. Futagbi, Leonard K. Amekudzi, Memuna M. Mattah, Dziedzorm K. de Souza, Worlasi D. Kartey-Attipoe, Langbong Bimi, and Michael D. Wilson. "Diversity in Breeding Sites and Distribution of Anopheles Mosquitoes in Selected Urban Areas of Southern Ghana." *Parasites & Vectors* 10 (2017): 1–15. https://doi.org/10/gf42hj.

Matter, Stephen F., Jessica R. Brzyski, Christopher J. Harrison, Sara Hyams, Clement Loo, Jessica Loomis, and Hannah R. Lubbers, et al. "Invading from the Garden? A Comparison of Leaf Herbivory for Exotic and Native Plants in Natural and Ornamental Settings." *Insect Science* 19 (2012): 677–82. https://doi.org/10/gf42ck.

Matteson, Kevin C., and Gail A. Langellotto. "Determinates of Inner City Butterfly and Bee Species Richness." *Urban Ecosystems* 13 (2010): 333–47. https://doi.org/10/cqnmdf.

Matteson, Kevin C., and Gail A. Langellotto. "Bumble bee abundance in New York City community gardens implications for urban agriculture." *Cities and the Environment* 2 (2009): 1–12. https://doi.org/10/gf4z72.

Matthys, Barbara, Eliézer K. N'Goran, Moussa Kone, Benjamin G. Koudou, Penelope Vounatsou, Guéladio Cisse, Andres B. Tschannen, Marcel Tanner, and Jürg Utzinger. "Urban Agricultural Land Use and Characterization of Mosquito Larval Habitats in a Medium-Sized Town of Cote D'Ivoire." *Journal of Vector Ecology* 31 (2006): 319–33. https://doi.org/10.3376/1081-1710(2006)31[319:ualuac]2.0.co;2.

Mazzeo, Nadia M., and Juan P. Torretta. "Wild bees (Hymenoptera Apoidea) in an urban botanical garden in Buenos Aires, Argentina." *Studies on Neotropical Fauna and Environment* 50 (2015): 182–93. http://dx.doi.org/10.1080/01650521.2015.1093764

McDonald-Madden, Eve, E. Sabine G. Schreiber, David M. Forsyth, David Choquenot, and T. F. Clancy. "Factors Affecting Grey-Headed Flying-Fox (Pteropus Poliocephalus: Pteropodidae) Foraging in the Melbourne Metropolitan Area, Australia." *Austral Ecology* 30, no. 5 (2005): 600–608. https://doi.org/10/crshxg.

McLean, Phil, John R.U. Wilson, Mirijam Gaertner, Suzaan Kritzinger-Klopper, and David M. Richardson. "The Distribution and Status of Alien Plants in a Small South African Town." *South African Journal of Botany* 117 (2018): 71–78. https://doi.org/10/gdwwx5.

Melendez-Ackerman, Elvia J., Christopher J. Nytch, Luis E. Santiago-Acevedo, Julio C. Verdejo-Ortiz, Raul Santiago-Bartolomei, Luis E. Ramos-Santiago, and Tischa A. Munoz-Erickson. "Synthesis of Household Yard Area Dynamics in the City of San Juan Using Multi-Scalar Social-Ecological Perspectives." *Sustainability* 8 (2016): 1–21. https://doi.org/10/gf42kt.

Merlin-Uribe, Yair, Carlos E. Gonzalez-Esquivel, Armando Contreras-Hernandez, Luis Zambrano, Patricia Moreno-Casasola, and Marta Astier. "Environmental and Socio-Economic Sustainability of Chinampas (Raised Beds) in Xochimilco, Mexico City." *International Journal of Agricultural Sustainability* 11 (2013): 216–33. https://doi.org/10/gf42cr.

Moorhead, Leigh C., and Stacy M. Philpott. "Richness and Composition of Spiders in Urban Green Spaces in Toledo, Ohio." *Journal of Arachnology* 41 (2013): 356–63. https://doi.org/10/gf42dg.

Morales, Helda, Bruce G. Ferguson, Linda E. Marin, Dario N. Gutierrez, Peter Bichier, and Stacy M. Philpott. "Agroecological Pest Management in the City: Experiences from California and Chiapas." *Sustainability* 10 (2018): 1–17. https://doi.org/10/gd74tg.

Moro, Marcelo F., Christian Westerkamp, and Francisca S. de Araújo. "How Much Importance is Given to Native Plants in Cities' Treescape? A Case Study in Fortaleza, Brazil." *Urban Forestry & Urban Greening* 13, no. 2 (2014): 365–74. https://doi.org/10/f56cj4.

Mosina, Gabolwelwe K. E., and Alfred Maroyi. "Edible Plants of Urban Domestic Gardens in the Capricorn District, Limpopo Province, South Africa." *Tropical Ecology* 57 (2016): 181–91.

Mosina, Gabolwelwe K. E., Alfred Maroyi, and Martin J. Potgieter. "Useful Plants Grown and Maintained in Domestic Gardens of the Capricorn District, Limpopo Province, South Africa." *Studies on Ethno-Medicine* 9 (2015): 43–58. https://doi.org/10/gf42kq.

Mukwevho, Vuledzani O., James S. Pryke, and Francois Roets. "Habitat Preferences of the Invasive Harlequin Ladybeetle Harmonia Axyridis (Coleoptera Coccinellidae) in the Western Cape Province, South Africa." *African Entomology* 25 (2017): 86–97. https://doi.org/10/f93kgw.

Mumaw, Laura, and Sarah Bekessy. "Wildlife Gardening for Collaborative Public-Private Biodiversity Conservation." *Australasian Journal of Environmental Management* 24 (2017): 242–60. https://doi.org/10/gfsp86.

Narango, Desirée L., Douglas W. Tallamy, and Peter P. Marra. "Nonnative Plants Reduce Population Growth of an Insectivorous Bird." *Proceedings of the National Academy of Sciences* 115, no. 45 (2018): 11549–54. https://doi.org/10/gfkgdt.

Norfolk, Olivia, Markus P. Eichhorn, and Francis Gilbert. "Culturally Valuable Minority Crops Provide a Succession of Floral Resources for Flower Visitors in Traditional Orchard Gardens." *Biodiversity and Conservation* 23 (2014): 3199–3217. https://doi.org/10/f6qjjz.

Ockendon, Nancy, Sarah E. Davis, Teresa Miyar, and Mike P. Toms. "Urbanization and Time of Arrival of Common Birds at Garden Feeding Stations." *Bird Study* 56 (2009): 405–10. https://doi.org/10/bt5skn.

Oh, Rachel R. Y., Daniel R. Richards, and Alex T. K. Yee. "Community-Driven Skyrise Greenery in a Dense Tropical City Provides Biodiversity and Ecosystem Service Benefits." *Landscape and Urban Planning* 169 (2018): 115–23. https://doi.org/10/gf42f6.

Orros, Melanie E., and Mark D. E. Fellowes. "Supplementary Feeding of Wild Birds Indirectly Affects the Local Abundance of Arthropod Prey." *Basic and Applied Ecology* 13 (2012): 286–93. https://doi.org/10/f365jj.

Orros, Melanie E., Rebecca L. Thomas, Graham J. Holloway, and Mark D. E. Fellowes. "Supplementary Feeding of Wild Birds Indirectly Affects Ground Beetle Populations in Suburban Gardens." *Urban Ecosystems* 18 (2015): 465–75. https://doi.org/10/f7cq6n.

Orsini, Francesco, Daniela Gasperi, Livia Marchetti, Chiara Piovene, Stefano Draghetti, Solange Ramazzotti, Giovanni Bazzocchi, and Giorgio Gianquinto. "Exploring the Production Capacity of Rooftop Gardens (RTGs) in Urban Agriculture: The Potential Impact on Food and Nutrition Security, Biodiversity and Other Ecosystem Services in the City of Bologna." *Food Security* 6 (2014): 781–92. https://doi.org/10/gf4z97.

Ortiz-Perea, Natali, Rebecca Gander, Oliver Abbey, and Amanda Callaghan. "The Effect of Pond Dyes on Oviposition and Survival in Wild UK Culex Mosquitoes." *Plos One* 13 (2018): 1–15. https://doi.org/10/gc9d7m.

Ossola, Alessandro, Dexter Locke, Brenda Lin, and Emily S. Minor. "Greening in Style Urban form, Architecture and the Structure of Front and Backyard Vegetation." *Landscape and Urban Planning* 185 (2019): 141–57. https://doi.org/10/gf42gc.

Otoshi, Michelle D., Peter Bichier, and Stacy M. Philpott. "Local and Landscape Correlates of Spider Activity Density and Species Richness in Urban Gardens." *Environmental Entomology* 44 (2015): 1043–51. https://doi.org/10/f7pdsj.

Palheta, Ivanete C., Ana C. C. Tavares-Martins, Flavia C. A. Lucas, and Mario A. G. Jardim. "Ethnobotanical Study of Medicinal Plants in Urban Home Gardens in the City of Abaetetuba, Para State, Brazil." *Boletin Latinoamericano Y Del Caribe De Plantas Medicinales Y Aromaticas* 16 (2017): 206–62.

Palliwoda, Julia, Ingo Kowarik, and Moritz von der Lippe. "Human-Biodiversity Interactions in Urban Parks: The Species Level Matters." *Landscape and Urban Planning* 157 (2017): 394–406. https://doi.org/10/f9g5bb.

Parris, Kirsten M., and Donna L. Hazell. "Biotic Effects of Climate Change in Urban Environments: The Case of the Grey-Headed Flying-Fox (Pteropus Poliocephalus) in Melbourne, Australia." *Biological Conservation* 124, no. 2 (2005): 267–76. https://doi.org/10/dwrpzx.

Parsons, Holly, Richard E. Major, and Kris French. "Species Interactions and Habitat Associations of Birds Inhabiting Urban Areas of Sydney, Australia." *Austral Ecology* 31 (2006): 217–27. https://doi.org/10/bhdhtq.

Patterson, Lindsay, Riddhika Kalle, and Collen Downs. "Factors Affecting Presence of Vervet Monkey Troops in a Suburban Matrix in KwaZulu-Natal, South Africa." *Landscape and Urban Planning* 169 (2018): 220–28. https://doi.org/10/gf42gd.

———. "Predation of Artificial Bird Nests in Suburban Gardens of KwaZulu-Natal, South Africa." *Urban Ecosystems* 19 (2016): 615–30. https://doi.org/10/f83dzk.

Pauw, Anton, and Kirsten Louw. "Urbanization Drives a Reduction in Functional Diversity in a Guild of Nectar-Feeding Birds." *Ecology and Society* 17 (2012): 1–8. https://doi.org/10/f9989m.

Pawelek, Jaime C., Gordon W. Frankie, Robbin W. Thorp, and Maggie Przybylski. "Modification of a Community Garden to Attract Native Bee Pollinators in Urban San Luis Obispo, California." *Cities and the Environment* 2, no. 1 (2009): 1–21. https://doi.org/10/ggkjxg.

Peck, Hannah L., Henrietta E. Pringle, Harry H. Marshall, Ian P. F. Owens, and Alexa M. Lord. "Experimental Evidence of Impacts of an Invasive Parakeet on Foraging Behavior of Native Birds." *Behavioral Ecology* 25 (2014): 582–90. https://doi.org/10/f548hv.

Peralta, Guadalupe, Maria S. Fenoglio, and Adriana Salvo. "Physical Barriers and Corridors in Urban Habitats Affect Colonisation and Parasitism Rates of a Specialist Leaf Miner." *Ecological Entomology* 36 (2011): 673–79. https://doi.org/10/dqbc9j.

Peterson, M. Nils, Brandi Thurmond, Melissa McHale, Shari Rodriguez, Howard D. Bondell, and Merril Cook. "Predicting Native Plant Landscaping Preferences in Urban Areas." *Sustainable Cities and Society* 5 (2012): 70–76. https://doi.org/10/gf42kz.

Philpott, Stacy M., Simone Albuquerque, Peter Bichier, Hamutahl Cohen, Monika H. Egerer, Claire Kirk, and Kipling W. Will. "Local and Landscape Drivers of Carabid Activity, Species Richness, and Traits in Urban Gardens in Coastal California." *Insects* 10 (2019): 1–14. https://doi.org/10/gf42cp.

Philpott, Stacy M., and Peter Bichier. "Local and Landscape Drivers of Predation Services in Urban Gardens." *Ecological Applications* 27, no. 3 (2017): 966–76. https://doi.org/10/gf4z8c.

Pierart, Antoine, Camille Dumat, Arthur Q. Maes, and Nathalie Sejalon-Delmas. "Influence of Arbuscular Mycorrhizal Fungi on Antimony Phyto-Uptake and Compartmentation in Vegetables Cultivated in Urban Gardens." *Chemosphere* 191 (2018): 272–79. https://doi.org/10/gf4z7x.

Pincetl, Stephanie, Setal S. Prabhu, Thomas W. Gillespie, G. Darrel Jenerette, and Diane E. Pataki. "The Evolution of Tree Nursery Offerings in Los Angeles County Over the Last 110 Years." *Landscape and Urban Planning* 118 (2013): 10–17. https://doi.org/10/f5bhg8.

Pinilla-Gallego, Mario S., Valentina N. Fernandez, and Guiomar Nates-Parra. "Pollen Resource and Seasonal Cycle of Thygater Aethiops (Hymenoptera Apidae) in an Urban Environment (Bogota-Colombia)." *Revista de Biología Tropical* 64 (2016): 1247–57. https://doi.org/10/gf42jn.

Plummer, Kate E., Kate Risely, Mike P. Toms, and Gavin M. Siriwardena. "The Composition of British Bird Communities is Associated with Long-Term Garden Bird Feeding." *Nature Communications* 10 (2019): 1–8. https://doi.org/10/c55h.

Plummer, Kate E., Gavin M. Siriwardena, Greg J. Conway, Kate Risely, and Mike P. Toms. "Is Supplementary Feeding in Gardens a Driver of Evolutionary Change in a Migratory Bird Species?" *Global Change Biology* 21 (2015): 4353–63. https://doi.org/10/gf42b2.

Poot-Pool, Wilbert S., Hans van der Wal, Salvador Flores-Guido, Juan M. Pat-Fernandez, and Ligia Esparza-Olguin. "Home Garden Agrobiodiversity Differentiates Along a Rural-Peri-Urban Gradient in Campeche, Mexico." *Economic Botany* 69 (2015): 203–17. https://doi.org/10/gf4z8r.

Pradeiczuk, Aline, Mayra T. Eichemberg, and Camila Kissmann. "Urban Ethnobotany: A Case Study in Neighborhoods of Different Ages in Chapeco, Santa Catarina State." *Acta Botanica Brasilica* 31 (2017): 276–85. https://doi.org/10/gf4z4q.

Prevot, Anne-Caroline, Hélène Cheval, Richard Raymond, and Alix Cosquer. "Routine Experiences of Nature in Cities can Increase Personal Commitment Toward Biodiversity Conservation." *Biological Conservation* 226 (2018): 1–8. https://doi.org/10/gfd6wm.

Prosser, Christine, Simon Hudson, and Michael B. Thompson. "Effects of Urbanization on Behavior, Performance, and Morphology of the Garden Skink, Lampropholis Guichenoti." *Journal of Herpetology* 40 (2006): 151–59. https://doi.org/10/fv8d22.

Quistberg, Robyn D., Peter Bichier, and Stacy M. Philpott. "Landscape and Local Correlates of Bee Abundance and Species Richness in Urban Gardens." *Environmental Entomology* 45 (2016): 592–601. https://doi.org/10/f8zqcm.

R Core Team. R: A Language and Environment for Statistical Computing. *R Foundation for Statistical Computing, Vienna, Austria.* (version 3.6.1), 2019. http://www.r-project.org/index.html.

Raymond, Christopher M., Alan P. Diduck, Arjen Buijs, Morrissa Boerchers, and Robert Moquin. "Exploring the Co-Benefits (and Costs) of Home Gardening for Biodiversity Conservation." *Local Environment* 24 (2019): 258–73. https://doi.org/10/gf42gr.

Rayol, Breno P., Igor D. Do Vale, and Izildinha S. Miranda. "Tree and Palm Diversity in Homegardens in the Central Amazon." *Agroforestry Systems* 93 (2019): 515–29. https://doi.org/10/gf4z5j.

Reid, S. Karrie, and Lorence R. Oki. "Field Trials Identify More Native Plants Suited to Urban Landscaping." *California Agriculture* 62 (2008): 104–97. https://doi.org/10/bvz2sf.

Ricogray, Victor, Jose G. Garciafranco, Alexandra Chemas, Armando Puch, and Paulino Sima. "Species Composition, Similarity, and Structure of Mayan Homegardens in Tixpeual and Tixcacaltuyub, Yucatan, Mexico." *Economic Botany* 44 (1990): 470–87. https://doi.org/10/fj44gw.

Robert, Vincent, H. P. Awono-Ambene, and Jean Thioulouse. "Ecology of Larval Mosquitoes, with Special Reference to Anopheles Arabiensis (Diptera Culcidae) in Market-Garden Wells in Urban Dakar, Senegal." *Journal of Medical Entomology* 35 (1998): 948–55. https://doi.org/10/gf42d4.

Rodewald, Amanda D. "Urban Agriculture as Habitat for Birds." In *Sowing Seeds in the City*, pp. 229–233. Springer, Dordrecht, 2016.

Rodriguez, Shari L., M. Nils Peterson, and Christopher J. Moorman. "Does Education Influence Wildlife Friendly Landscaping Preferences?" *Urban Ecosystems* 20 (2017): 489–96. https://doi.org/10/f95vmn.

Rodriguez-Rodriguez, Susana E., Karen P. Solis-Catalan, and Alejandro Valdez-Mondragon. "Diversity and Seasonal Abundance of Anthropogenic Spiders (Arachnida Araneae) in Different Urban Zones of the City of Chilpancingo, Guerrero, Mexico." *Revista Mexicana de Biodiversidad* 86 (2015): 962–71. https://doi.org/10/gf42jr.

Rolf, Werner, Stephan Pauleit, and Hubert Wiggering. "A Stakeholder Approach, Door Opener for Farmland and Multifunctionality in Urban Green Infrastructure." *Urban Forestry & Urban Greening* 40 (2019): 73–83. https://doi.org/10/gf42m5.

Šantrůčková, Markéta, Katarína Demkova, Jiří Dostalek, and Tomáš Frantik. "Manor Gardens: Harbors of Local Natural Habitats?" *Biological Conservation* 205 (2017): 16–22. https://doi.org/10/f9n6p2.

Schneider, Annemarie, Mark A. Friedl, and David Potere. "Mapping Global Urban Areas Using MODIS 500-m Data: New Methods and Datasets Based on 'Urban Ecoregions'." *Remote Sensing of Environment* 114, no. 8 (2010): 1733–1746.

Seburanga, Jean L., Beth A. Kaplin, Qixiang Zhang, and Theodette Gatesire. "Amenity Trees and Green Space Structure in Urban Settlements of Kigali, Rwanda." *Urban Forestry & Urban Greening* 13 (2014): 84–93. https://doi.org/10/gf42m3.

Shackleton, Charlie M., J. Guild, Byron Bromham, Sian Impey, M. Jarrett, Sduduzo Ngubane, and Kirsty Steijl. "How Compatible are Urban Livestock and Urban Green Spaces and Trees? An Assessment in a Medium-Sized South African Town." *International Journal of Urban Sustainable Development* 9 (2017): 243–52. https://doi.org/10/gf42cw.

Shwartz, Assaf, Helene Cheval, Laurent Simon, and Romain Julliard. "Virtual Garden Computer Program for Use in Exploring the Elements of Biodiversity People Want in Cities." *Conservation Biology* 27 (2013b): 876–86. https://doi.org/10/f4494v.

Shwartz, Assaf, Audrey Muratet, Laurent Simon, and Romain Julliard. "Local and Management Variables Outweigh Landscape Effects in Enhancing the Diversity of Different Taxa in a Big Metropolis." *Biological Conservation* 157 (2013a): 285–92. https://doi.org/10/f4s8t3.

Shwartz, Assaf, Anne Turbe, Laurent Simon, and Romain Julliard. "Enhancing Urban Biodiversity and its Influence on City-Dwellers: An Experiment." *Biological Conservation* 171 (2014): 82–90. https://doi.org/10/f528gw.

Sierra-Guerrero, María C., and Angela R. Amarillo-Suarez. "Socioecological Features of Plant Diversity in Domestic Gardens in the City of Bogota, Colombia." *Urban Forestry & Urban Greening* 28 (2017): 54–62. https://doi.org/10/gf42mz.

Siviero, Amauri, Thiago A. Delunardo, Moacir Haverroth, Luis C. de Oliveira, and Angela M. S. Mendonca. "Cultivation of Food Species in Urban Gardens in Rio Branco, Acre, Brazil." *Acta Botanica Brasilica* 25 (2011): 549–56. https://doi.org/10/fcbchm.

Smith, Richard M., Kevin J. Gaston, Philip H. Warren, and Ken Thompson, John G. Hodgson, Philip H. Warren, and Kevin J. Gaston. "Urban domestic gardens (IX) Composition and richness of the vascular plant flora, and implications for native biodiversity." *Biological Conservation* 129 (2006): 312–22. https://doi.org/10/ckj6d9.

Standish, Rachel J., Richard J. Hobbs, and James R. Miller. "Improving City Life Options for Ecological Restoration in Urban Landscapes and how these Might Influence Interactions between People and Nature." *Landscape Ecology* 28 (2013): 1213–21. https://doi.org/10/f44vbg.

Stary, Petr. "Philadelphus Coronarius L. as a Reservoir of Aphids and Parasitoids." *Journal of Applied Entomology* 112 (1991): 1–10. https://doi.org/10/d9ndw2.

Taheri, Ahmed, James K. Wetterer, and Joaquin L. Reyes-Lopez. "Tramp Ants of Tangier, Morocco." *Transactions of the American Entomological Society* 143 (2017): 299–304. https://doi.org/10/gf42k3.

Taylor, John R., and Sarah T. Lovell. "Urban Home Gardens in the Global North: A Mixed Methods Study of Ethnic and Migrant Home Gardens in Chicago, IL." *Renewable Agriculture and Food Systems* 30 (2014): 1–11. https://doi.org/10.1017/S1742170514000180.

Taylor, John R., Sarah T. Lovell, Sam E. Wortman, and Michelle Chan. "Ecosystem Services and Tradeoffs in the Home Food Gardens of African American, Chinese-Origin and Mexican-Origin Households in Chicago, IL." *Renewable Agriculture and Food Systems* 32 (2017): 69–86. https://doi.org/10/gf42jf.

Thabethe, Vuyisile, and Colleen T. Downs. "Citizen Science Reveals Widespread Supplementary Feeding of African Woolly-Necked Storks in Suburban Areas of KwaZulu-Natal, South Africa." *Urban Ecosystems* 21 (2018): 965–73. https://doi.org/10/gfd52j.

Thiengo, Silvana C., Fabio A. Faraco, Norma C. Salgado, Robert H. Cowie, and Monica A. Fernandez. "Rapid Spread of an Invasive Snail in South America: the Giant African Snail, Achatina Fulica, in Brasil." *Biological Invasions* 9 (2007): 693–702. https://doi.org/10/c82w9w.

Threlfall, Caragh G., Ken Walker, Nicholas S. G. Williams, Amy K. Hahs, Luis Mata, Nigel E. Stork, and Stephen J. Livesley. "The Conservation Value of Urban Green Space Habitats for Australian Native Bee Communities." *Biological Conservation* 187 (2015): 240–48. https://doi.org/10/f7jntp.

Tonietto, Rebecca, Jeremie Fant, John S. Ascher, Katherine Ellis, and Daniel Larkin. "A Comparison of Bee Communities of Chicago Green Roofs, Parks and Prairies." *Landscape and Urban Planning* 103 (2011): 102–8. https://doi.org/10/ckqxv3.

Tratalos, Jamie, Richard A. Fuller, Philip H. Warren, Richard G. Davies, and Kevin J. Gaston. "Urban form, Biodiversity Potential and Ecosystem Services." *Landscape and Urban Planning* 83 (2007): 308–17. https://doi.org/10/dqhj8n.

Tresch, Simon, David Frey, Renée-Claire Le Bayon, Andrea Zanetta, Frank Rasche, Andreas Fliessbach, and Marco Moretti. "Litter Decomposition Driven by Soil Fauna, Plant Diversity and Soil Management in Urban Gardens." *Science of the Total Environment* 658 (2019): 1614–29. https://doi .org/10/gf42j4.

Tresch, Simon, Marco Moretti, Renée-Claire Le Bayon, Paul Mader, Andrea Zanetta, David Frey, and Andreas Fliessbach. "A Gardener's Influence on Urban Soil Quality." *Frontiers in Environmental Science* 6 (2018): 1–17. https://doi.org/10/gf42bh.

Tryjanowski, Piotr, Anders P. Moller, Federico Morelli, Piotr Indykiewicz, Piotr Zduniak, and Łukasz Myczko. "Food Preferences by Birds Using Bird-Feeders in Winter a Large-Scale Experiment." *Avian Research* 9 (2018): 1–6. https://doi.org/10/gdnv6w.

United Nations, Department of Economics and Social Affairs, Population Divison (2015). *World Urbanization Prospects: The 2014 Revision.* New York, NY.

van Heezik, Yolanda M., Katharine J. M. Dickinson, and Claire Freeman. "Closing the Gap: Communicating to Change Gardening Practices in Support of Native Biodiversity in Urban Private Gardens." *Ecology and Society* 17 (2012): 1–9. https://doi.org/10/f99x9g.

van Heezik, Yolanda M., Katharine J. M. Dickinson, Claire Freeman, Stefan Porter, Janine M. Wing, and Barbara I. P. Barratt. "To What Extent does Vegetation Composition and Structure Influence Beetle Communities and Species Richness in Private Gardens in New Zealand?" *Landscape and Urban Planning* 151 (2016): 79–88. https://doi.org/10/gf42f7.

van Heezik, Yolanda M., and Karin Ludwig. "Proximity to Source Populations and Untidy Gardens Predict Occurrence of a Small Lizard in an Urban Area." *Landscape and Urban Planning* 104 (2012): 253–59. https://doi.org/10/d569nf.

van Heezik, Yolanda M., Amber Smyth, Amy Adams, and Joanna Gordon. "Do Domestic Cats Impose an Unsustainable Harvest on Urban Bird Populations?" *Biological Conservation* 143 (2010): 121–30. https://doi.org/10/d22csh.

Verderame, Mariailaria, and Rosaria Scudiero. "Health Status of the Lizard Podarcis Siculus (Rafinesque-Schmaltz, 1810) Subject to Different Anthropogenic Pressures." *Comptes Rendus Biologies* 342 (2019): 81–89. https://doi.org/10/gf4z75.

Walter, Theresa, Richard Zink, Gregor Laaha, Johann G. Zaller, and Florian Heigl. "Fox Sightings in a City are Related to Certain Land Use Classes and Sociodemographics Results from a Citizen Science Project." *Bmc Ecology* 18 (2018): 18–50. https://doi.org/10/gfrdwc.

Walters, S. Alan, and Karen S. Midden. "Sustainability of Urban Agriculture: Vegetable Production on Green Roofs." *Agriculture* 8, no. 168 (2018). https://doi.org/10/gf4z46.

Walton, Kerry. "Hygromia Cinctella (Draparnaud, 1801) (Mollusca: Gastropoda: Hygromiidae): A New Adventive Land Snail for New Zealand." *New Zealand Journal of Zoology* 44 (2017): 13–19. https://doi. org/10/gf42g9.

Wang, James W., Choon H. Poh, Chloe Y. T. Tan, Vivien N. Lee, Anuj Jain, and Edward L. Webb. "Building Biodiversity Drivers of Bird and Butterfly Diversity on Tropical Urban Roof Gardens." *Ecosphere* 8 (2017): 1–22. https://doi.org/10/gf4z8v.

Ward, Scott G., and Kathryn L. Amatangelo. "Suburban Gardening in Rochester, New York Exotic Plant Preference and Risk of Invasion." *Landscape and Urban Planning* 180 (2018): 161–65. https://doi.org/10/gfp324.

Warren, Paige S., Susannah B. Lerman, Riley Andrade, Kelli L. Larson, and Heather L. Bateman. "The More Things Change: Species Losses Detected in Phoenix Despite Stability in Bird-Socioeconomic Relationships." *Ecosphere* 10 (2019): 1–22. https://doi.org/10/gfxfqz.

Westgate, Martin J. "Revtools: An R Package to Support Article Screening for Evidence Synthesis." *Research Synthesis Methods* 10, no. 4 (2019). https://doi.org/10/ggbssh.

Wezel, Alexander, and Julia Ohl. "Does Remoteness from Urban Centres Influence Plant Diversity in Homegardens and Swidden Fields? A Case Study from the Matsiguenka in the Amazonian Rain Forest of Peru." *Agroforestry Systems* 65 (2005): 241–51. https://doi.org/10/chdw2j.

Whelan, Robert J., David G. Roberts, Philip R. England, and David J. Ayre. "The Potential for Genetic Contamination vs. Augmentation by Native Plants in Urban Gardens." *Biological Conservation* 128 (2006): 493–500. https://doi.org/10/bp3m2r.

Widows, Steffenie A., and David Drake. "Evaluating the National Wildlife Federation's Certified Wildlife Habitat (TM) Program." *Landscape and Urban Planning* 129 (2014): 32–43. https://doi.org/10/f6fgfg.

Williams, Rachel L., Richard Stafford, and Anne E. Goodenough. "Biodiversity in Urban Gardens: Assessing the Accuracy of Citizen Science Data on Garden Hedgehogs." *Urban Ecosystems* 18 (2015): 819–33. https://doi.org/10/f7pdvj.

Wilson, Sandra B., Rosanna Freyre, Gary W. Knox, and Zhanao Deng. "Characterizing the Invasive Potential of Ornamental Plants." *Acta Horticulturae* 937 (2012): 1183–92. https://doi.org/10/ggpwsg.

Woods, Matthew E., Rehman Ata, Zachary Teitel, Nishara M. Arachchige, Yi Yang, Brian E. Raychaba, James Kuhns, and Lesley G. Campbell. "Crop Diversity and Plant-Plant Interactions in Urban Allotment Gardens." *Renewable Agriculture and Food Systems* 31 (2016): 540–49. https://doi.org/10/gf42jg.

Wojcik, Victoria A., Gordon W. Frankie, Robbin W. Thorp, and Jennifer L. Hernandez. "Seasonality in bees and their floral resource plants at a constructed urban bee habitat in Berkeley, California." *Journal of the Kansas Entomological Society* 81 (2008): 15–28. https://doi.org/10/b994tp.

Yadav, Priyanka, Kathy Duckworth, and Parwinder S. Grewal. "Habitat Structure Influences below Ground Biocontrol Services: A Comparison between Urban Gardens and Vacant Lots." *Landscape and Urban Planning* 104 (2012): 238–44. https://doi.org/10/fp49xm.

Yadouleton, Anges W. M., Alex Asidi, Rousseau F. Djouaka, James Braima, Christian D. Agossou, and Martin C. Akogbeto. "Development of Vegetable Farming a Cause of the Emergence of Insecticide Resistance in Populations of Anopheles Gambiae in Urban Areas of Benin." *Malaria Journal* 8 (2009): 1–8. https://doi.org/10/fgprnf.

Zuin, M C, Anna Lante, F Zocca, Giampaolo Zanin, and Giusepe Zanin. "A Phytoalimurgic Garden to Promote Wild Edible Plants." *Acta Horticulturae* 881 (2010): 855–58. https://doi.org/10/ggpwsk.

For additional online resources and the accompanying data and code set, please see the Online Resources tab at https://www.routledge.com/9780367260019

2

Complex Ecological Interactions and Ecosystem Services in Urban Agroecosystems

Stacy M Philpott[1, §], Shalene Jha[2], Azucena Lucatero[1], Monika Egerer[1], and Heidi Liere[3]
[1] Department of Ecology and Ecosystem Management, School of Life Sciences - Weihenstephan, Technische Universität München
[2] Integrative Biology Department, University of Texas at Austin, Texas, USA
[3] Environmental Studies Department, Seattle University, Washington, USA
[§] Corresponding author: Email - sphilpot@ucsc.edu, 1156 High Street, Santa Cruz, CA 95064, USA

CONTENTS

KEY WORDS: *ecological network, garden, pest control, pollination, herbivore, natural enemy, functional traits*

2.1 Introduction

The local and landscape conditions that support biodiversity are fundamental features of a functioning ecosystem and its ecosystem services (Loreau et al. 2001, Hooper et al. 2005, Cardinale et al. 2012), including urban agroecosystems. Urban agroecosystems such as community gardens and urban farms are increasingly recognized for their contribution to biodiversity conservation and ecosystem service provision (Clinton et al. 2018, Clucas et al. 2018). Local features of urban agroecosystems include patch size, soil quality, ground cover, and vegetation diversity and structure, among other factors, while landscape characteristics include quantifications of natural habitat cover, development intensity, and landscape composition and complexity at larger spatial scales. Ecosystem services, processes provided by ecosystems that contribute to human well-being (Daily 1997, Millennium Ecosystem Assessment 2005), include pollination, pest control, and food production, among others, and are worth >$18 trillion globally (Costanza 1997, IPBES 2016). In rural agroecosystems, ecosystem services boost US crop production value by $57 billion per year (Losey and Vaughan 2006) providing strong financial motivation for optimizing service provision. While increasing biodiversity (e.g., of pollinators and natural enemies) is often associated with increases in ecosystem services (e.g., Tscharntke et al. 2012), this is not always the case (e.g., Cardinale et al. 2006). One potential explanation is that ecosystem services depend on species traits and representation in ecological networks (Lavorel and Garnier 2002, Schleuning et al. 2015, Perovič et al. 2018), not biodiversity alone. While many factors have been documented to affect biodiversity, ecological interactions, and networks in rural agroecosystems, less is known about urban agroecosystems; this system presents a new research context ripe for developing general urban agroecological principles. In this chapter, we will examine how a mechanistic understanding of the relationship between local and landscape factors, functional traits, and ecological networks may provide key information for reducing species or ecosystem function loss, better informing management, and optimizing overall ecosystem service acquisition in urban agroecosystems. In doing so, we contribute a new perspective on the future of urban agroecology studies, highlighting key research questions and methods.

Functional ecology and network ecology provide an ideal framework for understanding the impact of species loss and its implications for crop production and related ecosystem services. The functional ecology literature assesses which traits make species more sensitive to environmental change or impact provisioning of ecosystem services (Lavorel and Garnier 2002, Suding and Goldstein 2008, Lavorel et al. 2013). For ecosystem service providers, like pollinators and natural enemies, whose presence and functional effectiveness depends on ability to colonize, forage, and reproduce (Kremen et al. 2007), a broader investigation of species traits could yield valuable insight into function and resilience in agroecosystems. Networks depict and quantify interactions between species across trophic levels (Bascompte et al. 2003, Ings et al. 2009) and thus are a powerful tool to assess how species within a community respond to environmental change and how this potentially impacts the delivery of ecosystem services. Given that the loss of species interactions potentially weakens ecosystem services (Bohan et al. 2013), interaction metrics can serve as key indicators of ecosystem service provision and agroecosystem response to perturbation (Blüthgen 2010).

Urban agroecosystems are an ideal system for exploring the local and landscape management impacts on functional traits, ecological networks, and ecosystem services and may also alleviate local food security challenges. In this chapter, we consider urban agroecosystems as community, allotment, backyard, and rooftop gardens, urban farms, and other spaces within and at the edges of cities dedicated to cultivation of vegetables, medicinal plants, fruit trees, ornamental plants, and associated products (Lovell 2010, Zezza and Tasciotti 2010, Lin et al. 2015). Specifically, we focus on the ecological dimensions of these systems. By 2030, 80–90% of the global population will live in cities (United Nations 2005, Seto et al. 2012) and many urban residents lack access to fresh produce (e.g. Larson et al. 2009). Urban agroecosystems provide 15–20% of the global food supply (Smit et al. 1996, Hodgson et al. 2011), are an important source of vitamin-rich vegetables and fruits (Wakefield et al. 2007, Gregory et al. 2016), and promote gardener health and well-being (Brown and Jameton 2000, Classens 2015). Despite documented negative impacts of urban sprawl and increased urban developed cover, urban agroecosystems often host high biodiversity (Lin et al. 2015); however, the species assemblages in these disturbed habitats may be different from natural habitats, likely due to the different traits and species interactions that allow for their persistence in urban areas (e.g., Cane et al. 2006, Faeth et al. 2005). In addition, gardeners often express challenges in food production in these urban agriculture environments, such managing pests (Oberholtzer et al. 2014). In order to provide needed technical assistance with greater certainty, we need to understand how garden management practices are best implemented to optimize pest control, pollination, and crop production (Lin et al. 2015). This is vital given the increased importance of urban agriculture for food security, especially in underserved communities (Alig et al. 2004, Pothukuchi and Thomas 2004, Ver Ploeg et al. 2009, Chappell and LaValle 2011).

In this chapter, we summarize current knowledge of local and landscape drivers of herbivore, natural enemy, and pollinator abundance, richness, and community composition, and resulting impacts on functional traits of service-providing organisms, the structure of natural enemy-herbivore and pollinator-plant networks, ecological interactions, and pest control and pollination services in urban agroecosystems. We highlight different observational and experimental approaches that have been used in ecological studies in urban agroecosystems, propose a unification of functional and network ecology, and end with an analysis of some challenges and future opportunities in ecological research in urban agroecosystems.

2.2 Local and Landscape Drivers of Herbivore, Natural Enemy, and Pollinator Communities and Traits

Local and landscape management strongly impacts the abundance, richness and composition of herbivores, natural enemies, and pollinators in agroecosystems. Local-scale vegetation diversity and complexity and higher floral abundance and richness boost abundance and richness of natural enemies (e.g., Andow 1991, Langellotto and Denno 2004) and pollinators (e.g., Baldock et al. 2015, Ballare et al. 2019). Landscapes with more natural habitat cover offer more resources (Landis et al. 2000, Ricketts et al. 2008, Schellhorn et al. 2014) and support a higher density and diversity of arthropods, even in sites with low vegetation diversity (Bianchi et al. 2006, Chaplin-Kramer et al. 2011). Although these relationships are well established for rural agroecosystems, less is known, especially in terms of landscape context, for urban agroecosystems, which may be different given heating effects of concrete, and perhaps different conceptualization of landscape diversity when multiple habitat types are highly degraded rather than a mix of natural and semi-natural habitats.

Knowledge of the functional traits of organisms, such as herbivores, natural enemies, and pollinators, occurring within urban agroecosystems is vital to understanding community structure and composition (Fountain-Jones et al. 2015) and can increase the predictive power of community studies beyond taxonomic analyses alone (Barton et al. 2011). A functional trait approach also reduces context dependency and allows for generalization across communities and ecosystems (Moretti et al. 2017). For natural enemies and pollinators, life history traits like voltinism or nesting habits are linked to fitness but are also sensitive to environmental stress, making them useful traits for assessing how species and their interactions might respond to global change (Moretti et al. 2017). Morphological and consumption-related traits, such as body size, diet, and hunting mode may influence colonization and likelihood of extinction, and effective delivery of ecosystem services (Peters and Wassenberg 1983, Woodward et al. 2005, Stang et al. 2009, Ibañez 2012, Ibañez et al. 2013). Increases in urban cover (e.g., concrete and other impermeable surfaces) change the physical and biotic properties of habitats, resulting in changes in soil and vegetation characteristics that act as environmental filters for arthropod communities via their functional traits (Aronson et al. 2016, Maisto et al. 2017). In particular, species (and their interactions) that exist in urban areas are present after passing through extra environmental 'filters' distinct from those in rural and especially natural habitats, such as influences of urban form (e.g., buildings, density, sprawl), history of development and land use (e.g., toxic wastes, soil compaction), and human facilitation both at the regional and local site scales (Aronson et al. 2016). Here, we review the local and landscape drivers of herbivore, natural enemy, and pollinator abundance, richness, community composition and traits in urban agroecosystems.

2.2.1 Herbivores

Herbivorous insects are potentially detrimental to crop yield and quality, and managing pests is challenging for gardeners, who may lack pest management knowledge or avoid using chemical control due to health concerns and community garden regulations (Oberholtzer et al. 2014, Gregory et al. 2016). Herbivore populations can be regulated by intrinsic factors such as growth rates, survival, reproduction, dispersal capacity, as well as by bottom-up and top-down forces (Figure 2.1).

Major pest species differ in life history strategies and dispersal ranges which inform their responsiveness to habitat and management changes at different scales (Raupp et al. 2009, Mazzi and Dorn 2012). Thus, understanding variable pest responses to local and landscape features of urban systems requires consideration of pest traits and life histories in order to prevent pest outbreaks and minimize

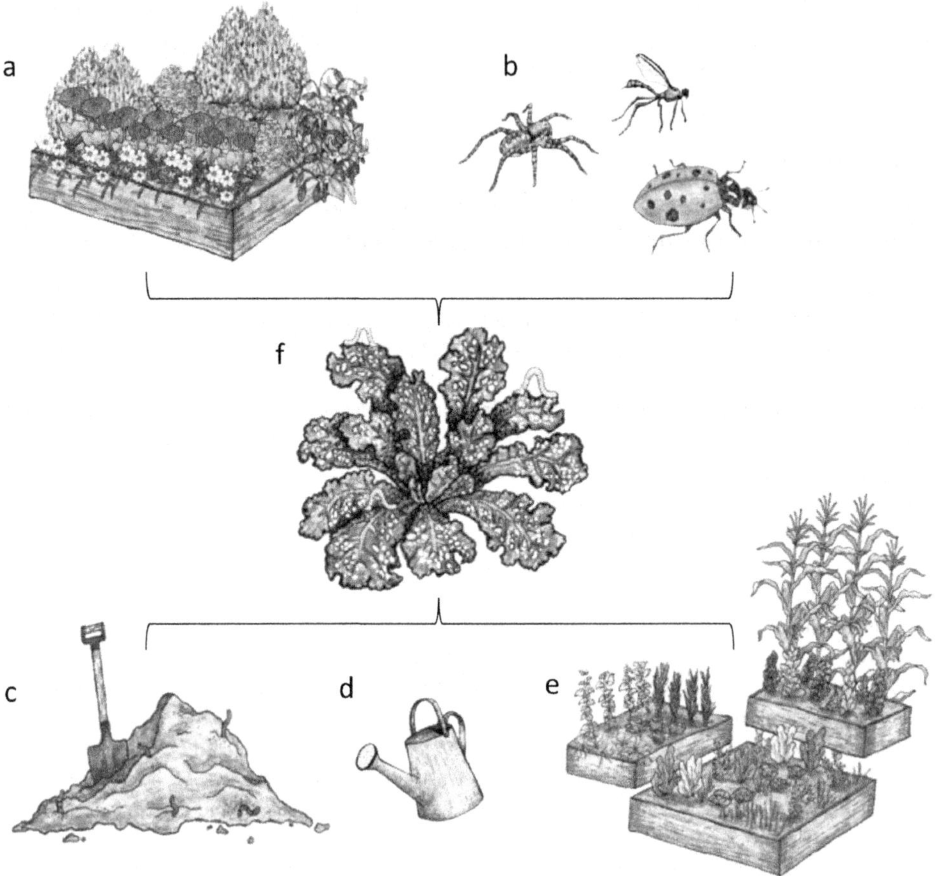

FIGURE 2.1 Depiction of selected top-down (a, b) and bottom-up (c-e) impacts on herbivore insect pests and crop damage (f) in urban agroecosystems. Top-down impacts can result from planting flower strips (a) to support, or direct release of natural enemies (b). Bottom-up impacts result from changes to soil characteristics (c), water management (d) crop species identity and diversity (e). Drawings by Charlotte Grenier.

crop damage. For example, herbivorous species with life-histories that depend on specific host plants benefit when their host plants are cultivated and these insects may then become pests (Raupp et al. 2009). Further, pest traits such as tolerance to pollutants and urban heat may influence pest density in urban settings. For example, scale insects in urban trees surrounded by impervious landscapes experience higher temperatures resulting in higher fecundity and population growth (Dale and Frank 2014). Similarly, aphids exposed to polluted urban air have higher population growth rates and shorter development times due to greater availability of amino acids in plant phloem (Bolsinger and Flückiger 1989). However, no studies have specifically examined the distribution of pest functional traits in urban agroecosystems.

Both local and landscape changes can alter bottom-up effects on herbivores. At the local scale, gardeners alter plant communities and add soil amendments and water; all of these activities control the quality of plant resources for herbivores (Faeth et al. 2011, Čeplová et al. 2017). For example, in California, garden soils with greater water holding capacity support larger plants, and thus higher cabbage aphid densities (Egerer et al. 2018b). Similarly, herbivorous bugs and aphids are more abundant in urban green spaces with higher vegetation volume and host plant density (Mata et al. 2017, Egerer et al. 2018b). Urban agroecosystems also support high crop diversity (compared with rural farms), and evidence from urban green spaces shows species-specific variation in insect responses to plant diversity, with some herbivorous species positively affected by plant species diversity and others negatively affected (Raupp et al. 2009, Mata et al. 2017). This suggests that gardeners hoping to manipulate the composition of the crops they cultivate as a pest management strategy will likely face trade-offs in controlling different herbivore species, and they may have to prioritize the most destructive or populous pests over more innocuous herbivores. Urban landscapes are characterized by a variety of land-use types, and the composition of landscapes surrounding urban agroecosystems may also influence pest populations. In rural agroecosystems, pest densities increase in more complex, well-connected landscapes (Kruess and Tscharntke 1994, Martin et al. 2015). In urban settings, few studies focus on the influence of landscape composition on herbivore pests, and the existing evidence on how urban impervious land cover (a component that strongly affects connectivity) affects pests is inconclusive. One study found a positive correlation between herbivore abundance and impervious land cover (Parsons and Frank 2019) while others have found no effect (Lowenstein and Minor 2018, Egerer et al. 2018b).

Similar to what rural farmers have done for many years, urban gardeners can also manage herbivore populations by manipulating top-down control (Figure 2.1). A common strategy used to enhance biological pest control in rural systems is planting flower strips or borders to prevent pest population build-up (Uyttenbroeck et al. 2016). This practice could be easily adopted by urban gardeners and farmers (Altieri and Nicholls 2018). Prey removal experiments in urban agroecosystems suggest that herbivore control by natural enemies can be very effective, with predators removing 15–100% of sentinel prey within 24 hours (Morales et al. 2018). But understanding more details about the interactions between herbivorous insects and their natural enemies in urban agriculture settings may help further enhance biological pest control and minimize crop losses to herbivory.

2.2.2 Natural Enemies

Natural enemies, animals that prey upon or parasitize crop pests, are important ecosystem service providers in agroecosystems and often maintain pest populations below damage thresholds without the use of chemical control. Agroecological principles, where external inputs are replaced by ecological interactions, may optimize natural pest control via natural enemy support in urban agriculture (Altieri and Nicholls 2018). For example, planting hedgerows of nectar and pollen-rich plants or setting aside beetle refuges, can benefit natural enemies (Bolger et al. 2000, Philpott et al. 2014, Altieri and Nicholls 2018). Even so, due to small size, isolation from natural areas, and constant disturbance, urban agroecosystems may not sustain populations of natural enemies necessary for pest control needs. It is therefore important to understand which local and landscape factors drive diversity, abundance, communities and traits of critical natural enemies as well as to understand which natural enemies in fact provide better control of urban agroecosystem herbivores.

Landscape context is likely important in urban systems, and the degree to which local resource availability increases natural enemy abundance and diversity depends on the composition of habitats in the surrounding areas (Bennett and Gratton 2012, Philpott et al. 2014, Egerer et al. 2017a). Yet unique city characteristics may lead to different effects of landscape context on beneficial organisms than expected from rural systems (Cadenasso et al. 2007, Pickett and Cadenasso 2009). For example, in contrast to patterns for rural agroecosystems, beneficial insect abundance and diversity in community gardens does not necessarily decrease with reduced proportion of natural habitats in the surrounding landscape (Egerer et al. 2017b).

Urban agroecosystems have unique local and landscape characteristics that influence natural enemy abundance, diversity, and species composition (Bennett and Gratton 2012, Burkman and Gardiner 2014, Otoshi et al. 2015, Burks and Philpott 2017, Lowenstein and Minor 2018). Larger gardens with more plant species may support higher abundance and richness of natural enemies (Burkman and Gardiner 2014, Lowenstein and Minor 2018). Garden soils vary in physical properties (e.g., texture, organic matter content), soil water retention, and water holding capacity (Dexter 2004, Lin et al. 2018), all of which can directly and indirectly affect ground dwelling natural enemies (Grewal et al. 2011). Ground covers such as concrete, rocks, and mulch are used in gardens for aesthetic and utilitarian reasons and can positively or negatively influence natural enemies. For example, impervious cover limits predators and parasitoids (Sattler et al. 2010, Bennett and Gratton 2012), while mulch cover increases the activity, abundance, and richness of spiders (Otoshi et al. 2015), carabid beetles (Philpott et al. 2019) and ladybeetles (Egerer et al. 2017b). Increased vegetation diversity and complexity boosts abundance and richness of natural enemies. For example, floral richness benefits spiders and parasitoids (Bennett and Gratton 2012, Burks and Philpott 2017), while ladybeetles benefit from increased abundance of ornamental plants and crop diversity (Egerer et al. 2017b).

Landscape connectivity, landscape diversity, proportion of impervious cover, and position along the rural to urban gradient all affect natural enemies within urban agroecosystems. The movement and colonization of natural enemies in urban landscapes is negatively affected by increased impervious cover and fragmentation (McKinney 2002, 2008, Faeth et al. 2005, Williams 2009). These factors can affect a wide array of natural enemies including non-flying spiders (Ponge 2003, Sattler et al. 2010), parasitic wasps (Bennett and Gratton 2012, Burks and Philpott 2017), and strong fliers such as ladybeetles (Egerer et al. 2018a, Liere et al. 2019). But impacts of landscape changes on natural enemies may be context dependent. For example, abundance and diversity of ladybeetles increases with urbanization in urban agroecosystems in California, but decreases in Michigan—potentially due to regional differences in water availability in the landscape or urbanization history (Egerer et al. 2018d).

Compared to species abundance, richness, and diversity, we know less about changes in community composition of natural enemies in urban agroecosystems. Impervious surface area and greenspace connectivity affect the community assembly of natural enemies within a patch (Burkman and Gardiner 2014). Ground cover features of urban gardens such as bare soil availability, woodchip mulch cover, or grass cover affect both ground-dwelling (ground beetles) and flying (lady beetles) natural enemy community assembly (Liere et al. 2019, Philpott et al. 2019). However, in urban community gardens in California, ground beetle species composition is solely driven by leaf-litter cover (Philpott et al. 2019), while lady-beetle composition responds to both local and landscape environmental drivers (Liere et al. 2019).

Natural enemy functional traits influence their responses to local and landscape features. Highly mobile arthropods such as ladybeetles may not perceive the urban matrix as a barrier to movement, thus urban agroecosystems may host numerous native species that vary in size, diet, and diet breadth (Liere et al. 2019). Natural enemies with low dispersal capabilities, such as ground-dwelling beetles (Carabidae and Staphylinidae), are abundant in urban agroecosystems better connected to other green spaces (e.g. in Paris, France; Vergnes et al. 2012) but these beetles also benefit from local gardens features, such as mulch and leaf litter, that provide them shelter and refuges from predation (e.g. in California, US; Philpott et al. 2019). Distinct garden microhabitats created by irrigation, vegetation, mulch, or bare patches may provide resources for organisms with a variety of life history strategies. For example, urban farms in Cleveland, Ohio support beetles with a variety of habitat preferences including xerophilous (dry/hot environment species), hygrophilous (wet environment species), open habitat beetles, as well as species associated with anthropogenic activity and environmental disturbance (Delgado de la Flor et al. 2017). Yet, more could be learned about which suites of traits are supported by which types of local or landscape features for a variety of organisms (see section IV for specific research suggestions on this topic).

2.2.3 Pollinators

The unique local and landscape characteristics of urban agroecosystems also strongly affect pollinator abundance, diversity, and traits. Animal pollinators are essential for reproduction of >80% of flowering plant species (Winfree et al. 2011), and are even more critical in urban agroecosystems (Lin et al. 2015). Because global pollinator populations are declining (Potts et al. 2010), there is great interest in better understanding and documenting the response of pollinators to urban habitat management practices occurring at local and landscape scales (Quistberg et al. 2016, Baldock et al. 2019, Ballare et al. 2019). Within urban agroecosystems, both local floral and nesting resources influence native pollinator abundance and composition within urban agroecosystems. For example, the most common positive predictor of bee abundance/visitation and bee diversity in urban habitats is local floral abundance or diversity (Ahrne et al. 2009, Hennig and Ghazoul 2012, Pardee and Philpott 2014, Quistberg et al. 2016,

Ballare et al. 2019). Similarly, within urban systems, floral area or abundance can enhance adult butterfly richness (e.g., Hardy and Dennis 1999, Matteson and Langellotto 2011) and hoverfly abundance and richness (e.g., Bates et al. 2011, Hennig and Ghazoul 2012) and a recent meta-analysis of multiple pollinator taxa (Hymenoptera, Lepidoptera, Diptera, Coleoptera) showed that within-garden features typically had stronger effect sizes than landscape-level attributes (Majewska and Altizer 2018). Non-floral vegetation features can have contrasting impacts on pollinator abundance and diversity, often depending on how they interact with floral resources. For example, in urban green spaces, canopy cover from trees can negatively impact flower-visiting insects due to reductions in floral resource levels (Matteson et al. 2013) while in urban agroecosystems, floral abundance, vegetation height, greater woody plant cover, and less grass cover all correlate positively with pollinator visitation (Pardee and Philpott 2014).

Landscape characteristics of urban agroecosystems also strongly influence pollinators. Given that pollinators are highly mobile, with small and large bodied pollinators often foraging hundreds of meters in a single day (Castilla et al. 2017), habitat composition can also critically influence their abundance and composition in urban settings. While some urban agroecosystem studies have not detected landscape-level effects on pollinator communities (e.g. Matteson and Langellotto 2011, Quistberg et al. 2016), others document positive effects of semi-natural habitat on bee species richness in urban landscapes (Eremeeva and Sushchev 2005, Cane et al. 2006, Matteson and Langellotto 2011). Outside of urban agroecosystems, but within urban landscapes, butterfly abundance (Blair and Launer 1997) and richness declines with landscape-level urbanization (Dennis and Hardy 2001, Bergerot et al. 2011) as does hoverfly abundance and richness (Bates et al. 2011). Both landscape heterogeneity and spatial connectivity indices are relatively understudied in urban systems, despite evidence that vegetation-level heterogeneity influences butterfly diversity in rural systems (Maccherini et al. 2009, Ferrer-Paris et al. 2013) and that spatial connectivity of urban agroecosystems may facilitate the movement of native bees (Colding et al. 2006).

There is also growing interest in potential interactions between local and landscape conditions within urban agroecosystems. In the urban context, one past study has found that locally simplified habitats (urban agroecosystems) exhibited greater bee abundance and diversity in response to landscape-level natural habitat cover compared to locally complex habitats (urban grasslands) (Ballare et al. 2019). Further, habitat heterogeneity (defined as the total number of land-use patches within a 2 km buffer) interacted with local habitat management type to differentially impact bee evenness across urban areas. Specifically, while heterogeneity negatively impacted evenness in both urban grasslands and urban agroecosystems, it only did so via increased rare species in the urban grassland sites. These results indicate that landscape-level heterogeneity may play a particularly important role in regulating pollinator dispersal across urban areas, with positive impacts on rare species if local habitat requirements are met.

Because pollinators require distinct nesting and food resources (Westrich 1996, Steffan-Dewenter 2002, Franzen and Nilsson 2010) and vary dramatically in size and time of emergence and senescence (Michener 2000, Danforth 2019), pollinator functional traits may differentially mediate responses to local and landscape change. For example, urban green spaces support a greater proportion of generalist (utilize a greater number of host-plant species) relative to specialist butterflies (Blair and Launer 1997, Niell et al. 2007, Bergerot et al. 2011). Sensitivity to nesting and sheltering resources is also seen in bees within urban agroecosystems, where ground-nesting bees are less abundant in sites with less bare ground or more mulch (Quistberg et al. 2016, Ballare et al. 2019). Both ground-nesting and cavity-nesting bees are more abundant in urban agroecosystems surrounded by greater forest cover (Pardee and Philpott 2014); indeed, cavity-nesting bees are more abundant in urban habitat with greater impervious cover (Fortel et al. 2014) and wood-nesting bees are likely more abundant in cities due to increased nesting resources found in built structures (Cane et al. 2006, Matteson et al. 2008). Apart from nesting substrates, traits related to body size may impact dispersal ability given that size predicts genetic structure for bees (López-Uribe et al. 2019) and given that larger Hymenoptera forage greater distances (Greenleaf et al. (2007); this pattern has also been proposed in Diptera, Lepidoptera, and Coleoptera in Baldock et al. (2019), but see Castilla et al. (2017)), suggesting that a pollinator's ability to colonize a site may depend on its body size. Indeed, in rural agricultural landscapes, pollinator composition may shift towards larger-bodied species with less natural cover (Steffan-Dewenter and Tscharntke 1999) but see (Larsen et al. 2005), while pollinator species with greater dispersal abilities may persist longer in fragmented habitats (Wood and Pullin 2002, Jauker et al. 2009). Interestingly, in urban agroecosystems, some find no correlation between landscape context and bee body size (Cane et al. 2006, Fortel et al. 2014), while others demonstrate that small-bodied pollinators respond positively to greater semi-natural habitat cover, and that larger bodied bees respond positively to landscapes with greater pasture or crop cover (Ballare et al. 2019) or greater impervious cover (Bennett and Lovell 2019). These studies suggest that larger-bodied bees may be less dependent on local nesting resources than smaller-bodied bees, but more research is needed to fully understand the species-level traits that render bees more or less sensitive to habitat composition at local and landscape scales within urban agroecosystems.

2.3 Local and Landscape Impacts on Ecological Networks and Ecosystem Services

Local and landscape features of urban agroecosystems also affect interactions such as herbivory, predation and parasitism, and pollination. In this section, we outline interactions between plants and arthropods with a specific focus on pest control and pollination and also examine local and landscape drivers of ecological networks, given their importance in predicting urban agroecosystem function. Networks depict ecological interactions between species across adjacent trophic levels (Bascompte et al. 2003, Ings et al. 2009) and network metrics used to quantify interactions can predict how ecosystem services respond to perturbations (Blüthgen 2010). Ecological networks can increase our understanding of how management and community characteristics relate to ecosystem services (Perrings et al. 2010, Bohan et al. 2013), but to our knowledge, only two studies have explicitly linked changes in natural enemy-herbivore or plant-pollinator networks to ecosystem services in urban agroecosystems (Geslin et al. 2013, Theodorou et al. 2017, discussed below). Analysis of networks in agroecosystems is thus an important analytical tool (Woodward and Bohan 2013, Tylianakis and Binzer 2014) but underexplored in the urban agroecosystem context. Trait-function relationships may strongly impact the outcome of ecosystem services (Poisot et al. 2013, Peralta et al. 2014) but these relationships are poorly understood, especially for urban systems (Lin et al. 2015).

2.3.1 Natural Enemy and Herbivore Interactions and Networks

Differences in local and landscape features of urban agroecosystems mediate ecological interactions and likely impact ecological network structure with critical implications for biological pest control. In urban agroecosystems, local-scale factors such as floral richness enhance natural enemies and pest suppression services (Bennett and Gratton 2012, Gardiner et al. 2014, Burks and Philpott 2017). However, predator exclusion experiments in urban agroecosystems suggest that there may be discrepancies in factors that promote different natural enemy groups at different scales. For example, pest control levels increase with local vegetation complexity but not with landscape diversity (Philpott and Bichier 2017). Further, management factors and their influence may differ regionally. Prey removal rates in predator exclusion experiments were generally higher in California, USA than in Chiapas, Mexico, and the factors that influenced the removal rates of different kinds of prey (aphids, moth eggs, and moth larvae) at times deviated between the regions (Morales et al. 2018). These differences may be due to differences in regional filters or differences in the particular urban landscape context (e.g., Aronson et al. 2016). For instance, shifts along an urban to rural gradient can impact natural enemy-herbivore interactions in urban green spaces, and the relative strength of bottom-up and top-down regulatory forces. In the Central Arizona Phoenix area, bird exclusion led to higher abundances of insect herbivores in urban areas but not rural desert areas, and birds in urban areas experienced lower predation risk thereby exerting higher top-down control of arthropods (Faeth et al. 2005). Turrini et al. (2016) ran predator exclusion experiments with syrphid larvae, vetch aphids, and bean plants and found differences between urban and rural areas. In particular, predator control of aphids was stronger in rural sites, whereas aphid impacts on plant biomass were stronger in urban sites. Thus, urbanization negatively affected both available plant biomass and top-down control of aphids.

Although several studies examine how ecological networks involving natural enemies and herbivores shift with rural agroecosystem management, this is relatively unexplored in urban agroecosystems. Past research on antagonistic networks of herbivores and their natural enemies has largely focused on understanding network structure and network dynamics (Pascual and Dunne 2006), with less work on investigating responses to global change forces as has been done for mutualistic networks (see Pollinator and Plant Interactions and Networks below). Specifically, recent reviews of antagonistic networks indicate that antagonistic network structure is often modular, with distinct modules, or compartments, of interacting species making up the larger network. Interestingly, studies of antagonistic network dynamics have indicated that antagonistic interactions are quite variable across time points and that environmental factors and landscape gradients can drive high rates of interaction turnover (Tylianakis and Morris 2017). Host-parasitoid networks (for bee and wasp parasitoids) in tropical agricultural landscapes strongly respond to shifts in land use, even though species richness and evenness do not (Tylianakis et al. 2007). Contrary to expectations, given similar species richness in modified and unmodified habitats, food web interaction evenness declined, the ratio of parasitoids to hosts increased, and the specialization of the most abundant parasitoid increased in more simple agroecosystems compared with complex agroecosystems in the landscape (Tylianakis et al. 2007). Along a regional gradient of land-use intensity, tropical host-parasitoid networks are more homogenous in their interaction composition in more simplified, deforested habitat types compared to forested habitat types (Laliberté and Tylianakis 2010). However, changes in network structure associated with land-use may not always have obvious effects on ecosystem services. In aphid-parasitoid-hyperparasitoid networks, agricultural intensification increased food web complexity and temporal variability, while also decreasing parasitism rates (an ecosystem service), and increasing hyperparasitism rates (an ecosystem disservice) (Gagic et al. 2012). Studies on natural enemy-herbivore networks in urban agroecosystems (Philpott et al. 2020) reveal that interaction richness (assumed to boost pest control) increases with floral richness and declines with landscape-level agricultural cover whereas trophic complementarity (assumed to negatively impact pest control) increases with agricultural cover. But a more detailed examination of the specific impacts of shifts in ecological networks on pest suppression is warranted. By examining more about the details of networks, the number of links between natural enemy and herbivore taxa, and shifts in shared herbivores for different natural enemies, we may learn more about why loss of certain species along local or landscape management gradients may disproportionately affect pest control function within the urban agroecosystem.

2.3.2 Pollinator and Plant Interactions and Networks

Plant-pollinator network observation has long existed (e.g., Darwin 1862), but more rigorous quantification and utilization of networks has recently bloomed, especially across natural and rural agricultural landscapes (reviewed in Bascompte and Jordano 2007, Heleno et al. 2014). Within these landscapes, pollinator network ecology has focused on three main themes: 1) network structure, 2) network dynamics, and 3) network response to global change (Valdovinos 2019). Past studies on network structure reveal that most mutualistic networks tend to have a similar distribution or number of links (Jordano et al. 2003), that networks are often nested and asymmetric, where interactions involving specialists are subsets of interactions involving generalists (Bascompte et al. 2003, Vázquez et al. 2007), and that networks are frequently modular with certain groups of interacting species (Vázquez et al. 2007, Valdovinos 2019). In terms of network dynamics, plant-pollinator networks are highly plastic (CaraDonna et al. 2017, Ponisio et al. 2017), with frequent partner switching (Kaiser-Bunbury et al. 2010). Studies on network response to global change is growing and often focuses on simulating species loss by removing pollinator or plant species (Memmott et al. 2004) or simulating climate change (Memmott et al. 2007), habitat loss (Fortuna and Bascompte 2006), or species invasion (Valdovinos 2019), and then measuring impacts on network metrics. These reviews broadly reveal that highly nested networks are more resilient to extinction events and that specialized pollinators (with narrow diet breadths) are particularly prone to extinction under global change scenarios (Valdovinos 2019), but relatively few studies have examined whether these general patterns exist in urban systems.

One study conducted across urban green spaces in Bangkok, Thailand found that the most common urban pollinators were bees and that urban plant-pollinator networks exhibited high pollinator generalism (Stewart et al. 2018). Moreover, the most abundant pollinators were also the most generalist; the authors propose this feature as particularly beneficial for persistence in an urban landscape where the floral environment is highly managed and frequently altered. In non-garden urban green spaces,

Mukherjee et al. (2018) and Maruyama et al. (2019) found similar patterns of higher generalization in more urbanized areas for butterfly-plant networks in Kolkata, India and for hummingbird-plant networks in southeastern Brazil, respectively; however, Mukherjee et al. (2018) also found greater nestedness with increasing urbanization levels. Similarly, a study conducted in the UK found that urban sites had lower overall network specialization but that pollinators exhibited both greater generality (foraging on a greater number of plant species) and also greater specialization (visited a lower proportion of available plant species) compared to pollinators in agricultural and natural sites (Baldock et al. 2015).

A handful of studies have explicitly examined plant-pollinator interactions within urban agricultural sites, and these studies have found contrasting patterns with respect to levels of robustness and specialization. In a massive study comparing pollinator communities and plant-pollinator networks across a wide range of urban land use types, Baldock and colleagues found that residential and allotment gardens served as pollinator network 'hotspots' via high pollinator diversity and dispersal into other urban sites, which increased city-scale network robustness (resilience to secondary extinctions following an initial loss of species) (Baldock et al. 2019). In other words, urban gardens may play a particularly critical role in the persistence of plant-pollinator networks across distinct urban land uses. Other urban garden network studies have focused primarily on patterns of specialization within urban gardens themselves (not at city-scales) and have found distinct patterns. For example, in an urban garden system in Chicago, USA, Lowenstein and colleagues found that garden pollinators were quite generalist, with a substantial fraction of the plant community experiencing limited pollinator visitation and a smaller subset of plants experiencing very high levels of visitation (Lowenstein et al. 2019). In contrast, in a study comparing urban residential gardens with semi-natural areas in Montreal, Canada, Martins (2017) found that gardens supported more specialized bees that visited a smaller fraction of the plant community compared with semi-natural sites. Overall, these studies suggest that urban gardens play a critical role in supporting city-level plant-pollinator robustness; but that within the focal urban garden systems itself, it is possible that both specialized and generalized pollinator groups dominate, depending on the local, landscape, and biogeographic context. More work examining network dynamics or potential response of networks to global change explicitly within urban systems is needed.

While the utilization of plant-pollinator network studies in urban areas has clearly been growing, networks are rarely linked to pollination function (Kaiser-Bunbury and Blüthgen 2015). To our knowledge, only two studies have investigated the explicit links between plant-pollinator networks and pollination services in urban systems. The first study evaluated 'open' and 'tubular' experimental flowers in different landscapes, and found that the number of interactions performed by flower-visitors decreased in urban habitat compared to semi-natural and agricultural habitat, and specifically the number of Syrphidae and solitary bees visits were lower in urban habitat, leading to lower reproductive success of the 'open' flower functional group (Geslin et al. 2013). In the second study connecting urban plant-pollinator networks with pollination functions, the authors found that flower visitation metrics (e.g., linkage density) increased with urban land cover, flower visitors were more generalized in urban compared to agricultural areas, and bees were more specialized in urban areas; despite these patterns, no network metrics were predictive of plant reproductive success (Theodorou et al. 2017). Instead, urban cover, bee richness, and flying insect abundance better predicted plant reproductive success (Theodorou et al. 2017). Overall, these studies indicate an urgent need to understand the impacts of plant-pollinator networks on plant reproduction and to determine the importance of network metrics relative to other biotic and abiotic factors within urban systems.

2.4 Challenges and Opportunities for Ecological Research in Urban Agroecology

Although a flurry of ecological research has emerged from urban agroecosystems in recent years, several challenges remain in linking biodiversity and species interactions with ecosystem function. These challenges present distinct opportunities for future research. We propose an urban agroecology framework that integrates more experimental work and that considers species response and effect traits.

We then examine how ecologists could do more manipulative experimental work in urban agroecosystems to assess causal mechanisms of delivery of ecosystem services. We discuss how the use of trait and network analysis may better link biodiversity and ecosystem function and services in urban agroecosystems. Finally, we discuss limitations in assessing linkages between ecological interactions and food production.

2.4.1 The Urban Agroecology Framework

One challenge in urban agroecology research is that the field fits into two separate bodies of literature with distinct histories and goals—the urban ecology literature on the one hand, and the rural agroecology literature, on the other (Table 2.1). Those working in between, in urban agroecology, often grapple with how to frame their research. The urban ecology literature has historically focused on biodiversity, nutrient cycling, and other ecological processes (McDonnell 2011), since the main goal is to determine how urbanization affects biodiversity and ecosystem processes. One hallmark of this research branch examines shifts in biodiversity and ecological interactions along an urban-to-rural gradient documenting changes in everything from pollution, resources, and forest composition to bird abundance and ground beetle diversity along this gradient (McDonnell and Pickett 1990, Clergeau et al. 1998, Niemelä et al. 2002, McDonnell et al. 2008). On the other hand, the rural agroecology research has largely focused on impacts of farming practices, agricultural intensification, and landscape composition and configuration on biodiversity and ecosystem services; one of the main goals of this research branch is to determine how to enhance biodiversity-mediated food production within the agricultural systems. The rural agroecology literature also clearly demonstrates a relationship between local habitat management and both natural enemy and pollinator abundance and richness (e.g. higher floral abundance, crop diversity, hedgerows leads to greater abundance and richness) (Klein et al. 2003, Bianchi et al. 2006, Letourneau et al. 2011, Morandin and Kremen 2013); however, the strength of local habitat enhancements on biodiversity depends on changes in landscape composition, the proportion of certain habitat types (Thies et al. 2003, Tscharntke et al. 2005, Chaplin-Kramer et al. 2011, Jha and Kremen 2013), heterogeneity, the diversity of habitat types (Klein et al. 2003, Winfree et al. 2007, Liere et al. 2015), or the configuration, such as distance from a larger fragment or natural habitat (Perfecto and Vandermeer 2002, Armbrecht and Perfecto 2003, Klein et al. 2006).

Another difference between rural agroecology and urban ecology literatures is that landscape composition and diversity are assessed differently. Urban ecologists often focus on the amount of developed area (e.g., impervious, paved, concrete cover) within the landscape due to the prevalence of this habitat, and its strong impact on abiotic conditions, such as temperature, in the urban environment (e.g., McKinney 2002, 2008, Yuan and Bauer 2007, Egerer et al. 2018a). In contrast, rural agroecologists more often use the proportion of natural or semi-natural habitat as the landscape variable of interest (e.g., Perfecto and Vandermeer 2002, Klein et al. 2003, Tscharntke et al. 2005) due to its importance for providing alternative resources and refuges for mobile ecosystem service providers (Landis et al. 2000). Urban agroecologists usually choose one of these two, often correlated, landscape variables making comparisons across the two bodies of literature difficult. Furthermore, based on patterns found in rural agroecosystems, urban agroecologists often assume that increases in natural habitat in the landscape will benefit urban agroecosystems, but that is not always the case (e.g. Egerer et al. 2017b) and far fewer studies have examined landscape impacts on ecosystem services in urban agroecosystems. A recent review concluded that urban farmers and gardeners rely on information and extension geared towards rural agroecosystems, but that management recommendations built on rural agroecology studies may have different, unknown, or unexpected impacts in urban systems (Arnold et al. 2019). These authors also note that unique urban conditions, such as heat island effects, temperature changes, and impervious surface may differentially affect beneficial insects in urban vs. rural agroecosystems.

The urban agroecology literature can serve as a bridge between these two distinct, but not critically linked research areas. Some studies successfully bridge these two research areas by including both urban and rural sites in their analyses. For example, Pereira-Peixoto et al. (2014) examined trap nesting bees and wasps both within city-center gardens and urban-edge gardens; their results suggest that mass flowering resources in rural agricultural areas “spillover” to affect bees, but not wasps, specifically in urban-edge gardens. Clarke et al. (2014) compared plant diversity in food gardens situated along a socio-economic gradient as well as in a suburban, peri-urban, and exurban gradient in Beijing. They found impacts of economic status and landscape position on plant composition and documented a shift from ornamental plants in the city center to more food plants with increasing distance from the city. Other urban agroecology studies that are more representative of the rural agroecology literature focus on comparing urban habitat types – community or allotment gardens, vacant lots, city parks, or restored prairies – for their value in supporting biodiversity and associated ecosystem services (e.g., Gardiner et al. 2013, 2014, Philpott et al. 2014, Speak et al. 2015, Clucas et al. 2018). We suggest that urban agroecology researchers should better acknowledge linkages between these two subdisciplines of ecology. Specifically, future studies could focus more on combining approaches by examining the role of natural and semi-natural habitat (typically explored in rural systems) in urban landscape, but also layering on finer-scale analyses of the urban-to-rural gradient or examining abiotic influences (typically explored in urban systems) to how urban agroecosystems differ from rural agroecosystems. Generally, we stress that the field of urban agroecology can importantly act as a bridge between these two disciplines because it focuses both on biodiversity conservation in cities (a main goal of urban ecology) and the biodiversity-mediated ecosystem services (a main goal of rural agroecology) (Table 2.1).

2.4.2 Experimental Manipulation

Another challenge in urban agroecology research is a lack of experimental manipulations examining local or landscape impacts on biodiversity, interactions, and ecosystem services. Most urban agroecosystem studies are correlational, examining differences across local management or landscape gradients (e.g., Arnold et al. 2019). This approach fails to assess causal effects of urban agroecosystem characteristics on herbivore, natural enemy, or pollinator communities; as a result, management recommendations based on these studies may be inaccurate. Manipulating variables at landscape-scales is logistically difficult, yet manipulative experiments of local habitat features are easier and often successful. For example, researchers have converted vacant lots into urban prairies (Turo and Gardiner 2019) and have added floral resources into urban green spaces (Simao et al. 2018), private gardens (Gaston et al. 2005) or public

TABLE 2.1

Description of the research goals, themes, and landscape characteristics primarily used in the distinct, but overlapping, fields of urban ecology, rural agroecology, and urban agroecology.

Research Feature	Urban Ecology	Rural Agroecology	Urban Agroecology
Major research goals	Determine how urbanization affects biodiversity and ecosystem function	Enhance biodiversity-mediated ecosystem services for food production within agricultural systems	Promote biodiversity conservation and biodiversity-mediated ecosystem services and food production in cities
Major research themes	Impact of city specific factors (i.e. heat island effects, urban to rural gradients) on biodiversity, nutrient cycling, and ecosystem functions	Impacts of farming practices, agricultural intensification gradients, landscape composition and configuration on biodiversity-mediated ecosystem services for agriculture	Impacts of garden/farm local characteristics and landscape composition on biodiversity and biodiversity-mediated ecosystem services for food production
Landscape focus	Developed area (impervious, paved, concrete cover); connectivity between green spaces	Natural or semi-natural habitat; landscape diversity/ complexity	Natural, semi-natural habitat; developed; agriculture; open spaces, etc.
Local focus	Characteristics of urban green spaces (e.g., fragment size, age, land use history, vegetation structure, irrigation and soil moisture)	Farm management intensity (e.g., soil and agrochemical inputs, crop diversity, weed management, tillage practices, presence of hedgerows or native vegetation strips)	Agroecosystem management (e.g., crop and ornamental plant richness, presence of trees and shrubs, soil inputs) and site characteristics (e.g., age, land use history, garden size).

garden spaces (Shwartz et al. 2014). Few have conducted manipulations in urban agroecosystems (see Chapter 12 in the book (Ong and Fitch) in this volume for further discussions on this topic).

The manipulative studies conducted in urban agroecosystems can be categorized into three groups: (1) manipulation of local garden habitat features, (2) introductions of sentinel pests to assess predation, and (3) introductions of potted plants to assess pollination or pollinator visitation. Some have manipulated local features like floral abundance and richness, nesting habitat, and bare soil. One study maintained a multi-year manipulation of floral abundance and richness to assess impacts on bee pollinators (Pawelek et al. 2009). This study is impressive in timescale, but limited in the spatial extent and replication needed to evaluate how local management changes impact pollinators in the context of landscape conditions. Matteson and Langelloto (2011) added native wildflowers to gardens and monitored for changes in pollinator (bee, butterfly) and predatory wasp communities for two years. They found that floral additions did not alter the community of floral visitors, and concluded that more floral resource addition (beyond the ~70 plants of seven species that they added) would be necessary to influence beneficial insect communities. Egerer et al. (2018a) added floral resources of five species to garden sites in highly urban and less urban landscapes in California, USA and found that floral addition augmented parasitoid abundance and changed parasitoid composition but did not affect ladybeetle abundance, richness, or site fidelity regardless of landscape context. A few authors have manipulated insect nesting resources. One study added 'bee hotels' for cavity-nesting bees, and found that although abundance increased with use of hotels, so did abundance of bee parasites (MacIvor and Packer 2015). A second study added cavity-nesting resources for ants and found that nest occupation was very low in gardens (Friedrich and Philpott 2009), but argued that nests could be added to increase opportunistic nest use by ants. Finally, one study introduced microcosms (i.e. large pots) with soil or with soil and grass into gardens to assess colonization by insects, seeds, and vegetation growth and found that both types of microcosms augmented insects and vegetation (Sperling and Lortie 2010). Additions of sentinel pests and exclosures have been used to assess pest suppression in a variety of studies (Burkman and Gardiner 2014, Mace-Hill 2015, Lowenstein et al. 2017, Philpott and Bichier 2017, Frey et al. 2018, Morales et al. 2018). Likewise, researchers have introduced potted crop and ornamental plant species (e.g., sungold tomatoes, jalapeño peppers, purple coneflowers) to test how local and landscape features influence pollination services (Cohen et al. 2020, Lowenstein et al. 2014, Potter and LeBuhn 2015). At least one study introduced ornamental plants to assess changes in pollinator visitation and to construct pollinator-plant networks across a gradient of garden systems (Udy 2017).

There are likely many reasons that few manipulations have been done in urban agroecosystem research, but we posit that many of these obstacles can be overcome, creating a promising experimental opportunity for ecologists. For one, researchers may not have access to garden plots in a large enough number of gardens in order to test manipulation effects across existing local or landscape gradients; many urban garden sites have long waiting lists inhibiting researcher access. Researchers may also perceive that manipulations might dramatically affect production or plant composition in the garden plots that gardeners rent and manage, and that gardeners may be unwilling to accept research into their gardens. There are at least three opportunities presented here. First, many gardeners may be willing to accept changes that may affect the garden, but not their specific garden plots. For instance, adding mulch to pathways, adding potted ornamental, flowering, or crop plant species to pathways or common spaces, or adding nesting resources may all be easily accepted manipulations that can tell us about how resources affect garden insects. Second, we lack information regarding how changes in soil properties influence secondary compounds, herbivore loads, and related bottom-up influences on beneficial insects. Although soil manipulations may dramatically alter growing conditions, gardeners, and especially gardeners with little access to soil amendments might be willing to engage in research projects. Third, ecologists may consider expanding their research to include more community engaged, participatory, or citizen science research methodologies (e.g. (Dickinson et al. 2012, Chevalier and Buckles 2019) so that knowledge is co-produced with gardeners, and gardeners and their families are active participants in developing research questions useful for both ecological science and for management. Working directly with gardeners will also give ecologists (and social science collaborators) the opportunity to focus more on how human perceptions of management practices or their willingness to incorporate new techniques may shape ecological communities.

2.4.3 Incorporating Traits into Urban Ecological Networks

Although many urban ecology studies examine either impacts of local and landscape changes on species traits or on natural enemy-herbivore and plant-pollinator networks, there is little emphasis on how those traits link to changes in network structure or ecosystem service provision, despite the potential management implications for conservation and food production. There is thus a tremendous opportunity to develop studies linking functional traits of organisms to their ecological functions and services within urban agroecosystems. While all functional traits affect reproduction, growth, and survival (Violle et al. 2007), response traits strongly influence how species respond to environmental changes, and effect traits have a greater impact on how species affect ecosystem services (Lavorel and Garnier 2002, Suding and Goldstein 2008, Dias et al. 2013, Lavorel et al. 2013, Astor et al. 2014, Woodcock et al. 2014). Examples of natural enemy and pollinator response traits include voltinism and nesting habits whereas effect traits may include traits such as body size, diet breadth, hunting or dispersal mode, mouth type, and pollen carrying structures (Figure 2.2).

We argue that by integrating trait based and network approaches we can dually improve our mechanistic understanding of how biodiversity relates to ecosystem services and improve our ability to predict how those services will deteriorate in the face of environmental stressors (Schleuning et al. 2015). Given species-based networks do not necessarily reflect functional redundancy (Diaz et al. 2007, 2013) or proneness to extinction (Suding et al. 2008), trait-based networks may better predict ecosystem services. Under this integrated approach, we would expect that a community with high functional diversity and high interaction richness would promote the provisioning and stability of ecosystem services (Schleuning et al. 2015). By examining the identity, function, and network interactions of species that are lost and those that remain, a trait- and network-based approach could expose mechanisms that drive ecosystem services. Past studies in urban agroecosystems (reviewed above) demonstrate that taxa with certain traits (e.g. wood nesting behavior) may be likely to experience declines under specific local or landscape changes (e.g., loss of semi-natural habitat); but no studies have integrated a trait-based approach to better predict ecosystem services in urban agroecosystems.

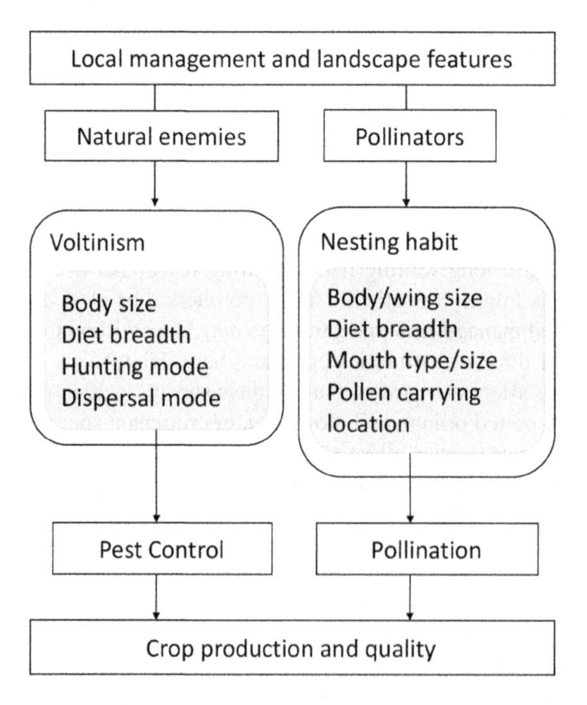

FIGURE 2.2 Schematic of response-effect traits showing how ecosystem service delivery is mediated through local garden management and landscape surroundings. Response-effect traits of the arthropod community will determine the delivery of pollination and pest control services. Traits in larger white boxes represent response traits while those in smaller shaded boxes represent both response and effect traits.

Further, we propose using the traits of organisms within urban networks to understand ecological and evolutionary processes creating and maintaining species interactions. Research suggests that there are two main ecological and evolutionary processes guiding species interaction patterns in ecological networks: one that is more 'neutral' (Vázquez et al. 2009), where we might expect that the interaction frequencies are largely explained by relative abundance (Vázquez et al. 2007), and one that is more morphologically restricted (e.g., more restrictive floral shape, Stang et al. 2009), or temporally and spatially limited, where traits may have a greater influence on interaction patterns (Kaiser-Bunbury et al. 2014). Of course, this approach is limited by a lack of information on species traits, and especially those related to diet breadth for predatory arthropods, but new molecular techniques (De Barba et al. 2014, Macias-Hernandez et al. 2018) and data compilations (Williams et al. 2010, Garibaldi et al. 2015) are making such studies increasingly more feasible. For traits that can be measured, we propose generating specific predictions regarding network neutrality versus networks with morphological, spatial and temporal constraints as well as predictions related to network resilience to perturbation and ecosystem function. For example, floral complexity (e.g., corolla depth) and body/mouth-part size are likely linked to interaction constraints (Eklöf et al. 2013, Kaiser-Bunbury et al. 2014), although these plant traits have never been examined as traits guiding network structure, dynamics, and pollination function in urban systems. In another example, one might hypothesize that hoverflies and beetles may be less influenced by the floral community traits or the loss of specialized plants compared to bees and butterflies, due to a less specialized diet (Faegri and van der Pijl 1979, Kaiser-Bunbury et al. 2010), but this has not been evaluated in urban garden systems. Overall, we suggest linking response and effect traits to ecosystem service provision by investigating the following questions: Which traits are linked to effective pest suppression? Which traits of parasitoids augment parasitism in gardens? Which traits increase con-specific pollen deposition on animal-pollinated crops? By determining the local and landscape features that boost abundance of the most effective ecosystem service providers or support communities more resilient to disturbance, we can determine how to best manage gardens for optimal services.

2.4.4 Biodiversity Influences on Food Production

Estimates of the food production capacity of urban garden systems at local and global scales exist, but few make the links between increased biodiversity, provisioning of ecosystem services, and boosts in food production from urban agroecosystems. Urban agroecosystems and farms may supply 15–20% of global food (Hodgson et al. 2011), but values vary dramatically depending on country or global region (Lin et al. 2015, Clinton et al. 2018). Nonetheless, a robust measurement of how urban agroecosystems can improve food security at the global scale is still needed (Siegner et al. 2018). Although much is assumed about links between biodiversity and food production in gardens, few, if any studies have quantified this relationship. This is in part an ecological issue—it is hard to make links from habitat management to the richness and abundance of beneficial insects, to plant damage and yield of crops in any agroecosystem (Letourneau and Bothwell 2008, Chaplin-Kramer et al. 2011, Karp et al. 2018). In natural systems and agroecosystems, there are time lags in food web interactions (e.g., Polis and Holt 1992) that can make it difficult to assess impacts of biodiversity on herbivory. Herbivory can induce plant defenses (Karban and Myers 1989) or overcompensation (Agrawal 2000), potentially masking negative effects of natural enemies on herbivore populations. The connection between pollination and yield is also difficult to track as pollen deposition happens at one time, but harvest is often much later, and fruit ripening and size is not always correlated with pollination-mediated seed number (e.g., Marcelis and Baan Hofman-Eijer 1997). A lack of direct links between biodiversity and yield is also in part a logistical and methodological issue. Even if researchers intend to collect information on yield, gardeners may harvest ripe fruits or leaves before the researcher can collect this data; citizen science or gardener reporting of weight of harvested food presents significant challenges (Gittleman et al. 2012). It is also not always possible to have the same varieties or cultivars in a large number of gardens due to specific preferences of gardeners for certain plants (e.g. Taylor et al. 2017). Clearly, there are many other biotic and abiotic factors that can affect plant yields in gardens (e.g., soil amendments, irrigation, rainfall, temperature), and urban gardeners are constantly changing their management practices to match new local conditions.

Many studies reviewed here point to benefits of natural enemies and pollinators for food production, but only a handful directly examine pollination or pest control (Lin et al. 2015, Arnold et al. 2019, Bennett and Lovell 2019), and none specifically link pest control and pollination to yields in urban agroecosystems. Some suggest positive linkages between biodiversity, ecosystem services and food production. For example, Werrell et al. (2009) examined cucumber pollination and found that conspecific pollen deposition increased in larger gardens, and gardens with more flowering plants and thus suggest that spatial arrangement of plants within a garden can positively influence fruit and vegetable yields in gardens. Jha et al. (in prep), found that tomato pollen deposition in urban agroecosystems is predominantly heterospecific, and that the proportion of conspecific pollen correlates positively with the abundance and diversity of pollinators, especially when garden-level floral resources are high. In other words, pollinator abundance and diversity are essential for conspecific pollen deposition, which is higher in gardens with abundant floral resources. The study further shows that the positive value of both pollinator diversity and garden-level floral abundance also exists for overall fruit production rates, but the significance of these factors disappears at later stages in the ecosystem service cascade, specifically when evaluating fruit volume. Moreover, a recent global analysis suggests that pest control and pollination services in urban agroecosystems are valued at hundreds of thousands of dollars (Clinton et al. 2018). Clearly more work is needed in this area to more closely link biodiversity and food production in urban garden systems.

Finally, although increasing food production in urban areas is often proposed as an important outcome to improve urban food security, many gardeners do not simply garden for food production, but spent time in gardens for a diverse array of reasons and well-being benefits derived (Egerer et al. 2018c). Although we here focus on ecological aspects of urban agroecosystem studies, urban agroecosystems are rich spaces for sociological, anthropological, economic, and interdisciplinary research (for example, see Chapters 7 (Morales et al) and 14 (Caswell et al) in this volume). Thus, identifying how changes in management, biodiversity, traits, and ecological interactions and networks influences food production is important from an ecological research standpoint, but may not be the only way to view benefits of biodiversity in garden spaces.

2.5 Conclusions

Studies published in the past decade have highlighted a lack of natural science research in urban agroecosystems (Guitart et al. 2012), and have called for additional research into assessing how habitat management impacts biodiversity conservation and food production for urban farmers and gardeners. Many researchers have answered this call by documenting a wide array of local management and landscape features that affect abundance, richness, community composition and trait values for herbivores, natural enemies, and pollinators and the ecosystem services that they provide in urban agroecosystems. Through our summary, we highlight an impressive research foundation, but also point to future directions calling for stronger linkages between the urban ecology and rural agroecology literature, more manipulative research, greater integration of trait and network biology to better elucidate biodiversity-ecosystem service linkages, as well as more research that explicitly examines how biodiversity can benefit food production.

REFERENCES

Agrawal, A. A. 2000. Overcompensation of plants in response to herbivory and the by-product benefits of mutualism. *Trends in Plant Science* 5:309–313.

Ahrne, K., J. Bengtsson, and T. Elmqvist. 2009. Bumble bees (*bombus* spp) along a gradient of increasing urbanization. *Plos One* 4:9.

Alig, R. J., J. D. Kline, and M. Lichtenstein. 2004. Urbanization on the US landscape: looking ahead in the 21st century. *Landscape and Urban Planning* 69:219–234.

Altieri, M. A., and C. I. Nicholls. 2018. Urban agroecology: designing biodiverse, productive and resilient city farms. *Agro Sur* 46:49–60.

Andow, D. 1991. Vegetational diversity and arthropod population response. *Annual Review of Entomology* 36:561–586.

Armbrecht, I., and I. Perfecto. 2003. Litter-twig dwelling ant species richness and predation potential within a forest fragment and neighboring coffee plantations of contrasting habitat quality in Mexico. *Agriculture, Ecosystems & Environment* 97:107–115.

Arnold, J. E., M. Egerer, and K. M. Daane. 2019. Local and landscape effects to biological controls in urban agriculture—a review. *Insects* 10:215.

Aronson, M. F., C. H. Nilon, C. A. Lepczyk, T. S. Parker, P. S. Warren, S. S. Cilliers, M. A. Goddard, A. K. Hahs, C. Herzog, and M. Katti. 2016. Hierarchical filters determine community assembly of urban species pools. *Ecology* 97:2952–2963.

Astor, T., J. Strengbom, M. P. Berg, L. Lenoir, B. Marteinsdottir, and J. Bengtsson. 2014. Underdispersion and overdispersion of traits in terrestrial snail communities on islands. *Ecology and Evolution* 4:2090–2102.

Baldock, K. C. R., M. A. Goddard, D. M. Hicks, W. E. Kunin, N. Mitschunas, H. Morse, L. M. Osgathorpe, S. G. Potts, K. M. Robertson, A. V. Scott, P. P. A. Staniczenko, G. N. Stone, I. P. Vaughan, and J. Memmott. 2019. A systems approach reveals urban pollinator hotspots and conservation opportunities. *Nature Ecology & Evolution* 3:363–373.

Baldock, K. C. R., M. A. Goddard, D. M. Hicks, W. E. Kunin, N. Mitschunas, L. M. Osgathorpe, S. G. Potts, K. M. Robertson, A. V. Scott, G. N. Stone, I. P. Vaughan, and J. Memmott. 2015. Where is the UK's pollinator biodiversity? The importance of urban areas for flower-visiting insects. *Proceedings of the Royal Society of London. Series B: Biological Sciences* 282:20142849.

Ballare, K. M., J. L. Neff, R. Ruppel, and S. Jha. 2019. Multi-scalar drivers of biodiversity: local management mediates wild bee community response to regional urbanization. *Ecological Applications* 29:e01869.

Barton, P. S., H. Gibb, A. D. Manning, D. B. Lindenmayer, and S. A. Cunningham. 2011. Morphological traits as predictors of diet and microhabitat use in a diverse beetle assemblage. *Biological Journal of the Linnaean Society* 102:301–310.

Bascompte, J., and P. Jordano. 2007. Plant-animal mutualistic networks: The architecture of biodiversity. *Annual Review of Ecology Evolution and Systematics* 38:567–593.

Bascompte, J., P. Jordano, C. J. Melian, and J. M. Olesen. 2003. The nested assembly of plant-animal mutualistic networks. *Proceedings of the National Academy of Sciences of the United States of America* 100:9383–9387.

Bates, A. J., J. P. Sadler, A. J. Fairbrass, S. J. Falk, J. D. Hale, and T. J. Matthews. 2011. Changing bee and hoverfly pollinator assemblages along an urban-rural gradient. *Plos One* 6:e23459.

Bennett, A. B., and C. Gratton. 2012. Local and landscape scale variables impact parasitoid assemblages across an urbanization gradient. *Landscape and Urban Planning* 104:26–33.

Bennett, A. B., and S. Lovell. 2019. Landscape and local site variables differentially influence pollinators and pollination services in urban agricultural sites. *Plos One* 14:e0212034.

Bergerot, B., B. Fontaine, R. Julliard, and M. Baguette. 2011. Landscape variables impact the structure and composition of butterfly assemblages along an urbanization gradient. *Landscape Ecology* 26:83–94.

Bianchi, F. J, C. J. Booij, and T. Tscharntke. 2006. Sustainable pest regulation in agricultural landscapes: a review on landscape composition, biodiversity and natural pest control. *Proceedings of the Royal Society of London. Series B: Biological Sciences* 273:1715–1727.

Blair, R. B., and A. E. Launer. 1997. Butterfly diversity and human land use: Species assemblages along an urban gradient. *Biological Conservation* 80:113–125.

Blüthgen, N. 2010. Why network analysis is often disconnected from community ecology: A critique and an ecologist's guide. *Basic and Applied Ecology* 11:185–195.

Bohan, D. A., A. Raybould, C. Mulder, G. Woodward, A. Tamaddoni-Nezhad, N. Bluthgen, M. J. O. Pocock, S. Muggleton, D. M. Evans, J. Astegiano, F. Massol, N. Loeuille, S. Petit, and S. Macfadyen. 2013. Networking Agroecology: Integrating the Diversity of Agroecosystem Interactions. Pages 1–67 *in* G. Woodward and D. A. Bohan, editors. *Advances in Ecological Research, Vol 49: Ecological Networks in an Agricultural World.* Academic Press.

Bolger, D. T., A. V. Suarez, K. R. Crooks, S. A. Morrison, and T. J. Case. 2000. Arthropods in urban habitat fragments in Southern California: area, age, and edge effects. *Ecological Applications* 10:1230–1248.

Bolsinger, M., and W. Flückiger. 1989. Ambient air pollution induced changes in amino acid pattern of phloem sap in host plants-relevance to aphid infestation. *Environmental Pollution* 56:209–216.

Brown, K., and A. Jameton. 2000. Public health implications of urban agriculture. *Journal of Public Health Policy* 21:20–39.

Burkman, C., and M. Gardiner. 2014. Urban greenspace composition and landscape context influence natural enemy community composition and function. *Biological Control* 75:58–67.

Burks, J. M., and S. M. Philpott. 2017. Local and landscape drivers of parasitoid abundance, richness, and composition in urban gardens. *Environmental Entomology* 46:201–209.

Cadenasso, M. L., S. T. A. Pickett, and K. Schwarz. 2007. Spatial heterogeneity in urban ecosystems: reconceptualizing land cover and a framework for classification. *Frontiers in Ecology and the Environment* 5:80–88.

Cane, J. H., R. L. Minckley, L. J. Kervin, T. H. Roulston, and N. M. Williams. 2006. Complex responses within a desert bee guild (Hymenoptera : Apiformes) to urban habitat fragmentation. *Ecological Applications* 16:632–644.

CaraDonna, P. J., W. K. Petry, R. M. Brennan, J. L. Cunningham, J. L. Bronstein, N. M. Waser, and N. J. Sanders. 2017. Interaction rewiring and the rapid turnover of plant-pollinator networks. *Ecology Letters* 20:385–394.

Cardinale, B. J., J. E. Duffy, A. Gonzalez, D. U. Hooper, C. Perrings, P. Venail, A. Narwani, G. M. Mace, D. Tilman, D. A. Wardle, A. P. Kinzig, G. C. Daily, M. Loreau, J. B. Grace, A. Larigauderie, D. S. Srivastava, and S. Naeem. 2012. Biodiversity loss and its impact on humanity. *Nature* 486:59–67.

Cardinale, B. J., D. S. Srivastava, J. E. Duffy, J. P. Wright, A. L. Downing, M. Sankaran, and C. Jouseau. 2006. Effects of biodiversity on the functioning of trophic groups and ecosystems. *Nature* 443:989–992.

Castilla, A. R., N. S. Pope, M. O'Connell, M. F. Rodriguez, L. Treviño, A. Santos, and S. Jha. 2017. Adding landscape genetics and individual traits to the ecosystem function paradigm reveals the importance of species functional breadth. *Proceedings of the National Academy of Sciences of the United States of America* 114:12761–12766.

Čeplová, N., V. Kalusová, and Z. Lososová. 2017. Effects of settlement size, urban heat island and habitat type on urban plant biodiversity. *Landscape and Urban Planning* 159:15–22.

Chaplin-Kramer, R., K. Tuxen-Bettman, and C. Kremen. 2011. Value of wildlands habitat for supplying pollination services to Californian agriculture. *Rangelands* 33:33–41.

Chappell, M., and L. LaValle. 2011. Food security and biodiversity: can we have both? An agroecological analysis. *Agriculture and Human Values* 28:3–26.

Chevalier, J. M., and D. J. Buckles. 2019. *Participatory Action Research: Theory and Methods for Engaged Inquiry*. Routledge.

Clarke, L. W., L. Li, G. D. Jenerette, and Z. Yu. 2014. Drivers of plant biodiversity and ecosystem service production in home gardens across the Beijing municipality of China. *Urban Ecosystems* 17:741–760.

Classens, M. 2015. The nature of urban gardens: toward a political ecology of urban agriculture. *Agriculture and Human Values* 32:229–239.

Clergeau, P., J.-P. L. Savard, G. Mennechez, and G. Falardeau. 1998. Bird abundance and diversity along an urban-rural gradient: a comparative study between two cities on different continents. *The Condor* 100:413–425.

Clinton, N., M. Stuhlmacher, A. Miles, N. U. Aragon, M. Wagner, M. Georgescu, C. Herwig, and P. Gong. 2018. A Global geospatial ecosystem services estimate of urban agriculture. *Earth's Future* 6:40–60.

Clucas, B., I. D. Parker, and A. M. Feldpausch-Parker. 2018. A systematic review of the relationship between urban agriculture and biodiversity. *Urban Ecosystems* 21:635–643.

Cohen, H., S. M. Philpott, H. Liere, B. B. Lin, and S. Jha. (in review). *Delivery of Pollination Services in Urban Gardens is Mediated by Pollinator Biodiversity and the Interaction Between Local and Landscape Habitat Factors*. Urban Ecosystems.

Cohen, H., Philpott S., Liere, H., Lin, B., Jha, S. 2020. The relationship between pollinator community and pollination services is mediated by floral abundance in urban landscapes. Urban Ecosystems. In press. 10.1007/s11252-020-01024-z.

Colding, J., J. Lundberg, and C. Folke. 2006. Incorporating green-area user groups in urban ecosystem management. *Ambio* 35:237–244.

Costanza, R. 1997. The value of the world's ecosystem services and natural capital. *Nature* 387:253–260.

Daily, G. C. 1997. *Nature's Services: Societal Dependence on Natural Ecosystems*. Island Press, Washington, D.C., USA.

Dale, A. G., and S. D. Frank. 2014. Urban warming trumps natural enemy regulation of herbivorous pests. *Ecological Applications* 24:1596–1607.

Danforth, B. N., R. L. Minckley, J. L. Neff, F. Fawcett. 2019. *The Solitary Bees: Biology, Evolution, Conservation*. http://www.danforthlab.entomology.cornell.edu/2019/07/24/the-solitary-bees/.

Darwin, C. 1862. *On the Various Contrivances by Which British and Foreign Orchids are Fertilised by Insects, and On the Good Effects of Intercrossing*. Second Edition. John Murray. London.

De Barba, M., C. Miquel, F. Boyer, C. Mercier, D. Rioux, E. Coissac, and P. Taberlet. 2014. DNA metabarcoding multiplexing and validation of data accuracy for diet assessment: application to omnivorous diet. *Molecular Ecology Resources* 14:306–323.

Delgado de la Flor, Y. A., C. E. Burkman, T. K. Eldredge, and M. M. Gardiner. 2017. Patch and landscape-scale variables influence the taxonomic and functional composition of beetles in urban greenspaces. *Ecosphere* 8:Article e02007.

Dennis, R. L. H., and P. B. Hardy. 2001. Loss rates of butterfly species with urban development. A test of atlas data and sampling artefacts at a fine scale. *Biodiversity and Conservation* 10:1831–1837.

Dexter, A. R. 2004. Soil physical quality - part I. Theory, effects of soil texture, density, and organic matter, and effects on root growth. *Geoderma* 120:201–214.

Dias, A. T. C., E. J. Krab, J. Marien, M. Zimmer, J. H. C. Cornelissen, J. Ellers, D. A. Wardle, and M. P. Berg. 2013. Traits underpinning desiccation resistance explain distribution patterns of terrestrial isopods. *Oecologia* 172:667–677.

Diaz, S., S. Lavorel, F. de Bello, F. Quetier, K. Grigulis, and M. Robson. 2007. Incorporating plant functional diversity effects in ecosystem service assessments. *Proceedings of the National Academy of Sciences of the United States of America* 104:20684–20689.

Diaz, S., A. Purvis, J. H. C. Cornelissen, G. M. Mace, M. J. Donoghue, R. M. Ewers, P. Jordano, and W. D. Pearse. 2013. Functional traits, the phylogeny of function, and ecosystem service vulnerability. *Ecology and Evolution* 3:2958–2975.

Dickinson, J. L., J. Shirk, D. Bonter, R. Bonney, R. L. Crain, J. Martin, T. Phillips, and K. Purcell. 2012. The current state of citizen science as a tool for ecological research and public engagement. *Frontiers in Ecology and the Environment* 10:291–297.

Egerer, M. H., C. Arel, M. D. Otoshi, R. D. Quistberg, P. Bichier, and S. M. Philpott. 2017a. Urban arthropods respond variably to changes in landscape context and spatial scale. *Journal of Urban Ecology* 3:1–10.

Egerer, M. H., P. Bichier, and S. M. Philpott. 2017b. Landscape and local habitat correlates of lady beetle abundance and species richness in urban agriculture. *Annals of the Entomological Society of America* 110:97–103.

Egerer, M. H., H. Liere, P. Bichier, and S. M. Philpott. 2018a. Cityscape quality and resource manipulation affect natural enemy biodiversity in and fidelity to urban agroecosystems. *Landscape Ecology* 33:985–998.

Egerer, M. H., H. Liere, B. B. Lin, S. Jha, P. Bichier, and S. M. Philpott. 2018b. Herbivore regulation in urban agroecosystems: Direct and indirect effects. *Basic and Applied Ecology* 29:44–54.

Egerer, M. H., S. M. Philpott, P. Bichier, S. Jha, H. Liere, and B. B. Lin. 2018c. Gardener well-being along social and biophysical landscape gradients. *Sustainability* 10:96

Egerer, M., K. Li, and T. W. Y. Ong. 2018d. Context matters: Contrasting ladybird beetle responses to urban environments across two US Regions. *Sustainability* 10:1829.

Eklöf, A., U. Jacob, J. Kopp, J. Bosch, R. Castro-Urgal, N. P. Chacoff, B. Dalsgaard, C. de Sassi, M. Galetti, P. R. Guimarães, S. B. Lomáscolo, A. M. M. González, M. A. Pizo, R. Rader, A. Rodrigo, J. M. Tylianakis, D. P. Vázquez, and S. Allesina. 2013. The dimensionality of ecological networks. *Ecology Letters* 16:577–583.

Eremeeva, N. I., and D. V. Sushchev. 2005. Structural changes in the fauna of pollinating insects in urban landscapes. *Russian Journal of Ecology* 36:259–265.

Faegri, K., and L. van der Pijl. 1979. The Principles of Pollination Ecology. Pergamon, Oxford.

Faeth, S. H., C. Bang, and S. Saari. 2011. Urban biodiversity: patterns and mechanisms. *Annals of the New York Academy of Sciences* 1223:69–81.

Faeth, S. H., W. A. Marussich, E. Shochat, and P. S. Warren. 2005. Trophic dynamics in urban communities. *BioScience* 55:399–407.

Ferrer-Paris, J. R., A. Sánchez-Mercado, Á. L. Viloria, and J. Donaldson. 2013. Congruence and diversity of butterfly-host plant associations at higher taxonomic levels. *Plos One* 8:e63570.

Fortel, L., M. Henry, L. Guilbaud, A. L. Guirao, M. Kuhlmann, H. Mouret, O. Rollin, and B. E. Vaissière. 2014. Decreasing abundance, increasing diversity and changing structure of the wild bee community (Hymenoptera: Anthophila) along an urbanization gradient. *Plos One* 9:e104679.

Fortuna, M. A., and J. Bascompte. 2006. Habitat loss and the structure of plant-animal mutualistic networks. *Ecology Letters* 9:278–283.

Fountain-Jones, N. M., S. C. Baker, and G. J. Jordan. 2015. Moving beyond the guild concept: developing a practical functional trait framework for terrestrial beetles. *Ecological Entomology* 40:1–13.

Franzen, M., and S. G. Nilsson. 2010. Both population size and patch quality affect local extinctions and colonizations. *Proceedings of the Royal Society of London. Series B: Biological Sciences* 277:79–85.

Frey, D., K. Vega, F. Zellweger, J. Ghazoul, D. Hansen, and M. Moretti. 2018. Predation risk shaped by habitat and landscape complexity in urban environments. *Journal of Applied Ecology* 55:2343–2353.

Friedrich, R. L., and S. M. Philpott. 2009. Nest-site limitation and nesting resources of ants (Hymenoptera: Formicidae) in urban green spaces. *Environmental Entomology* 38:600–607.

Gagic, V., S. Hänke, C. Thies, C. Scherber, Ž. Tomanović, and T. Tscharntke. 2012. Agricultural intensification and cereal aphid-parasitoid-hyperparasitoid food webs: Network complexity, temporal variability and parasitism rates. *Oecologia* 170:1099–1109.

Gardiner, M. M., C. E. Burkman, and S. P. Prajzner. 2013. The Value of urban vacant land to support arthropod biodiversity and ecosystem services. *Environmental Entomology* 42:1123–1136.

Gardiner, M. M., S. P. Prajzner, C. E. Burkman, S. Albro, and P. S. Grewal. 2014. Vacant land conversion to community gardens: influences on generalist arthropod predators and biocontrol services in urban greenspaces. *Urban Ecosystems* 17:101–122.

Garibaldi, L. A., I. Bartomeus, R. Bommarco, A. M. Klein, S. A. Cunningham, M. A. Aizen, V. Boreux, M. P. D. Garratt, L. G. Carvalheiro, C. Kremen, C. L. Morales, C. Schuepp, N. P. Chacoff, B. M. Freitas, V. Gagic, A. Holzschuh, B. K. Klatt, K. M. Krewenka, S. Krishnan, M. M. Mayfield, I. Motzke, M. Otieno, J. Petersen, S. G. Potts, T. H. Ricketts, M. Rundlof, A. Sciligo, P. A. Sinu, I. Steffan-Dewenter, H. Taki, T. Tscharntke, C. H. Vergara, B. F. Viana, and M. Woyciechowski. 2015. Trait matching of flower visitors and crops predicts fruit set better than trait diversity. *Journal of Applied Ecology* 52:1436–1444.

Gaston, K. J., R. M. Smith, K. Thompson, and P. H. Warren. 2005. Urban domestic gardens (II): experimental tests of methods for increasing biodiversity. *Biodiversity and Conservation* 14:395–413.

Geslin, B., B. Gauzens, E. Thebault, and I. Dajoz. 2013. Plant pollinator networks along a gradient of urbanisation. *Plos One* 8:e63421.

Gittleman, M., K. Jordan, and E. Brelsford. 2012. Using citizen science to quantify community garden crop yields. *Cities and the Environment* 5:4.

Greenleaf, S. S., N. M. Williams, R. Winfree, and C. Kremen. 2007. Bee foraging ranges and their relationship to body size. *Oecologia* 153:589–596.

Gregory, M. M., T. W. Leslie, and L. E. Drinkwater. 2016. Agroecological and social characteristics of New York city community gardens: contributions to urban food security, ecosystem services, and environmental education. *Urban Ecosystems* 19:763–794.

Grewal, S. S., Z. Cheng, S. Masih, M. Wolboldt, N. Huda, A. Knight, and P. S. Grewal. 2011. An assessment of soil nematode food webs and nutrient pools in community gardens and vacant lots in two post-industrial American cities. *Urban Ecosystems* 14:181–194.

Guitart, D., C. Pickering, and J. Byrne. 2012. Past results and future directions in urban community gardens research. *Urban Forestry & Urban Greening* 11:364–373.

Hardy, P. B., and R. L. H. Dennis. 1999. The impact of urban development on butterflies within a city region. *Biodiversity and Conservation* 8:1261–1279.

Heleno, R., C. Garcia, P. Jordano, A. Traveset, J. M. Gómez, N. Blüthgen, J. Memmott, M. Moora, J. Cerdeira, S. Rodríguez-Echeverría, H. Freitas, and J. M. Olesen. 2014. Ecological networks: delving into the architecture of biodiversity. *Biology Letters* 10:20131000.

Hennig, E. I., and J. Ghazoul. 2012. Pollinating animals in the urban environment. *Urban Ecosystems* 15:149–166.

Hodgson, K., M. Campbell, and M. Bailkey. 2011. *Urban Agriculture: Growing Healthy, Sustainable Places.* American Planning Association.

Hooper, D. U., F. S. Chapin, J. J. Ewel, A. Hector, P. Inchausti, S. Lavorel, J. H. Lawton, D. M. Lodge, M. Loreau, S. Naeem, B. Schmid, H. Setala, A. J. Symstad, J. Vandermeer, and D. A. Wardle. 2005. Effects of biodiversity on ecosystem functioning: A consensus of current knowledge. *Ecological Monographs* 75:3–35.

Ibañez, S. 2012. Optimizing size thresholds in a plant-pollinator interaction web: towards a mechanistic understanding of ecological networks. *Oecologia* 170:233–242.

Ibañez, S., O. Manneville, C. Miquel, P. Taberlet, A. Valentini, S. Aubert, E. Coissac, M. P. Colace, Q. Duparc, S. Lavorel, and M. Moretti. 2013. Plant functional traits reveal the relative contribution of habitat and food preferences to the diet of grasshoppers. *Oecologia* 173:1459–1470.

Ings, T. C., J. M. Montoya, J. Bascompte, N. Bluthgen, L. Brown, C. F. Dormann, F. Edwards, D. Figueroa, U. Jacob, J. I. Jones, R. B. Lauridsen, M. E. Ledger, H. M. Lewis, J. M. Olesen, F. J. F. van Veen, P. H. Warren, and G. Woodward. 2009. Ecological networks - beyond food webs. *Journal of Animal Ecology* 78:253–269.

IPBES. 2016. The Assessment Report of the Intergovernmental Science-Policy Platform on Biodiversity and Ecosystem Services on Pollinators, Pollination and Food Production. Bonn, Germany.

Jauker, F., T. Diekötter, F. Schwarzbach, and V. Wolters. 2009. Pollinator dispersal in an agricultural matrix: opposing responses of wild bees and hoverflies to landscape structure and distance from main habitat. *Landscape Ecology* 24:547–555.

Jha, S., and C. Kremen. 2013. Urban land use limits regional bumble bee gene flow. *Molecular Ecology* 22:2483–2495.

Jha, S., E. Tran, Z. Jordan, H. Liere, B. Lin, H. Cohen, M. H. Egerer, and S. M. Philpott. (in prep). Pollinator and floral abundance interact to increase conspecific pollen deposition in urban gardens.

Jordano, P., J. Bascompte, and J. M. Olesen. 2003. Invariant properties in coevolutionary networks of plant-animal interactions. *Ecology Letters* 6:69–81.

Kaiser-Bunbury, C. N., and N. Blüthgen. 2015. Integrating network ecology with applied conservation: a synthesis and guide to implementation. *Aob Plants* 7.

Kaiser-Bunbury, C. N., S. Muff, J. Memmott, C. B. Muller, and A. Caflisch. 2010. The robustness of pollination networks to the loss of species and interactions: a quantitative approach incorporating pollinator behaviour. *Ecology Letters* 13:442–452.

Kaiser-Bunbury, C. N., D. P. Vázquez, M. Stang, and J. Ghazoul. 2014. Determinants of the microstructure of plant–pollinator networks. *Ecology* 95:3314–3324.

Karban, R., and J. H. Myers. 1989. Induced plant responses to herbivory. *Annual Review of Ecology and Systematics* 20:331–348.

Karp, D. S., R. Chaplin-Kramer, T. D. Meehan, E. A. Martin, F. DeClerck, H. Grab, C. Gratton, L. Hunt, A. E. Larsen, A. Martínez-Salinas, M. E. O'Rourke, A. Rusch, K. Poveda, M. Jonsson, J. A. Rosenheim, N. A. Schellhorn, T. Tscharntke, S. D. Wratten, W. Zhang, A. L. Iverson, L. S. Adler, M. Albrecht, A. Alignier, G. M. Angelella, M. Zubair Anjum, J. Avelino, P. Batáry, J. M. Baveco, F. J. J. A. Bianchi, K. Birkhofer, E. W. Bohnenblust, R. Bommarco, M. J. Brewer, B. Caballero-López, Y. Carrière, L. G. Carvalheiro, L. Cayuela, M. Centrella, A. Ćetković, D. C. Henri, A. Chabert, A. C. Costamagna, A. De la Mora, J. de Kraker, N. Desneux, E. Diehl, T. Diekötter, C. F. Dormann, J. O. Eckberg, M. H. Entling, D. Fiedler, P. Franck, F. J. Frank van Veen, T. Frank, V. Gagic, M. P. D. Garratt, A. Getachew, D. J. Gonthier, P. B. Goodell, I. Graziosi, R. L. Groves, G. M. Gurr, Z. Hajian-Forooshani, G. E. Heimpel, J. D. Herrmann, A. S. Huseth, D. J. Inclán, A. J. Ingrao, P. Iv, K. Jacot, G. A. Johnson, L. Jones, M. Kaiser, J. M. Kaser, T. Keasar, T. N. Kim, M. Kishinevsky, D. A. Landis, B. Lavandero, C. Lavigne, A. Le Ralec, D. Lemessa, D. K. Letourneau, H. Liere, Y. Lu, Y. Lubin, T. Luttermoser, B. Maas, K. Mace, F. Madeira, V. Mader, A. M. Cortesero, L. Marini, E. Martinez, H. M. Martinson, P. Menozzi, M. G. E. Mitchell, T. Miyashita, G. A. R. Molina, M. A. Molina-Montenegro, M. E. O'Neal, I. Opatovsky, S. Ortiz-Martinez, M. Nash, Ö. Östman, A. Ouin, D. Pak, D. Paredes, S. Parsa, H. Parry, R. Perez-Alvarez, D. J. Perović, J. A. Peterson, S. Petit, S. M. Philpott, M. Plantegenest, M. Plećaš, T. Pluess, X. Pons, S. G. Potts, R. F. Pywell, D. W. Ragsdale, T. A. Rand, L. Raymond, B. Ricci, C. Sargent, J.-P. Sarthou, J. Saulais, J. Schäckermann, N. P. Schmidt, G. Schneider, C. Schüepp, F. S. Sivakoff, H. G. Smith, K. Stack Whitney, S. Stutz, Z. Szendrei, M. B. Takada, H. Taki, G. Tamburini, L. J. Thomson, Y. Tricault, N. Tsafack, M. Tschumi, M. Valantin-Morison, M. Van Trinh, W. van der Werf, K. T. Vierling, B. P. Werling, J. B. Wickens, V. J. Wickens, B. A. Woodcock, K. Wyckhuys, H. Xiao, M. Yasuda, A. Yoshioka, and Y. Zou. 2018. Crop pests and predators exhibit inconsistent responses to surrounding landscape composition. *Proceedings of the National Academy of Sciences of the United States of America* 115:E7863–E7870.

Klein, A., I. Steffan–Dewenter, and T. Tscharntke. 2003. Fruit set of highland coffee increases with the diversity of pollinating bees. *Proceedings of the Royal Society of London. Series B: Biological Sciences* 270:955–961.

Klein, A.-M., I. Steffan-Dewenter, and T. Tscharntke. 2006. Rain forest promotes trophic interactions and diversity of trap-nesting hymenoptera in adjacent agroforestry. *Journal of Animal Ecology* 75:315–323.

Kremen, C., N. M. Williams, M. A. Aizen, B. Gemmill-Herren, G. LeBuhn, R. Minckley, L. Packer, S. G. Potts, T. Roulston, I. Steffan-Dewenter, D. P. Vazquez, R. Winfree, L. Adams, E. E. Crone, S. S. Greenleaf, T. H. Keitt, A. M. Klein, J. Regetz, and T. H. Ricketts. 2007. Pollination and other ecosystem services produced by mobile organisms: a conceptual framework for the effects of land-use change. *Ecology Letters* 10:299–314.

Kruess, A., and T. Tscharntke. 1994. Habitat fragmentation, species loss, and biological control. *Science* 264:1581–1584.

Laliberté, E., and J. M. Tylianakis. 2010. Deforestation homogenizes tropical parasitoid — host networks homogenizes tropical parasitoid host. *Ecology* 91:1740–1747.

Landis, D. A., S. D. Wratten, and G. M. Gurr. 2000. Habitat management to conserve natural enemies of arthropod pests in agriculture. *Annual Review of Entomology* 45:175–201.

Langellotto, G. A., and R. F. Denno. 2004. Responses of invertebrate natural enemies to complex-structured habitats: a meta-analytical synthesis. *Oecologia* 139:1–10.

Larsen, T. H., N. Williams, and C. Kremen. 2005. Extinction order and altered community structure rapidly disrupt ecosystem functioning. *Ecology Letters* 8:538–547.

Larson, N., Story, M., and Nelson, M. (2009). Neighborhood Environments: Disparities in Access to Healthy Foods in the US. *American Journal of Preventive Medicine.* 36: 74–81.

Lavorel, S., and E. Garnier. 2002. Predicting changes in community composition and ecosystem functioning from plant traits: revisiting the holy grail. *Functional Ecology* 16:545–556.

Lavorel, S., J. Storkey, R. D. Bardgett, F. de Bello, M. P. Berg, X. Le Roux, M. Moretti, C. Mulder, R. J. Pakeman, S. Diaz, and R. Harrington. 2013. A novel framework for linking functional diversity of plants with other trophic levels for the quantification of ecosystem services. *Journal of Vegetation Science* 24:942–948.

Letourneau, D. K., I. Armbrecht, B. S. Rivera, J. M. Lerma, E. J. Carmona, M. C. Daza, S. Escobar, V. Galindo, C. Gutiérrez, S. D. López, J. L. Mejía, A. M. A. Rangel, J. H. Rangel, L. Rivera, C. A. Saavedra, A. M. Torres, and A. R. Trujillo. 2011. Does plant diversity benefit agroecosystems? A synthetic review. *Ecological Applications* 21:9–21.

Letourneau, D. K., and S. G. Bothwell. 2008. Comparison of organic and conventional farms: challenging ecologists to make biodiversity functional. *Frontiers in Ecology and the Environment* 6:430–438.

Liere, H., M. H. Egerer, and S. M. Philpott. 2019. Environmental and spatial filtering of ladybird beetle community composition and functional traits in urban landscapes. *Journal of Urban Ecology* 5:1–12.

Liere, H., T. N. Kim, B. P. Werling, T. D. Meehan, D. A. Landis, and C. Gratton. 2015. Trophic cascades in agricultural landscapes: indirect effects of landscape composition on crop yield. *Ecological Applications* 25:652–661.

Lin, B. B., M. H. Egerer, H. Liere, S. Jha, and S. M. Philpott. 2018. Soil management is key to maintaining soil moisture in urban gardens facing changing climatic conditions. *Scientific Reports* 8:17565.

Lin, B. B., S. M. Philpott, and S. Jha. 2015. The future of urban agriculture and biodiversity-ecosystem services: challenges and next steps. *Basic and Applied Ecology* 16:189–201.

López-Uribe, M. M., S. Jha, and A. Soro. 2019. A trait-based approach to predict population genetic structure in bees. *Molecular Ecology* 28:1919–1929.

Loreau, M., S. Naeem, P. Inchausti, J. Bengtsson, J. P. Grime, A. Hector, D. U. Hooper, M. A. Huston, D. Raffaelli, B. Schmid, D. Tilman, and D. A. Wardle. 2001. Ecology - biodiversity and ecosystem functioning: Current knowledge and future challenges. *Science* 294:804–808.

Losey, J. E., and M. Vaughan. 2006. The economic value of ecological services provided by insects. *Bioscience* 56:311–323.

Lovell, S. T. 2010. Multifunctional urban agriculture for sustainable land use planning in the united states. *Sustainability* 2:2499–2522.

Lowenstein, D. M., M. Gharehaghaji, and D. H. Wise. 2017. Substantial mortality of cabbage looper (Lepidoptera: Noctuidae) from predators in urban agriculture is not influenced by scale of production or variation in local and landscape-level factors. *Environmental Entomology* 46:30–37.

Lowenstein, D. M., K. C. Matteson, and E. S. Minor. 2019. Evaluating the dependence of urban pollinators on ornamental, non-native, and 'weedy' floral resources. *Urban Ecosystems* 22:293–302.

Lowenstein, D. M., K. C. Matteson, I. Xiao, A. M. Silva, and E. S. Minor. 2014. Humans, bees, and pollination services in the city: the case of Chicago, IL (USA). *Biodiversity and Conservation* 23:2857–2874.

Lowenstein, D. M., and E. S. Minor. 2018. Herbivores and natural enemies of brassica crops in urban agriculture. *Urban Ecosystems* 21:519–529.

Maccherini, S., G. Bacaro, L. Favilli, S. Piazzini, E. Santi, and M. Marignani. 2009. Congruence among vascular plants and butterflies in the evaluation of grassland restoration success. *Acta Oecologica* 35:311–317.

Mace-Hill, K. C. 2015. Understanding, using, and promoting biological control: from commercial walnut orchards to school gardens. PhD Dissertation. UC Berkeley.

Macias-Hernandez, N., K. Athey, V. Tonzo, O. S. Wangensteen, M. Arnedo, and J. D. Harwood. 2018. Molecular gut content analysis of different spider body parts. *Plos One* 13:e0196589.

MacIvor, J. S., and L. Packer. 2015. 'Bee hotels' as tools for native pollinator conservation: A premature verdict? *Plos One* 10:e0122126.

Maisto, G., V. Milano, and L. Santorufo. 2017. Relationships among site characteristics, taxonomical structure and functional trait distribution of arthropods in forest, urban and agricultural soils of Southern Italy. *Ecological Research* 32:511–521.

Majewska, A. A., and S. Altizer. 2018. Planting gardens to support insect pollinators. *Conservation Biology* 34:15–25.

Marcelis, L. F. M., and L. R. Baan Hofman-Eijer. 1997. Effects of seed number on competition and dominance among fruits in *capsicum annuum* L. *Annals of Botany* 79:687–693.

Martin, E. A., B. Reineking, B. Seo, and I. Steffan-Dewenter. 2015. Pest control of aphids depends on landscape complexity and natural enemy interactions. *PeerJ* 3:e1095.

Martins, K. T., A. Gonzalez, and M. J. Lechowicz. 2017. Patterns of pollinator turnover and increasing diversity associated with urban habitats. *Urban Ecosystems* 20:1359–1371.

Maruyama, P. K., C. Bonizário, A. P. Marcon, G. D'Angelo, M. M. da Silva, E. N. da Silva Neto, P. E. Oliveira, I. Sazima, M. Sazima, J. Vizentin-Bugoni, L. dos Anjos, A. M. Rui, and O. Marçal Júnior. 2019. Plant-hummingbird interaction networks in urban areas: Generalization and the importance of trees with specialized flowers as a nectar resource for pollinator conservation. *Biological Conservation* 230:187–194.

Mata, L., C. G. Threlfall, N. S. G. Williams, A. K. Hahs, M. Malipatil, N. E. Stork, and S. J. Livesley. 2017. Conserving herbivorous and predatory insects in urban green spaces. *Scientific Reports* 7:1–12.

Matteson, K. C., J. S. Ascher, and G. A. Langellotto. 2008. Bee richness and abundance in New York City urban gardens. *Annals of the Entomological Society of America* 101:140–150.

Matteson, K. C., J. B. Grace, and E. S. Minor. 2013. Direct and indirect effects of land use on floral resources and flower-visiting insects across an urban landscape. *Oikos* 122:682–694.

Matteson, K. C., and G. A. Langellotto. 2011. Small scale additions of native plants fail to increase beneficial insect richness in urban gardens. *Insect Conservation and Diversity* 4:89–98.

Mazzi, D., and S. Dorn. 2012. Movement of insect pests in agricultural landscapes. *Annals of Applied Biology* 160:97–113.

McDonnell, M. J. 2011. The History of urban ecology: An ecologist's perspective. Pages 5–13 *in* J. Niemelä, J. Breuste, T. Elmqvist, G. Guntenspergen, P. James and N. McIntyre, editors. *Urban Ecology: Patterns, Processes, and Applications.* Oxford University Press, Oxford.

McDonnell, M. J., and S. T. A. Pickett. 1990. Ecosystem structure and function along urban-rural gradients: an unexploited opportunity for ecology. *Ecology* 71:1232–1237.

McDonnell, M. J., S. T. A. Pickett, P. Groffman, P. Bohlen, R. V. Pouyat, W. C. Zipperer, R. W. Parmelee, M. M. Carreiro, and K. Medley. 2008. Ecosystem Processes Along an Urban-to-Rural Gradient. Pages 299–313 *in* J. M. Marzluff, E. Shulenberger, W. Endlicher, M. Alberti, G. Bradley, C. Ryan, U. Simon, and C. ZumBrunnen, editors. *Urban Ecology: An International Perspective on the Interaction Between Humans and Nature.* Springer US, Boston, MA.

McKinney, M. L. 2002. Urbanization, biodiversity, and conservation. *Bioscience* 52:883–890.

McKinney, M. L. 2008. Effects of urbanization on species richness: A review of plants and animals. *Urban Ecosystems* 11:161–176.

Memmott, J., P. G. Craze, N. M. Waser, and M. V. Price. 2007. Global warming and the disruption of plant-pollinator interactions. *Ecology Letters* 10:710–717.

Memmott, J., N. M. Waser, and M. V. Price. 2004. Tolerance of pollination networks to species extinctions. *Proceedings of the Royal Society of London. Series B: Biological Sciences* 271:2605–2611.

Michener, C. D. 2000. Bees of the World. Johns Hopkins University Press, Baltimore, MD.

Millennium Ecosystem Assessment. 2005. Full Reports. Island Press, Washington, D.C.

Morales, H., B. G. Ferguson, L. E. Marín, D. N. Gutiérrez, P. Bichier, and S. M. Philpott. 2018. Agroecological pest management in the city: experiences from California and Chiapas. *Sustainability* 10:2068.

Morandin, L., and C. Kremen. 2013. Hedgerow restoration promotes pollinator populations and exports native bees to adjacent fields. *Ecological Applications* 23:829–839.

Moretti, M., A. T. C. Dias, F. de Bello, F. Altermatt, S. L. Chown, F. M. Azcarate, J. R. Bell, B. Fournier, M. Hedde, J. Hortal, S. Ibanez, E. Ockinger, J. P. Sousa, J. Ellers, and M. P. Berg. 2017. Handbook of protocols for standardized measurement of terrestrial invertebrate functional traits. *Functional Ecology* 31:558–567.

Mukherjee, S., S. Banerjee, P. Basu, G. K. Saha, and G. Aditya. 2018. Butterfly-plant network in urban landscape: Implication for conservation and urban greening. *Acta Oecologica* 92:16–25.

Niell, R. S., P. F. Brussard, and D. D. Murphy. 2007. Butterfly community composition and oak woodland vegetation response to rural residential development. *Landscape and Urban Planning* 81:235–245.

Niemelä, J., D. J. Kotze, S. Venn, L. Penev, I. Stoyanov, J. Spence, D. Hartley, and E. M. De Oca. 2002. Carabid beetle assemblages (Coleoptera, Carabidae) across urban-rural gradients: an international comparison. *Landscape Ecology* 17:387–401.

Oberholtzer, L., C. Dimitri, and A. Pressman. 2014. Urban agriculture in the United States: characteristics, challenges, and technical assistance needs. *Journal of Extension* 52:6FEA1.

Otoshi, M., P. Bichier, and P. SM. 2015. Local and landscape correlates of spider activity density and species richness in urban gardens. *Environmental Entomology* 44:1043–1051.

Pardee, G. L., and S. M. Philpott. 2014. Native plants are the bee's knees: local and landscape predictors of bee richness and abundance in backyard gardens. *Urban Ecosystems* 17:641–659.

Parsons, S. E., and S. D. Frank. 2019. Urban tree pests and natural enemies respond to habitat at different spatial scales. *Journal of Urban Ecology* 5:1–15.

Pascual, M., and J. A. Dunne. 2006. *Ecological Networks: Linking Structure to Dynamics in Food Webs.* Oxford University Press, Oxford.

Pawelek, J. C., G. W. Frankie, R. W. Thorp, and M. Przybylski. 2009. Modification of a community garden to attract native bee pollinators in urban San Luis Obispo, California. *Cities and the Environment* 2:20.

Peralta, G., C. M. Frost, T. A. Rand, R. K. Didham, and J. M. Tylianakis. 2014. Complementarity and redundancy of interactions enhance attack rates and spatial stability in host-parasitoid food webs. *Ecology* 95:1888–1896.

Pereira-Peixoto, M. H., G. Pufal, C. F. Martins, and A.-M. Klein. 2014. Spillover of trap-nesting bees and wasps in an urban–rural interface. *Journal of Insect Conservation* 18:815–826.

Perfecto, I., and J. Vandermeer. 2002. Quality of agroecological matrix in a tropical montane landscape: ants in coffee plantations in Southern Mexico. *Conservation Biology* 16:174–182.

Perovič, D. J., S. Gamez-Virues, D. A. Landis, F. Wackers, G. M. Gurr, S. D. Wratten, M. S. You, and N. Desneux. 2018. Managing biological control services through multi-trophic trait interactions: review and guidelines for implementation at local and landscape scales. *Biological Reviews* 93:306–321.

Perrings, C., S. Naeem, F. Ahrestani, D. E. Bunker, P. Burkill, G. Canziani, T. Elmqvist, R. Ferrati, J. A. Fuhrman, F. Jaksic, Z. Kawabata, A. Kinzig, G. M. Mace, F. Milano, H. Mooney, A. H. Prieur-Richard, J. Tschirhart, and W. Weisser. 2010. Ecosystem services for 2020. *Science* 330:323–324.

Peters, R., and K. Wassenberg. 1983. The effect of body size on animal abundance. *Oecologia* 60:89–96.

Philpott, S. M., S. Albuquerque, P. Bichier, H. Cohen, M. H. Egerer, C. Kirk, and K. W. Will. 2019. Local and landscape drivers of carabid activity, species richness, and traits in urban gardens in coastal California. *Insects* 10:112.

Philpott, S. M., and P. Bichier. 2017. Local and landscape drivers of predation services in urban gardens. *Ecological Applications* 27:966–976.

Philpott, S. M., J. Cotton, P. Bichier, R. L. Friedrich, L. C. Moorhead, S. Uno, and M. Valdez. 2014. Local and landscape drivers of arthropod abundance, richness, and trophic composition in urban habitats. *Urban Ecosystems* 17:513–532.

Philpott, S. M., A. Lucatero, P. Bichier, M. H. Egerer, B. B. Lin, S. Jha, and H. Liere. (2020). Changes in natural enemy-herbivore networks along an urban agroecosystem management gradient. *Ecological Applications.* In press. https://doi.org/10.1002/eap.2201

Pickett, S. T. A., and M. L. Cadenasso. 2009. Altered resources, disturbance, and heterogeneity: A framework for comparing urban and non-urban soils. *Urban Ecosystems* 12:23–44.

Poisot, T., N. Mouquet, and D. Gravel. 2013. Trophic complementarity drives the biodiversity-ecosystem functioning relationship in food webs. *Ecology Letters* 16:853–861.

Polis, G. A., and R. D. Holt. 1992. Intraguild predation: The dynamics of complex trophic interactions. *Trends in Ecology & Evolution* 7:151–154.

Ponge, J.-F. 2003. Humus forms in terrestrial ecosystems: a framework to biodiversity. *Soil Biology and Biochemistry* 35:935–945.

Ponisio, L. C., M. P. Gaiarsa, and C. Kremen. 2017. Opportunistic attachment assembles plant–pollinator networks. *Ecology Letters* 20:1261–1272.

Pothukuchi, K., and B. Thomas. 2004. Food deserts and access to retail grocery in inner cities. *Community News & Views* 17:6–7.

Potter, A., and G. LeBuhn. 2015. Pollination service to urban agriculture in San Francisco, CA. *Urban Ecosystems* 18:885–893.

Potts, S. G., Biesmeijer, J. C., Kremen, C., Neumann, P., Schweiger, O., & Kunin, W. E. (2010). Global pollinator declines: trends, impacts and drivers. *Trends in Ecology & Evolution*, 25(6), 345–353.

Quistberg, R. D., P. Bichier, and S. M. Philpott. 2016. Landscape and local correlates of bee abundance and species richness in urban gardens. *Environmental Entomology* 45:592–601.

Raupp, M. J., P. M. Shrewsbury, and D. A. Herms. 2009. Ecology of herbivorous arthropods in urban landscapes. *Annual Review of Entomology* 55:19–38.

Ricketts, T. H., J. Regetz, I. Steffan-Dewenter, S. A. Cunningham, C. Kremen, A. Bogdanski, B. Gemmill-Herren, S. S. Greenleaf, A. M. Klein, M. M. Mayfield, L. A. Morandin, A. Ochieng, and B. F. Viana. 2008. Landscape effects on crop pollination services: are there general patterns? *Ecology Letters* 11:499–515.

Sattler, T., P. Duelli, M. Obrist, R. Arlettaz, and M. Moretti. 2010. Response of arthropod species richness and functional groups to urban habitat structure and management. *Landscape Ecology* 25:941–954.

Schellhorn, N. A., F. Bianchi, and C. L. Hsu. 2014. Movement of entomophagous arthropods in agricultural landscapes: Links to pest suppression. *Annual Review of Entomology* 59:559–581.

Schleuning, M., J. Frund, and D. Garcia. 2015. Predicting ecosystem functions from biodiversity and mutualistic networks: an extension of trait-based concepts to plant-animal interactions. *Ecography* 38:380–392.

Seto, K. C., B. Guneralp, and L. R. Hutyra. 2012. Global forecasts of urban expansion to 2030 and direct impacts on biodiversity and carbon pools. *Proceedings of the National Academy of Sciences of the United States of America* 109:16083–16088.

Shwartz, A., A. Turbé, L. Simon, and R. Julliard. 2014. Enhancing urban biodiversity and its influence on city-dwellers: An experiment. *Biological Conservation* 171:82–90.

Siegner, A., J. Sowerwine, and C. Acey. 2018. Does urban agriculture improve food security? Examining the nexus of food access and distribution of urban produced foods in the United States: a systematic review. *Sustainability* 10:2988.

Simao, M.-C. M., J. Matthijs, and I. Perfecto. 2018. Experimental small-scale flower patches increase species density but not abundance of small urban bees. *Journal of Applied Ecology* 55:1759–1768.

Smit, J., J. Nasr, and A. Ratta. 1996. Urban Agriculture: Food, Jobs and Sustainable Cities. New York, USA.

Speak, A. F., A. Mizgajski, and J. Borysiak. 2015. Allotment gardens and parks: provision of ecosystem services with an emphasis on biodiversity. *Urban Forestry & Urban Greening* 14:772–781.

Sperling, C. D., and C. J. Lortie. 2010. The importance of urban backgardens on plant and invertebrate recruitment: a field microcosm experiment. *Urban Ecosystems* 13:223–235.

Stang, M., P. G. L. Klinkhamer, N. M. Waser, I. Stang, and E. van der Meijden. 2009. Size-specific interaction patterns and size matching in a plant-pollinator interaction web. *Annals of Botany* 103:1459–1469.

Steffan-Dewenter, I. 2002. Landscape context affects trap-nesting bees, wasps, and their natural enemies. *Ecological Entomology* 27:631–637.

Steffan-Dewenter, I., and T. Tscharntke. 1999. Effects of habitat isolation on pollinator communities and seed set. *Oecologia* 121:432–440.

Stewart, A. B., T. Sritongchuay, P. Teartisup, S. Kaewsomboon, and S. Bumrungsri. 2018. Habitat and landscape factors influence pollinators in a tropical megacity, Bangkok, Thailand. *PeerJ* 6:e5335.

Suding, K. N., and L. J. Goldstein. 2008. Testing the Holy Grail framework: using functional traits to predict ecosystem change. *New Phytologist* 180:559–562.

Suding, K. N., S. Lavorel, F. S. Chapin, J. H. C. Cornelissen, S. Diaz, E. Garnier, D. Goldberg, D. U. Hooper, S. T. Jackson, and M. L. Navas. 2008. Scaling environmental change through the community-level: a trait-based response-and-effect framework for plants. *Global Change Biology* 14:1125–1140.

Taylor, J. R., S. T. Lovell, S. E. Wortman, and M. Chan. 2017. Ecosystem services and tradeoffs in the home food gardens of African American, Chinese-origin and Mexican-origin households in Chicago, IL. *Renewable Agriculture and Food Systems* 32:69–86.

Theodorou, P., K. Albig, R. Radzevičiūtė, J. Settele, O. Schweiger, T. E. Murray, and R. J. Paxton. 2017. The structure of flower visitor networks in relation to pollination across an agricultural to urban gradient. *Functional Ecology* 31:838–847.

Thies, C., I. Steffan-Dewenter, and T. Tscharntke. 2003. Effects of landscape context on herbivory and parasitism at different spatial scales. *Oikos* 101:18–25.

Tscharntke, T., Y. Clough, T. C. Wanger, L. Jackson, I. Motzke, I. Perfecto, J. Vandermeer, and A. Whitbread. 2012. Global food security, biodiversity conservation and the future of agricultural intensification. *Biological Conservation* 151:53–59.

Tscharntke, T., A. M. Klein, A. Kruess, I. Steffan-Dewenter, and C. Thies. 2005. Landscape perspectives on agricultural intensification and biodiversity – ecosystem service management. *Ecology Letters* 8:857–874.

Turo, K. J., and M. M. Gardiner. 2019. From potential to practical: conserving bees in urban public green spaces. *Frontiers in Ecology and the Environment* 17:167–175.

Turrini, T., D. Sanders, and E. Knop. 2016. Effects of urbanization on direct and indirect interactions. *Ecological Applications* 26:664–675.

Tylianakis, J. M., and A. Binzer. 2014. Effects of global environmental changes on parasitoid-host food webs and biological control. *Biological Control* 75:77–86.

Tylianakis, J. M., and R. J. Morris. 2017. Ecological networks across environmental gradients. *Annual Review of Ecology, Evolution, and Systematics* 48:25–48.

Tylianakis, J. M., T. Tscharntke, and O. T. Lewis. 2007. Habitat modification alters the structure of tropical host–parasitoid food webs. *Nature* 445:202–205.

Udy, K. 2017. Scaling of Animal Communities: From local and landscape to global processes. PhD Dissertation. Georg-August-University Göttingen, Göttingen, Germany.

United Nations. 2005. *United Nations, World Urbanization Prospects: The 2005 Revision*. United Nations, New York.

Uyttenbroeck, R., S. Hatt, A. Paul, F. Boeraeve, J. Piqueray, F. Francis, S. Danthine, M. Frederich, M. Dufrêne, B. Bodson, and A. Monty. 2016. Pros and cons of flowers strips for farmers. A review. *Biotechnology, Agronomy, Society and Environment* 20:225–235.

Valdovinos, F. S. 2019. Mutualistic networks: moving closer to a predictive theory. *Ecology Letters* 22:1517–1534.

Vázquez, D. P., N. Bluthgen, L. Cagnolo, and N. P. Chacoff. 2009. Uniting pattern and process in plant-animal mutualistic networks: a review. *Annals of Botany* 103:1445–1457.

Vázquez, D. P., C. J. Melián, N. M. Williams, N. Blüthgen, B. R. Krasnov, and R. Poulin. 2007. Species abundance and asymmetric interaction strength in ecological networks. *Oikos* 116:1120–1127.

Ver Ploeg, M., V. Breneman, T. Farrigan, K. Hamrick, D. Hopkins, P. Kaufman, B. Lin, M. Nord, T. Smith, R. Williams, and K. Kinnison. 2009. Access to affordable and nutritious food: measuring and understanding food deserts and their consequences. In Report to Congress (pp. 1–150).

Vergnes, A., I. L. Viol, and P. Clergeau. 2012. Green corridors in urban landscapes affect the arthropod communities of domestic gardens. *Biological Conservation* 145:171–178.

Violle, C., M. L. Navas, D. Vile, E. Kazakou, C. Fortunel, I. Hummel, and E. Garnier. 2007. Let the concept of trait be functional! *Oikos* 116:882–892.

Wakefield, S., F. Yeudall, C. Taron, J. Reynolds, and A. Skinner. 2007. Growing urban health: community gardening in South-East Toronto. *Health Promotion International* 22:92–101.

Werrell, P. A., G. A. Langellotto, S. U. Morath, and K. C. Matteson. 2009. The Influence of garden size and floral cover on pollen deposition in urban community gardens. *Cities and the Environment* 2:16.

Westrich, P. 1996. Habitat Requirements of Central European Bees and Problems of Partial Habitats. Pages 1–16 *in* K. Matheson, P. O'Toole, P. Westrich and I. Williams, editors. *The Conservation of Bees*. Academic Press, London.

Williams, M. R. 2009. Butterflies and day-flying moths in a fragmented urban landscape, south-west Western Australia: patterns of species richness. *Pacific Conservation Biology* 15:32–46.

Williams, N. M., E. E. Crone, T. H. Roulston, R. L. Minckley, L. Packer, and S. G. Potts. 2010. Ecological and life-history traits predict bee species responses to environmental disturbances. *Biological Conservation* 143:2280–2291.

Winfree, R., I. Bartomeus, and D. Cariveau. 2011. Native pollinators in anthropogenic habitats. *Annual Review of Ecology, Evolution and Systematics* 42:1–22.

Winfree, R., T. Griswold, and C. Kremen. 2007. Effect of human disturbance on bee communities in a forested ecosystem. *Conservation Biology* 21:213–223.

Wood, B. C., and A. S. Pullin. 2002. Persistence of species in a fragmented urban landscape: the importance of dispersal ability and habitat availability for grassland butterflies. *Biodiversity and Conservation* 11:1451–1468.

Woodcock, B. A., C. Harrower, J. Redhead, M. Edwards, A. J. Vanbergen, M. S. Heard, D. B. Roy, and R. F. Pywell. 2014. National patterns of functional diversity and redundancy in predatory ground beetles and bees associated with key UK arable crops. *Journal of Applied Ecology* 51:142–151.

Woodward, G., and D. Bohan. 2013. Ecological Networks in an Agricultural World. 1st Edition. Academic Press.

Woodward, G., B. Ebenman, M. Emmerson, J. M. Montoya, J. M. Olesen, A. Valido, and P. H. Warren. 2005. Body size in ecological networks. *Trends in Ecology & Evolution* 20:402–409.

Yuan, F., and M. E. Bauer. 2007. Comparison of impervious surface area and normalized difference vegetation index as indicators of surface urban heat island effects in landsat imagery. *Remote Sensing of Environment* 106:375–386.

Zezza A., and L. Tasciotti. 2010. Urban agriculture, poverty, and food security: Empirical evidence from a sample of developing countries. *Food Policy* 35:265–273.

3

Climate Factors and Climate Change in Urban Agroecosystems

Monika Egerer[1*]

[1] *Department of Ecology and Ecosystem Management, School of Life Sciences - Weihenstephan, Technische Universität München*

[*] *Corresponding Author: Email: monika.egerer@tum.de Address: Hans Carl-von-Carlowitz-Platz 2, 85354 Freising, Germany*

CONTENTS

KEY WORDS: *urban agriculture, climate change, ecosystem services, environmental management*

3.1 Introduction

Climate change is impacting the resilience of agriculture and food systems, affecting food security and nutrition, and threatening the adaptive capacity of agricultural livelihoods to climate variability and extremes (FAO et al. 2018). Climate change is expected to bring more frequent and extreme heat and drought events in some agricultural regions (Pathak et al. 2018), and more extreme precipitation and flooding events in others (Wassmann et al. 2004). Such climate-related events are generally detrimental to plant growth, associated biodiversity and ecosystem functions, resource availability, and crop production in agricultural systems (Daryanto et al. 2017). Thus, agricultural systems at all scales of production, from industrialized monoculture farms to diverse polycultures, are vulnerable to changing and increasingly unpredictable temperature and precipitation patterns (Tubiello et al. 2007, Ray et al. 2015).

At the same time, changes in climate are impacting the environmental integrity of urban regions and the comfort and health of the world's growing urban populations, affecting how people use and manage urban environments and natural resources (UN-Habitat 2016). With over two-thirds of the world's population living in cities by the turn of the century (UN-Habitat 2016), recent urbanization is associated with concentrated human populations and urban land cover (Seto et al. 2012), and with social changes at the population- and society-level (World Health Organization 2010). The transition to urban living can both positively or negatively impact social isolation, public health and nutrition through increased or decreased access to social amenities, fresh food and community (World Health Organization 2010). Urban environments are also characterized by altered biogeochemical cycles, urban heat effects and changes in hydrological flows and functions (Grimm et al. 2008). Urban temperatures are higher than in rural areas (e.g. by 5 to 11 degrees) due to reduced heat reflectance and transport from built materials, reduced evaporation, and air pollution and sky obstruction that traps heat (Oke 1988). Alterations in temperature and precipitation in combination with land use change in urban areas affect biodiversity and associated ecological functions through effects on the thermal physiology and functioning of organisms, including urban plants (Duffy and Chown 2016, Kendal et al. 2018) and animals (Diamond et al. 2015). The current and forecasted effects of changing temperature and precipitation patterns associated with climate change are predicted to become stronger, impacting ecosystem properties and processes at the local ecosystem to regional spatial scales.

Temperature variability and more extreme heat also affect how human populations experience everyday urban life and are prompting behavioral adaptation strategies to perceived and expected changes in climate. The use and design of urban green and blue spaces for recreation or socializing varies with temperature variability and high heat (Nikolopoulou and Steemers 2003, Thorsson et al. 2004). How people respond to perceived current and forecasted temperature variability or extremes in behaviors and actions depends on people's social-environmental context, perceptions, values, and experience (Adger et al. 2005, Grothmann and Patt 2005). Larger-scale adaptation through societal adaptative behaviors is also occurring through governance and policy changes. Adaptive actions are in turn intimately tied to the form and functions of natural systems and landscapes (Berkes and Folke 1998), ultimately affecting system resilience—the capacity of a social-ecological system to self-organize and persist under changing conditions (Carpenter et al. 2001, Folke 2006). Behavioral adaptation of individuals and of larger-scale society-level governance bodies to change will be necessary to reduce social and environmental vulnerability (i.e., the susceptibility to be harmed; (Adger 2006) as climate change increasingly affects urban residents and urban society.

Urban agriculture is a widespread land use within cities that supports biodiversity of plants and animals, enhances fresh food access, and builds social cohesion among urban communities to deliver ecosystem goods and services (Lin et al. 2015). Urban agroecosystems including home gardens, community gardens, and market farms provide: habitat and resources for associated beneficial arthropods and birds involved in complex ecological networks (see Chapter 1–2); opportunity for human-nature interactions by mimicking structurally and functionally complex natural systems that promote human engagement with plant and animal biodiversity (see Chapter 6); and an opportunity to foster understanding of environmental processes such as soil functioning and climate patterns (Lin, Egerer, and Ossola 2018). The intensive management of urban agroecosystems can yield high productivity – nearly double that of rural agriculture (Mcdougall et al. 2018) – to improve access to fresh foods key to human health and most essential to food-insecure populations (Zezza and Tasciotti 2010). The social nature of urban agriculture also produces many perceived cultural services, including social cohesion, knowledge, and education (Glover et al. 2005, Yotti et al. 2006). Thus, urban agriculture engenders both high ecological and social functions that make urban agroecosystems foundational novel social-ecological systems in cities where human-environment interactions affect ecosystem properties and functions (Egerer, Ossola, et al. 2018), and the social benefits reaped from them (Ossola et al. 2018).

Similar to rural agriculture, recent global environmental change factors stemming from climate changes are impacting urban agriculture and the services that it delivers (Webster et al. 2017, Nogeire-McRae et al. 2018). Unprecedented urban heat and drought events can damage crops, reduce soil functioning, and prompt strict water reduction policies that limit the availability of necessary irrigation to maintain crops (Wortman and Lovell 2013). Extreme flooding events may leave soils depleted of nutrients and uncultivatable (Trenberth 2008). People play a decisive role in urban agriculture's resilience and sustainability through adaptive water, soil, and vegetation management under different and changing climate conditions. Urban agroecosystems may be sites of intensive management of system biophysical properties to revert past legacies to increase ecosystem functions (Egerer, Ossola, et al. 2018); they may also be sites where people individually and collectively learn through experience over time to build adaptive capacity and resilience to social-environmental disturbances (Barthel and Isendahl 2013).

Despite knowing the substantial impacts of climate factors in agriculture and increasingly also in cities on plants, animals, and ecosystem functions, it is still unclear how climate-related factors and climate change factors impact: (i) urban crop production systems; (ii) many ecosystem functions in urban agroecosytems such as pollination and pest-regulation, and the biodiversity of animals that underly these functions (i.e. "ecosystem service providers"); (iii) peoples adaptive urban agricultural management actions in response to changes in short-term weather patterns or longer-term climate cycles; and (iv) the provisioning, regulating, supporting and cultural services consequently derived from urban agriculture under changing climate conditions. These knowledge gaps stem from the lack of urban agriculture ecological and social investigations that incorporate climate-related and climate change factors. Such gaps are concerning because urban agriculture is a proposed potential adaptation strategy to climate change stressors (Clarke et al. 2019). Thus, furthering our understanding of how climate-related factors and climate change impacts biodiversity of plants and animals, ecosystem functions and services, management decisions, and human-centered benefits can better inform climate resilient urban agriculture production, policy and planning.

This chapter draws from knowledge of climate-related factors and climate change impacts in agroecosystems and urban ecosystems to discuss how these factors – particularly increasing temperature variability and high temperature extremes associated with climate change and exacerbated by urban heat – impact ecological and social facets of urban agroecosystems. Here, "climate" relates to the prevailing meteorological conditions of a region, including temperature, precipitation, humidity, solar radiation, etc.; "climate change" relates to the change in these long-term weather conditions at an unprecedented rate, particularly in regards to the rise and variability in global temperatures and more frequent weather-related extreme events (IPCC 2014). Rather than boxing "climate" related impacts within categories, this chapter considers weather patterns and climate cycles to vary across spatial and temporal scales; the impacts of weather and climate event frequency, intensity, duration, seasonality, and reoccurrence on urban agriculture and peoples' ability to adapt to them will occur at different spatial and temporal scales. This review first discusses impacts to the ecological dimensions of urban agroecosystems, including cropping systems, ecosystem functions, and ecosystem service providing animals that underly these functions. It then discusses impacts to social dimensions of urban agriculture, including human-centered benefits, agricultural management and climate adaptation behaviors.

3.2 Impacts of Climate Factors on Ecological Dimensions of Urban Agroecosystems: Cropping Systems, Ecosystem Functions, and Ecosystem Service Providing Animals

Urban agroecosystems support high crop diversity and animal diversity that are essential to crop production, including soil-dwelling organisms, arthropods and avifauna that provide key ecosystem functions of soil decomposition, pollination, and pest regulation (Lin et al. 2015, Liere et al. 2017). As temperatures in cities increase through climate change and urban heat, it is unclear how plant and animal communities and ecosystem functions, including pollination and pest regulation will change. The following sections discuss climate-related effects in urban agroecosystems on: (i) planted crops, which are the basis for food provisioning and cultural services; (ii) key regulating and supporting ecosystem functions, including pollination and pest-regulation and their associated animals; and (iii) soil-based properties and processes as an important site of human management and generator of ecosystem services. First, the sections draw from known climate impacts to agroecosystems from the agricultural literature to discuss how those findings may or may not apply to the urban agroecosystems, and known impacts to urban ecosystems from the urban literature are used to discuss how those findings may or may not be applied to urban agroecosystems. Second, the sections highlight research findings on climate-related or climate change impacts specifically in urban agroecology research. Third, the sections underline research gaps where climate has not been considered, and where considering climate may change our understanding of urban agrecosystems.

3.2.1 Impacts on Urban Crop Systems

In agriculture, crop survivorship and productivity are affected by the climatic characteristics of the environment through direct effects on plant ecophysiology and tolerances, or by indirect effects through effects on associated biodiversity and functions (pests, diseases). Crop exposure to extremely high temperatures can scorch leaves, desiccate fruit, affect phenology, and damage pollen (Teixeira et al. 2013). Changing weather patterns and climate-related factors can influence crop species composition, vulnerability of crops to pests and disease, and ultimately crop yield and food production. In both agricultural and urban systems, the interactions between local weather patterns, regional climate cycles, and human management facilitate plant species range shifts, allowing for plants to grow outside of their fundamental climatic niche. Environmental conditions, socioeconomic factors, and agricultural policies (e.g., the Colombian Exchange, NAFTA) influence crop selection and diversification across regions (Martin et al. 2019). In urban areas, warmer urban temperatures due to urban heat effects and climate change can be thermal physiological filters (Chown and Duffy 2015, Aronson et al. 2016), selecting for thermophilic native plants and allowing for non-native plants of warmer and tropical regions introduced by people to thrive (Williams et al. 2015). Urban residents' planting and intensive management (e.g., irrigation) shapes and maintains urban plant composition under different climate contexts (Jenerette et al. 2016). However, there is still a climate signature on urban plant composition due to species climatic niches at a longer-term and regional-scale (Kendal et al. 2018), and on species hydrological niches on the shorter-term seasonal scale and at the habitat-scale (Lewis et al. 2019).

Thus, the maintenance of crop diversity in an urban agroecosystem is similarly related to local to regional climate factors, including extremely high temperatures and temperature variability that occur within and over cropping seasons. One experimental study found that common varieties of garden ornamentals are more or less resilient to the effects of extreme drought conditions (Lewis et al. 2019). This work demonstrates that even highly-related garden cultivars vary in their stress tolerance and stress recovery likely due to narrow ecological niches of garden ornamentals. One empirical study in Melbourne, Australia found correlative evidence that urban allotment garden plots exhibiting higher interday temperature variability had fewer plant species including varieties (Egerer, Lin, Threlfall, et al. 2019). This work suggests that garden plot-level temperatures, local seasonal weather patterns, and regional climate changes can affect the diversity of crop species that survive and thrive in urban agriculture either directly through plant physiology or indirectly through gardener decision making related to weather and climate. Though topographic heterogeneity through greater vegetation cover and structural diversity in urban agriculture may allow for many different plants to grow by mediating heat loading and radiation to stabilize temperatures, or by creating diverse microclimates that may be unique to these systems, conditions at the regional scale due to extended high heat events and drought conditions exacerbated by urban heat island could still affect crop productivity and survivorship if temperatures reach above crop thermal tolerances and if varieties have low stress tolerances. Urban heat may also change the plant species able to grow if temperatures increase and interday temperature variability decreases due to nighttime warming. The sensitivity of cultivated garden plants to extreme weather events and climate change effects even in heavily managed systems further suggests that the diversity of species as we know today may be largely impacted (Bisgrove and Hadley 2002, Webster et al. 2017). Crop physiologies and phenology could change, hardiness zones could shift, extreme temperatures could demand more irrigation, stress tolerances could be reached, and we may see high declines in the diversity or shifts in species composition of cultivated crops in urban agriculture (Wortman and Lovell 2013).

Measuring the mechanisms behind crop survival and productivity under changing local microclimate conditions of urban environments (e.g. urban heat) will inform how future climate scenarios in cities will affect plant species taxonomic, functional and phylogenetic diversity supported in urban agroecosystems. Evidence for the signature of regional climate on crop productivity is demonstrated in simulations of global urban agricultural food production (Clinton et al. 2018), but empirical food-yield studies in urban agriculture have not measured local ecosystem temperature variability, regional urban heat effects, and natural precipitation (e.g. (Mcdougall et al. 2018). Thus there is little empirical basis for how to assess the role of temperature and precipitation patterns on crop yield, or how climate change will affect the sustainability of urban food production in rapidly developing cities that grapple with water limitation and exacerbated urban temperatures (Eriksen-Hamel and Danso 2010, Nogeire-McRae et al. 2018). Field-based studies should measure crop damage and fruit volume of plants in relation to humidity, temperature extremes and variability in the urban agroecosystem along with regional urban heat. This combination of information would provide a more mechanistic understanding of how urban climate variability and change at both the plot scale and at the regional scale affect primary production, pest and pathogen resistance that affect food delivery – a key motivation for and benefit to many urban farmers and gardeners.

3.2.2 Pollinators and Pollination

Pollinator communities are essential to the pollination of crops in agriculture, contributing to over one-third of crop production (Millennium Ecosystem Assessment 2005). The effects of climate-related factors on pollinators in urban agriculture can be gleaned from work in systems including rural agriculture and urban grasslands. In rural agricultural systems, changing temperatures influence bee foraging behavior (Harrison and Winfree 2015) and seasonal emergence with implications for pollination and crop production (Fründ et al. 2013). In urban environments, thermal physiological filters of arthropods predict that pollinators should have more narrow thermal tolerance ranges and earlier peak foraging times than in rural environments (Diamond et al. 2015). Empirical work shows that urban bee communities are sensitive to combined effects of climate change and urban heat because of these physiological limits that determine heat tolerance (i.e., critical thermal maximums, CT_{max}) (Hamblin et al. 2017), desiccation tolerance (i.e., critical water content, CWC), and thermal and hygric (moisture) safety margins (Burdine and McCluney 2019). Community assembly of wild pollinators can be influenced by urban temperature extremes due to lower CT_{max} that make species more sensitive to urban heat (Hamblin et al. 2017). The temporal aspect of community assembly predicts lag effects of urban heat and climatic change on current and future pollinator community composition. Temperature increases, urban heat, and drought will extirpate vulnerable yet functionally important pollinator species from urban areas even with habitat management that provides local resource availability (Hamblin et al. 2018) if novel conditions push species to the edge of their thermal and hygric tolerances (Burdine and McCluney 2019). On the other hand, thermal regimes may likely favor pollinator species that are currently at the low end of their thermal/hygric tolerances.

The effects of urban temperatures and extreme abiotic conditions on pollinator community assembly processes further suggest implications for plant pollination. Urban heat can modify plant phenology by affecting interactions between temperature, photoperiod and soil moisture that trigger flowering (Neil and Wu 2006). Phenological change such as earlier flowering can affect urban pollinator behavior and subsequent plant-pollinator interactions (Harrison and Winfree 2015) to negatively effect urban crop production (Werrell et al. 2009). For example, the temporal overlap of plants and their pollinators may shift with temperature to lower pollination effectiveness, and impacts of high temperature and increasing urban land cover on pollinators (Fitch et al. 2019) may reduce pollinator foraging and movement. While urban gardeners and farmers can support pollinator populations with floral additions (Werrell et al. 2009, Quistberg et al. 2016, Plascencia and Philpott 2017), these may be short term tactics whereas climate change will drive long term community assembly processes due to effects on pollinator physiology and behavior. Yet because most pollinator or pollination investigations in urban agroecology do not include local agroecosystem temperature variability and extremes or regional climate (but see Fitch et al. 2019), we do not know exactly how crop physiological changes will affect pollination function and fruit production as forecasted in rural agroecosystems. Predictions in rural agroecosystems for pollinators may not apply because urban areas generally have higher landscape heterogeneity with many patchy floral resources, and likely more extreme temperature events than in rural areas due to, for example combined effects of extreme heat and urban heat island. Furthermore, though most crop plants are generalist in their pollination relationships which could suggest little effect of phenological mismatching (Bartomeus et al. 2011), urban agroecosystems often harbor high diversity of crop species that may rely on specialist pollinators (Galluzzi et al. 2010, Baldock et al. 2019). Thus, future urban agroecology research should examine combined effects of temperature variability and urban heat, along with traditional metrics of land management on community assembly and pollination function to determine if shifts in pollinator species composition negatively affect pollination services. This research is necessary to manage for pollination services supporting urban food security in the future.

3.2.3 Pests, Natural Enemies, and Pest Regulation

Natural enemies and natural pest regulation of crop insect pests and disease are necessary for sustainable, pesticide-free urban agriculture (Oberholtzer et al. 2014). In agriculture, insect pests and crop diseases are diverse in form, function, and magnitude of impact on crop production. Chewing, sap-feeding, and mining insect pests reduce crop health and productivity through direct plant consumption or plant indirect responses. Bacterial, viral, and fungal diseases infect plants through soil, water or air transport and lead to vascular tissue wilt, decay, mold, and fruit mold and rot. The severity of attack differs by pest and disease species, but also by local temperature, humidity, and heat because these factors also drive disease spread and cycles (Colhoun 1973). Temperature generally directs herbivore development, survival, range and abundance (Bale et al. 2002). Herbivores may increase in abundance with changes in temperatures and precipitation that prolong their mating, prevent diapause, or enhance the availability or quality of hosts (Bale et al. 2002). In contrast, natural enemies may decrease in success in locating their hosts or regulating herbivores because of distributional shifts in plants, hosts and natural enemies. With warming in some latitudes, herbivore pests may expand their range as they track crop hosts, and escape their natural enemies (Thomson et al. 2010).

Studies in urban agroecology illustrate the predicted fluctuations in pest populations over time over the growing season likely due to temperature and precipitation changes. For example, cabbage aphid populations on *Brassica* in urban community gardens in California were at their peak in the colder, wetter months than in the hotter and drier months (Egerer, Liere, et al. 2018); and urban farms in Illinois, cabbage aphid densities peaked later in the growing season (August), and seasonality significantly affected natural enemy abundance (parasitoids, predators) (Lowenstein and Minor 2018). These cases in two different climate regimes (arid/Mediterranean versus humid continental) suggest the potential effects of broader climate context on pest regulation, and the role of temperature variability in potentially influencing pest dynamics. Such examples of pest fluctuations are often highly challenging for urban farmers and gardeners because most urban agricultural systems are diverse in their local vegetation (thus farmers may deal with a more diverse set of pests) and are managed organically either in principle or by requirement (Oberholtzer et al. 2014). Although urban farmers use a suite of integrated pest management approaches and biological control tactics to combat crop pests (e.g., insecticidal plant, natural enemy, bacteria biological controls; (Altieri et al. 1999), information is lacking on how to control pests given the high humidity, altered solar irradiance, and lack of air movement in cities that can increase crop pest and disease incidence and impact severity in urban agriculture (Eriksen-Hamel and Danso 2010). Furthermore, it is unclear how climate change may influence pest populations and their life history strategies that in turn influence pest outbreaks in urban agriculture. For example, in urban agriculture in California, the Bagrada Bug (*Bagrada hilaris*) has expanded its northerly range in part due to warming temperatures, dispersal across a fragmented urban landscape, and few natural predators to cause severe crop losses (UC IPM 2013). There is unfortunately little urban agroecological knowledge to manage this pest in urban contexts.

Regulation of pests in urban agroecosystems can be provisioned through natural enemy communities (e.g. beetles, spiders, parasitoid wasps, insectivorous birds), yet climate factors similarly affect these communities with implications for pest regulation. Urban ecology research shows that temperature variability and habitat temperature extremes can filter for arthropod natural enemy species with high temperature tolerances that are more thermophilic (e.g., ants, (Menke et al. 2011); carabid beetles, (Piano et al. 2017). The water balance of natural enemies (i.e. arthropods) within urban habitats is also affected by both temperature and natural precipitation events (McCluney et al. 2017). In urban agroecosystems, temperature variability, natural precipitation patterns, and interactions between urban form and urban heat could explain inconsistencies in local and landscape drivers of natural enemies and pest regulation in investigations across geographic regions and drastically different climate regimes (e.g., Mediterranean versus humid continental), though often climate features were not measured. For example, Coccinellids thrive in urban community gardens in highly urbanized landscapes in arid climates versus continental climates likely due to the local availability of water resources provided through irrigation (Egerer, Li, et al. 2018). Subsequently, pest control may be higher in these systems in comparison to those in other climate regions (e.g., sub-tropical) where water resources are unlimited in the landscape (Morales et al. 2018), and vegetation management factors affecting natural enemies and pest control may consequently be very different. Thus while variability and extremes in temperature and precipitation will have a range of effects on natural enemies and pest regulation in rural agriculture (Thomson et al. 2010), such comparative examples illustrate that lessons from rural agroecosystems are not always readily applied to urban agroecosystems. This is particularly likely under changing climate conditions.

In sum, in most urban agroecological research, climate-related factors are only tangentially mentioned as potential factors driving pest and natural enemy populations, and are generally not explicitly tested as explanatory variables. Climate change forecasts of pest outbreaks and crop losses worldwide (Deutsch et al. 2018) should motivate research on the unknown climate-related effects on pests, natural enemies, and pest regulation in urban agroecosystems. Work in rural agroecosystems and from greenhouse simulations can certainly inform how pests respond to climate-related factors, yet to provide an understanding in situ, research in urban agroecosystems should measure crop damage, pest and predator populations in tandem with local temperature fluctuations, precipitation availability, and humidity over the duration of the season.

3.2.4 Soil-Based Ecosystem Functions

Climate-related effects on abiotic biophysical properties in agroecosystems widely impact soil-based functions essential to crop production. Predicted by Jenny's state factor framework, agricultural soil-based processes, including nutrient cycling, hydrological flows and evaporation are affected by climate because temperature controls rates of chemical processes (e.g., chemical reactions, nutrient cycles) and biological processes (e.g., organismal metabolism) (Jenny 1941). Agricultural land use transforms hydrological flows and interacts with external climate factors to change water flows (Gordon et al. 2008). Thus, urban agroecosystem soil processes and water balance results from the combination of and interactions among climate factors, stressors, and agroecosystem management (Beniston and Lal 2012, Edmondson et al. 2014). Biological and geochemical processes in urban agroecosystem soils are intensively managed through soil and ground cover amendments that distinguish them from agricultural and natural soils through, for example, higher soil organic matter (Egerer, Ossola, et al. 2018, Tresch et al. 2018). However, both large scale climate-related factors (urban heat) and local scale temperature variability still strongly affect urban agroecosystem soil organic matter accumulation, decomposition, nutrient cycling and moisture retention (Tresch et al. 2018). One study in Zurich, Switzerland found that urban agroecosystem soils increase in basal respiration, decomposition rates, and labile SOM compounds in systems experiencing more extreme urban heat (Tresch et al. 2018). Soil moisture loss rates – an indication of water storage – are faster under more extreme high temperatures measured at the local plot-scale (Lin, Egerer, Liere, et al. 2018b). Soil hydrological flows are also a function of increasing ambient temperatures through irrigation practices (Lin, Egerer, Liere, et al. 2018a). In sum, the combination of macroclimate and nuances in local temperatures and water balance can affect soil physical, chemical, and biological properties and processes that all influence other above-ground agroecosystem processes (pollination, pest regulation). More variable temperatures and exacerbated urban heat are already challenging urban gardeners and farmers due to rapid losses in soil moisture under high heat (Egerer, Lin, et al. 2018, Lin, Egerer, Liere, et al. 2018b), but urban water availability regulated through urban policy mechanisms will further necessitate adaptive management and investigations on temperature variability and soil water conservation in the context of policy-level changes (Wortman and Lovell 2013).

3.3 Impacts of Climate Change on Social Dimensions of Urban Agriculture: Human-Centered Benefits, Agroecosystem Management, and Adaptation

Intersections between climate-related factors and agroecosystem management will grow in relevance as temperature variability and high heat in urban areas increase in normalcy under climate change. Changes in temperature and precipitation will have broad implications for urban food production, urban food security and the provisioning of related human benefits. Ecosystem goods and services are derived from the translation of ecosystem properties and processes into ecosystem functions that benefit human wellbeing (De Groot et al. 2002), and urban agriculture is considered "natural capital" for producing goods and services in cities (Van Veenhuizen and Danso 2007). In a recent global geospatial assessment, Clinton and colleagues estimated the value of all ecosystem services urban agriculture provides to range from $88 to $164 billion (USD), with $78 to $150 million estimated for urban food production alone (Clinton et al. 2018). However, the authors warn about the likely loss of these services under climate change. The following sections discuss the human-centered benefits provided by urban agriculture considering current and forecasted climate changes.

3.3.1 Changes in Cultural Services Related to Climate Factors

Extreme heat and temperature variability associated with climate change will affect how people use and benefit from green spaces in cities (Gill et al. 2007, Jennings et al. 2017). Changes in space usage have implications for the ability of urban agriculture to generate its numerous social benefits. For example, social interactions are a fundamental component of community gardens (Yotti et al. 2006, Firth et al. 2011) as is the opportunity for people to connect with natural elements and learn about the natural world, including weather patterns and climate change processes (Lin, Egerer, and Ossola 2018). Many people perceive mental and physical health improvements through urban agricultural activity (Kingsley et al. 2009, Egerer, Ordóñez-Barona, et al. 2019). The habitat-scale microclimates that urban agroecosystems create through vegetation management can differ from the surrounding urban matrix to also deliver cooling benefits to local garden users and surrounding neighborhoods (Bolund and Hunhammer 1999, Oliveira et al. 2011). Cooling benefits and microclimate regulation can further provide substantial health benefits to urban residents if people perceive these as colder, more preferred and valued spaces than other areas in the urban landscape. However, urban warming and increasing temperature variability could reduce these benefits if more extended and more frequent extreme heat events limits time and physical activity within gardens, and if gardeners opt to stay indoors instead. Less time spent in these spaces could reduce cultural service provision and wellbeing benefits while food production becomes more variable under changing weather conditions. Thus the alteration of health benefits is likely under different climate scenarios and should be considered in wellbeing and ecosystem service assessments through, for example measuring how self-reported personal well-being scores and motivations to participate in urban agriculture change after extreme weather events, or using semi-structured interviews and participant observation to elucidate how social interactions and community cohesion in urban agriculture change during climate-related events.

3.3.2 Adaptation to Changes in Climate through Agroecosystem Management

Urban agroecosystem management in response to variable weather patterns, climate cycles, and perceived climate changes is a social-ecological process involving complex decisions and agricultural practices, and also influenced by local and regional agricultural and environmental policies (e.g., water access, availability). Urban gardeners and farmers quickly respond to temperature extremes by increasing the frequency and duration of crop irrigation to maintain crops (Lin et al. 2018a; Egerer et al. 2019). Under climate change, the individual's behavioral response to temperature and precipitation variability are adaptation decisions shaped by perceptions of climate-related changes, beliefs about the potential risk, and the ability to adapt given the social-environmental context. People may enact precautionary behaviors to protect against perceived imminent risk (e.g., increasing plant care through watering), or

BOX 3.1 RESILIENCY OF URBAN AGROECOSYSTEMS TO CLIMATE CHANGE

As intensively managed ecosystems and highly social systems, human-environment interactions in urban agroecosystems affect ecosystem properties and functions. The local knowledge produced through interactions moves across social networks within the community over time, engendering ecological legacies on ecosystem properties and function. **Ecological legacies** are how past events affect the current structure and functioning of ecosystems (Vogt et al. 2013); for example, soil organic matter is a legacy of past primary production or changes in hydrological regimes. Simply, ecological legacies functionally link past, present and future ecosystems. **Social memories** are how experiences in the past affect peoples' current behaviors and practices (Misztal 2003), and include learning about climate change effects via lived experiences and associated learning processes. Social memories provide experience for dealing with change, and similarly link past experience with current and future decision-making (Folke et al. 2005). **Learning** relates social memory, management and resilience in agroecosystems under changing conditions (Tosey et al. 2012); new ideas and strategies emerge through continuous reflection on experiences that shape management action. These processes in agroecosystems, in communities contribute to system **resiliency** to change (Folke et al. 2005). From an ecological perspective, agroecosystems are made resilient through biological diversification, enhanced nutrient flows, and closed loop cycling (Altieri et al. 2017). Indicators of resiliency include soil enrichment, enhanced nutrient cycling, and enhanced hydrological functioning. From a social perspective, agroecosystems are made resilient through collective memory of a community that facilitate adaptive responses, and social structures that facilitate organized action (van Oudenhoven et al. 2011). Indicators of resiliency include diverse social interactions, group participation, knowledge sharing, and learning through experience. **Social-ecological resilience** emerges from adaptive management through which people change the system (van Oudenhoven et al. 2011). In the context of climate change, adaptive management grounded in resilience theory can conceptualize management actions and learning processes that build urban agriculture's resiliency to change.

may enact mitigation behaviors in response to predicted change (e.g., deciding not to replant a species due to prior weather-related crop mortality). Trial-and-error with growing crop species in a given climate context may drive crop management and crop species composition in these systems, ultimately shaping the plant and animal biodiversity that underlie urban agroecosystem functions. Although studies reveal that people are highly responsive in water and plant management to changes in temperature or precipitation (Egerer, Lin, and Kendal 2019, Egerer, Lin, Threlfall, et al. 2019), it is unclear how people are making long-term adaptive management decisions in response to changing conditions through learning processes. Ideas around social-ecological memories and legacies are relevant to understanding peoples' agricultural management and maintaining agroecosystem function resiliency under climate change (Box 3.1). Adaptive management through learning both at the individual level and community level, and the influence of past climate-related events on urban agroecosystem functions through impacts to soil properties or plants' ecological memory are mostly unknown. Future investigations should link learning about climate events and climate change through past and current experience to agroecosystem management change to functional properties and processes of the system.

3.4 Conclusion: Climate, Climate Change, and Urban Agroecology

Climate change will increasingly impact the ecological features, biophysical processes, and social dimensions of urban agroecosystems across multiple spatial and temporal scales. In urban agroecology, we must better investigate how climate-related factors and climate change will impact plant and animal biodiversity, ecosystem functions critical to urban agro-food system productivity, and ecosystem services that contribute to urban residents' well-being. Temperature extremes exacerbated by urban warming are shifting pollinator communities, and this will impact crop-pollinator interactions and pollination function. Crop pest range expansion with warming temperatures will prompt new challenges for urban gardeners and farmers grappling with how to optimize pest regulation function in unpredictable urban environments already prone to pest outbreaks. The variability and increasing unpredictability in temperatures and water availability influence agroecosystem management and the adaptive capacity of urban gardeners and farmers. Thus, we must also understand the social dimensions of climate change in combination with the ecological or biophysical. Though climate features of cities do and increasingly will impact the functions and management of urban agroecosystems, climate-related factors are often overlooked in ecological and social urban agroecological investigations, meaning that urban agriculture's sustainability is incredibly vulnerable to climate change. The unique social and biophysical context of urban agroecosystems as the intersection of urban and agricultural land use requires integrating climatology and climate change research in urban agroecology research to elucidate the impacts on urban agroecosystem functioning and the benefits provided to urban communities (Figure 3.1).

Urban agriculture is a proposed potential adaptation strategy to climate change stressors (Clarke et al. 2019). It is essential to know how urban agroecological management responds to climate-related factors and climate change to improve sustainable resource use and the resilience of urban agroecosystems to climate-related perturbations and stressors. Long-term social-ecological research in urban agroecology could parse out biotic and abiotic mechanisms driving ecosystem functions and management under changing conditions to expand theories in ecology, agroecology, climatology, and urban ecology to urban agroecosystems. Urban agroecological research must therefore measure climate-related factors and integrate the very real effects of climate change in research frameworks and investigations. By integrating climate in research, urban agroecology can ultimately best support ecosystem functioning and peoples' adaptive management for sustainable crop production in the world's future cities.

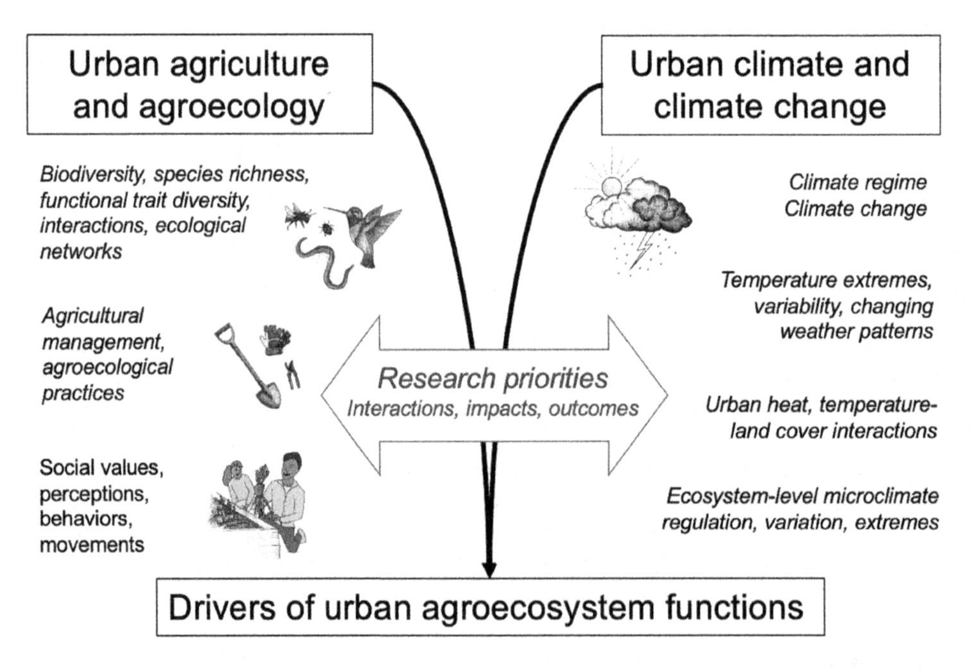

FIGURE 3.1 Under climate change, the biodiversity of plants and animals that maintain urban agroecosystem functions for urban food production can be better understood by integrating climate-related factors and the context of climate change. This figure illustrates integrating agroecological research with understanding of climate change specifically in urban areas to better promote food production and other ecosystem services and human benefits through urban agriculture.

Acknowledgments

The author wishes to thank Hamutahl Cohen, Brenda Lin, Dave Kendal, Stacy Philpott, and Alessandro Ossola for helpful suggestions and insight about the work presented in this chapter. This work was in part supported by National Science Foundation Graduate Research Fellowship under Grant No. DGE1339067.

REFERENCES

Adger, W.N. 2006. Vulnerability. *Global Environmental Change* 16(3):268–281.

Adger, W.N., Arnell, N.W., and Tompkins, E.L. 2005. Successful adaptation to climate change across scales. *Global Environmental Change* 15(2):77–86.

Altieri, M.A., Companioni, N., Cañizares, K., Murphy, C., Rosset, P., Bourque, M., and Nicholls, C.I. 1999. *The Greening of the 'Barrios': Urban Agriculture for Food Security in Cuba* 16(2):131.

Altieri, M.A., Nicholls, C.I., and Lana, M.A. 2017. Agroecology: Using functional biodiversity to design productive and resilient polycultural systems. *In*: D. Hunter, L. Guarino, C. Spillane, and P.C. McKeown, eds. *Routledge Handbook of Agricultural Biodiversity*. London and New York: Routledge, 692.

Aronson, M.F.J., Nilon, C.H., Lepczyk, C.A., Parker, T.S., Warren, P.S., Cilliers, S.S., Goddard, M.A., Hahs, A.K., Herzog, C., Katti, M., La Sorte, F.A., Williams, N.S.G., and Zipperer, W. 2016. Hierarchical filters determine community assembly of urban species pools. *Ecology* 97(11):2952–2963.

Baldock, K.C.R., Goddard, M.A., Hicks, D.M., Kunin, W.E., Mitschunas, N., Morse, H., Osgathorpe, L.M., Potts, S.G., Robertson, K.M., Scott, A. V., Staniczenko, P.P.A., Stone, G.N., Vaughan, I.P. & Memmott, J. (2019) A systems approach reveals urban pollinator hotspots and conservation opportunities. *Nature Ecology & Evolution*, 3, 363–373.

Bale, J.S., Masters, G.J., Hodkinson, I.D., Awmack, C., Bezemer, T.M., Brown, V.K., Butterfield, J., Buse, A., Coulson, J.C., Farrar, J., Good, J.E.G., Harrington, R., Hartley, S., Jones, T.H., Lindroth, R.L., Press, M.C., Symrnioudis, I., Watt, A.D., and Whittaker, J.B. 2002. Herbivory in global climate change research: Direct effects of rising temperature on insect herbivores. *Global Change Biology* 8(1):1–16.

Barthel, S. and Isendahl, C. 2013. Urban gardens, agriculture, and water management: sources of resilience for long-term food security in cities. *Ecological Economics* 86, 224–234.

Bartomeus, I., Ascher, J.S., Wagner, D., Danforth, B.N., Colla, S., Kornbluth, S., and Winfree, R. 2011. Climate-associated phenological advances in bee pollinators and bee-pollinated plants. *Proceedings of the National Academy of Sciences of the United States of America* 108(51):20645–20649.

Beniston, J. and Lal, R. 2012. Improving soil quality for urban agriculture in the North Central U.S. *In*: *Carbon sequestration in urban ecosystems*. Springer, 279–314.

Berkes, F. and Folke, C. 1998. *Linking sociological and ecological systems for resilience and sustainability. id. Linking Sociological and Ecological Systems: Management practices and social mechanisms for building resilience, Cambridge University Press, New York*, 1–25.

Bisgrove, R. and Hadley, P. 2002. *Gardening in the Global Greenhouse: the Impacts of Climate Change on Gardens in the UK. The UK Climate Impacts Programme.*

Bolund, P. and Hunhammer, S. 1999. Ecosystem services in urban areas. *Ecological Economics* 29(29):293–301.

Burdine, J.D. and McCluney, K.E. 2019. Interactive effects of urbanization and local habitat characteristics influence bee communities and flower visitation rates. *Oecologia* 1–9.

Carpenter, S., Walker, B., Anderies, J.M., and Abel, N. 2001. From Metaphor to Measurement: Resilience of What to What? *Ecosystems* 4(8):765–781.

Chown, S.L. and Duffy, G.A. 2015. Thermal physiology and urbanization: Perspectives on exit, entry and transformation rules. *Functional Ecology* 29(7):902–912.

Clarke, M., Davidson, M., Egerer, M., Anderson, E., and Fouch, N. 2019. The underutilized role of community gardens in improving cities' adaptation to climate change: a review. *People, Place and Policy* 12(3):241–251.

Clinton, N., Stuhlmacher, M., Miles, A., Aragon, N.U., Wagner, M., Georgescu, M., Herwig, C., and Gong, P. 2018. A global geospatial ecosystem services estimate of urban agriculture. *Earth's Future* 6, 40–60.

Colhoun, J. 1973. Effects of environmental factors on plant disease. *Annual Review of Phytopathology*.

Daryanto, S., Wang, L., and Jacinthe, P.A. 2017. Global synthesis of drought effects on cereal, legume, tuber and root crops production: A review. *Agricultural Water Management* 179, 18–33.

Deutsch, C.A., Tewksbury, J.J., Tigchelaar, M., Battisti, D.S., Merrill, S.C., Huey, R.B., and Naylor, R.L. 2018. Increase in crop losses to insect pests in a warming climate. *Science* 361(6405):916–919.

Diamond, S.E., Dunn, R.R., Frank, S.D., Haddad, N.M., and Martin, R.A. 2015. Shared and unique responses of insects to the interaction of urbanization and background climate. *Current Opinion in Insect Science* 11, 71–77.

Duffy, G.A. and Chown, S.L. 2016. Urban warming favours C4 plants in temperate European cities. *Journal of Ecology* 104(6):1618–1626.

Edmondson, J.L., Davies, Z.G., Gaston, K.J., and Leake, J.R. 2014. Urban cultivation in allotments maintains soil qualities adversely affected by conventional agriculture. *Journal of Applied Ecology* 51(4):880–889.

Egerer, M., Li, K., and Ong, T.Y.W. 2018. Context matters: contrasting ladybird beetle responses to urban environments across two US regions. *Sustainability* 10(1829):1–17.

Egerer, M., Ordóñez-Barona, C., Lin, B.B., and Kendal, D. 2019. Multicultural gardeners and park users benefit from and attach diverse values to urban nature spaces. *Urban Forestry & Urban Greening* 126445.

Egerer, M., Ossola, A., and Lin, B.B. 2018. Creating socioecological novelty in urban agroecosystems from the ground up. *BioScience* 68(1):25–34.

Egerer, M.H., Liere, H., Lin, B.B., Jha, S., Bichier, P., and Philpott, S.M. 2018. Herbivore regulation in urban agroecosystems: Direct and indirect effects. *Basic and Applied Ecology* 29, 44–54.

Egerer, M.H., Lin, B.B., and Kendal, D. 2019. Temperature variability differs in urban agroecosystems across two metropolitan regions. *Climate* 7(50):1–18.

Egerer, M.H., Lin, B.B., and Philpott, S.M., 2018. Water use behavior, learning, and adaptation to future change in urban gardens. *Frontiers in Sustainable Food Systems* 2(71):1–14.

Egerer, M.H., Lin, B.B., Threlfall, C.G., and Kendal, D., 2019. Temperature variability influences urban garden plant richness and gardener water use behavior, but not planting decisions. *Science of the Total Environment* 646, 111–120.

Eriksen-Hamel, N. and Danso, G. 2010. Agronomic considerations for urban agriculture in southern cities. *International Journal of Agricultural Sustainability* 8, 86–93.

FAO, IFAD, UNICEF, WFP, and WHO 2018. The State of Food Security and Nutrition in the World 2018. Building climate resilience for food security and nutrition. The State of the World. Rome.

Firth, C., Maye, D., and Pearson, D. 2011. Developing "community" in community gardens. *Local Environment* 16(6):555–568.

Fitch, G., Glaum, P., Simao, M.C., Vaidya, C., Matthijs, J., Iuliano, B., and Perfecto, I. 2019. Changes in adult sex ratio in wild bee communities are linked to urbanization. *Scientific Reports* 9(1):1–10.

Folke, C., 2006. Resilience: The emergence of a perspective for social–ecological systems analyses. *Global Environmental Change* 16(3):253–267.

Folke, C., Hahn, T., Olsson, P., and Norberg, J., 2005. Adaptive governance of social-ecological systems. *Annual Review of Environment and Resources* 30, 441–73.

Fründ, J., Zieger, S.L., and Tscharntke, T. 2013. Response diversity of wild bees to overwintering temperatures. *Oecologia* 173(4):1639–1648.

Galluzzi, G., Eyzaguirre, P., and Negri, V. 2010. Home gardens: neglected hotspots of agro-biodiversity and cultural diversity. *Biodiversity and Conservation* 19(13):3635–3654.

Gill, S.E., Handley, J.F., Ennos, A.R., and Pauleit, S. 2007. Adapting cities for climate change: The role of green infrastructure. *Built Environment* 33(1):115–133.

Glover, T.D., Parry, D.C., and Shinew, K.J. 2005. Building relationships, accessing resources: mobilizing social capital in community garden contexts. *Journal of Leisure Research* 37(4):450–474.

Gordon, L.J., Peterson, G.D., and Bennett, E.M. 2008. Agricultural modifications of hydrological flows create ecological surprises. *Trends in Ecology & Evolution* 23(4):211–219.

Grimm, N.B., Faeth, S.H., Golubiewski, N.E., Redman, C.L., Wu, J., Bai, X., and Briggs, J.M., 2008. Global change and the ecology of cities. *Science* 319(5864):756–760.

De Groot, R.S., Wilson, M.A., and Boumans, R.M.J. 2002. A typology for the classification, description and valuation of ecosystem functions, goods and services. *Ecological Economics* 41(3):393–408.

Grothmann, T. and Patt, A. 2005. Adaptive capacity and human cognition: The process of individual adaptation to climate change. *Global Environmental Change* 15(3):199–213.

Hamblin, A.L., Youngsteadt, E., and Frank, S.D. 2018. Wild bee abundance declines with urban warming, regardless of floral density. *Urban Ecosystems* 21, 419–428.

Hamblin, A.L., Youngsteadt, E., López-Uribe, M.M., and Frank, S.D. 2017. Physiological thermal limits predict differential responses of bees to urban heat-island effects. *Biology Letters* 13:20170125.

Harrison, T. and Winfree, R. 2015. Urban drivers of plant-pollinator interactions. *Functional Ecology* 29(7):879–888.

IPCC, 2014. *Climate Change 2014 Synthesis Report Summary for Policymakers.*

Jenerette, G.D., Clarke, L.W., Avolio, M.L., Pataki, D.E., Gillespie, T.W., Pincetl, S., Nowak, D.J., Hutyra, L.R., McHale, M., McFadden, J.P., and Alonzo, M. 2016. Climate tolerances and trait choices shape continental patterns of urban tree biodiversity. *Global Ecology and Biogeography* 25(11):1367–1376.

Jennings, V., Floyd, M.F., Shanahan, D., Coutts, C., and Sinykin, A. 2017. Emerging issues in urban ecology: Implications for research, social justice, human health, and well-being. Population Environment.

Jenny, H. 1941. Factors of soil formation: A system of quantitative pedology. *Courier Corporation.*

Kendal, D., Dobbs, C., Gallagher, R. V, Beaumont, L.J., Baumann, J., Williams, N.S.G., and Livesley, S.J. 2018. A global comparison of the climatic niches of urban and native tree populations. *Global Ecology and Biogeography* 1–9.

Kingsley, J. 'Yotti', Townsend, M., and Henderson-Wilson, C. 2009. Cultivating health and wellbeing: members' perceptions of the health benefits of a Port Melbourne community garden. *Leisure Studies* 28(2):207–219.

Lewis, E., Phoenix, G.K., Alexander, P., David, J., and Cameron, R.W.F. 2019. Rewilding in the Garden: are garden hybrid plants (cultivars) less resilient to the effects of hydrological extremes than their parent species? A case study with Primula. *Urban Ecosystems* 841–854.

Liere, H., Jha, S., and Philpott, S.M. 2017. Intersection between biodiversity conservation, agroecology, and ecosystem services. *Agroecology and Sustainable Food Systems* 41(7):723–760.

Lin, B.B., Egerer, M.H., Liere, H., Jha, S., Bichier, P., and Philpott, S.M. 2018a. Local- and landscape-scale land cover affects microclimate and water use in urban gardens. *Science of The Total Environment* 610–611, 570–575.

Lin, B.B., Egerer, M.H., Liere, H., Jha, S., and Philpott, S.M. 2018b. Soil management is key to maintaining soil moisture in urban gardens facing changing climatic conditions. *Scientific Reports* 8(1):17565.

Lin, B.B., Egerer, M.H., and Ossola, A. 2018. Urban gardens as a space to engender biophilia: Evidence and ways forward. *Frontiers in Built Environment* 4(December), 1–10.

Lin, B.B., Philpott, S.M., and Jha, S. 2015. The future of urban agriculture and biodiversity-ecosystem services: challenges and next steps. *Basic and Applied Ecology* 16(3):189–201.

Lowenstein, D.M. and Minor, E.S. 2018. Herbivores and natural enemies of brassica crops in urban agriculture. *Urban Ecosystems* (March), 1–11.

Martin, A.R., Cadotte, M.W., Isaac, M.E., Milla, R., Vile, D., and Violle, C. 2019. Regional and global shifts in crop diversity through the Anthropocene. *PLoS ONE* 14(2):1–18.

McCluney, K.E., Burdine, J.D., and Frank, S.D. 2017. Variation in arthropod hydration across US cities with distinct climate. *Journal of Urban Ecology* 3(1).

Mcdougall, R., Kristiansen, P., and Rader, R. 2018. Small-scale urban agriculture results in high yields but requires judicious management of inputs to achieve sustainability. *Proceedings of the National Academy of Science*, 1–6.

Menke, S.B., Guénard, B., Sexton, J.O., Weiser, M.D., Dunn, R.R., and Silverman, J. 2011. Urban areas may serve as habitat and corridors for dry-adapted, heat tolerant species; an example from ants. *Urban Ecosystems* 14(2):135–163.

Millennium Ecosystem Assessment, 2005. *Ecosystems and Human Well-being: Synthesis.* Ecosystems.

Misztal, B. 2003. *Theories of social remembering.* McGraw-Hill Education (UK).

Morales, H., Ferguson, B., Marin, L.E., Gutierrez, D.N., Bichier, P., and Philpott, S.M. 2018. Agroecological pest management in the city: Experiences from California and Chiapas. *Sustainability* 10, 1–17.

Neil, K. and Wu, J. 2006. Effects of urbanization on plant flowering phenology: A review. *Urban Ecosystems* 9(3):243–257.

Nikolopoulou, M. and Steemers, K. 2003. Thermal comfort and psychological adaptation as a guide for designing urban spaces. *Energy and Buildings* 35(1):95–101.

Nogeire-McRae, T., Ryan, Elizabeth, P., Jablonski, B.B.R., Carolan, M., Arathi, H.S., Brown, C.S., Saki, H.H., McKeen, S., Lapansky, E., and Schipanski, M.E. 2018. The role of urban agriculture in a secure, healthy, and sustainable food system. *BioScience* 1–12.

Oberholtzer, L., Dimitri, C., and Pressman, A. 2014. Organic Agriculture in U.S. *Urban Areas: Building Bridges between Organic Farms and Education. Istanbul, Turkey.*

Oke, T.R. 1988. Street design and urban canopy layer climate. *Energy and Buildings* 11(1–3):103–113.

Oliveira, S., Andrade, H., and Vaz, T. 2011. The cooling effect of green spaces as a contribution to the mitigation of urban heat: A case study in Lisbon. *Building and Environment* 46(11):2186–2194.

Ossola, A., Egerer, M.H., Lin, B.B., Rook, G., and Setälä, H. 2018. Lost food narratives can grow human health in cities. *Frontiers in Ecology and the Environment* 16(10):560–562.

van Oudenhoven, F.J.W., Mijatovic, D., and Eyzaguirre, P.B. 2011. Social-ecological indicators of resilience in agrarian and natural landscapes. *Management of Environmental Quality: An International Journal* 22(2):154–173.

Pathak, T., Maskey, M., Dahlberg, J., Kearns, F., Bali, K., Zaccaria, D., Pathak, T.B., Maskey, M.L., Dahlberg, J.A., Kearns, F., Bali, K.M., and Zaccaria, D. 2018. Climate change trends and impacts on California agriculture: A detailed review. *Agronomy* 8(3):25.

Piano, E., De Wolf, K., Bona, F., Bonte, D., Bowler, D.E., Isaia, M., Lens, L., Merckx, T., Mertens, D., van Kerckvoorde, M., De Meester, L., and Hendrickx, F. 2017. Urbanization drives community shifts towards thermophilic and dispersive species at local and landscape scales. *Global Change Biology* 23(7):2554–2564.

Plascencia, M. and Philpott, S.M. 2017. Floral abundance, richness, and spatial distribution drive urban garden bee communities. *Bulletin of Entomological Research.*

Quistberg, R.D., Bichier, P., and Philpott, S.M. 2016. Landscape and local correlates of bee abundance and species richness in urban gardens. *Environmental Entomology* 0(0):1–10.

Ray, D.K., Gerber, J.S., Macdonald, G.K., and West, P.C. 2015. Climate variation explains a third of global crop yield variability. *Nature Communications* 6, 1–9.

Seto, K.C., Guneralp, B., and Hutyra, L.R. 2012. Global forecasts of urban expansion to 2030 and direct impacts on biodiversity and carbon pools. *Proceedings of the National Academy of Sciences* 109(40):16083–16088.

Teixeira, E.I., Fischer, G., van Velthuizen, H., Walter, C., and Ewert, F. 2013. Global hot-spots of heat stress on agricultural crops due to climate change. *Agricultural and Forest Meteorology* 170, 206–215.

Thomson, L.J., Macfadyen, S., and Hoffmann, A.A. 2010. Predicting the effects of climate change on natural enemies of agricultural pests. *Biological Control* 52(3):296–306.

Thorsson, S., Lindqvist, M., and Lindqvist, S. 2004. Thermal bioclimatic conditions and patterns of behaviour in an urban park in Goteborg, Sweden. *International Journal of Biometeorology* 48(3):149–156.

Tosey, P., Visser, M., and Saunders, M.N. 2012. The origins and conceptualizations of 'triple-loop' learning: A critical review. *Management Learning* 43(3):291–307.

Trenberth, K.E. 2008. The Impact of Climate Change and Variability on Heavy Precipitation, Floods, and Droughts. *In*: *Encyclopedia of Hydrological Sciences.* Chichester, UK: John Wiley & Sons, Ltd.

Tresch, S., Moretti, M., Bayon, R. Le, Mäder, P., Zanetta, A., Frey, D., and Fliessbach, A. 2018. A gardener's influence on urban soil quality. *Frontiers in Environmental Science* 6:1–18.

Tubiello, F.N., Soussana, J.-F., and Mark Howden, S. 2007. Crop and pasture response to climate change. *PNAS* 104(50):19626–19690.

UC IPM 2013. Pest Alert: Bagrada Bug [online]. Exotic and Invasive Pests. Available from: http://ipm.ucanr.edu/pestalert/pabagradabug.html [Accessed 19 May 2019].

UN-Habitat 2016. World Cities Report 2016: Urbanization and Development: Emerging Futures. *International Journal.* United Nations Human Settlements Programme (UN-Habitat).

Van Veenhuizen, R. and Danso, G. 2007. *Profitability and sustainability of urban and periurban agriculture.* Food & Agriculture Org.

Vogt, K., Gordon, J., Wargo, J., Vogt, D., Asbjornsen, H., Palmiotto, P.A., Clark, H.J., O'Hara, J.L., Keeton, W.S., and Patel-Weynand, T. 2013. *Ecosystems: balancing science with management.* Springer Science & Business Media.

Wassmann, R., Hien, N.X., Hoanh, C.T., and Tuong, T.P. 2004. Sea level rise affecting the Vietnamese Mekong Delta: Water elevation in the flood season and implications for rice production. *Climatic Change* 66(1/2):89–107.

Webster, E., Cameron, R., and Culham, A. 2017. *Gardening in a Changing Climate*. UK.
Werrell, P.A., Langellotto, G.A., Morath, S.U., and Matteson, K.C. 2009. The Influence of Garden Size and Floral Cover on Pollen Deposition in Urban Community Gardens, 2(1):1–16.
Williams, N.S.G., Hahs, A.K., and Vesk, P.A. 2015. Urbanisation, plant traits and the composition of urban floras. *Perspectives in Plant Ecology, Evolution and Systematics* 17(1):78–86.
World Health Organization, 2010. *Hidden cities : unmasking and overcoming health inequities in urban settings*. Word Health Organization, Centre for Health Development.
Wortman, S.E. and Lovell, S.T. 2013. Environmental challenges threatening the growth of urban agriculture in the United States. *Journal of Environmental Quality* 42(5):1283–94.
Yotti, Kingsley, J., and Townsend, M. 2006. 'Dig In'' to Social Capital: Community Gardens as Mechanisms for Growing Urban Social Connectedness'. *Urban Policy and Research* 24(4):525–537.
Zezza, A. and Tasciotti, L. 2010. Urban agriculture, poverty, and food security: Empirical evidence from a sample of developing countries. *Food Policy* 35(4):265–273.

4

Restoring Soil and Supporting Food Sovereignty across Urban–Rural Landscapes: An Interdisciplinary Perspective

Coleman Rainey[1], Monika Egerer[2], Dustin Herrmann[3], and Timothy Bowles[1]
[1] Department of Environmental Science, Policy and Management, University of California Berkeley, Berkeley, California, USA
[2] Department of Ecology, Ecosystem Science/Plant Ecology, Technische Universität Berlin, Germany
[3] Department of Botany and Plant Sciences, University of California, Riverside, California, USA
Corresponding author: Email- cwrainey@berkeley.edu, Department of Environmental Science, Policy and Management, University of California Berkeley, Berkeley, California, USA

CONTENTS

KEY WORDS: *soil health, food sovereignty, urbanization, urban–rural systems, food systems, ecosystem services, critical geography*

4.1 Introduction

The health of soil is under threat. Around the world, soils are being degraded through desertification, acidification, pollution, biodiversity loss, compaction, erosion, and organic matter loss (FAO & ITPS 2015; D'Odorico et al. 2013; Lal 2007; Gomiero 2016). Soil degradation is largely a syndrome of environmental, socio-political, and cultural paradigms that drive an exploitative and unsustainable global food system (Vandermeer et al. 2018; Hunter et al. 2017; Heller and Keoleian 2003; McMichael 2011). Industrial agriculture has altered global soil carbon fluxes (Ellis 2011; Sanderman et al. 2017), disturbed soil hydrology (Vorosmarty 2000; Postel et al. 1996), polluted water resources (Bowles et al. 2018; Bennett et al. 2001), driven deforestation and desertification (D'Odorico et al. 2013), and diminished both ecosystem and soil biodiversity (Geisen et al. 2019; Isbell et al. 2019; Kremen et al. 2012). Despite undermining soil resources, intensive agriculture manages to produce more calories per capita than the world needs (Cassidy et al. 2013), largely through dependence on non-renewable inputs. Meanwhile nearly a billion people face hunger and another billion face nutrient deficiencies, creating a global agrarian crisis where underconsumption meets overconsumption, and overproduction is weaponized to engineer hunger for poor urban dwellers and rural populations that face deteriorating livelihoods (Ruel et al. 2017; Lal 2007; Chappell 2019).

Current threats to soil resources and food security are spatial by nature: they link distinct geographies through processes of capital exchange, industrialization, urban densification, and land development. A dominant paradigm motivating the spatial differentiation of land and the destruction of soil ecosystems is the urban–rural divide (Champion and Graeme 2004). Concepts of "urban" and "rural" are reinforced through modifications to both the ecological and built environment, creating strong feedbacks between land use, ecological systems, and the provisioning of food, fiber, and medicine. The socio-ecological and spatial organization of human-dominated landscapes, paired with the increasing density and size of human settlements, we will call "urbanization." Urbanization is interwoven with changes to soil, ecology, and agriculture through a number of processes including migration from food-growing communities, decreased farming labor forces, increased demand for imported foods, and material flows between cities and hinterlands (Krugman and Elizondo 1996; Ruel et al. 2017; Satterthwaite et al. 2010; Davis and Henderson 2003; Gozgor and Kablamaci 2015). Over two-thirds of the global population is projected to live in cities by 2050 and urban land use is expected to triple by 2030, primarily on former agricultural land (Bren d'Amour et al. 2017; United Nations 2014; Glaeser 2014). Following industrial agriculture, urbanization is the second leading cause of acute and accelerated soil degradation, with damages extending far beyond city boundaries through waste disposal (Marcotullio et al. 2008), air and waterborne pollutants (Pouyat and McDonnell 1991; Marsalek et al. 2006; Chen 2007), mining and resource extraction (Richards and VanWey 2015), urban–rural trade (Vorley and Lançon 2016), and urban sprawl. Across urban and rural landscapes, these patterns of ecological degradation disproportionately affect Black, Indigenous, and People of Color (BIPOC) as well as low-income communities in the rural Northern hemisphere and global South.

Urban–rural divisions also reflect shifts in diets and precarity in food security. Food systems become increasingly vulnerable to socioeconomic and public health crises when they are embedded within the built environment and diminished ecological systems. Urban development is often coupled with increased obesity and other diet-related illnesses, shifting access towards diets rich in fat, sugar, and refined carbohydrates, while contributing to hunger among poor populations that increasingly rely on food purchases (Popkin 1999; Treuhaft and Karpyn 2010; Ruel et al. 2017; Ruel and Garrett 2004). Yet poor, rural communities often face food shortages of their own as crops are exported to higher payer urban centers and rural livelihoods languish under urban-centric economies.

Taken together, threats to soil and food resources materially link urban–rural systems. Urbanization creates a "rift" in what Marx called our "social metabolism," breaking the cycling of nutrients between soil and society that ultimately supports the health of both (Foster 1999; Foster and Magdoff 1998; Marx 1959). Soil nutrients, food, capital, and labor flow across urban–rural interfaces and never return, creating a material imbalance exhibited by the stark inequalities in soil degradation and food insecurity. This overwhelmingly pits urbanization against efforts towards food sovereignty, defined as "the right of peoples to healthy and culturally appropriate food produced through ecologically sound and sustainable methods, and their right to define their own food and agriculture systems" (Nyéléni 2007).

Food sovereignty initiatives must contend with urbanization and address malnourishment as urban populations grow and increasingly rely on imported foods. To accomplish this goal, a theoretical and practical framework is needed that guides equitable soil regeneration and agricultural self-determination across urban–rural landscapes.

Discourses on food systems are increasingly calling for a holistic, action-oriented, and integrated approach to address crises such as food insecurity and soil degradation—an approach that acknowledges the complex political and cultural factors driving socio-ecological unsustainability (McMichael 2011; Vandermeer et al. 2018; Béné et al. 2019a; Béné et al. 2019b; Oteros-Rozas et al. 2019; Holt Giménez and Shattuck 2011). In this chapter, we propose that a dialectic among critical geography, soil science, and agroecology can create a starting point for how to: 1) Restore a rich suite of soil-based ecosystem services to environments impacted by urbanization; and 2) Build more equitable food systems in a rapidly urbanizing world. Our objective is to place these disciplines in conversation to inform research and action aimed at addressing the twin challenges of soil degradation and diminished food sovereignty across urban–rural systems. We approach this as interdisciplinary researchers working mainly in the United States, and we do not all have direct experiences as farmers or practitioners. For this reason, where possible, we have chosen case studies and examples from our experiences working in North America. We recognize these biases limit the scope of our work and can skew our perspectives, and thus recommend that future work on this topic should include a more inclusive range of experiences, knowledge-bases, and geographies.

4.2 Urbanization, Soil, and Food Systems

In the following sections, we contextualize the biophysical and social processes working across urban–rural landscapes that impact soil and food sovereignty. First, we provide a brief history of urban soil classification, outlining how views on their biophysical properties and ecosystem services have shifted over time, ultimately defining urban soils as socio-ecological systems. Second, taking cues from critical geography and urban studies, we briefly define urbanization and outline how it aptly describes transformations of soil resources and food systems. Third, we explain how soil and food systems link urban and rural. Finally, we outline how food sovereignty is impacted by urbanization. This lays the foundation for how critical geography can problematize and inform the work of soil science and agroecology.

4.2.1 History and Classification of "Urban" Soils

Urban landscapes were historically neglected by soil investigations. This led to early descriptions of "urban" soils as those from highly disturbed sites with substantial human constructed surface layers (Craul 1985; Patterson et al. 1980). Urban soils were narrowly defined as any unconsolidated, nonagricultural, and man-made surface layer produced by mixing, filling, or contamination of land surface in urban areas (Hollis 1991). This led to the concept of anthropedogenesis, or the human-driven formation of soils. Urban soils were generally considered degraded and characterized by several common attributes, including compaction; limited root penetration and water infiltration; novel spatial variability with abrupt changes in vertical profile properties; elevated pH; and modified regimes of biogeochemistry, temperature, and biota (Craul 1985). These understandings were largely motivated for their use in creating ornamental, aesthetic, and designed landscapes (Craul 1992).

More recent investigations placed soils in urban environments along a continuum. At one extreme are anthropogenic soils where human activity is the dominant soil-forming factor, enacted primarily through grading, filling, and terracing (Trammell et al. 2020; Jones et al. 2014) and governed by technical requirements of the built environment. These soils, or Technosols, have been classified as a novel soil type (Rossiter 2007). On the other extreme are "natural" soils, fragmented but common within cities, that are largely unaltered directly by human activity yet impacted by indirect effects such as altered hydrologic flow paths, particulate deposition, and the urban heat island effect (Kaye et al. 2006; Hope et al. 2005; Pickett and Cadenasso 2009).

A growing appreciation for the global importance of urban ecosystems and the development of urban ecology as a discipline shifted the focus on urban soils towards their ecological functioning

(Pickett et al. 2001) and ecosystem services (Morel et al. 2015; Herrmann et al. 2017; Pavao-Zuckerman 2008 Ziter and Turner 2018). Soils in urbanized landscapes are often still shaped by legacies of natural history and can support ecosystem structure and function (Herrmann et al. 2017). Soils can provide cities with a full suite of ecosystem services, including: 1) supporting services like maintaining or enhancing biodiversity (Johnson et al. 2015); 2) regulating services like sequestering carbon (Pouyat, Yesilonis, and Nowak 2006), buffering contaminants (Paterson et al. 1996), promoting hydrologic functioning (Stewart et al. 2019), and nutrient cycling (Herrmann, Shuster, and Garmestani 2017); 3) provisioning services like food production; and 4) cultural services like recreation, community building, and education (Nahuelhual et al. 2013; Grossman et al. 2012; Armstrong 2000). Soils can also play a large role in improving the response capacity of cities to climate change, potentially improving services such as flood storage, infiltration capacity, and mitigating heat island effects through supporting plant growth that provides cooling and shading (Gill et al. 2007).

Increasingly, however, societies shape and alter soils through social and ecological processes that diminish their capacity to perform these services. The dominant land-use paradigms of both "urban" and "rural" space (soil capping and industrial agriculture, purportedly) together tend to diminish soil health, and are being introduced to ecological systems in ways that reinforce the role of markets and capital in shaping landscapes. The increasing complexity of human settlements and the encroachment of the built environment have created mixed ecosystems in a patchy mosaic, all linked by the flow of soil nutrients, food supplies, and capital. This has led the fields of urban studies and critical geography to largely shift away from viewing "urban" as a fixed category, and towards an understanding of "urbanization" as a process that can work along urban–rural gradients (Brenner 2013; Brenner and Schmid 2012; Lefebvre 2003).

4.2.2 Defining Urbanization

Urbanization can be understood as the creation of human-dominated landscapes, primarily through the increasing density, agglomeration, and contiguity of human populations (Angel et al. 2018; Martino et al. 2016; Ritchie and Roser 2020; Brenner 2013). Urbanization is accelerating around the world, exhibited by similar shifts toward transnational markets, foreign direct investment, diminished local agriculture, and access to mass media that characterize globalization (Davis and Henderson 2003; Krugman and Elizondo 1996; Glaeser 2014; Gozgor and Kablamaci 2015). These processes are unfolding with sufficient dynamism and scale so as to blur the boundary between urban and non-anthropogenic environments. Urbanization is accompanied by land use change for economic and material gain, where structures of capital commodify land resources for human development to the exclusion of ecological processes (Williams and Holt-Giménez 2017; Lefebvre 2003). In the twenty-first century, urbanization and globalization connect even the most remote places to and through market forces (Krause 2013), expressing a materiality of capital and resource extraction that underpin social reproduction (Krzywoszynska and Marchesi 2020; Tornaghi and Dehaene 2020).

Insights into the material transformations during urbanization better situate soil and food resources in their wider geographic, ecological, and social context. In Oakland, CA, USA, the excavation and importation of soil from city margins and natural lands supported the industrialization of the city's economy and created novel Technosols with elevated levels of lead contamination (Nathan McClintock 2011; 2012). High levels of metals and pollutants were concentrated in fill dredged from estuaries and marshlands due to decades of accumulation from industrial effluent, landfill leachate, and hydraulic mining refuse carried down the Sacramento River (Nathan McClintock 2015). The material process of urbanization and building the "industrial garden" of Oakland brought with it trace metals from across its watershed and rural hinterlands. These patterns of contamination are accompanied by food insecurity and poverty, as a history of red-lining and other racial discrimination policies led to impoverishment and malnourishment of Black neighborhoods where industrial pollution is most concentrated (Alkon et al. 2013).

The case of soil contamination in Oakland exemplifies how urbanization works across an urban–rural continuum and along axes of perceived social difference. This continuum motivated the urban–rural ecological gradient concept, a framework that investigates urban ecology along socio-ecological axes such as population density or land cover type (Pouyat and McDonnell 1991; McDonnell et al. 2008). Studies across urban–rural gradients have often revealed non-linear, heterogeneous, or confounding soil ecosystems that obscure any coherent distinction between urban and rural. In forest soils that spanned from New York City, NY to its rural outskirts, soils closer to the city (counterintuitively) exhibited increased earthworm populations, nitrogen decomposition rates, and nitrification (McDonnell et al. 2008). In a study of 30 metropolitan areas in the United States, habitat quality and biotic homogenization were constant across urban and rural ecosystems, largely due to the homogenizing effect of croplands (Warren et al. 2018). And in Leipzig, Germany, industrial decline, migration, and de-economization have created a distinct patchwork of land use types that diminish measurable differences in soil cover between urban and rural poles of its urban land use (Haase and Nuissl 2010).

This brief outline shows how the study of urban soils has increasingly blurred the distinction between urban and rural soil ecosystems. Urban studies, critical geography, and urban political ecology have largely moved beyond false dualisms of city and nature, urban and rural, or society and nature that often carry implicit capitalist or colonial notions of modernity (Haller 2019), and towards an understanding of cities as co-produced socio-ecological systems.

4.2.3 Soil and Food Systems: Linking Urban and Rural

Urban and rural, so often seen in opposition to one another, are in fact linked by the flow of capital and material between cities and their rural hinterlands. The movement of people and resources across urban–rural interfaces creates and sustains urbanization through three primary processes: socioeconomic change, spatial relationship, and material flows.

4.2.3.1 Socioeconomic Links

The socioeconomic relationship between rural-agricultural and urban settlements is well documented. In *Capital*, Marx clearly outlined how primitive accumulation, or the systematic dispossession of rural lands, imposed wage-labor and commodity value relations upon the agrarian countryside to fuel the creation of industrial cities in the 18th and 19th centuries (Marx 1959). Urbanization is often caused by migration from rural, predominantly agricultural, settlements to more densely populated areas (Davis and Henderson 2003; Knorr, Khoo, and Augustin 2018; Condon et al. 2010; Satterthwaite, McGranahan, and Tacoli 2010). Labor shifts from agriculture to industrial and service sectors also strongly predict patterns of urbanization, to the degree that some demographers use this as a proxy measurement for urbanization (Davis and Henderson 2003). The intensification and industrialization of agriculture has allowed for fewer individuals to engage in farming: in low-income nations, agriculture comprises a third of the labor force and 40% of the GDP, but only 1–2% of the labor force and 2% of the GDP in high-income nations (Satterthwaite et al. 2010; World Bank 2000).

4.2.3.2 Spatial Relationships

There is also a direct spatial relationship between rural-agricultural ecosystems and urbanization. Urbanization, especially in the poor megacities of the global south, are taking on increasingly fractal and spider-like formations where loss of agricultural land and landscape fragmentation directly threaten food production (Makse et al. 1995; Angel et al. 2012). Future urban development will compete with soil conservation and agriculture for land resources, as projections show that the majority of urbanization will occur on former agricultural land (Bren d'Amour et al. 2017), and more than 60% of the world's irrigated croplands are located near cities (Thebo, Drechsel, and Lambin 2014). In California, USA, approximately 40,000 acres per year of prime farm and rangeland have been converted to urban land uses since 1984 (DOC 2016), coming with much higher per unit area greenhouse gas emissions (Jackson et al. 2012). These changes are likely greatest in the global south, particularly Asia and Africa, where over 90% of projected increases in the global urban population will occur. These are also the regions with high levels of poverty and high rates of food insecurity (Satterthwaite et al. 2010). Urbanization and industrial agriculture are the dominant modes of land transformation (and thus soil alteration) on Earth (Vitousek et al. 1997). Croplands, rangelands, and dense human settlements (>300 people/km^2) now cover 55% of the global land surface area, with another 30% of unused lands embedded within or fragmented by those anthropogenic ecosystems (Ellis et al. 2010). The scope and magnitude of these changes are vast and only increasing.

4.2.3.3 Metabolism and Material Flow

Often overlooked are the *material* links that urbanization forms between city and hinterland. These linkages invite new understandings of the metabolism of cities and reveal the effects that urbanization has on soil resources and food sovereignty. In a study of Lisbon, Portugal, the material flows of food imported to the city each year was 2.06 tons per capita, and excavated soil and rock leaving the city reached 6–8 tons per capita (Rosado et al. 2014). For a city with 10 million people or more, over 6,000 tons of food has to be imported every day, traveling an average of 1,000 miles (De Zeeuw et al. 2011). In California, USA, recent regulations mandate diversion of 75% of organic waste from landfills by 2025, much of which will go from cities to composting facilities located in rural areas (Harrison et al. 2020). While compost provides an important source of organic matter and nutrients for soil, composting facilities themselves create air quality concerns in rural areas of California already facing some of the worst air quality in the country (Kumar et al. 2011). And rural, largely undocumented, agricultural labor in the United States face disproportionately high rates of hunger as valuable food crops are exported to high-paying urban customers (Brown and Getz 2011).

The metabolic patterns of society that impact soil health and food sovereignty—the material exchange of water, nutrients, food, waste, and people across space and time—are best understood in relation to urban–rural systems. Capital production and industrialization disconnect individuals from the natural resources that support their livelihoods, and impedes the return of nutrients derived from food and fiber to the soils that produced them (Foster 1999). The scale of urbanization's impact on soil and food systems, and its increasing fragmentation and complexity, invites a more critical perspective that views urban–rural systems as a dynamic membrane of material interfaces and linkages. Urban–rural perspectives have gained traction among some scholars and activists who see the urgency in translating agrarian reform and food systems analysis to urban environments (Tornaghi 2017; N. McClintock 2010; Holt Giménez and Shattuck 2011; Nyéléni 2018). The impact of urbanization and agriculture on diverse ecosystems demonstrates strong linkages and interfaces across urban–rural systems. Food, fertilizer, soil, and people flow between "city" and "countryside" through processes of capital accumulation, labor and natural resource exploitation, and consumerism.

4.2.3.4 Urbanization and Food Sovereignty

The rapid growth of cities and resultant soil degradation is also interwoven with dramatic changes in urban food access and increasingly vulnerable urban food systems. Patterns of food consumption in cities are converging around the world, moving from more local or traditional foods towards increased dietary diversity, but diets rich in fat, sugar, and refined carbohydrates (Mendez and Popkin 2004). These dietary shifts reflect increased rates of obesity and other diet-related chronic diseases (Dixon et al. 2007), while economic inequalities cause malnutrition among low-income urban populations that rely on food purchases (Alkon et al. 2013; Ruel and Garrett 2004; Ruel et al. 2017; Handbury et al. 2015).

China is an illustrative case study of the materiality that connects soil resources and food sovereignty in urban–rural systems. Between 1979 and 1998, urban populations rose by 222 million in China, 75% of which was due to rural-urban migration (Zhang and Song 2003). This has led to the dramatic abandonment of farmland in those regions of China (Shi et al. 2016), with serious impacts on both soil health and food sovereignty in urbanized areas (Mendez and Popkin 2004; Chen 2007). Pollution from China's megacities has extended to peri-urban and agricultural soils, with one study finding over 50% of sites in the Pearl River Delta suffering from heavy metal and petroleum hydrocarbon pollutants (Li and Jia 2018; Chen 2007). And these pollutants wind up back in cities: a study in 2000 found that heavy metal concentrations exceeded permissible values of the National Standard for Food Safety in over 30% of the agricultural products surveyed (Chen 2007). At the same time, urban dietary preferences had put increasing pressure on agricultural production. Obesity among Chinese adults in urban areas closely reflected the availability of imported and processed foods, and the number of adults consuming high-fat (>30%

calories from fat) diets increased from 33.0 to 60.8% in urban areas over just 20 years (Mendez and Popkin 2004). These changes are expanding beyond large urban centers and into smaller cities and towns, as western-style supermarkets are now found in smaller cities and towns along the Eastern coast and interior (Reardon Timmer and Berdegue 2003). Importantly, the largest increases in obesity were found in rural areas with high levels of urban development and infrastructure, indicating how urbanization works across urban–rural systems and allows consumers to access new food supplies that negatively impact their health.

A "nutrition transition" towards processed, high-fat, high sugar diets is driven by a host of factors linked to both urbanization and globalization: integration of the global marketplace into food systems, corporate consolidation of grocery supplies and outlets, mass media advertising, and diminishing agricultural self-sufficiency (Dixon et al. 2007; Popkin 1999; Hawkes 2006). Like soil degradation, these changes tend to have adverse and inequitable effects, disproportionately affecting developing countries in the global south and low-income communities in the global north (Dixon et al. 2007). In the United States and other wealthy nations, this often means BIPOC due to long perpetuated racial and class segregation (Eisenhauer 2002; Treuhaft and Karpyn 2010; Alkon et al. 2013).

Food sovereignty is positioned largely in opposition to these inequalities, which are overwhelmingly caused by industrial agriculture, transnational food regimes, capital exploitation, and urbanization. Food sovereignty emerged from agrarian social movements predominantly in Central America, taking hold in international peasant organizations like *La Via Campesina*, and ultimately breaking onto the international stage in the 2000s (Edelman 2014; Hannah Wittman 2011). Since then, food sovereignty has become a mobilizing frame for social movements around the world. While the term has taken on diverse and at times contradictory meanings, food sovereignty broadly describes a set of ideologies and practices aimed at transforming food and agriculture systems in ways that restore agency, self-determination, and self-sufficiency (Patel 2009). In order to realize food sovereignty in urbanized landscapes, however, a better understanding of how urbanization alters soil ecosystems is required.

4.3 How Urbanization Changes Soil Properties and Ecosystem Services

The common forces driving urbanization, despite regionally distinct economic or political outcomes, cause direct and often irreversible consequences for the physical, chemical, and biological properties of soils (Lehman and Stahr, 2007), which in turn tend to diminish soil ecosystem services (Hermann et al. 2017; Pavao-Zuckerman 2008 Ziter and Turner 2018). These negative impacts of urbanization do not affect all communities equally: soil degradation and contamination tend to be concentrated in the global south (Lal 2007) and in the poor, often non-white, neighborhoods of wealthier nations (Ernstson 2013; Nathan McClintock 2012; Wilson et al. 2017). These inequalities reaffirm the socio-ecological nature of urbanization. Through processes of land commodification, capital production, and landscape fragmentation, urbanization enacts human-driven disturbances, heterogeneities, altered fluxes, and feedbacks which in turn determine the ecological trajectories of soils across urban–rural systems. Urbanization is a dominant structuring force on soils. Several outcomes of this force are considered below to highlight major pathways for how urbanization works on soils, but it is not an exhaustive coverage.

4.3.1 Land Commodification, Soil Heterogeneity, and Convergence

Human-driven soil formation and the commodification of land resources are central drivers of urbanization and its negative impacts on soil. The fictitious view that land and soil resources can simply be exchanged for one another as commodities is driven by a logic of substitution that ignores the immense ecological, environmental, and cultural value tied to particular places (Naybor 2015). This raises a central tension between soil's two major functions in cities: first, as support for built infrastructures and human use, and second, for ecological functioning and ecosystem services. These functions are often in direct opposition. City-making activities are dominantly undertaken to create built infrastructures, and as a result, soils have been evaluated and reworked for their usefulness to this role. Environmental functions of soils have historically been evaluated in relation to this role as well (Craul 1985). Soils are intentionally altered in urban landscapes to function as a component of engineered systems. Soil properties are selected for serving as a stable geotechnical base or matrix to support building foundations, paved surfaces, and pipe networks. Ideal soil properties for supporting and protecting built infrastructures without the potential for subsidence include a uniform profile of low to intermediate soil particle coarseness (i.e., fine silty) as well as low and stable organic matter levels (ICC 2018; ASTM 2017). These properties are not well suited for biologically active, dynamic soil ecosystems.

The dominating logic of the built environment produces sharp gradients of horizontal and vertical differentiation that correlate positively with site age, degree of alteration, and land use history (D. L. Herrmann et al. 2018; A. Lehmann and Stahr 2007; Pickett et al. 2017). Urbanization creates local heterogeneity greater than that typically exerted on by natural soil forming factors. This heterogeneity is reflected in metrics such as waste and rubble admixture, inversion of soil horizons, soil sealing, compaction, nutrient cycling, and contamination. The mixing of soil with waste and rubble tends to increase calcium carbonate concentrations in soils, which can have a cementing effect, limiting infiltration and aeration (A. Lehmann and Stahr 2007) The inversion or removal of soil horizons can decrease rooting zone depth, reduce soil water reserves, limit contaminant buffering, and reduce capacity to store or cycle infiltrated storm water runoff (D. L. Herrmann et al. 2018) And soil sealing, or the capping of soil with an impermeable or semi-permeable surface layer, leads to a cascade of negative effects for soil ecosystems, most notably biodiversity loss and an increased heat island effect (Scalenghe and Marsan 2009).

The requirements of the built environment create local heterogeneity within urbanized landscapes, carving the soil into a patchwork of states determined by legacies of land use. But when all lands become the same—when land becomes a commodity—these causes and conditions are created everywhere. By externalizing ecological and cultural values from capital land exchange, soil becomes a passive medium for development that is carved the same everywhere. Paradoxically, then, the ubiquity and similarity of soil-forming decisions during urbanization tends to converge the physical, chemical, and biological properties of soils in cities across climatic zones and continents (Trammell et al. 2020; Pouyat et al. 2015; Groffman et al. 2017; Kaye et al. 2006). This is also true for managed landscapes, where soils can be shaped by such practices as plant selection, irrigation, and fertilization (Trammell et al. 2020; Groffman et al. 2017; Pouyat et al. 2015). For example, soil organic carbon levels across six climatically distinct U.S. cities converged on a surface soil (0–10 cm) content of ~2.5-3.0 kg/m^2 (Trammell et al. 2020). Common landscaping choices in urbanized areas drive biotic homogenization in plant communities and generate soils with similar moisture contents, nitrogen pools, and microbial biomass, converging the properties of reference "natural" soils to a common urbanized condition (Groffman et al. 2017). Contamination from both point and non-point sources also converge soil trajectories as altered atmospheric chemistry and deposition of pollutants and nutrients affects all city soils simultaneously (Wilson et al. 2017; Kaye et al. 2006). Driven by the logic of substitution and commodity, soil ecosystems converge through the reshaping of land with heavy equipment, similar characteristics in the choice of fill soils, soil sealing, industrial processes, automotive activities, and common residential landscaping choices along with nutrient and water inputs. Anthropedogenesis and common attitudes towards soil resources create patchworked soil landscapes with socially-determined local heterogeneities that are found across cities and continents, indicating the need for improved methods pairing social and biophysical metrics (Effland and Pouyat 1997, Swidler 2009).

4.3.2 Case Study: Community Gardens in the California Central Coast

A case study from 25 community gardens in the California Central Coast provides an example of the dynamic interplay of heterogeneity and convergence in urban soil ecosystems (Egerer et al. 2018). Through soil surveys and geographic analysis of historic land use, Egerer et al. found that community gardeners' consistent and deliberate management of their soils in the quest to optimize soil-based ecosystem services for crop production can create similar soils across a heterogeneous landscape. This case study represents an example of how 1) soil management can wash out land use legacies, and 2) novel agroecosystems emerge within cities through soil biotic homogenization. Urban allotment community gardens are pockets of green space in cities where species diversity, ecological processes, and ecosystem services may be maintained through strong social and ecological interactions. Urban gardeners carefully select crops for food, ornamental, and medicinal purposes (Baker 2004), but they also heavily cultivate the soil through purchased soil media, soil amendments, and irrigation that support the cultivated crop plant diversity of these systems. Community gardens also provide a place where people can interact with one another and co-produce ecological and biophysical outcomes, collectively learning and enacting management practices that shape soil ecologies.

The California Central Coast landscape comprises intensive agriculture, dense urban development, and forested mountains that create biophysical and climatic distinctions within and across cities. These biophysical gradients are mirrored by socioeconomic gradients that create an unjust and inequitable social landscape. Both low-income farm workers that experience high rates of food insecurity and limited land access (Brown and Getz 2011) and high-income technology workers in Silicon Valley (Pellow and Park 2002) populate the region. Land-use and demographic differences shape the diverse physical and social composition of Central Coast cities. But urban community gardens across the region embody how intensive soil management can mediate the effects of land-use histories and converge soil properties in less than 15 years. In these spaces, gardeners rent individual beds that they manage to their preference (e.g., crop choice, ornamental plantings, or levels of amendments). Variation in these soil systems was expected: diverse social and environmental factors were thought to yield a range of management decisions, and legacy effects were expected to differentiate garden soils of previous land use types in the region (e.g. agriculture or pasture, forest, housing, tree orchard, and vacant lot or open space). Despite this, however, no significant differences among sites of different types or age were found in soil nutrients (C, N), soil structure (bulk density), or soil organic matter. The urban garden soils represented emergent systems where intensive, transformative soil management progressively washed out legacy effects typical of urban soil characteristics (Raciti et al. 2012; Pouyat et al. 2015). In comparison with soils in urban parks and recreational areas, this increased management intensity may be advantageous by ameliorating poor soil quality and increasing soil fertility in unfavourable urban areas for food production.

Furthermore, the soil properties of the gardens demonstrated how raised beds can further drive soil homogenization and novelty. Raised beds with high organic matter provide a gardening system in which soils can be completely cultivated outside of the original soil profile, thereby allowing gardeners to create novel soils to their liking. Gardeners often bring base soils from other locations, such as gardening stores, which produce soils of higher soil organic matter, carbon, and nitrogen stocks. The similarity of garden bed soils across regions—each of which has different microclimate characteristics, sociodemographics, and degree of urbanization— demonstrate how soil systems converge towards very similar properties. This suggests similar agroecological practices, knowledge, and attitudes to soil cultivation across the region.

4.3.3 Capital Production and Patterns of Contamination

A second force of urbanization on soils is capital production, which impacts patterns of soil contamination. Urbanized landscapes emerge from historic processes of capital accumulation, interlinked through trade with other ecosystems around the world (Harvey 1996; Heynen, Kaika, and Swyngedouw 2006; Swyngedouw 2003). Since urbanization is governed by capital, soils and their potential ecosystem services are valued mainly by their potential return on investment (Ernstson 2013). Accumulation inevitably makes unequal distribution of soil contaminants and ecological degradation a fundamental component of urbanized landscapes (McClintock 2015). These differences tend to follow socioeconomic and racial differences. Historic processes of industrialization, zoning, and redlining in Oakland, California, have created contemporary lead contamination patterns that disproportionately affect low-income, BIPOC communities (Nathan McClintock 2011). In Charleston, South Carolina, concentration gradients of heavy metal pollutants were found to closely correlate with distance from industrial facilities, and with disparities in exposure primarily affecting Black neighborhoods (Wilson et al. 2017).

Urbanization makes material the historic cycles of capitalist development. Industrial, automotive, and construction activities associated with urbanization and construction of the built environment produce elevated levels of contamination (Wortman and Lovell 2013). Soils can retain and accumulate pollutants, notably heavy metals as legacy pollutants but also contemporary pollutants such as microplastic and per- and polyfluoroalkyl substances (PFAS) (Rodríguez Eugenio et al. 2018). Sources can be concentrated in the case of industrial sites or widely dispersed in the case of lead from car exhaust and exterior paint. Yet perceptions of, and potential interventions in, soil contamination are governed by prevailing political, economic, and social paradigms. For example, soils found to have lead pollution are less likely to be preserved for conservation or restoration (Cutts et al. 2017). This has direct implications for the communities already suffering from soil degradation and contamination. Agroecological approaches to amending soil contamination must be adopted with a lens towards equity and environmental justice, to ensure that patterns of contamination that result from racialized and discriminatory policies do not end up repeating legacies of harm (Ernstson 2013; Cook and Swyngedouw 2012).

4.3.4 Land Use and Landscape Fragmentation

The commodification of land, the loss of the commons, and the rise of private property rights within urbanized societies has led to land use decision making paradigms that divide and subdivide land resources (Williams and Holt-Giménez 2017; Haller 2019). High population densities and fragmented ecological interfaces produced by urbanization create heightened interactions between socio-cultural and biophysical components of ecosystems (Alberti et al. 2003; Yu et al. 2012; Andersson et al. 2014). These interlocking factors produce significant *landscape fragmentation*, with implications for soil ecosystem functioning through both ecological and physical processes (Alberti 2004; Donald and Evans 2006; Yu et al. 2012). Urbanization, particularly in former rural areas, drive soil landscapes to become more fragmented, more irregular, and more isolated (Xiao et al. 2016), typically altering the urban mosaic in small patches <4,000 m^2 in area (Ellis et al. 2006). This increasing patchwork and isolation of soil habitat has distinct implications for soil biological, physical, and chemical properties.

Biologically, species richness in soil microarthropod and faunal communities are negatively impacted by increased fragmentation and decreasing urban habitat size, in part due to altered plant species composition and limited soil typologies (Ooms et al. 2020; Byrne 2007). Fragmentation affects community assembly processes like dispersal that drive patterns of soil biodiversity, which can create isolated soil communities subject to high fluctuations from environmental variability (Caruso et al. 2017), especially for relatively dispersal-limited soil organisms like fauna. Soil nematode communities are impacted by landscape fragmentation and other urbanizing forces like soil contamination in distinct ways, often maintaining overall species richness but creating distinct functional guilds (Pavao-Zuckerman and Coleman 2007). It is important to note that most, if not all, of the studies on urbanization and soil biology sample from standing forests, parks, or gardens. Few evaluate the direct impacts of soil sealing, dredging, or other human-driven soil formation, limiting the scope of these conclusions.

The physical processes of soil landscape fragmentation have been demonstrated to directly impact hydrologic functioning (Stewart et al. 2019), with urbanization often causing disconnects with landscape position-wetness positions that are found in natural landscapes (Tenenbaum et al. 2006). In the past, it was argued that the change to landscape hydrology was a result of surface compaction and other infiltration restrictions. However, improving infiltration has not redressed altered catchment hydrology, and new work is showing that subsurface soil modifications are having a substantial effect by limiting drainage and soil water storage (Bhaskar, Hogan, and Archfield 2016).

Finally, landscape fragmentation increases the dispersal rate of chemical pollutants and fluxes from both point and nonpoint sources. The increasing encroachment of the built environment on soil ecosystems causes severe chemical alteration, making contamination a pressing and ubiquitous concern regarding urbanization. Altered nitrogen biogeochemical cycling of urban-adjacent deserts (Hall et al. 2009), pollution from severe densification of rural settlements and refugee camps (Qadeer 2004), legacies of contamination in declining of industrial cities (D. Herrmann et al. 2018), and heavy metal dispersal in the merging of multiple megacities (Li and Jia 2018) are all examples of the increasing spatial complexity of how urbanization impacts soil chemical properties.

4.3.5 Altered Ecosystem Services

Through the forces outlined above, urbanization directly impacts the capacity of soils to support ecosystem services, or those benefits that people derive from their environment. Typically ecosystem services are conceptualized as emanating from biodiversity, with the hypothesis that greater levels of biodiversity support delivery of a larger magnitude and stability of ecosystem services (Millennium Ecosystem Assessment (Program) 2005; Bardgett and van der Putten 2014). For soil-based ecosystem services, soil chemical and physical properties are also fundamental to their functioning (Dominati et al. 2010). Due to the negative impact of urbanizing forces on all aspects (biological, chemical, and physical) of soils, soil-based ecosystem services are generally diminished in urbanizing environments, or skewed toward just one service (D. L. Herrmann et al. 2017; Pavao-Zuckerman and Coleman 2007; Pavao-Zuckerman 2008; Shuster et al. 2005). For instance, capital development and industrial pollution diminish the ability for soil to effectively buffer contaminants and often increase the flux of pollutants from soils, endangering water resources (Ding et al. 2018). In addition, landscape discontinuity and altered hydrologic flow paths limit water retention and flood mitigation often performed by soil not altered by urbanization (Shuster et al. 2005; Bhaskar et al. 2016).

Because of these challenges, the provisioning ecosystem services of urban soils are often underestimated. However in certain places and contexts, the services provided by urban soils can be quite significant (Bon et al. 2010; Mok et al. 2014). A recent geospatial analysis of the ecosystem services provided by urban agriculture estimated that in addition to saving energy, fixing nitrogen, and reducing storm runoff on a global scale, agriculture in cities produces between 98 - 178 million tons of food each year, constituting as much as 10% of the world's food crops (Clinton et al. 2018). Developing countries have particularly robust provisioning services from urban soils, with significant portions of the urban population (approximately 40% in Africa and 50% in Latin America) engaging in agriculture for subsistence and employment (Bon, Parrot, and Moustier 2010; De Zeeuw et al. 2011). Agrarian reform famously allowed Cuba to recover from a precarious food supply and stunted agricultural production after the collapse of the Soviet Union, ultimately producing 4.2 million tons of food through urban agriculture (Hamilton et al. 2014; Altieri et al. 1999). In the United States, potential for meaningful urban food production is particularly dismissed. Yet when the social and political will was mobilized through the military-funded Victory Gardens program of World War II, over 50% of the country's fruits and vegetables were being produced by small-scale, suburban gardens (Mok et al. 2014). Further, urban systems have been shown to produce more food per unit area than conventional agricultural production (McDougall et al. 2019). Uncertainty remains, however, regarding the capacity of urban agriculture to address food insecurity in urbanized areas, especially in low-income and vulnerable communities (A. Siegner et al. 2018). This is questioned largely on account of urban agriculture's role in perpetuating gentrification, attracting venture capital investment, green washing, and forms of self-exploitation (Alkon and Agyeman 2014), patterns that often lead to a privileging of whiteness and white culture in urban food production even while BIPOC lead the resistance and reimagination of food sovereignty in cities (Ramírez 2015).

Practitioners in diverse social and geographic contexts cultivate soils in urbanizing environments for different cultural, political, and social reasons. Through urban farming and gardening, practitioners seek to improve nutrition (Alaimo et al. 2008), beautify urban spaces and elevate mental health (Wakefield et al. 2007), build community cohesion and empowerment (Armstrong 2000; Baker 2004), improve community food security (A. Siegner et al. 2018; Meenar and Hoover 2012), increase educational opportunities (Grossman et al. 2012), renew cultural sovereignty (Baker 2004; Hoover 2017a; Meenar and Hoover 2012), and establish a spiritual connection with the natural world (Lin, Egerer, and Ossola 2018). Increasingly, practitioners also see themselves as protecting the environment, addressing climate change, or alleviating ecological harms done by the industrial food system (Odewumi 2013; A. B. Siegner et al. 2020; Travaline and Hunold 2010). Diverse stakeholders, including urban farmers, gardeners, ecologists, activists, urban planners, landscape architects, and others engage in these activities to both sustain ecological functioning and bring about social change (Piso et al. 2019).

The social forces that drive urbanization and degrade soil resources result in soil-based ecosystem services remaining far below their potential. Urbanization is a process that determines the material, social, and ecological trajectories of the social metabolism. Without an interdisciplinary and action-oriented approach, urbanization's inequitable erosion of soil health and food sovereignty will likely only deepen as the world's populations expand and densify. Here we propose how a dialectic among agroecology, soil science, and critical geography might better address the twin challenges of soil degradation and diminished food sovereignty caused by urbanization.

4.4 Addressing the Twin Challenges of Soil Degradation and Diminished Food Sovereignty across Urban–Rural Systems

Soil management, restoration, and conservation have long been central to all three dimensions of agroecology (science, practice, movement). For instance, early scientific articles in the field investigated the soil-building practices of traditional cultures, measured key soil parameters, and found long-term soil fertility to be a desired outcome of practitioners (Gliessman, Garcia, and Amador 1981; Altieri 1995). Practitioners of agroecology across the world emphasize the importance of soil to agroecology (Nyéléni 2015). Yet a number of recent developments in soil science have not been widely integrated into agroecological research or practice, with implications for restoring ecosystem services. On the other hand, since soil science as a scientific discipline focuses mainly on biophysical understanding of soils and technical aspects of soil management, it is not well-poised to consider the motivations and knowledge of practitioners, or to motivate the action needed to change socio-political systems that constrain alternative futures for farmers. And both fields have much to gain from engaging more deeply with critical geography to address growing urbanization and the social inequities brought with it. In the following sections, we begin a series of conversations among these three disciplines, in which solutions from one is contextualized or *problematized* by the others. In doing so, we hope to foster deep, critical engagement with the social and ecological issues that span urban–rural landscapes.

4.4.1 Building Healthy Soils in Urbanized Landscapes

The idea of "soil health" has rapidly gained traction among researchers, practitioners, and even the general public as a means of conceptualizing soil as an emergent, living system and suggesting links between soil and human health (Doran and Zeiss 2000). Soil health focuses on the extent to which the dynamic properties of soil are managed for multiple ecosystem services, given inherent constraints of climate, parent material, and soil type (Moebius-Clune 2016). In this way, soil health is understood in relation to human intervention, and is made meaningful only when translated into practice (Doran 2002). The focus on multiple ecosystem services also emphasizes that soil can do more than just provide a medium for plant growth and provisioning food, emphasizing the role of soils in supporting, regulating, and cultural services as well (Kibblewhite, Ritz, and Swift 2008).

But urbanization significantly alters perceptions of and engagement with soil through pollution, displacement, and physical disconnection (Kim et al. 2014; Armstrong 2000; Brevik et al. 2019), placing soil management practices within the increasingly political context of urban land use (Cutts et al. 2017). The contested history of community farms and gardens in Sacramento, CA and changing perceptions of soil health, provide a stark example. In the 1950s, a number of working-class homes and businesses located near Sacramento's capitol building were razed by the state government, but the site was never redeveloped (Cutts et al. 2016). By the 1970s, a robust, community-based garden serving over 100 families had taken root in those vacant lots, ultimately calling themselves the Ron Mandella Community Garden (Francis 1987). Over a protracted 30 year conflict with state and city governments, the issue of soil lead contamination and perceptions of the garden's soil health became central to the fate of the garden (Cutts et al. 2017). When proponents of the garden justified protecting the site because of its well-established soil health, state and city officials countered by planning to physically transport the topsoil to a new location. This led activists to decry the soil lead pollution at proposed alternative sites as prohibitive. Then, when high concentrations of lead were found at Mandella's original location, the logic of contamination suddenly cut the other way, discrediting the claims that Mandella's soil was healthy and worthy of cultivation (Cutts et al. 2017). Ultimately, the garden was razed. This demonstrates how perceptions of soil health—highly politicized and co-opted by stakeholders for distinct ends—is used to justify access and use of soil resources in urbanized areas.

This is a far cry from the "soil health" concept leveraged by rural farmers, soil scientists, and agronomists. Soils in cities are rarely viewed as provisioning food, fiber, or medicine, and therefore the soil health framework has not been as widely applied to urbanized landscapes as to agricultural soils in rural areas (Brevik et al. 2019). Perceiving soil as a living system, or connected to human and ecosystem

FIGURE 4.1 A volunteer harvests lettuce from the Oxford Tract in Berkeley, CA, where researchers and students established a model no-till farm, studying how small-scale agroecosystems can be managed to both improve soil health and address food insecurity

health, is in direct competition with urban development that sees land as a commodity to be bought, graded, and shaped (Cutts et al. 2017). Some notable examples of soil health analyses in urban areas exist in the literature, many of which develop a "soil quality" or soil health index to determine the most critical factors governing soil functioning in urbanized landscapes (Scharenbroch et al. 2005; Beniston et al. 2016; Tresch et al. 2018; Knight et al. 2013; Reeves et al. 2014). In applying an agroecological perspective, however, it is apparent that these studies do not capture the political, social, and cultural contexts that are necessary to sustain or enhance soil management in those urbanized landscapes.

Soil health is thus a useful framework for conceptualizing and assessing soils across urbanizing systems and for establishing benchmarks relevant to highly altered, politicized, and contested soils. By focusing on dynamic properties, soil health is also differentiated from soil quality, a concept that includes inherent soil properties in describing the suitability of soil for agricultural production or other uses. For instance, inherent properties (e.g. texture) of imported soils or soil blends often diverge strongly from locally-formed soils (Obrycki et al. 2017). This changes the absolute magnitude and array of soil-based ecosystem services that can be promoted from the filling and transport of soils, a common practice in the logic of urbanization that treats land as an exchangeable commodity. Soil health management that meaningfully addresses urbanization should try to promote holistic soil functioning while also considering the biophysical, social, and cultural constraints that occur in cities. As an example, a study in Berkeley, CA helps illustrate how concepts of soil health are problematized by these aspects of urbanization.

4.4.1.1 Case Study: Connecting Soil Health and Hunger Relief on Contested Land

In 2017, an interdisciplinary team of agroecologists with expertise in entomology and soil, along with city planners and cooperative extension specialists at the University of California, Berkeley (UCB) set out to evaluate: 1) how urban agroecosystems can be designed to enhance ecological processes; and 2) whether urban food production can meaningfully address food insecurity (A. Siegner et al. 2018). This work included input from a constellation of small farms across the San Francisco Bay Area and other regions of Northern California, including two model no-till farms on land owned by UCB: the Oxford Tract (1.2 acres) located in Berkeley, CA and the Gill Tract Community Farm (10 acres) located in Albany, CA.

Maximizing ecological processes in urban agroecosystems poses unique challenges regarding labor, land, capital, machinery, and amendments. By working with small urban and rural farmers across

California, researchers hoped to better understand how social and ecological contexts determined farmer's success in building soil health. These farms were united by their use of no-till management systems and all experienced land and capital limitations of some kind. However the farms also spanned an urban–rural gradient that mapped onto socioeconomic and racial differences: urban farmers were predominantly Black, Indigenous, or POC and operated non-profit or cooperative projects, while rural farmers were predominantly white with market-facing enterprises. At a symposium held on February 11, 2019 in Davis, California, practitioners from over 30 different farms and organizations discussed scientific, agronomic, economic, and cultural issues surrounding no-till farming for organic farmers with limited access to land and capital. At this symposium, the group helped outline the foundational principles included in "biointensive" no-till: 1) minimize soil disturbance, 2) maximize crop density in space and time, 3) maximize on-farm biodiversity above and belowground, 4) promote holistic management that increases biological activity, and 5) mitigate and adapt to climate change. These principles were shared widely amongst farmers but implemented in a variety of ways across California depending on their social context and site-specific needs and constraints. Crucially, biointensive no-till management requires little machinery, inputs, capital, or infrastructure, and can be implemented at both small and medium scales.

The research team at UCB implemented biointensive no-till principles on a model farm at the Oxford Tract, creating a factorially replicated and randomized block design to understand the impact of no-till management and cover cropping on key indicators of soil health. These indicators included soil carbon stocks and dynamics, water retention, biological activity and diversity, nutrient density and acquisition, and drought stress tolerance. After two years of diversified vegetable production and no-till management, the model farm was intensively sampled for these indicators in the summer of 2019. During that season, a model crop (*Phaseolus vulgaris;* black bean), grown as part of a polyculture, was subjected to simulated drought conditions and monitored over the growing season. The results demonstrated that the physical, chemical, and biological properties of soil ecosystems can shift dramatically in a short period of time under no-till management, much faster than the majority of the literature claims (Mitchell et al. 2017). First, no-till management dramatically increased soil moisture content under a drought scenario. Stem water potential, a proxy for plant drought stress, was also found to be lowest under no-till management and cover cropping, likely due in part to improvements in measured water retention and soil moisture. This has serious implications for urban agriculture facing drought conditions in California and high water costs due to residential rates (Pathak et al. 2018). Black bean yield was highest under no-till management and cover cropping, but showed no significant differences across treatments in nutrient acquisition or protein content. Still, high yields of a protein and calorie-rich food in the urban context demonstrates the capacity of urban food production to meaningfully contribute to nutrition. Plant performance and drought resilience showed no clear relationship with arbuscular mycorrhizal fungi (AMF) colonization of plant roots or with fungal community composition. Future research aims to resolve some of the intricacies of these plant-soil-microbe interactions. Management was also found to alter both the rate and location of soil enzymatic activity: biointensive no-till increased enzyme activity at all soil depths and shifted enzymatic activity to smaller particle fractions (clay particles, <20 μm). This has direct implications for the stability of soil carbon under no-till management (Cotrufo et al. 2013). Taken together, these measurements helped to capture the dynamic physical, chemical, and biological responses of soil ecosystems to a management system.

The study went beyond just evaluating soil biophysical properties, however. The research team also sought to understand whether urban farms meaningfully address food insecurity in the region. Through a series of interviews, mapping projects, surveys, and community partnerships, UCB researchers were able to track the movement of produce from the Oxford and Gill Tracts to dozens of organizations directly serving low-income residents (A. B. Siegner et al. 2020). This research is ongoing, but preliminary results demonstrate how informal networks, coalition building and cooperativism between organizations, anti-capitalist solidarity economies, and centralized distribution centers all help urban agriculture better address food insecurity. These findings are situated within a vital culture of urban agriculture in Oakland that draws inspiration from the city's radical food justice programs and politics, including the free breakfast program hosted by the Black Panthers starting in 1969 (Curran and González 2011).

Despite this history, UCB's agricultural research sites are heavily contested, both politically and culturally. The Gill Tract was the site of the Occupy the Farm (OTF) movement in 2012, a direct action mobilization to realize food sovereignty that stemmed from the broader Occupy movement of the previous year. In the late 2000s, the Gill Tract was slated to be developed into a parking lot, Whole Foods, senior living center, and baseball fields. OTF was a land occupation that opposed this development, establishing a community farming project at the Gill Tract and calling for anti-capitalist solidarity economies, land reform, and democratic or autonomous governance structures (Roman-Alcalá 2018). The occupation was a threat to UCB, who ultimately sent in the University of California Police Department (UCPD) and arrested the occupiers. But a protracted battle between UCB and community organizers delayed the development and OTF was successful in establishing the Gill Tract Community Farm (Roman-Alcalá 2018). The farm still operates today, growing 11–12,000 pounds of free produce a year for the community.

The Oxford Tract is the site of more recent controversy, as UCB's Chancellor and Capital Strategies office have announced their decision to develop student housing on the site. This poses threats to the research, teaching, student organizing, and farm-to-food pantry programs that currently operate on the site (Treffinger and Jacobson 2018). It also poses threats to the rich ecosystem services and biodiversity of the site, where, for example, 84 distinct native bee species were found over a two year period (Frankie et al. 2009). Two of the primary justifications made by students and activists for maintaining these sites for agricultural use are the fertile, alluvial soils (Treffinger and Jacobson 2018) and contribution of student farming projects to campus food security through regular donations to the Berkeley Student Food Pantry. These struggles situate concepts of soil health within a complex web of social, cultural, and ecological forces, and complicate how soil health is interpreted in urbanized landscapes.

4.4.2 Restoring Soil Organic Matter

Soil organic matter is often cited as the most critical determinant of soil health and the sustainability of agricultural production (Lehman et al. 2015; Bot and Benites 2005; Johnston, Poulton, and Coleman 2009), serving as one of the primary energy sources for soil microbes, a reservoir of fertility, and an important influence on soil structure through aggregation of soil particles, just to name a few functions. Further, since soil organic matter is approximately 50–58% carbon (Pribyl 2010), it is also an important source and sink of carbon dioxide and is repeatedly invoked as a target for negative emissions to offset carbon emissions from other sectors (Minasny et al. 2017). Building soil health through increasing soil organic matter is a high priority for many urban agriculturalists, in part motivated by an interest in climate change mitigation and adaptation. For instance, in a survey of 35 urban producers in the San Francisco Bay Area, USA, most respondents reported using soil organic matter-enhancing techniques like organic matter amendments, cover cropping, and reduced tillage (A. B. Siegner et al. 2020).

In considering how to increase soil organic matter, and especially how to sequester soil carbon, it is important to incorporate recent understanding of how soil organic matter forms and stabilizes in soil. This emerging view emphasizes the importance of microbial processing and mineral surface chemistry on the formation and long-term storage of soil carbon (Schmidt et al. 2011; Miltner et al. 2012). For many years it was hypothesized that plant-derived carbon, in the form of polymeric residues and diverse root exudates, itself formed complex and chemically recalcitrant forms of carbon called "humus." But improved organic matter characterization and accounting have revealed that the majority of soil organic matter is actually relatively simple macromolecules derived from microbial necromass, or the accumulated cellular residues of microorganisms (Kallenbach et al. 2016). To be stabilized, microbial necromass must either form associations with mineral surfaces of clay-sized particles or be physically protected inside aggregates; "chemical recalcitrance" is no longer widely-recognized as an important stabilization mechanism of organic matter (J. Lehmann and Kleber 2015). These changes in understanding have at least two important implications, particularly in addressing urbanization's impact on soil. First, the proportion of organic carbon that soil microbes respire vs. incorporate into their biomass, termed their "carbon use efficiency" is an important control on soil carbon accrual (Kallenbach et al. 2019). In contrast to received wisdom about the chemical recalcitrance of an organic input being important for increasing soil carbon, higher quality organic inputs (e.g. residues from a legume cover crop with low carbon:nitrogen ratios) can actually contribute more to stabilized organic matter (Cotrufo et al. 2013). Second, the mineral component of soil is increasingly recognized as crucial to forming long-lived soil organic matter, both through direct associations with organic matter and indirectly through its influence on soil aggregates. As an example of an implication, establishing raised beds using potting mixes or imported soil with low mineral content is not a means of stabilizing carbon for the long-term, even if the initial organic matter content is high.

The widespread use of compost in urban agricultural systems also deserves attention. While often necessary for amending highly degraded soils, building soil carbon through compost additions is not necessarily a sink for carbon; instead, it is a transfer of carbon from one ecosystem or source to another. Understanding the net energy cost and carbon dioxide emissions of compost additions in urbanized landscapes, which often require large organic amendments to restore degraded soils, requires a life cycle analysis (DeLonge et al. 2013). In addition, a study of small urban farms in Sydney, Australia demonstrated that while urban agriculture can have large yields relative to other agricultural sectors, high rates of amendment can over fertilize food crops and cause high energy costs of production (McDougall et al. 2019). This contradicts some practitioners desire to combat climate and environmental degradation through soil management. The source and quality of compost additions must also be closely evaluated, especially when feedstocks are municipal in origin. Even green waste feedstocks may contain substantial quantities of chemical and/or physical contaminants and that counter potential benefits of compost (Cattle et al. 2020; Montejo et al. 2010). Further, municipal compost production often occurs in rural areas that outsource the air quality issues associated with its production to other communities (Kumar et al. 2011).

This discussion challenges soil organic matter's role as a panacea for soil health—particularly in urbanized areas where purchased soil media, infill, amendments, and soil transport are common. Further research should better evaluate how local, abundant organic matter in cities can be better recycled and used to fertilize urban soils. This restoration of rich soil ecosystems and meaningful food sovereignty in the face of capital-driven urbanization will require the participation of farmers, practitioners, and communities.

4.4.3 Incorporating Practitioner or Traditional Knowledge

The incorporation of practitioner or traditional knowledge has been central to agroecology since its beginnings as a scientific discipline (Altieri, Miguel A. 2002; Altieri 1995). A focus on local farming wisdom has contributed to agroecology's championed role as both a set of agricultural principles and broad-based social movement (Wezel et al. 2009). This is perhaps best articulated by La Via Campesina in their six principles of food sovereignty, formalized at the Nyéléni International Forum for Food Sovereignty in 2007: "Food sovereignty builds on the skills and local knowledge of food providers and their local organizations that conserve, develop and manage localized food production and harvesting systems, developing appropriate research systems to support this and passing on this wisdom to future generations." (Nyéléni 2007) Despite this, local soil knowledge has not been reflected in the majority of soil science research or the historic development of the discipline (Warkentin, International Union of Soil Sciences, and Soil Science Society of America 2006).

What's more, the biophysical and ecological losses caused by urbanization are often accompanied by losses of cultural resources and traditional knowledge. In many cases, this is due to the forced assimilation, removal, or migration of people from the lands that supported their agricultural practices, cultures, and foodways (Daniel 2013; Balvanz et al. 2011; Wires and LaRose 2019; Carney and Rosomoff 2009; White 2018). In North America, this predominantly means the violent displacement of Black and Indigenous people, a central element of modern city-making that attempts to pave over sacred sites, traditional knowledge systems, and cultures (Mays 2015; Edmonds 2010). Urbanization, along with corresponding shifts towards globalized market economies, industrial agriculture, and industrialization, have severed intergenerational transfer of agricultural knowledge in regions as distinct as Spain, India, Vanuatu, and California, USA (Anderson 2013; Brodt 2001; McCarter et al. 2014; Gómez-Bagget et al. 2010). Nishnaabeg scholar Leanne Betasamosake Simpson has emphasized how a resurgence of Indigenous and traditional cultures will require generations of people to grow up "intimately and strongly connected to [their] homelands." (Simpson 2014) The call from advocates of food sovereignty and agroecology to incorporate practitioner knowledge is challenged in a capital-driven and urbanizing world that actively threatens the connection between societies and soil.

Yet scholars and activists across a range of disciplines are striving to protect the cultures, practices, and lifeways that contain within them traditional land and soil management strategies. In Detroit, Monica White has described how Black activists participated in urban agriculture to reconnect with their cultural traditions and reclaim personal power through food self-sufficiency (White 2011). These experiences were situated within a rich history of Black agrarianism in the United States, where both urban and rural agricultural production were a means of self determination, economic prosperity, cooperativism, and liberation (White 2018), ultimately reclaiming the deep agricultural knowledge of African cultures that domesticated much of the world's food crops (Carney and Rosomoff 2009).

Elizabeth Hoover's work is another striking example, chronicling the tension between food sovereignty initiatives in Native American communities and urban environmental pollution caused by discriminatory state, federal, and industry policies (Hoover 2010). Akwesasne is a Mohawk community that straddles the borders between New York State and the Canadian provinces of Quebec and Ontario. The territory is downwind, downriver, and downgradient from three Superfund sites, aluminum foundries that used fluoride and polychlorinated biphenyls (PCBs), contaminating the waterways, fish, soils, and sediments of the region (Hoover et al. 2012). This continues to have severe health impacts on Mohawk communities and has altered traditional foodways that relied on fishing, hunting, and farming in the region's alluvial soils to sustain their communities. Hoover's work in Akwesasne brilliantly catalogues how a rich culture of food production slowly shifted in the twentieth century towards wage-based economies, industrial farming, urban migration, and the abandonment of croplands. A significant driver of this shift was the pollution of Akwesasne soils and water resources, particularly the atmospheric deposition of fluoride in soils from aluminum foundry smokestacks (Hoover 2010). Yet many members of the Akwesasne Mohawk community persisted in maintaining gardens of traditional foods including corn, beans, squash, and medicinal herbs, in turn preserving their ceremonies that focused on the cycles of the harvest. This was an expression of their desire for food sovereignty, which has led in recent years to the development of programs like the Akwesasne Task Force on the Environment and Kanenhi:io Ionkwaienthon:hakie (We Are Planting Good Seeds) that establish gardens and support food production in the community (Hoover 2017b). And these trends are not isolated to Akwesasne. Across the country, Native American food sovereignty initiatives maintain connections with cultural foodways and cultivation practices to restore the lifeways and ecosystems lost to soil and water contamination, land dispossession, and damming (Hoover 2017a). This demonstrates how calls for the inclusion of traditional or practitioner knowledge in agroecology must also confront legacies of oppression, many of which are evident in the health of soils and tied to waves of industrialization and urbanization.

Growing interest in preserving traditional soil cultivation methods is also found in the burgeoning field of ethnopedology—a hybrid discipline incorporating aspects of ethnography, geography, agronomy, agroecology, and soil science that seeks to encompass the land and soil knowledge of local people (Barrera-Bassols and Zinck 2003). A review of ethnopedologic studies from around the world shows how 432 studies conducted of local soil knowledge systems represented over 217 ethno-linguistic groups from Africa (41%), the Americas (23%), Asia (23%), Europe (8%) and the Pacific (5%) (Barrera-Bassols and Zinck 2003). A common set of principles emerges in these traditional knowledge systems of soil: 1) the existence of a complex system of knowledge about the hierarchical organization of soil horizons, 2) the recognition and use of soil morphological attributes for classification, 3) comparisons made between soil ecosystems to construct classification systems, and 4) the existence of certain universal criteria for evaluating soils (color, texture, moisture, topography, drainage, etc.) (Barrera-Bassols and Zinck 1998). Understanding social theories of soil and land resources require integration of three primary domains: the symbolic (*kosmos*), the cognitive (*corpus*), and the practical (*praxis*) dimensions (Barrera-Bassols and Toledo 2005; Toledo 2002). The inclusion of practitioner's own symbolic *cosmovisions* or worldviews, and integration of those worldviews into biophysical or practical understandings of ecological health, is crucial to the adoption of new management practices that achieve sustainability and food sovereignty goals (WinklerPrins and Barrera-Bassols 2004). A limited number of studies, however, meaningfully integrate the symbolic worldview of practitioners. This must become a part of research agendas moving forward. Another limitation in preserving local soil knowledge is that most ethnopedological studies focus on rural landscapes and communities. Further work must be done to understand how experiential learning (learning by doing) and ecological management feedback with an urbanizing landscape to create new forms of knowledge.

4.4.4 City–Region and Urban–Rural: A Systems Approach

Another key advancement in agroecology is the realization that agroecological principles can be applied not just to individual producers, agroecosystems, or social movements, but whole food systems (Vaarst et al. 2018). As the discussion of urban–rural systems and linkages above indicates, activists and scholars are increasingly calling for an integration of urban–rural or city–regional analyses to address food system failures like hunger and soil degradation (Armendáriz, Armenia, and Atzori 2016; Karg et al. 2016; N. McClintock 2010; Vaarst et al. 2018; Chappell 2019; Nyéléni 2018; Tornaghi 2017). Such debates have generally called for a re-evaluation of the linkages between urban and rural landscapes (Nyéléni 2018), the economic revitalization of rural areas (Armendáriz, Armenia, and Atzori 2016), or the inclusion of the "urban" context as meaningful to agrarian issues (Tornaghi 2017). Recently, however, even these approaches have been questioned as reifying or reinforcing false dichotomies between urban and rural landscapes that must be deconstructed to create a just and sustainable food system.

Vaarst et al. established a framework for this emerging thread in food systems, food sovereignty, and critical geography literature by illustrating what an agroecological food system in city-regional contexts would require. They emphasize the importance of linking urban, peri-urban, and rural landscapes through food systems that embody agroecological principles: minimizing external inputs and encouraging resource recycling, promoting adaptive capacity, ensuring equity, and sustaining human nourishment (Vaarst et al. 2018). This means connecting places where food is grown and where food is consumed, and promoting ecological and cultural diversification, zoning urban–rural landscapes, planning for seasonality, producing at scale, and critically, challenging patterns of "urban consumption" and "rural production." This calls for a "radical shift" in urban–rural thinking and challenges the commodification, exploitation, and injustices that underpin the dichotomy in the first place (Vaarst et al. 2018). Monika Krause further complicates the urban–rural concept. (Krause 2013) She highlights that in the absence of such labels, the true nature of human settlements must be re-examined through networks of solidarity, physical places, governance structures, and human livelihoods that actually function in a given society. This, she claims, will reveal that societies are never completely "urban" and resist the zeitgeist of capital through the "persistence of informal social relations, the natural, the problem of food and livelihoods, basic need, and complex dependencies that nevertheless enable survival." (Krause 2013) This expands the idea of city-regional and urban–rural systems to encompass the entire set of social and ecological relationships that sustain human societies.

Expanded ideas of city-regional and urban–rural systems can improve real-world outcomes in ecological health and food sovereignty by operating across axes of social difference. The success of hunger abatement programs in Belo Horizonte, Brazil was achieved through strong policies and governing bodies whose jurisdiction spanned rural livelihoods, agrarian sustainability, and city food security initiatives (Chappell 2019). Urban gardens and allotments of Santarém, Brazil also build strong links between urban centers and nearby rural communities that support regional food sovereignty. These city-regions are connected by a rich set of soil management practices that includes the burning of organic waste to yield biochar amendments, the application of manure, and cultivation of diverse fruit trees, ornamentals, vegetables, and medicinal herbs (WinklerPrins 2002). And in Oakland, California, the history of food policy and advocacy reflects a central tension: between the anti-state and anti-capitalist politics of self determination championed by the Black Panthers in the city, and government programs and neo-liberal institutions promoting food security through regional food production (Curran and González 2011). Each of these examples demonstrates how the broken metabolism of society, and the production of food, soil, and culture, is created through material exchanges along lines of perceived social differences.

These examples challenge us to move beyond repairing city-regional or urban–rural connections and interrogate deeply the nature of those designations, ultimately reformulating the connection between human societies and ecological systems. As this article has outlined, understanding the role of soil and food resources in our social metabolism is central to such a project. Tornaghi and Dehaene affirm this view, calling for a "soil-based society":

> "If the urban is not just a geographical location whose oppressive power over the rural is to be reversed, but rather the reflection of specific social arrangements, collective inter-dependencies, value and exchange systems, in short, urban-'isms' reflecting specific forms of urbanis-'ations'

> … and if we acknowledge and aim to build on the work of critical geography that has unpacked the intimate link between capitalism and urbanization … we are yet to dig deeper into these dynamics to foresee ways to dismantle oppressive mechanisms." (Tornaghi and Dehaene 2020)

Tornaghi and Dehaene examine food systems through the lens of 'social reproduction,' derived from Marx and historic materialism, that encompasses practices through which social relations and the material basis for capitalism are renewed. This places food and soil at the heart of the social metabolism by which societies, cultures, and ecologies are linked. Fundamental injustices comprise these linkages, however, upholding capitalist societies through the devaluation of women's and farm labor, the loss of biocultural diversity, the commodification of food that leaves poor people hungry, and the externalization of ecological costs that allows for polluting industries (Tornaghi and Dehaene 2020). How might we change course?

Perhaps the nexus of change is the material, the body, itself (Ayres 2014). In the final analysis, theoretical strivings to comprehend the "urban" and the "rural" point us beyond the dichotomy and towards the material: the pollution of Indigenous lands, the starvation of the urban poor, the blossoming within the concrete, and the food of cultures. These material realities are experienced by real human beings and sustained by dynamic, living ecological systems. Transforming urban–rural food systems will require grounding individual actions and experiences in this materiality: how soil, food, and cultural exchange impacts human and non-human bodies, including our own. To restore soils and build food sovereignty requires a return to embodied experiences where food, soil, and culture root individuals back into the ecologies that sustain them.

4.5 Conclusions

In this article, we began a dialectic that may offer pathways toward addressing the social inequity and ecological unsustainability caused by urbanization. A meaningful dialectic between agroecology, critical geography, and soil science reveals how solutions from each discipline are contextualized and *problematized* by the others. Soil degradation and diminished food sovereignty are central to the material exchange that sustains urbanization and capital production across urban–rural landscapes. Integration of these disciplines leads to novel, transdisciplinary methods and perspectives on how social forces determine the ecological trajectories of soils. In this article we begin a number of conversations that span these disciplines. First, viewing soil as living systems whose health, services, and biological diversity can be finely understood in relation to social forces expands the horizons of how land is valued and managed. This can also help orient soil science towards improved transdisciplinarity, political engagement, and action. Second, incorporating new paradigms of soil carbon accrual and stabilization into land management strategies will help make climate intervention and advocacy more successful. Better understanding the role of organic inputs in carbon stabilization and nutrient management will help establish badly needed organic matter recycling across urban–rural systems. Third, the meaningful inclusion of traditional and Indigenous pedalogic knowledge is complicated by patterns of cultural loss, especially in urban areas. Cultural revitalization is ongoing, however, both in theory and practice. Supporting traditional knowledge systems threatened by patterns of contamination and land dispossession requires access to land and healthy soil. This is essential to uplifting practitioner livelihoods and food sovereignty among vulnerable communities. Finally, critical engagement with city-regional systems begins to deconstruct the boundary between urban and rural, creating improvements in soil health and food sovereignty by working along lines of perceived social difference. This ultimately points beyond urban–rural systems and towards the metabolic exchanges that create them, calling for a direct, embodied engagement with soil, food, and cultural reproduction.

Taken together, these methodological shifts begin to form a more robust engagement with the material injustices embedded within our social metabolism, posing a strong response to the harmful ecological and social patterns created by urbanization. The contamination, degradation, and loss of urban soil resources is governed by the same socioeconomic, racial, political, and cultural paradigms that drive food insecurity and disproportionally affects poor and BIPOC communities in urban areas of the global North while devastating rural livelihoods around the globe. In human-dominated landscapes, soil *materializes* prevailing social paradigms. This makes real the possibility that the health of our soils can be viewed as a proxy for our cultural and societal wellbeing. Restoring a rich suite of soil ecosystem services while ensuring equitable food sovereignty in both urban *and* rural landscapes will require the integration of agroecology and critical geography. Working across urban–rural landscapes challenges colonial notions of progress that reinforce soil and wage exploitation. In order to address the rapid and unequal expansion of soil degradation and food insecurity, scientists, practitioners, and communities must work together to support rich, dynamic soil ecosystems in ways that prioritize ecological vibrancy, cultural sovereignty, and social engagement over commodification and urbanization.

REFERENCES

Alaimo, Katherine, Elizabeth Packnett, Richard A. Miles, and Daniel J. Kruger. 2008. "Fruit and Vegetable Intake among Urban Community Gardeners." *Journal of Nutrition Education and Behavior* 40(2): 94–101. https://doi.org/10.1016/j.jneb.2006.12.003.

Alberti, Marina. 2004. "Ecological Resilience in Urban Ecosystems: Linking Urban Patterns to Human and Ecological Functions." *Urban Ecosystems* 7: 241–65.

Alberti, Marina, John M Marzluff, Eric Shulenberger, Gordon Bradley, Clare Ryan, and Craig Zumbrunnen. 2003. "Integrating Humans into Ecology: Opportunities and Challenges for Studying Urban Ecosystems." *BioScience* 53(12): 1169–1179.

Alkon, Alison Hope, and Julian Agyeman. 2014. *Cultivating Food Justice Race, Class, and Sustainability.* Cambridge: MIT Press.

Alkon, Alison Hope, Daniel Block, Kelly Moore, Catherine Gillis, Nicole DiNuccio, and Noel Chavez. 2013. "Foodways of the Urban Poor." *Geoforum* 48 (August): 126–35. https://doi.org/10.1016/j.geoforum.2013.04.021.

Altieri, Miguel A. 1995. *Agroecology: The Science of Sustainable Agriculture.* 2nd ed. Boulder, Colo. : London: Westview Press ; IT Publications.

Altieri, Miguel A. 2002. "Agroecology: The Science of Natural Resource Management for Poor Farmers in Marginal Environments." *Agriculture, Ecosystems & Environment* 1971: 1–24.

Altieri, Miguel A., Nelso Companioni, Kristina Cañizares, Catherine Murphy, Peter Rosset, Martin Bourque, and Clara I. Nicholls. 1999. "The Greening of the 'Barrios': Urban Agriculture for Food Security in Cuba." *Agriculture and Human Values* 16(2): 131–40. https://doi.org/10.1023/A:1007545304561.

Andersson, Erik, Stephan Barthel, Sara Borgström, Johan Colding, Thomas Elmqvist, Carl Folke, and Åsa Gren. 2014. "Reconnecting Cities to the Biosphere: Stewardship of Green Infrastructure and Urban Ecosystem Services." *AMBIO* 43(4): 445–53. https://doi.org/10.1007/s13280-014-0506-y.

Anderson, Kat. 2005. *Tending the Wild: Native American Knowledge and the Management of California's Natural Resources.* University of California Press.

Angel, Shlomo, Jason Parent, and Daniel L. Civco. 2012. "The Fragmentation of Urban Landscapes: Global Evidence of a Key Attribute of the Spatial Structure of Cities, 1990–2000." *Environment and Urbanization* 24(1): 249–83. https://doi.org/10.1177/0956247811433536.

Angel, Shlomo, Patrick Lamson-Hall, Bibiana Guerra, Yang Liu, Nicolás Galarza, and Alejandro M. Blei. 2018. *"Our Not So Urban World." 42.* The Marron Institute of Urban Management, New York University. https://marroninstitute.nyu.edu/uploads/content/Angel_et_al_Our_Not-So-Urban_World,_revised_on_22_Aug_2018_v2.pdf.

Armendáriz, Vanessa, Stefano Armenia, and Alberto Atzori. 2016. "Systemic Analysis of Food Supply and Distribution Systems in City-Region Systems—An Examination of FAO's Policy Guidelines towards Sustainable Agri-Food Systems." *Agriculture* 6(4): 65. https://doi.org/10.3390/agriculture6040065.

Armstrong, Donna. 2000. "A Survey of Community Gardens in Upstate New York: Implications for Health Promotion and Community Development." *Health & Place* 6(4): 319–27. https://doi.org/10.1016/S1353-8292(00)00013-7.

ASTM. 2017. *Standard Practice for Classification of Soils for Engineering Purposes (Unified Soil Classification System).* West Conshohocken, PA: ASTM International. https://doi.org/10.1520/D2487-17.

Ayres, Jennifer R. 2014. "Learning on the Ground: Ecology, Engagement, and Embodiment: Learning on the Ground." *Teaching Theology & Religion* 17(3): 203–16. https://doi.org/10.1111/teth.12202.

Baker, Lauren E. 2004. "Tending Cultural Landscapes and Food Citizenship in Toronto's Community Gardens." *Geographical Review* 94(3): 305–25. https://doi.org/10.1111/j.1931-0846.2004.tb00175.x.

Balvanz, Peter, Morgan L. Barlow, Lillianne M. Lewis, Kari Samuel, William Owens, Donna L. Parker, Molly De Marco, et al. 2011. "'The Next Generation, That's Why We Continue To Do What We Do': African American Farmers Speak About Experiences with Land Ownership and Loss in North Carolina." *Journal of Agriculture, Food Systems, and Community Development* 1(3): 67–88. https://doi.org/10.5304/jafscd.2011.013.011.

Bardgett, Richard D., and Wim H. van der Putten. 2014. "Belowground Biodiversity and Ecosystem Functioning." *Nature* 515(7528): 505–11. https://doi.org/10.1038/nature13855.

Barrera-Bassols, Narciso, and Víctor M. Toledo. 2005. "Ethnoecology of the Yucatec Maya: Symbolism, Knowledge and Management of Natural Resources." *Journal of Latin American Geography* 4(1): 9–41.

Barrera-Bassols, Narciso, and Alfred J. Zinck. 1998. "The Other Pedology: Empirical Wisdom of Local People." *Proceedings of the 16th World Congress of Soil Science*, 9.

Barrera-Bassols, Narciso, and J.A. Zinck. 2003. "Ethnopedology: A Worldwide View on the Soil Knowledge of Local People." *Geoderma* 111(3–4): 171–95. https://doi.org/10.1016/S0016-7061(02)00263-X.

Béné, Christophe, Peter Oosterveer, Lea Lamotte, Inge D. Brouwer, Stef de Haan, Steve D. Prager, Elise F. Talsma, and Colin K. Khoury. 2019. "When Food Systems Meet Sustainability – Current Narratives and Implications for Actions." *World Development* 113 (January): 116–30. https://doi.org/10.1016/j.worlddev.2018.08.011.

Béné, Christophe, Steven D. Prager, Harold A. E. Achicanoy, Patricia Alvarez Toro, Lea Lamotte, Camila Bonilla, and Brendan R. Mapes. 2019. "Global Map and Indicators of Food System Sustainability." *Scientific Data* 6(1): 279. https://doi.org/10.1038/s41597-019-0301-5.

Beniston, Joshua W., Rattan Lal, and Kristin L. Mercer. 2016. "Assessing and Managing Soil Quality for Urban Agriculture in a Degraded Vacant Lot Soil: Assessing and Managing Soil Quality for Urban Agriculture." *Land Degradation & Development* 27(4): 996–1006. https://doi.org/10.1002/ldr.2342.

Bennett, Elena M., Stephen R. Carpenter, and Nina F. Caraco. 2001. "Human Impact on Erodable Phosphorus and Eutrophication: A Global Perspective." *BioScience* 51(3): 227. https://doi.org/10.1641/0006-3568(2001)051[0227:HIOEPA]2.0.CO;2.

Bhaskar, Aditi S., Dianna M. Hogan, and Stacey A. Archfield. 2016. "Urban Base Flow with Low Impact Development: Urban Base Flow with Low Impact Development." *Hydrological Processes* 30(18): 3156–71. https://doi.org/10.1002/hyp.10808.

Bon, Hubert, Laurent Parrot, and Paule Moustier. 2010. "Sustainable Urban Agriculture in Developing Countries. A Review." *Agronomy for Sustainable Development* 30(1): 21–32. https://doi.org/10.1051/agro:2008062.

Bot, Alexandra, and José Benites. 2005. *The Importance of Soil Organic Matter: Key to Drought-Resistant Soil and Sustained Food Production*. FAO Soils Bulletin 80. Rome: Food and Agriculture Organization of the United Nations.

Bowles, Timothy M., Shady S. Atallah, Eleanor E. Campbell, Amélie C. M. Gaudin, William R. Wieder, and A. Stuart Grandy. 2018. "Addressing Agricultural Nitrogen Losses in a Changing Climate." *Nature Sustainability* 1(8): 399–408. https://doi.org/10.1038/s41893-018-0106-0.

Bren d'Amour, Christopher, Femke Reitsma, Giovanni Baiocchi, Stephan Barthel, Burak Güneralp, Karl-Heinz Erb, Helmut Haberl, Felix Creutzig, and Karen C. Seto. 2017. "Future Urban Land Expansion and Implications for Global Croplands." *Proceedings of the National Academy of Sciences* 114(34): 8939–44. https://doi.org/10.1073/pnas.1606036114.

Brenner, Neil. 2013. "Theses on Urbanization." *Public Culture* 25 (1): 85–114. https://doi.org/10.1215/08992363-1890477.

Brenner, Neil, and Christian Schmid. 2012. "Planetary Urbanisation." In *Urban Constellations*, edited by Matthew Gandy, 10–13. Berlin: Jovis.

Brevik, Eric C., Joshua J. Steffan, Jesús Rodrigo-Comino, Darrell Neubert, Lynn C. Burgess, and Artemi Cerdà. 2019. "Connecting the Public with Soil to Improve Human Health." *European Journal of Soil Science* 70(4): 898–910. https://doi.org/10.1111/ejss.12764.

Brodt, Sonja B. 2001. "A Systems Perspective on the Conservation and Erosion of Indigenous Agricultural Knowledge in Central India." *Human Ecology*, 29, 99–120. https://doi.org/10.1023/A:1007147806213

Brown, Sandy, and Christy Getz, eds. 2011. "Farmworker Food Insecurity and the Production of Hunger in California." In *Cultivating Food Justice: Race, Class, and Sustainability*. Cambridge, Massachusetts: The MIT Press. https://doi.org/10.7551/mitpress/8922.001.0001.

Byrne, Loren B. 2007. "Habitat Structure: A Fundamental Concept and Framework for Urban Soil Ecology." *Urban Ecosystems* 10(3): 255–74. https://doi.org/10.1007/s11252-007-0027-6.

Carney, Judith Ann, and Richard Nicholas Rosomoff. 2009. *In the Shadow of Slavery: Africa's Botanical Legacy in the Atlantic World.* Berkeley: University of California Press.

Caruso, Tancredi, Massimo Migliorini, Emilia Rota, and Roberto Bargagli. 2017. "Highly Diverse Urban Soil Communities: Does Stochasticity Play a Major Role?" *Applied Soil Ecology* 110 (February): 73–78. https://doi.org/10.1016/j.apsoil.2016.10.012.

Cassidy, Emily S, Paul C West, James S Gerber, and Jonathan A Foley. 2013. "Redefining Agricultural Yields: From Tonnes to People Nourished per Hectare." *Environmental Research Letters* 8(3): 034015. https://doi.org/10.1088/1748-9326/8/3/034015.

Cattle, Stephen R., Carl Robinson, and Mark Whatmuff. 2020. "The Character and Distribution of Physical Contaminants Found in Soil Previously Treated with Mixed Waste Organic Outputs and Garden Waste Compost." *Waste Management* 101 (January): 94–105. https://doi.org/10.1016/j.wasman.2019.09.043.

Champion, Tony, and Graeme Hugo, eds. 2004. *New Forms of Urbanization: Beyond the Urban–Rural Dichotomy.* Burlington, VT: Ashgate.

Chappell, M. Jahi. 2019. *Beginning to End Hunger: Food and the Environment in Belo Horizonte, Brazil, and Beyond.* Berkeley: University of California Press. https://doi.org/10.1525/9780520966338.

Chen, Jie. 2007. "Rapid Urbanization in China: A Real Challenge to Soil Protection and Food Security." *CATENA* 69(1): 1–15. https://doi.org/10.1016/j.catena.2006.04.019.

Clinton, Nicholas, Michelle Stuhlmacher, Albie Miles, Nazli Uludere Aragon, Melissa Wagner, Matei Georgescu, Chris Herwig, and Peng Gong. 2018. "A Global Geospatial Ecosystem Services Estimate of Urban Agriculture." *Earth's Future* 6(1): 40–60. https://doi.org/10.1002/2017EF000536.

Condon, Patrick M., Kent Mullinix, Arthur Fallick, and Mike Harcourt. 2010. "Agriculture on the Edge: Strategies to Abate Urban Encroachment onto Agricultural Lands by Promoting Viable Human-Scale Agriculture as an Integral Element of Urbanization." *International Journal of Agricultural Sustainability* 8(1–2): 104–15. https://doi.org/10.3763/ijas.2009.0465.

Cook, Ian R., and Erik Swyngedouw. 2012. "Cities, Social Cohesion and the Environment: Towards a Future Research Agenda." *Urban Studies* 49(9): 1959–79. https://doi.org/10.1177/0042098012444887.

Cotrufo, M. Francesca, Matthew D. Wallenstein, Claudia M. Boot, Karolien Denef, and Eldor Paul. 2013. "The Microbial Efficiency-Matrix Stabilization (MEMS) Framework Integrates Plant Litter Decomposition with Soil Organic Matter Stabilization: Do Labile Plant Inputs Form Stable Soil Organic Matter?" *Global Change Biology* 19(4): 988–95. https://doi.org/10.1111/gcb.12113.

Craul, Phillip J. 1985. "A Description of Urban Soils and Their Desired Characteristics." *Journal of Arboriculture* 11(11).

Craul, Phillip J. 1992. *Urban Soil in Landscape Design.* New York: Wiley.

Curran, Christopher J., and Marc-Tizoc González. 2011. "Food Justice as Interracial Justice: Urban Farmers, Community Organizations and the Role of Government in Oakland, California." *U. Miami Inter-American Law Review* 43: 207–32.

Cutts, Bethany B., Danqi Fang, Kaitlyn Hornik, Jonathan K. London, Kirsten Schwarz, and Mary L. Cadenasso. 2016. "Media Frames and Shifting Places of Environmental (In)Justice: A Qualitative Historical Geographic Information System Method." *Environmental Justice* 9(1): 23–28. https://doi.org/10.1089/env.2015.0027.

Cutts, Bethany B., Jonathan K. London, Shaina Meiners, Kirsten Schwarz, and Mary L. Cadenasso. 2017. "Moving Dirt: Soil, Lead, and the Dynamic Spatial Politics of Urban Gardening." *Local Environment* 22(8): 998–1018. https://doi.org/10.1080/13549839.2017.1320539.

Daniel, Pete. 2013. *Dispossession: Discrimination against African American Farmers in the Age of Civil Rights.* Chapel Hill: University of North Carolina Press.

Davis, James C., and J. Vernon Henderson. 2003. "Evidence on the Political Economy of the Urbanization Process." *Journal of Urban Economics* 53(1): 98–125. https://doi.org/10.1016/S0094-1190(02)00504-1.

De Zeeuw, H., R. Van Veenhuizen, and M. Dubbeling. 2011. "The Role of Urban Agriculture in Building Resilient Cities in Developing Countries." *The Journal of Agricultural Science* 149(S1): 153–63. https://doi.org/10.1017/S0021859610001279.

DeLonge, Marcia S., Rebecca Ryals, and Whendee L. Silver. 2013. "A Lifecycle Model to Evaluate Carbon Sequestration Potential and Greenhouse Gas Dynamics of Managed Grasslands." *Ecosystems* 16(6): 962–79. https://doi.org/10.1007/s10021-013-9660-5.

Ding, Kengbo, Qing Wu, Hang Wei, Wenjun Yang, Geoffroy Séré, Shizhong Wang, Guillaume Echevarria, et al. 2018. "Ecosystem Services Provided by Heavy Metal-Contaminated Soils in China." *Journal of Soils and Sediments* 18(2): 380–90. https://doi.org/10.1007/s11368-016-1547-6.

Dixon, Jane, Abiud M. Omwega, Sharon Friel, Cate Burns, Kelly Donati, and Rachel Carlisle. 2007. "The Health Equity Dimensions of Urban Food Systems." *Journal of Urban Health* 84 (S1): 118–29. https://doi.org/10.1007/s11524-007-9176-4.

DOC. 2016. "California Farmland Conversion Summary." Department of Conservation. www.conservation.ca.gov/dlrp/fmmp/trends/Pages/FastFacts.aspx.

D'Odorico, Paolo, Abinash Bhattachan, Kyle F. Davis, Sujith Ravi, and Christiane W. Runyan. 2013. "Global Desertification: Drivers and Feedbacks." *Advances in Water Resources* 51 (January): 326–44. https://doi.org/10.1016/j.advwatres.2012.01.013.

Dominati, Estelle, Murray Patterson, and Alec Mackay. 2010. "A Framework for Classifying and Quantifying the Natural Capital and Ecosystem Services of Soils." *Ecological Economics* 69(9): 1858–68. https://doi.org/10.1016/j.ecolecon.2010.05.002.

Donald, Paul F., and Andy D. Evans. 2006. "Habitat Connectivity and Matrix Restoration: The Wider Implications of Agri-Environment Schemes: Habitat Connectivity and Matrix Restoration." *Journal of Applied Ecology* 43(2): 209–18. https://doi.org/10.1111/j.1365-2664.2006.01146.x.

Doran, John W. 2002. "Soil Health and Global Sustainability: Translating Science into Practice." *Agriculture, Ecosystems & Environment* 88: 119–27.

Doran, John W., and Michael R. Zeiss. 2000. "Soil Health and Sustainability: Managing the Biotic Component of Soil Quality." *Applied Soil Ecology* 15(1): 3–11. https://doi.org/10.1016/S0929-1393(00)00067-6.

Edelman, Marc. 2014. "Food Sovereignty: Forgotten Genealogies and Future Regulatory Challenges." *The Journal of Peasant Studies* 41(6): 959–78. https://doi.org/10.1080/03066150.2013.876998.

Edmonds, Penelope. 2010. *Urbanizing Frontiers: Indigenous Peoples and Settlers in 19th-Century Pacific Rim Cities*. Vancouver: UBC Press.

Egerer, Monika, Alessandro Ossola, and Brenda B Lin. 2018. "Creating Socioecological Novelty in Urban Agroecosystems from the Ground Up." *BioScience* 68 (1): 25–34. https://doi.org/10.1093/biosci/bix144.

Eisenhauer, Elizabeth. 2002. "In Poor Health: Supermarket Redlining and Urban Nutrition." *GeoJournal* 53: 125–33.

Ellis, Erle C. 2011. "Anthropogenic Transformation of the Terrestrial Biosphere." *Philosophical Transactions of the Royal Society A: Mathematical, Physical and Engineering Sciences* 369(1938): 1010–35. https://doi.org/10.1098/rsta.2010.0331.

Ellis, Erle C., Kees Klein Goldewijk, Stefan Siebert, Deborah Lightman, and Navin Ramankutty. 2010. "AnthropogenicTransformationoftheBiomes,1700to2000:AnthropogenicTransformationoftheBiomes." *Global Ecology and Biogeography*, June, no-no. https://doi.org/10.1111/j.1466-8238.2010.00540.x.

Ellis, Erle C., Hongqing Wang, Hong Sheng Xiao, Kui Peng, Xin Ping Liu, Shou Cheng Li, Hua Ouyang, Xu Cheng, and Lin Zhang Yang. 2006. "Measuring Long-Term Ecological Changes in Densely Populated Landscapes Using Current and Historical High Resolution Imagery." *Remote Sensing of Environment* 100(4): 457–73. https://doi.org/10.1016/j.rse.2005.11.002.

Ernstson, Henrik. 2013. "The Social Production of Ecosystem Services: A Framework for Studying Environmental Justice and Ecological Complexity in Urbanized Landscapes." *Landscape and Urban Planning* 109(1): 7–17. https://doi.org/10.1016/j.landurbplan.2012.10.005.

FAO & ITPS. 2015. *Status of the World's Soil Resources: Main Report*. Rome: Food and Agriculture Organization of the United Nations & Intergovernmental Technical Panel on Soils.

Foster, John Bellamy. 1999. "Marx's Theory of Metabolic Rift: Classical Foundations for Environmental Sociology." *American Journal of Sociology* 105(2): 366–405. https://doi.org/10.1086/210315.

Foster, John Bellamy, and Fred Magdoff. 1998. "Liebig, Marx and the Depletion of Soil Fertility: Relevance for Today's Agriculture." *Monthly Review* 50 (3): 32. https://doi.org/10.14452/MR-050-03-1998-07_3.

Francis, Mark. 1987. "Meanings Attached to a City Park and a Community Garden in Sacramento." *Landscape Research* 12(1): 8–12. https://doi.org/10.1080/01426398708706216.

Frankie, Gordon W., Robbin W. Thorp, Jennifer Hernandez, Mark Rizzardi, Barbara Ertter, Jaime C. Pawelek, Sara L. Witt, Mary Schindler, Rollin Coville, and Victoria A. Wojcik. 2009. "Native Bees Are a Rich Natural Resource in Urban California Gardens." *California Agriculture* 63(3): 113–20. https://doi.org/10.3733/ca.v063n03p113.

Geisen, Stefan, Diana H. Wall, and Wim H. van der Putten. 2019. "Challenges and Opportunities for Soil Biodiversity in the Anthropocene." *Current Biology* 29(19): R1036–44. https://doi.org/10.1016/j.cub.2019.08.007.

Gill, S.E, J.F. Handley, A.R. Ennos, and S. Pauleit. 2007. "Adapting Cities for Climate Change: The Role of the Green Infrastructure." *Built Environment* 33(1): 115–33. https://doi.org/10.2148/benv.33.1.115.

Glaeser, Edward L. 2014. "A World of Cities: The Causes and Consequences of Urbanization in Poorer Countries." *Journal of the European Economic Association* 12(5): 1154–99.

Gliessman, S.R., R.E. Garcia, and M.A. Amador. 1981. "The Ecological Basis for the Application of Traditional Agricultural Technology in the Management of Tropical Agro-Ecosystems." *Agro-Ecosystems* 7(3): 173–85. https://doi.org/10.1016/0304-3746(81)90001-9.

Gómez-Bagget, Erik, Sara Mingorría, Victoria Reyes-García, Laura Calvet, and Carlos Montes. 2010. "Traditional Ecological Knowledge Trends in the Transition to a Market Economy: Empirical Study in the Doñana Natural Areas." *Conservation Biology* 24(3): 721–29.

Gomiero, Tiziano. 2016. "Soil Degradation, Land Scarcity and Food Security: Reviewing a Complex Challenge." *Sustainability* 8(3): 281. https://doi.org/10.3390/su8030281.

Gozgor, Giray, and Baris Kablamaci. 2015. "What Happened to Urbanization in the Globalization Era? An Empirical Examination for Poor Emerging Countries." *The Annals of Regional Science* 55(2–3): 533–53. https://doi.org/10.1007/s00168-015-0716-7.

Groffman, Peter M., Meghan Avolio, Jeannine Cavender-Bares, Neil D. Bettez, J. Morgan Grove, Sharon J. Hall, Sarah E. Hobbie, et al. 2017. "Ecological Homogenization of Residential Macrosystems." *Nature Ecology & Evolution* 1(7): 0191. https://doi.org/10.1038/s41559-017-0191.

Grossman, Julie, Maximilian Sherard, Seb M. Prohn, Lucy Bradley, L. Suzanne Goodell, and Katherine Andrew. 2012. "An Exploratory Analysis of Student- Community Interactions in Urban Agriculture." *Journal of Higher Education Outreach and Engagement* 16(2): 178–96.

Haase, Dagmar, and Henning Nuissl. 2010. "The Urban-to-Rural Gradient of Land Use Change and Impervious Cover: A Long-Term Trajectory for the City of Leipzig." *Journal of Land Use Science* 5(2): 123–41. https://doi.org/10.1080/1747423X.2010.481079.

Hall, S. J., B. Ahmed, P. Ortiz, R. Davies, R. A. Sponseller, and N. B. Grimm. 2009. "Urbanization Alters Soil Microbial Functioning in the Sonoran Desert." *Ecosystems* 12(4): 654–71. https://doi.org/10.1007/s10021-009-9249-1.

Haller. 2019. "The Different Meanings of Land in the Age of Neoliberalism: Theoretical Reflections on Commons and Resilience Grabbing from a Social Anthropological Perspective." *Land* 8(7): 104. https://doi.org/10.3390/land8070104.

Hamilton, Andrew J., Kristal Burry, Hoi-Fei Mok, S. Fiona Barker, James R. Grove, and Virginia G. Williamson. 2014. "Give Peas a Chance? Urban Agriculture in Developing Countries. A Review." *Agronomy for Sustainable Development* 34(1): 45–73. https://doi.org/10.1007/s13593-013-0155-8.

Handbury, Jessie, Ilya M. Rahkovsky, and Molly Schnell. 2015. "What Drives Nutritional Disparities? Retail Access and Food Purchases Across the Socioeconomic Spectrum." *SSRN Electronic Journal*. https://doi.org/10.2139/ssrn.2632216.

Hannah Wittman. 2011. "Food Sovereignty: A New Rights Framework for Food and Nature." *Environment and Society* 2(1). https://doi.org/10.3167/ares.2011.020106.

Harrison, Brendan P., Evan Chopra, Rebecca Ryals, and J. Elliott Campbell. 2020. "Quantifying the Farmland Application of Compost to Help Meet California's Organic Waste Diversion Law." *Environmental Science & Technology* 54(7): 4545–53. https://doi.org/10.1021/acs.est.9b05377.

Harvey, David. 1996. "Cities or Urbanization?" *City* 1 (1–2): 38–61. https://doi.org/10.1080/13604819608900022.

Hawkes, Corinna. 2006. "Uneven Dietary Development: Linking the Policies and Processes of Globalization with the Nutrition Transition, Obesity and Diet-Related Chronic Diseases." *Globalization and Health* 2 (1): 4. https://doi.org/10.1186/1744-8603-2-4.

Heller, Martin C., and Gregory A. Keoleian. 2003. "Assessing the Sustainability of the US Food System: A Life Cycle Perspective." *Agricultural Systems* 76(3): 1007–41. https://doi.org/10.1016/S0308-521X(02)00027-6.

Herrmann, Dustin, Wen-Ching Chuang, Kirsten Schwarz, Timothy Bowles, Ahjond Garmestani, William Shuster, Tarsha Eason, Matthew Hopton, and Craig Allen. 2018. "Agroecology for the Shrinking City." *Sustainability* 10(3): 675. https://doi.org/10.3390/su10030675.

Herrmann, Dustin L., Laura A. Schifman, and William D. Shuster. 2018. "Widespread Loss of Intermediate Soil Horizons in Urban Landscapes." *Proceedings of the National Academy of Sciences* 115(26): 6751–55. https://doi.org/10.1073/pnas.1800305115.

Herrmann, Dustin L., William D. Shuster, and Ahjond S. Garmestani. 2017. "Vacant Urban Lot Soils and Their Potential to Support Ecosystem Services." *Plant and Soil* 413(1–2): 45–57. https://doi.org/10.1007/s11104-016-2874-5.

Heynen, Nik, Maria Kaika, and Erik Swyngedouw, eds. 2006. *In the Nature of Cities: Urban Political Ecology and the Politics of Urban Metabolism*. 0 ed. Routledge. https://doi.org/10.4324/9780203027523.

Hollis, J. M. 1991. "The Classification of Soils in Urban Areas." In *Soils in the Urban Environment*, edited by Peter Bullock and Peter J. Gregory, 5–27. Oxford, UK: Blackwell Publishing Ltd. https://doi.org/10.1002/9781444310603.ch2.

Holt Giménez, Eric, and Annie Shattuck. 2011. "Food Crises, Food Regimes and Food Movements: Rumblings of Reform or Tides of Transformation." *Journal of Peasant Studies* 38(1): 109–44. https://doi.org/10.1080/03066150.2010.538578.

Hoover, Elizabeth. 2010. "*Local Food Production and Community Illness Narratives: Responses to Environmental Contaminatino and Health Studies in the Mohawk Community of Akwesasne*." Providence, RI: Brown University.

———. 2017a. "'You Can't Say You're Sovereign If You Can't Feed Yourself': Defining and Enacting Food Sovereignty in American Indian Community Gardening." *American Indian Culture and Research Journal* 41(3): 31–70. https://doi.org/10.17953/aicrj.41.3.hoover.

———. 2017b. *The River Is in Us: Fighting Toxics in a Mohawk Community*. University of Minnesota Press. https://doi.org/10.5749/j.ctt1pwt6mk.

Hoover, Elizabeth, Katsi Cook, Ron Plain, Kathy Sanchez, Vi Waghiyi, Pamela Miller, Renee Dufault, Caitlin Sislin, and David O. Carpenter. 2012. "Indigenous Peoples of North America: Environmental Exposures and Reproductive Justice." *Environmental Health Perspectives* 120(12): 1645–49. https://doi.org/10.1289/ehp.1205422.

Hope, Diane, Weixing Zhu, Corinna Gries, Jacob Oleson, Jason Kaye, Nancy B. Grimm, and Lawrence A. Baker. 2005. "Spatial Variation in Soil Inorganic Nitrogen across an Arid Urban Ecosystem." *Urban Ecosystems* 8 (3–4): 251–73. https://doi.org/10.1007/s11252-005-3261-9.

Hunter, Mitchell C., Richard G. Smith, Meagan E. Schipanski, Lesley W. Atwood, and David A. Mortensen. 2017. "Agriculture in 2050: Recalibrating Targets for Sustainable Intensification." *BioScience* 67(4): 386–91. https://doi.org/10.1093/biosci/bix010.

ICC. 2018. "International Building Code." International Code Council. https://codes.iccsafe.org/content/IBC2018/chapter-18-soils-and-foundations.

Isbell, Forest, David Tilman, Peter B. Reich, and Adam Thomas Clark. 2019. "Deficits of Biodiversity and Productivity Linger a Century after Agricultural Abandonment." *Nature Ecology & Evolution* 3(11): 1533–38. https://doi.org/10.1038/s41559-019-1012-1.

Jackson, Louise, Van R. Haden, Allan D. Hollander, Hyunok Lee, Mark Lubell, Vishal K. Mehta, Toby O'Geen, et al. 2012. "*Adaptation Strategies for Agricultural Sustainability in Yolo County, California*." California Energy Commission.

Johnson, Anna L., Erica C. Tauzer, and Christopher M. Swan. 2015. "Human Legacies Differentially Organize Functional and Phylogenetic Diversity of Urban Herbaceous Plant Communities at Multiple Spatial Scales." Edited by Norbert Hölzel. *Applied Vegetation Science* 18(3): 513–27. https://doi.org/10.1111/avsc.12155.

Johnston, A. Edward, Paul R. Poulton, and Kevin Coleman. 2009. "Soil Organic Matter: Its Importance in Sustainable Agriculture and Carbon Dioxide Fluxes." In *Advances in Agronomy*, 101:1–57. Elsevier. https://doi.org/10.1016/S0065-2113(08)00801-8.

Jones, Daniel K., Matthew E. Baker, Andrew J. Miller, S. Taylor Jarnagin, and Dianna M. Hogan. 2014. "Tracking Geomorphic Signatures of Watershed Suburbanization with Multitemporal LiDAR." *Geomorphology* 219 (August): 42–52. https://doi.org/10.1016/j.geomorph.2014.04.038.

Kallenbach, Cynthia M., Serita D. Frey, and A. Stuart Grandy. 2016. "Direct Evidence for Microbial-Derived Soil Organic Matter Formation and Its Ecophysiological Controls." *Nature Communications* 7(1): 13630. https://doi.org/10.1038/ncomms13630.

Kallenbach, Cynthia M., Matthew D. Wallenstein, Meagan E. Schipanksi, and A. Stuart Grandy. 2019. “Managing Agroecosystems for Soil Microbial Carbon Use Efficiency: Ecological Unknowns, Potential Outcomes, and a Path Forward.” *Frontiers in Microbiology* 10 (May): 1146. https://doi.org/10.3389/fmicb.2019.01146.

Karg, Hanna, Pay Drechsel, Edmund Akoto-Danso, Rüdiger Glaser, George Nyarko, and Andreas Buerkert. 2016. “Foodsheds and City Region Food Systems in Two West African Cities.” *Sustainability* 8(12): 1175. https://doi.org/10.3390/su8121175.

Kaye, J, P. Groffman, N. Grimm, L. Baker, and R. Pouyat. 2006. “A Distinct Urban Biogeochemistry.” *Trends in Ecology & Evolution* 21(4): 192–99. https://doi.org/10.1016/j.tree.2005.12.006.

Kibblewhite, M.G., K. Ritz, and M.J. Swift. 2008. “Soil Health in Agricultural Systems.” *Philosophical Transactions of the Royal Society B: Biological Sciences* 363(1492): 685–701. https://doi.org/10.1098/rstb.2007.2178.

Kim, Brent F., Melissa N. Poulsen, Jared D. Margulies, Katie L. Dix, Anne M. Palmer, and Keeve E. Nachman. 2014. “Urban Community Gardeners’ Knowledge and Perceptions of Soil Contaminant Risks.” Edited by Sunghun Park. *PLoS ONE* 9(2): e87913. https://doi.org/10.1371/journal.pone.0087913.

Knight, A., Z. Cheng, S. S. Grewal, K. R. Islam, M. D. Kleinhenz, and P. S. Grewal. 2013. “Soil Health as a Predictor of Lettuce Productivity and Quality: A Case Study of Urban Vacant Lots.” *Urban Ecosystems* 16(3): 637–56. https://doi.org/10.1007/s11252-013-0288-1.

Knorr, Dietrich, Chor San Heng Khoo, and Mary Ann Augustin. 2018. “Food for an Urban Planet: Challenges and Research Opportunities.” *Frontiers in Nutrition* 4 (January): 73. https://doi.org/10.3389/fnut.2017.00073.

Krause, M. 2013. “The Ruralization of the World.” *Public Culture* 25(2 70): 233–48. https://doi.org/10.1215/08992363-2020575.

Kremen, Claire, Alastair Iles, and Christopher Bacon. 2012. “Diversified Farming Systems: An Agroecological, Systems-Based Alternative to Modern Industrial Agriculture.” *Ecology and Society* 17(4): art44. https://doi.org/10.5751/ES-05103-170444.

Krugman, Paul, and Raul Livas Elizondo. 1996. “Trade Policy and the Third World Metropolis.” *Journal of Development Economics* 49: 137–50.

Krzywoszynska, Anna., and Greta Marchesi. 2020. “Towards a Relational Materiality of Soils.” *Environmental Humanities* 12 (1): 190–204.

Kumar, Anuj, Christopher P. Alaimo, Robert Horowitz, Frank M. Mitloehner, Michael J. Kleeman, and Peter G. Green. 2011. “Volatile Organic Compound Emissions from Green Waste Composting: Characterization and Ozone Formation.” *Atmospheric Environment* 45(10): 1841–48. https://doi.org/10.1016/j.atmosenv.2011.01.014.

Lal, Rattan. 2007. “Anthropogenic Influences on World Soils and Implications to Global Food Security.” In *Advances in Agronomy*, 93:69–93. Elsevier. https://doi.org/10.1016/S0065-2113(06)93002-8.

Lefebvre, Henri. 2003. *The Urban Revolution*. University of Minnesota Press.

Lehman, R., Cynthia Cambardella, Diane Stott, Veronica Acosta-Martinez, Daniel Manter, Jeffrey Buyer, Jude Maul, et al. 2015. “Understanding and Enhancing Soil Biological Health: The Solution for Reversing Soil Degradation.” *Sustainability* 7(1): 988–1027. https://doi.org/10.3390/su7010988.

Lehmann, Andreas, and Karl Stahr. 2007. “Nature and Significance of Anthropogenic Urban Soils.” *Journal of Soils and Sediments* 7(4): 247–60. https://doi.org/10.1065/jss2007.06.235.

Lehmann, Johannes, and Markus Kleber. 2015. “The Contentious Nature of Soil Organic Matter.” *Nature* 528(7580): 60–68. https://doi.org/10.1038/nature16069.

Li, Siyue, and Zhongmin Jia. 2018. “Heavy Metals in Soils from a Representative Rapidly Developing Megacity (SW China): Levels, Source Identification and Apportionment.” *CATENA* 163 (April): 414–23. https://doi.org/10.1016/j.catena.2017.12.035.

Lin, Brenda B., Monika H. Egerer, and Alessandro Ossola. 2018. “Urban Gardens as a Space to Engender Biophilia: Evidence and Ways Forward.” *Frontiers in Built Environment* 4 (December): 79. https://doi.org/10.3389/fbuil.2018.00079.

Makse, Hernán A., Shlomo Havlin, and H. Eugene Stanley. 1995. “Modelling Urban Growth Patterns.” *Nature* 377(6550): 608–12. https://doi.org/10.1038/377608a0.

Marcotullio, Peter J., Ademola K. Braimoh, and Takashi Onishi. 2008. “The Impact of Urbanization on Soils.” In *Land Use and Soil Resources*, edited by Ademola K. Braimoh and Paul L. G. Vlek, 201–50. Dordrecht: Springer Netherlands. https://doi.org/10.1007/978-1-4020-6778-5_10.

Marsalek, J., B.E. Jiménez-Cisneros, P.-A. Malmquist, M. Karamouz, J. Goldenfum, and B. Chocat. 2006. "Urban Water Cycle Processes and Interactions." *Technical Documents in Hydrology*. Paris: International Hydrological Programme, UNESCO.

Martino, Pesaresi, Melchiorri Michele, Siragusa Alice, and Kemper Thomas. 2016. "Atlas of the Human Planet 2016. Mapping Human Presence on Earth with the Global Human Settlement Layer." European Commission.

Marx, Karl. 1959. *Capital: A Critique of Political Economy, Vol. 3*. New York, NY: International Publishers.

Mays, Kyle T. 2015. *Indigenous Detroit: Indigeneity, Modernity, and Racial and Gender Formation in a Modern American City*, 1871–2000. Urbana, Illinois: University of Illinois at Urbana-Champaign.

McCarter, Joe, Michael C. Gavin, Sue Baereleo, and Mark Love. 2014. "The Challenges of Maintaining Indigenous Ecological Knowledge." *Ecology and Society* 19(3). https://www.jstor.org/stable/26269617.

McClintock, N. 2010. "Why Farm the City? Theorizing Urban Agriculture through a Lens of Metabolic Rift." *Cambridge Journal of Regions, Economy and Society* 3(2): 191–207. https://doi.org/10.1093/cjres/rsq005.

McClintock, Nathan. 2011. "From Industrial Garden to Food Desert: Demarcated Devaluation in the Flatlands of Oakland, California." In *Cultivating Food Justice: Race, Class, and Sustainability*. MIT Press. https://direct.mit.edu/books/book/4423/chapter-abstract/189390/From-Industrial-Garden-to-Food-Desert-Demarcated?redirectedFrom=fulltext.

———. 2012. "Assessing Soil Lead Contamination at Multiple Scales in Oakland, California: Implications for Urban Agriculture and Environmental Justice." *Applied Geography* 35(1–2): 460–73. https://doi.org/10.1016/j.apgeog.2012.10.001.

———. 2015. "A Critical Physical Geography of Urban Soil Contamination." *Geoforum* 65 (October): 69–85. https://doi.org/10.1016/j.geoforum.2015.07.010.

McDonnell, M. J., and S. T. A. Pickett. 1990. "Ecosystem Structure and Function along Urban–Rural Gradients: An Unexploited Opportunity for Ecology." *Ecology* 71 (4): 1232–37. https://doi.org/10.2307/1938259.

McDonnell, Mark J., Steward T.A. Pickett, Peter Groffman, Patrick Bohlen, Richard V. Pouyat, Wayne C. Zipperer, Robert W. Parmelee, Margaret M. Carreiro, and Kimberly Medley. 2008. "Ecosystem Processes Along an Urban-to-Rural Gradient." In *Urban Ecology*, edited by John M. Marzluff, Eric Shulenberger, Wilfried Endlicher, Marina Alberti, Gordon Bradley, Clare Ryan, Ute Simon, and Craig ZumBrunnen, 299–313. Boston, MA: Springer US. https://doi.org/10.1007/978-0-387-73412-5_18.

McDougall, Robert, Paul Kristiansen, and Romina Rader. 2019. "Small-Scale Urban Agriculture Results in High Yields but Requires Judicious Management of Inputs to Achieve Sustainability." *Proceedings of the National Academy of Sciences* 116(1): 129–34. https://doi.org/10.1073/pnas.1809707115.

McMichael, Philip. 2011. "Food System Sustainability: Questions of Environmental Governance in the New World (Dis)Order." *Global Environmental Change* 21(3): 804–12. https://doi.org/10.1016/j.gloenvcha.2011.03.016.

Meenar, Mahbubur, and Brandon Hoover. 2012. "Community Food Security via Urban Agriculture: Understanding People, Place, Economy, and Accessibility from a Food Justice Perspective." *Journal of Agriculture, Food Systems, and Community Development*, November, 143–60. https://doi.org/10.5304/jafscd.2012.031.013.

Mendez, Michelle A., and Barry M. Popkin. 2004. "*Globalization, Urbanization and Nutritional Change in the Developing World*." *Journal of Development and Agricultural Economics* 1(2): 220–241.

Millennium Ecosystem Assessment (Program), ed. 2005. *Ecosystems and Human Well-Being: Synthesis*. Washington, DC: Island Press.

Miltner, Anja, Petra Bombach, Burkhard Schmidt-Brücken, and Matthias Kästner. 2012. "SOM Genesis: Microbial Biomass as a Significant Source." *Biogeochemistry* 111(1–3): 41–55. https://doi.org/10.1007/s10533-011-9658-z.

Minasny, Budiman, Brendan P. Malone, Alex B. McBratney, Denis A. Angers, Dominique Arrouays, Adam Chambers, Vincent Chaplot, et al. 2017. "Soil Carbon 4 per Mille." *Geoderma* 292 (April): 59–86. https://doi.org/10.1016/j.geoderma.2017.01.002.

Mitchell, Jeffrey P., Anil Shrestha, Konrad Mathesius, Kate M. Scow, Randal J. Southard, Richard L. Haney, Radomir Schmidt, Daniel S. Munk, and William R. Horwath. 2017. "Cover Cropping and No-Tillage Improve Soil Health in an Arid Irrigated Cropping System in California's San Joaquin Valley, USA." *Soil and Tillage Research* 165 (January): 325–35. https://doi.org/10.1016/j.still.2016.09.001.

Moebius-Clune, Bianca Nadine. 2016. Comprehensive Assessment of Soil Health: The Cornell Framework Manual.

Mok, Hoi-Fei, Virginia G. Williamson, James R. Grove, Kristal Burry, S. Fiona Barker, and Andrew J. Hamilton. 2014. "Strawberry Fields Forever? Urban Agriculture in Developed Countries: A Review." *Agronomy for Sustainable Development* 34(1): 21–43. https://doi.org/10.1007/s13593-013-0156-7.

Montejo, C., P. Ramos, C. Costa, and M.C. Márquez. 2010. "Analysis of the Presence of Improper Materials in the Composting Process Performed in Ten MBT Plants." *Bioresource Technology* 101(21): 8267–72. https://doi.org/10.1016/j.biortech.2010.06.024.

Morel, Jean Louis, Claire Chenu, and Klaus Lorenz. 2015. "Ecosystem Services Provided by Soils of Urban, Industrial, Traffic, Mining, and Military Areas (SUITMAs)." *Journal of Soils and Sediments* 15 (8): 1659–66. https://doi.org/10.1007/s11368-014-0926-0.

Nahuelhual, Laura, Alejandra Carmona, Paola Lozada, Amerindia Jaramillo, and Mauricio Aguayo. 2013. "Mapping Recreation and Ecotourism as a Cultural Ecosystem Service: An Application at the Local Level in Southern Chile." *Applied Geography* 40 (June): 71–82. https://doi.org/10.1016/j.apgeog.2012.12.004.

Naybor, Deborah. 2015. "Land as Fictitious Commodity: The Continuing Evolution of Women's Land Rights in Uganda." *Gender, Place & Culture* 22(6): 884–900. https://doi.org/10.1080/0966369X.2014.917275.

Nyéléni. 2007. "Declaration of Nyéléni." In. Selingue, Mali. https://nyeleni.org/spip.php?article290.

———. 2015. "Declaration of the International Forum for Agroecology." Nyéléni, Mali. https://viacampesina.org/en/declaration-of-the-international-forum-for-agroecology/.

———. 2018. "Food Sovereignty at the Rural-Urban Interface." 35. Nyéléni Newsletter.

Obrycki, John F., Nicholas T. Basta, and Steven W. Culman. 2017. "Management Options for Contaminated Urban Soils to Reduce Public Exposure and Maintain Soil Health." *Journal of Environmental Quality* 46(2): 420–30. https://doi.org/10.2134/jeq2016.07.0275.

Odewumi, Odewumi. 2013. "Farmers Perception on the Effect of Climate Change and Variation on Urban Agriculture in Ibadan Metropolis, South-Western Nigeria." *Journal of Geography and Regional Planning* 6(6): 209–17. https://doi.org/10.5897/JGRP2013.0370.

Ooms, A., A.T.C. Dias, A.R. van Oosten, J.H.C. Cornelissen, J. Ellers, and M.P. Berg. 2020. "Species Richness and Functional Diversity of Isopod Communities Vary across an Urbanisation Gradient, but the Direction and Strength Depend on Soil Type." *Soil Biology and Biochemistry*, May, 107851. https://doi.org/10.1016/j.soilbio.2020.107851.

Oteros-Rozas, Elisa, Adriana Ruiz-Almeida, Mateo Aguado, José A. González, and Marta G. Rivera-Ferre. 2019. "A Social–Ecological Analysis of the Global Agrifood System." *Proceedings of the National Academy of Sciences* 116(52): 26465–73. https://doi.org/10.1073/pnas.1912710116.

Patel, Raj. 2009. "Food Sovereignty." *The Journal of Peasant Studies* 36(3): 663–706. https://doi.org/10.1080/03066150903143079.

Paterson, E., M. Sanka, and L. Clark. 1996. "Urban Soils as Pollutant Sinks — a Case Study from Aberdeen, Scotland." *Applied Geochemistry* 11(1–2): 129–31. https://doi.org/10.1016/0883-2927(95)00081-X.

Pathak, Tapan, Mahesh Maskey, Jeffery Dahlberg, Faith Kearns, Khaled Bali, and Daniele Zaccaria. 2018. "Climate Change Trends and Impacts on California Agriculture: A Detailed Review." *Agronomy* 8(3): 25. https://doi.org/10.3390/agronomy8030025.

Patterson, J.C., J.J. Murray, and J.R. Short. 1980. "The Impact of Urban Soils on Vegetation." *Proceedings of the Third Conference of the Metropolitan Tree Improvement Alliance* 3: 33–56.

Pavao-Zuckerman, Mitchell A. 2008. "The Nature of Urban Soils and Their Role in Ecological Restoration in Cities." *Restoration Ecology* 16(4): 642–49. https://doi.org/10.1111/j.1526-100X.2008.00486.x.

Pavao-Zuckerman, Mitchell A., and David C. Coleman. 2007. "Urbanization Alters the Functional Composition, but Not Taxonomic Diversity, of the Soil Nematode Community." *Applied Soil Ecology* 35(2): 329–39. https://doi.org/10.1016/j.apsoil.2006.07.008.

Pellow, David N., and Lisa Sun-Hee Park. 2002. *The Silicon Valley of Dreams: Environmental Injustice, Immigrant Workers, and the High-Tech Global Economy. Critical America.* New York: New York University Press.

Pickett, S. T. A., and M. L. Cadenasso. 2009. "Altered Resources, Disturbance, and Heterogeneity: A Framework for Comparing Urban and Non-Urban Soils." *Urban Ecosystems* 12 (1): 23–44. https://doi.org/10.1007/s11252-008-0047-x.

Pickett, S. T. A., M. L. Cadenasso, J. M. Grove, C. H. Nilon, R. V. Pouyat, W. C. Zipperer, and R. Costanza. 2001. "Urban Ecological Systems: Linking Terrestrial Ecological, Physical, and Socioeconomic Components of Metropolitan Areas." *Annual Review of Ecology and Systematics* 32(1): 127–57. https://doi.org/10.1146/annurev.ecolsys.32.081501.114012.

Pickett, S. T. A., M. L. Cadenasso, E. J. Rosi-Marshall, K. T. Belt, P. M. Groffman, J. M. Grove, E. G. Irwin, et al. 2017. "Dynamic Heterogeneity: A Framework to Promote Ecological Integration and Hypothesis Generation in Urban Systems." *Urban Ecosystems* 20(1): 1–14. https://doi.org/10.1007/s11252-016-0574-9.

Piso, Zachary, Lissy Goralnik, Julie C. Libarkin, and Maria Claudia Lopez. 2019. "Types of Urban Agricultural Stakeholders and Their Understandings of Governance." *Ecology and Society* 24(2): art18. https://doi.org/10.5751/ES-10650-240218.

Popkin, Barry M. 1999. "Urbanization, Lifestyle Changes and the Nutrition Transition." *World Development* 27(11): 1905–16. https://doi.org/10.1016/S0305-750X(99)00094-7.

Postel, S. L., G. C. Daily, and P. R. Ehrlich. 1996. "Human Appropriation of Renewable Fresh Water." *Science* 271(5250): 785–88. https://doi.org/10.1126/science.271.5250.785.

Pouyat, Richard. V., and M. J. McDonnell. 1991. "Heavy Metal Accumulations in Forest Soils along an Urban–Rural Gradient in Southeastern New York, USA." *Water, Air, and Soil Pollution* 57–58(1): 797–807. https://doi.org/10.1007/BF00282943.

Pouyat, Richard V., Ian D. Yesilonis, Miklós Dombos, Katalin Szlavecz, Heikki Setälä, Sarel Cilliers, Erzsébet Hornung, D. Johan Kotze, and Stephanie Yarwood. 2015. "A Global Comparison of Surface Soil Characteristics Across Five Cities: A Test of the Urban Ecosystem Convergence Hypothesis." *Soil Science* 180(4/5): 136–45. https://doi.org/10.1097/SS.0000000000000125.

Pouyat, Richard V., Ian D. Yesilonis, and David J. Nowak. 2006. "Carbon Storage by Urban Soils in the United States." *Journal of Environmental Quality*, 1566–1575.

Pribyl, Douglas W. 2010. "A Critical Review of the Conventional SOC to SOM Conversion Factor." *Geoderma* 156(3–4): 75–83. https://doi.org/10.1016/j.geoderma.2010.02.003.

Qadeer, Mohammad A. 2004. "Urbanization by Implosion." *Habitat International* 28(1): 1–12.

Raciti, Steve M., Lucy R. Hutyra, Preeti Rao, and Adrien C. Finzi. 2012. "Inconsistent Definitions of 'Urban' Result in Different Conclusions about the Size of Urban Carbon and Nitrogen Stocks." *Ecological Applications* 22 (3): 1015–35. https://doi.org/10.1890/11-1250.1.

Ramírez, Margaret Marietta. 2015. "The Elusive Inclusive: Black Food Geographies and Racialized Food Spaces." *Antipode* 47(3): 748–69. https://doi.org/10.1111/anti.12131.

Reardon, Thomas, C. Peter Timmer, Christopher B. Barrett, and Julio Berdegué. 2003. "The Rise of Supermarkets in Africa, Asia, and Latin America." *American Journal of Agricultural Economics* 85 (5): 1140–46. https://doi.org/10.1111/j.0092-5853.2003.00520.x.

Reeves, J., Z. Cheng, J. Kovach, M. D. Kleinhenz, and P. S. Grewal. 2014. "Quantifying Soil Health and Tomato Crop Productivity in Urban Community and Market Gardens." *Urban Ecosystems* 17(1): 221–38. https://doi.org/10.1007/s11252-013-0308-1.

Richards, Peter, and Leah VanWey. 2015. "Where Deforestation Leads to Urbanization: How Resource Extraction Is Leading to Urban Growth in the Brazilian Amazon." *Annals of the Association of American Geographers* 105(4): 806–23. https://doi.org/10.1080/00045608.2015.1052337.

Ritchie, Hannah, and Max Roser. 2018. "Urbanization." *Our World In Data*. https://ourworldindata.org/urbanization.

Rodríguez Eugenio, Natalie, M. J. McLaughlin, Daniel Pennock, and Global Soil Partnership. 2018. *Soil Pollution: A Hidden Reality*.

Roman-Alcalá, Antonio. 2018. "(Relative) Autonomism, Policy Currents and the Politics of Mobilisation for Food Sovereignty in the United States: The Case of Occupy the Farm." *Local Environment* 23(6): 619–34. https://doi.org/10.1080/13549839.2018.1456516.

Rosado, Leonardo, Samuel Niza, and Paulo Ferrão. 2014. "A Material Flow Accounting Case Study of the Lisbon Metropolitan Area Using the Urban Metabolism Analyst Model." *Journal of Industrial Ecology* 18(1): 84–101. https://doi.org/10.1111/jiec.12083.

Rossiter, David G. 2007. "Classification of Urban and Industrial Soils in the World Reference Base for Soil Resources (5 Pp)." *Journal of Soils and Sediments* 7 (2): 96–100. https://doi.org/10.1065/jss2007.02.208.

Ruel, Marie T., and James L. Garrett. 2004. "Features of Urban Food and Nutrition Security and Considerations for Successful Urban Programming." *Journal of Development and Agricultural Economics* 1(2): 242–271.

Ruel, Marie T., James L.Garrett, Sivan Yosef, and Meghan Olivier. 2017. "Urbanization, Food Security and Nutrition." *In Nutrition and Health in a Developing World*, edited by Saskia de Pee, Douglas Taren, and Martin W. Bloem, 705–35. Cham: Springer International Publishing. https://doi.org/10.1007/978-3-319-43739-2_32.

Sanderman, Jonathan, Tomislav Hengl, and Gregory J. Fiske. 2017. "Soil Carbon Debt of 12,000 Years of Human Land Use." *Proceedings of the National Academy of Sciences* 114(36): 9575–80. https://doi.org/10.1073/pnas.1706103114.

Satterthwaite, David, Gordon McGranahan, and Cecilia Tacoli. 2010. "Urbanization and Its Implications for Food and Farming." *Philosophical Transactions of the Royal Society B: Biological Sciences* 365(1554): 2809–20. https://doi.org/10.1098/rstb.2010.0136.

Scalenghe, Riccardo, and Franco Ajmone Marsan. 2009. "The Anthropogenic Sealing of Soils in Urban Areas." *Landscape and Urban Planning* 90(1–2): 1–10. https://doi.org/10.1016/j.landurbplan.2008.10.011.

Scharenbroch, Bryant C., John E. Lloyd, and Jodi L. Johnson-Maynard. 2005. "Distinguishing Urban Soils with Physical, Chemical, and Biological Properties." *Pedobiologia* 49(4): 283–96. https://doi.org/10.1016/j.pedobi.2004.12.002.

Schmidt, Michael W. I., Margaret S. Torn, Samuel Abiven, Thorsten Dittmar, Georg Guggenberger, Ivan A. Janssens, Markus Kleber, et al. 2011. "Persistence of Soil Organic Matter as an Ecosystem Property." *Nature* 478(7367): 49–56. https://doi.org/10.1038/nature10386.

Shi, Kaifang, Yun Chen, Bailang Yu, Tingbao Xu, Linyi Li, Chang Huang, Rui Liu, Zuoqi Chen, and Jianping Wu. 2016. "Urban Expansion and Agricultural Land Loss in China: A Multiscale Perspective." *Sustainability* 8(8): 790. https://doi.org/10.3390/su8080790.

Shuster, W. D., J. Bonta, H. Thurston, E. Warnemuende, and D. R. Smith. 2005. "Impacts of Impervious Surface on Watershed Hydrology: A Review." *Urban Water Journal* 2(4): 263–75. https://doi.org/10.1080/15730620500386529.

Siegner, Alana Bowen, Charisma Acey, and Jennifer Sowerwine. 2020. "Producing Urban Agroecology in the East Bay: From Soil Health to Community Empowerment." *Agroecology and Sustainable Food Systems* 44(5): 566–93. https://doi.org/10.1080/21683565.2019.1690615.

Siegner, Alana, Jennifer Sowerwine, and Charisma Acey. 2018. "Does Urban Agriculture Improve Food Security? Examining the Nexus of Food Access and Distribution of Urban Produced Foods in the United States: A Systematic Review." *Sustainability* 10(9): 2988. https://doi.org/10.3390/su10092988.

Simpson, Leanne Betasamosake. 2014. "Land as Pedagogy: Nishnaabeg Intelligence and Rebellious Transformation." *Decolonization: Indigeneity, Education & Society* 3(3): 1–25.

Stewart, Ryan D., Aditi S. Bhaskar, Anthony J. Parolari, Dustin L. Herrmann, Jinshi Jian, Laura A. Schifman, and William D. Shuster. 2019. "An Analytical Approach to Ascertain Saturation-excess versus Infiltration-excess Overland Flow in Urban and Reference Landscapes." *Hydrological Processes* 33(26): 3349–63. https://doi.org/10.1002/hyp.13562.

Swidler, Eva-Maria. 2009. "The Social Production of Soil." *Soil Science* 174 (1): 2–8. https://doi.org/10.1097/SS.0b013e318194274d.

Swyngedouw, Erik, and Nikolas C Heynen. 2003. "Urban Political Ecology, Justice and the Politics of Scale." *Antipode* 35 (5): 898–918. https://doi.org/10.1111/j.1467-8330.2003.00364.x.

Tenenbaum, D. E., L. E. Band, S. T. Kenworthy, and C. L. Tague. 2006. "Analysis of Soil Moisture Patterns in Forested and Suburban Catchments in Baltimore, Maryland, Using High-Resolution Photogrammetric and LIDAR Digital Elevation Datasets." *Hydrological Processes* 20(2): 219–40. https://doi.org/10.1002/hyp.5895.

Thebo, A L, P Drechsel, and E F Lambin. 2014. "Global Assessment of Urban and Peri-Urban Agriculture: Irrigated and Rainfed Croplands." *Environmental Research Letters* 9 (11): 114002. https://doi.org/10.1088/1748-9326/9/11/114002.

Toledo, Víctor M. 2002. "Ethnoecology: A Conceptual Framework for the Study of Indigenous Knowledges of Nature." In *International Congress of Ethnobiology: Ethnobiology and Biocultural Diversity*, 511–22. International Society of Ethnobiology.

Tornaghi, Chiara. 2017. "Urban Agriculture in the Food-Disabling City: (Re)Defining Urban Food Justice, Reimagining a Politics of Empowerment: (Re)Defining Urban Food Justice." *Antipode* 49(3): 781–801. https://doi.org/10.1111/anti.12291.

Tornaghi, Chiara, and Michiel Dehaene. 2020. "The Prefigurative Power of Urban Political Agroecology: Rethinking the Urbanisms of Agroecological Transitions for Food System Transformation." *Agroecology and Sustainable Food Systems* 44(5): 594–610. https://doi.org/10.1080/21683565.2019.1680593.

Trammell, Tara L. E., Brad P. Schneid, and Margaret M. Carreiro. 2011. "Forest Soils Adjacent to Urban Interstates: Soil Physical and Chemical Properties, Heavy Metals, Disturbance Legacies, and Relationships with Woody Vegetation." *Urban Ecosystems* 14 (4): 525–52. https://doi.org/10.1007/s11252-011-0194-3.

Trammell, Tara L. E., Diane E. Pataki, Richard V. Pouyat, Peter M. Groffman, Carl Rosier, Neil Bettez, Jeannine Cavender-Bares, et al. 2020. "Urban Soil Carbon and Nitrogen Converge at a Continental Scale." *Ecological Monographs* 90(2). https://doi.org/10.1002/ecm.1401.

Travaline, Katharine, and Christian Hunold. 2010. "Urban Agriculture and Ecological Citizenship in Philadelphia." *Local Environment* 15(6): 581–90. https://doi.org/10.1080/13549839.2010.487529.

Treffinger, Grace, and Tyler Jacobson. 2018. *Oxford Tract Planning Committee Minority Report*. Berkeley, CA: University of California, Berkeley. https://nature.berkeley.edu/sites/edu.oxford-facility/files/Oxford%20Tract%20Planning%20Committee%20Minority%20Report.pdf.

Tresch, Simon, Marco Moretti, Renée-Claire Le Bayon, Paul Mäder, Andrea Zanetta, David Frey, and Andreas Fliessbach. 2018. "A Gardener's Influence on Urban Soil Quality." *Frontiers in Environmental Science* 6 (May): 25. https://doi.org/10.3389/fenvs.2018.00025.

Treuhaft, Sarah, and Allison Karpyn. 2010. *The Grocery Gap: Who Has Access to Healthy Food and Why It Matters*. Oakland, CA: PolicyLink.

United Nations. 2014. "World Urbanization Prospects." New York: United Nations. https://population.un.org/wup/Publications/Files/WUP2014-Methodology.pdf.

Vaarst, Mette, Arthur Getz Escudero, M. Jahi Chappell, Catherine Brinkley, Ravic Nijbroek, Nilson A.M. Arraes, Lise Andreasen, et al. 2018. "Exploring the Concept of Agroecological Food Systems in a City-Region Context." *Agroecology and Sustainable Food Systems* 42(6): 686–711. https://doi.org/10.1080/21683565.2017.1365321.

Vandermeer, John, Aniket Aga, Jake Allgeier, Catherine Badgley, Regina Baucom, Jennifer Blesh, Lilly F. Shapiro, et al. 2018. "Feeding Prometheus: An Interdisciplinary Approach for Solving the Global Food Crisis." *Frontiers in Sustainable Food Systems* 2 (July): 39. https://doi.org/10.3389/fsufs.2018.00039.

Vitousek, Peter M, Harold A Mooney, Jane Lubchenco, and Jerry M Melillo. 1997. "Human Domination of Earth's Ecosystems" 277: 6.

Vorley, Bill, and Fréderic Lançon. 2016. *Food Consumption, Urbanisation and Rural Transformation The Trade Dimensions*. London: International Institute for Environment and Development.

Vorosmarty, C. J. 2000. "Global Water Resources: Vulnerability from Climate Change and Population Growth." *Science* 289(5477): 284–88. https://doi.org/10.1126/science.289.5477.284.

Wakefield, Sarah, Fiona Yeudall, Carolin Taron, Jennifer Reynolds, and Ana Skinner. 2007. "Growing Urban Health: Community Gardening in South-East Toronto." *Health Promotion International* 22(2): 92–101. https://doi.org/10.1093/heapro/dam001.

Warkentin, Benno P., International Union of Soil Sciences, and Soil Science Society of America, eds. 2006. *Footprints in the Soil: People and Ideas in Soil History*. 1st ed. Amsterdam ; Boston: Elsevier.

Warren, Robert J., Katelyn Reed, Michael Olejnizcak, and Daniel L. Potts. 2018. "Rural Land Use Bifurcation in the Urban–Rural Gradient." *Urban Ecosystems* 21(3): 577–83. https://doi.org/10.1007/s11252-018-0734-1.

Wezel, A., S. Bellon, T. Doré, C. Francis, D. Vallod, and C. David. 2009. "Agroecology as a Science, a Movement and a Practice. A Review." *Agronomy for Sustainable Development* 29(4): 503–15. https://doi.org/10.1051/agro/2009004.

White, Monica M. 2011. "Sisters of the Soil: Urban Gardening as Resistance in Detroit." *Race/Ethnicity: Multidisciplinary Global Contexts* 5(1): 13–28. https://doi.org/10.2979/racethmulglocon.5.1.13.

White, Monica M. 2018. Freedom Farmers: Agricultural Resistance and the Black Freedom Movement. http://search.ebscohost.com/login.aspx?direct=true&scope=site&db=nlebk&db=nlabk&AN=1930136.

Williams, Justine M., and Eric Holt-Giménez, eds. 2017. *Land Justice: Re-Imagining Land, Food, and the Commons in the United States*. Oakland, CA: Food First Books: Institute for Food and Development Policy.

Wilson, Sacoby, Aaron Aber, Vivek Ravichandran, Lindsey Wright, and Omar Muhammad. 2017. "Soil Contamination in Urban Communities Impacted by Industrial Pollution and Goods Movement Activities." *Environmental Justice* 10(1): 16–22. https://doi.org/10.1089/env.2016.0040.

WinklerPrins, Antoinette M. G. A., and Narciso Barrera-Bassols. 2004. "Latin American Ethnopedology: A Vision of Its Past, Present, and Future." *Agriculture and Human Values* 21(2/3): 139–56. https://doi.org/10.1023/B:AHUM.0000029405.37237.c8.

WinklerPrins, Antoinette M.G.A. 2002. "House-Lot Gardens in Santarém, Pará, Brazil: Linking Rural with Urban." *Urban Ecosystems* 6(1/2): 43–65. https://doi.org/10.1023/A:1025914629492.

Wires, K. Nicole, and Johnella LaRose. 2019. "Sogorea Te' Land Trust and Indigenous Food Sovereignty in the San Francisco Bay Area." *Journal of Agriculture, Food Systems, and Community Development*, November, 1–4. https://doi.org/10.5304/jafscd.2019.09B.003.

World Bank. 2000. *World Development Report 2000/2001: Attacking Poverty*. New York: Oxford University Press.

Wortman, Sam E., and Sarah Taylor Lovell. 2013. "Environmental Challenges Threatening the Growth of Urban Agriculture in the United States." *Journal of Environmental Quality* 42(5): 1283–94. https://doi.org/10.2134/jeq2013.01.0031.

Xiao, Rui, Diwei Jiang, George Christakos, Xufeng Fei, and Jiaping Wu. 2016. "Soil Landscape Pattern Changes in Response to Rural Anthropogenic Activity across Tiaoxi Watershed, China." Edited by Ben Bond-Lamberty. *PLOS ONE* 11(11): e0166224. https://doi.org/10.1371/journal.pone.0166224.

Yu, Deyong, Bin Xun, Peijun Shi, Hongbo Shao, and Yupeng Liu. 2012. "Ecological Restoration Planning Based on Connectivity in an Urban Area." *Ecological Engineering* 46 (September): 24–33. https://doi.org/10.1016/j.ecoleng.2012.04.033.

Zhang, Kevin Honglin, and Shunfeng Song. 2003. "Rural–Urban Migration and Urbanization in China: Evidence from Time-Series and Cross-Section Analyses." *China Economic Review* 14(4): 386–400. https://doi.org/10.1016/j.chieco.2003.09.018.

Ziter, Carly, and Monica G. Turner. 2018. "Current and Historical Land Use Influence Soil-Based Ecosystem Services in an Urban Landscape." *Ecological Applications* 28 (3): 643–54. https://doi.org/10.1002/eap.1689.

5

Urban Foraging: Where Cultural Knowledge and Local Biodiversity Meet

Leonie K Fischer[1,2,3*], **Jonah Landor-Yamagata**[2,4], **and Ingo Kowarik**[2,3]
[1] *University of Stuttgart, Institute of Landscape Planning and Ecology, Germany*
[2] *Technische Universität, Berlin Department of Ecology, Ecosystem Science/Plant Ecology, Germany*
[3] *Berlin-Brandenburg Institute of Advanced Biodiversity Research (BBIB), Berlin, Germany*
[4] *Urban Tilth, Richmond, CA USA*
[*] *Corresponding author: leonie.fischer@ilpoe.uni-stuttgart.de, Keplerstraße 11, D-70174 Stuttgart, Germany*

CONTENTS

5.1 Introduction

> A woman, possibly in her mid-50s, hair pulled back, walks amidst the roadside vegetation in a park in the city of Oakland (U.S.) picking low-growing plants. She wears rubber gloves and carries a plastic shopping bag, which holds her 6-inch-long harvests. She belongs to the Mien ethnic group, possibly from Southeast Asia, and is gathering young flower stalks of a plant familiar to her from this region. It grows in open areas, she explains, and the shoots should be picked before they become too old then cooked lightly so they are still crunchy, with beef or bacon. Later investigation identified the plant as *Hypochaeris radicata*, or hairy catsear, a plant native to Europe with wide global distribution.[1]

The collection of wild plants is a traditional human–nature interaction that spans from prehistoric times through to modern days (Moffett 1991, Kubiak-Martins 1999, Hummer 2013, Antolín et al. 2016). In contemporary urban areas, this practice adds to other aspects of multifaceted urban space planning, the "edible green infrastructure" (Russo et al. 2017, Russo & Cirella 2019)—a range of private and public places where plants are deliberately cultivated for consumption such as domestic, community and market gardens. While urban gardening and farming present highly visible interactions between people and plants and are often the focus within urban agroecology research, we address a rarely studied form of

[1] Own observation J. Landor-Yamagata

PHOTOS 5.1A, B: Collecting edible flowers, greens, berries etc. is an outdoor activity that connects people to urban nature and enables them to prepare and eat fresh, local, healthy food.

urban agronomy: the collection and harvesting of plants in places where they have not been explicitly cultivated for the purpose of consumption.

Today, the use of wild-gathered plants has been studied in rural communities around the globe (Pieroni et al. 2009, Ladio & Lozada 2003, Bharucha & Pretty 2010, Sujarwo et al. 2016), where gathering non-cultivated plants can contribute importantly to food security and nutrition (Lockett & Calvert 2000, Arnold et al. 2011), medical care (Willcox 1999, Pouliot 2011) and economic livelihoods (Shackleton et al. 2008). In urban areas, on the other hand, the collection of wild-growing plants has been largely overlooked by researchers and planners as an important part of the urban food system and of green infrastructure (McLain et al. 2012, Shackleton et al. 2017). A range of recent studies from North America, however, have revealed the significance of this practice for urban food access, green space planning and environmental stewardship (McLain et al. 2014, Poe et al. 2014, McLain et al. 2017, Synk et al. 2017). This research, and others, have begun to shed light on foraging practices of people in cities around the globe, in both the Global South (van Andel & Carvalheiro 2013, Unnikrishnan & Nagendra 2015, Mollee et al. 2017) as well as North (Wehi & Wehi 2010, Poe et al. 2013, Palliwoda et al. 2017). This still small but growing body of work has shown that gathering wild plants can be surprisingly prevalent in urban areas: practiced by over half of respondents in three South African cities (Kaoma & Shackleton 2014);

47% of respondents in Kampala, Uganda (Mollee et al. 2017); one-quarter of urban and rural residents in New England, U.S. (Robbins et al. 2008), and 5% of park visitors in five European cities (Fischer et al. 2018).

Generally, urban foraging involves the collection of plants growing outside areas where they have been purposefully cultivated for consumption, i.e. the domestic gardens, community gardens and agricultural fields where most work on urban food systems and urban agroecology is focused. We define urban foraging as the gathering of raw biological resources (e.g., plants and plant parts, fungi) in urban and peri-urban areas for food, medicine, crafts, small-scale sale, or other purposes (adapted from Shackleton et al. 2017). These "forageables" include wild and domesticated species that occur spontaneously, those that spread or persist without human intervention, and those that are introduced primarily for non-edible/material purposes—such as landscaping (Poe et al. 2013). Urban foraging can occur in a wide variety of spaces, both managed and unmanaged, public and private, including parks and forests, abandoned lots, alongside rights-of-ways and in nature preserves (McLain et al. 2014). It thus relies on the role of various urban greenspace types in harboring considerable biological richness (e.g., Kowarik & von der Lippe 2018) which, in addition to fulfilling habitat functions, also include a wealth of forageable species for humans that ultimately become a component of urban residents' daily nutrition and the circular economy of the urban food system as a whole.

While it is likely that foraging occurs in most cities around the world, the particulars, including species and quantities gathered, harvesting environments, socio-demographics and motivations of foragers are varied and related to specific times and places (Synk et al. 2017, Landor-Yamagata et. al 2018). Intriguingly, over the past decade or so in the Global North, foraging in general has enjoyed some time in the limelight, even becoming trendy (Reyes-García 2015). Explanations for this rise in popularity could include increased interest in locally-grown foods and urban agriculture, desire of urban populations to connect with nature, as well as influences of 'tastemakers' such as well-known restaurants that showcase regionally-foraged ingredients (Łuczaj et al. 2012). On the other hand, and in stark contrast, foraged foods have also provided essential nutrition and added variety during times of great hardship, such as recent war and famine (Redžić 2010, Łuczaj & Pieroni 2016). Research has also shown that some species collected in times of crisis were later abandoned and stigmatized as "famine foods" (Svanberg 2012, Reyes-García 2015). Thus, though urban foraging has certainly experienced waves of popularity in both times of prosperity and hardship, overall it has remained an important practice in urban food-based movements.

This chapter investigates urban foraging research broadly, presenting it as a growing edge of the evolving scholarship concerning the ability of diverse elements of the urban green infrastructure to provide food and nature-based benefits—both material and non-material, social and ecological—to people living in cities. We discuss existing urban foraging research approaches, including approaches such as interviews and questionnaire studies (e.g., Chipeniuk 1995, Poe et al. 2014), vegetation-related methods (e.g., Fischer et al. 2019, Stark et al. 2019), and research based on user-based data (Arrington et al. 2017, Hurley & Emery 2018). Drawing from these studies, we summarize the health, economic, social and ecological benefits as well as the connections between its cultural and ecological dimensions of urban life that urban foraging reveals. A clearer understanding of the benefits and challenges that urban foraging presents could lead to a greater understanding—and perhaps legitimization—of this multi-faceted practice, which in many places is officially discouraged, if acknowledged at all. Ultimately, this chapter aims to expand perspectives on nature- and food-based practices in urban areas that contribute to the ecology of cities and their food systems. While foraging generally is often not discussed in agroecology, its role and prevalence in cities for nutrition and its contribution to urban food systems requires that we consider it in the urban agroecology movement—and in regard to the benefits and challenges bound to it.

5.2 Urban Biodiversity as a Prerequisite for Collecting Edible Species

Urban foraging directly depends on a city's biodiversity—primarily its diversity in ecosystems that harbor a range of plant communities and its richness in plant species, which include wild-growing spontaneous species as well as cultivated plants. Thus, collecting edible plants simply requires that a certain

PHOTOS 5.2A, B: Elderberries (*Sambucus nigra*) and rose hips (*Rosa canina*) at the Schöneberger Südgelände, a former vacant land in Berlin that was transformed to a "Nature-Park". Here, even though given the comparably low legal regulations in Germany, picking the fruits is prohibited.

species pool is available in an area and the edible species within this pool and their occurrences within the area are known to people. Several studies have now assessed the species collected by urban foragers (e.g., Wehi & Wehi 2010, Synk et al. 2017, Palliwoda et al. 2017), and also revealed the potential of urban floras to provide edible plant parts that *could* be collected (Wang et al. 2015, Hurley & Emery 2018, Klein et al. 2019). To know which plants to collect and where, when and how to prepare them is part of the local ecological knowledge that is involved in urban foraging. Local ecological knowledge about foraging is maintained in many ways; for example, studies reported that knowledge is passed from older to younger people and from experienced foragers to less experienced beginners (Hurley et al. 2015, Landor-Yamagata et al. 2018). The role of local ecological knowledge and its intergenerational transfer in urban foraging mirrors traditional knowledge often discussed in relation to cultivated plants in urban home gardens and community gardens (Corlett et al. 2003, Taylor et al. 2015, Glowa et al. 2019).

Often, people involved in urban foraging report incorporating voluntary codes of conduct to avoid overharvesting and practices aimed to encourage desired species to persist (Hurley et al. 2015, Charnley et al. 2018). This is an important point when relating urban foraging to nature conservation issues in cities. Up till now, some studies indicate that urban foraging largely does not threaten species of conservation concern; for example, red-listed species in Berlin (Germany) were not reported as collected species by the general public, and rarely from people that had a more profound background in urban foraging (Landor-Yamagata et al. 2018, Fischer & Kowarik 2020). In this regard it is also of interest in which surroundings, and in which urban ecosystems people collect wild edible plants, and whether these may play a role for the conservation of specific plants or populations. After all, potential threats of foraging to urban biodiversity are not entirely unwarranted—especially when foraged products are destined for the cash economy (Petersen et al. 2012). Longer-term in-depth research about how foragers interact with and potentially affect harvested species and the ecosystems in which they occur would be instrumental in understanding how foraging could contribute to biodiversity goals while minimizing threats to urban ecosystems. However, the benefits it provides on multiple levels indicate its potential to be a valuable tool for both urban biodiversity and human well-being.

5.3 Urban Foraging: A Practice That Offers Multiple Social-Ecological Benefits

The benefits connected to the practice of collecting edible plants in cities are manifold (Figure 5.1). They include aspects such as the conservation of ecological knowledge about (wild) edible plants and identification skills (McLain et al. 2014), health benefits such as improved diet (Synk et al. 2017), positive social and cultural implications like feelings of belonging and identification with certain places (Poe et al. 2014), and positive economic outcomes such as the procurement of fruits and vegetables at little or no cost (Kaoma & Shackleton 2014). These social, health and economic benefits are similar to those often found through urban crop cultivation in gardens, orchards and farms (Barthel et al. 2010, Guitart et al. 2014, Poulsen et al. 2015).

Within the realm of health benefits (see health-related benefits and challenges in Figure 5.1), the contribution of fresh vegetables, fruits, and greens to complement a healthy diet may be among the most prominent benefit cited by foragers (Poe et al. 2013, Palliwoda et al. 2017, Synk et al. 2017). Indeed, common urban edible greens—AKA weeds—can be part of the edible species pool (Frazee et al. 2016) and have high levels of essential nutrients (Stark et al. 2019). Urban foraging can also have an important role to play in promoting healthy food literacy in urban populations. Identifying, harvesting and consuming wild-growing food can be a powerful pathway by which city dwellers reclaim agency in their participation in the food system as well as their own health maintenance and well-being (Poe et al. 2013).

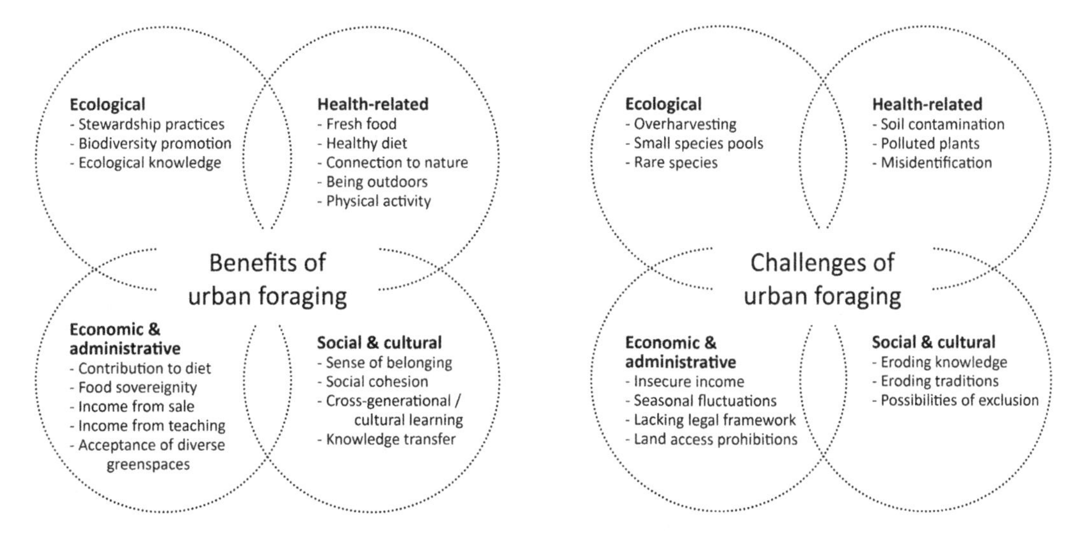

FIGURE 5.1 Urban foraging incorporates numerous benefits, but also incorporates challenges from different fields such as ecological, economic, administrative, health-related, social and cultural aspects.

PHOTOS 5.3A, B: In the dense urban city area of Valencia, Spain, orange trees are a common part of the green infrastructure in streets and parks. In the rural areas surrounding the city, citriculture is part of the agricultural tradition.

The availability of wild edible plants in urban areas can also help citizens weather times of crisis or hardship. For example, Redžić (2010) reports that foraged plants were used to mitigate food scarcity during the siege of Sarajevo in Bosnia and Herzegovina during the mid-1990s. From a more everyday health perspective, urban foraging involves being outside while collecting plants and thereby walking, wandering and moving within greenspaces. Physical outdoor activities in nature have positive implications for people's physical and mental health and wellbeing, at any age (Hartig et al. 2014, Kabisch et al. 2017). Thus, foraging can support positive health outcomes for city dwellers both in situations of acute need as well as during their normal day-to-day routines, becoming an important part of urban food systems.

Urban foraging includes economic aspects as well (see economic & administrative benefits and challenges in Figure 5.1). Some people report that the practice contributes to their livelihood, not only by supplementing diets, but also because they can sell forageables to bolster their income (Poe et al. 2013, Synk et al. 2017). Others lead courses and workshops about wild edible plants and their uses to supplement their income (Landor-Yamagata et al. 2018). There are few studies that explicitly focus on the economic dimensions of urban foraging. Because the economic climate may influence how foraging is perceived, the lack of research is a missed opportunity for understanding the use of foraging during times of economic dearth. For example, there was an increase in urban foraging practice and its promotion during the Great Recession in the U.S. (Sachdeva et al. 2018), but we do not know if this pattern is seen globally.

Urban foraging includes diverse aspects that relate to the social and cultural background and the personal setting of urban people (see social & cultural benefits and challenges in Figure 5.1). For example, traditions for collecting edible plant species at specific times of year are often based both on the regional and seasonal availability of fruits, nuts, greens, etc. (McLain et al. 2014), as well as cultural and family traditions (Landor-Yamagata et al. 2018). Passing on traditions to others helps to maintain both the practices and the knowledge involved. In the long term, such knowledge transfer may help to improve stewardship for urban nature and its diversity (Chipeniuk 1995, Maller et al. 2019) as increased personal experience of nature can translate to a higher willingness to promote and protect nature (Soga & Gaston 2016).

Along with the local ecological knowledge associated with urban foraging, further social and cultural aspects are also involved; for many foragers these attributes may be what attracts them to the practice. For example, collecting plants in groups allows for direct person-to-person contact, and thus contributes to social cohesion on a local basis. At the same time, urban foraging may assist people in their personal place-making, contributing to their identification with and connection to a place (Poe et al. 2014). Relatedly, people with migrant backgrounds from abroad may have chances to interact with species familiar from their homelands, which represent familiar plants—and food—from home (McLain et al. 2014). Urban agroecological literature has investigated how immigrants maintain cultural ties by cultivating plants in gardens (e.g. Glowa et al. 2019); similarly, the maintenance of traditions has been cited as a reason that foraging is practiced in cities (Mollee et al. 2017, Garekae & Shackleton 2020). It is also important to recognize that foraging practices and traditions are dynamic and diverse. For example, several European countries demonstrate shifts in the species collected by people privately or sold at local markets, with some becoming more popular and others less so—for example, due to changes in traditional land use practices (Luczaj et al. 2012), new trends in contemporary cuisine, or because species became a source of income for immigrants (Svanberg 2012).

That people with migrant backgrounds "import" traditions in plant use and collection has been mentioned in several studies. For example, one study uncovered that Vietnamese immigrants in Hawai'i were aware of more food uses for certain traditional Vietnamese (albeit cultivated) species than Vietnamese people in their home country due to adopting the large variety of food in the new country (Nguyen 2003). Another cross-country insight is that wild edible plants that were traditionally collected and eaten in South Korea were popular for exporting and sale in Korean food markets abroad (Pemberton 1996). In sum, learning from others (cross-generational, across different groups of society) may mean that traditions are passed on as well as mixed with those of others. Thus, this specific aspect of urban foraging adds to the role of gardening and community gardening—important and common urban agroecosystems – for promoting ecological knowledge and its multicultural benefits in broader society.

5.4 Approaches of Urban Foraging Studies

Interest in the topic of urban foraging has arisen from different research fields, connecting social and ecological disciplines. Likewise, publications from several research areas are available that involve diverse methods from the disciplines engaged (Shackleton et al. 2017). Initially, urban foraging was often researched in ethnographic studies and studies from the social sciences, with the urban setting being a novel field of work as opposed to rural areas and communities (see especially the pioneering work of Poe et al. 2013, McLain et al. 2014 for U.S. cities). These studies especially utilize interviewing as a research method, and this method is often combined with other methods that are unique to a specific field of research. Other methods to collect data include field surveys and observation studies (Poe et al. 2014, Palliwoda et al. 2017) and—more recently—analyzed user data of online platforms (Arrington et al. 2017, Sachdeva et al. 2018).

Below, Table 5.1 illustrates the range of thematic approaches and topics brought to the wide field of urban foraging research (column 1), with research methods employed for the selected corresponding studies (columns 2 to 4). Urban foraging studies include topics such as

i. which edible species people collect,
ii. why people collect,
iii. where they collect, and
iv. the assessment of services and disservices of collecting edible plants.

TABLE 5.1

Examples for topics, approaches to, and methods used in urban foraging (UF) studies. We differentiate methods involved in past studies on how they assessed the data, for example if interviews (qualitative, quantitative) helped to gather information or if vegetation-ecological methods (e.g., vegetation mapping) or other, specific research methods (e.g., comparison of gathered taxa to the full local species pool) were used to answer research questions.

	Methods involved				
Topic / Approach	**Interview, questionnaire**	**Vegetation-related methods**	**User-based or Internet data**	**Specific / other methods***	**References**
Assessment of UF and its relation to urban society as a specific of urban non timber forest produce	x				Poe et al. 2013, Mollee et al. 2017, Garekae & Shackleton 2020
Assessment of UF and its relation to the local flora of the study region	x			Links to local flora	Wehi & Wehi 2010, Landor-Yamagata et al. 2018
Education about edible species; assessment of species collected and their uses during times of shortage	x			+ Field demonstrations	Redžić 2010
Assessment of UF and its nature–society relationships	x			Ethnographically -grounded framework	Poe et al. 2014
Assessment of UF and its relation to planning and management	x			Geography-grounded framework	McLain et al. 2014
Assessment of UF practice in children	x			+ Quiz	Chipeniuk 1995
Assessment of UF practice in urban and rural areas of several U.S. states	x			Random-digit dial telephone survey	Robbins et al. 2008
Assessment of UF and linkages to the local flora of the study region	x	x			Fischer & Kowarik 2020
Assessment of local UF practice and its relation to the area's species pool		x		Observation	Palliwoda et al. 2017

TABLE 5.1 (Continued)

	Methods involved				
Topic / Approach	**Interview, questionnaire**	**Vegetation-related methods**	**User-based or Internet data**	**Specific / other methods***	**References**
Assessment of edible species within a case study area to reveal UF potential		x		+ stakeholder involvement	Fischer et al. 2019
Assessment of the presence of edible species in an urban flora		x		Plant database of a city	Wang et al. 2015
Analyses of the toxicological and nutritional content of forageables		x		Nutrient and toxin analysis of plants	Stark et al. 2019
Quantification of edible forest yield			x	Tree database of a city as basis for choosing sample trees	Bunge et al. 2019
Analyses of the changes in UF practice through time			x	Automated content analysis of media articles	Sachdeva et al. 2018
Assessment of edible tree species within an urban area			x	GIS analyses of official tree inventory	Hurley & Emery 2018
Assessment of edible tree species within an urban area in relation to the city's historic development			x	GIS analyses of UF platform data, land use data	Larondelle & Strohbach 2016
Analyses of the composition and distribution of web-entries on UF locations			x	Primary analyses of UF platform data in relation to urban and rural areas	Klein et al. 2019
Analyses of foraging locations and foragers' background data by the use of citizen science data			x	Web data, mobile app usage	Arrington et al. 2017

* In this column, a "+" indicates if the described method or approach is combined with another method or approach.

In the ecological literature, studies focus on different levels of biodiversity. For example, some studies present information on the ecosystems in which urban foraging takes place, and others on the specific species that are collected by foragers. Ecological studies have largely focused on the "foraging potential" within a flora, i.e., the proportion of edible or medicinal species within the study area, regardless of whether they are actively being gathered by foragers (Hurley & Emery 2018, Wang et al. 2015). More recently, studies have also assessed or discussed the local species pool or nature conservation issues, for example, in relation to potential benefits or disservices of urban foraging towards biodiversity (Wehi & Wehi 2010, Landor-Yamagata et al. 2018, Fischer & Kowarik 2020).

BOX 5.1 CASE STUDY: UNCOVERING LINKAGES BETWEEN EDIBLE PLANT KNOWLEDGE, PLANT OCCURRENCE AND FORAGER BACKGROUNDS.

Given that we still know little about how knowledge about wild edible plants relates to the general public—and not to the specific group of urban foragers *per se*—here we present a study where we asked both foragers and non-foragers about their relation to the practice. Within the urban area of Berlin, Germany, we conducted vegetation surveys at a wild natural site (for details see Fischer et al. 2019), and determined which species were edible, and whether they were frequently or infrequently found in the Berlin city area. We then assembled a questionnaire with embedded photos of two herb species and two woody species, one of each which was frequently and infrequently found in the study area. We asked 535 people if they could identify any of the four species and whether they knew if any of them were edible. The results show that a person's botanical knowledge about a species' identity does not necessarily parallel their knowledge about the plant's edibility and its local abundance (Figure 5.2). More likely, it is that easy-to-recognize edible plant parts are better known, especially if these are also available in stores and markets in a cultivated form or variety—such as blackberry fruits. At the same time, another part of this survey (see Fischer & Kowarik 2020) showed that people with different sociocultural backgrounds collect edible plants outside of gardens but that some barriers prevent people from foraging, such as fear of contamination or mistaking potentially toxic species for edible species. Here, our results show that the practice of urban foraging may be seen as a tool for connecting urban people to nature, but that this requires integrated approaches addressing environmental policy, environmental education and greenspace management.

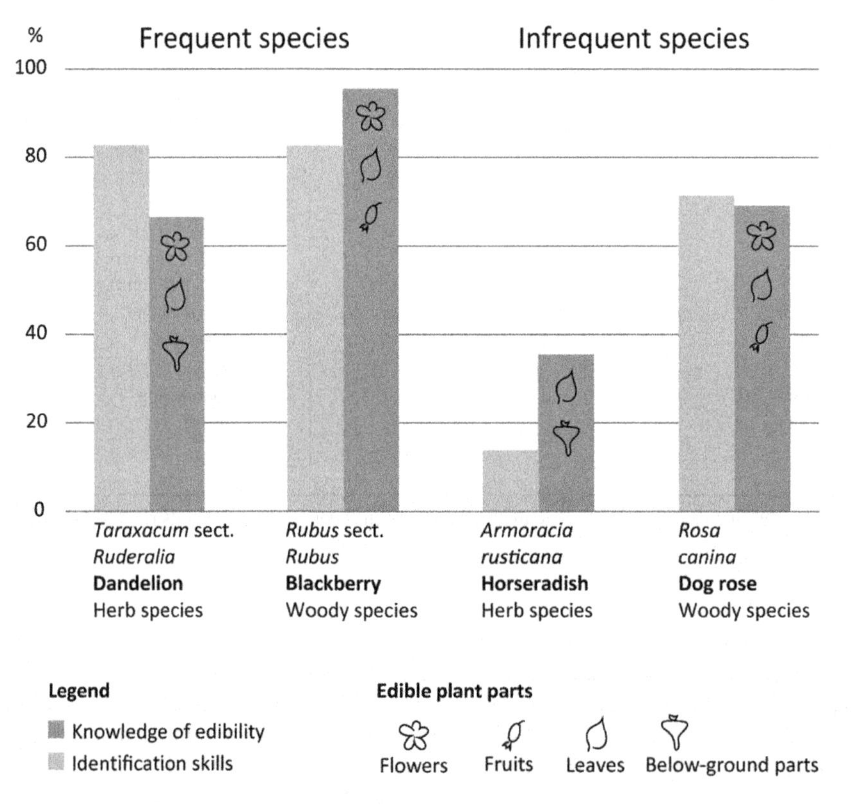

FIGURE 5.2 Identification skills and knowledge of edibility of four plant species that were illustrated by photographic stimuli material in a field survey in Berlin. Bars represent the proportions in which the respondents named each species correctly and identified their edibility; N=535. Symbols illustrate the main edible plant parts of each species (cf. Fleischhauer 2003).

5.5 Practical Implications

Today, within many cities around the world, urban foragers are confronted with various challenges, including biodiversity conservation and prohibitive city policy (Figure 5.1). For many of these challenges, urban green planning and greenspace management can provide approaches to support urban foraging as a sustainable and socially responsible way to connect to urban nature.

First, **ecological challenges** such as overharvesting of certain plant species or harvesting (even small quantities of) rare species have been reported before (e.g., orchid tubers; Molnar et al. 2017). Such practices can potentially have negative outcomes for urban biodiversity, and may contradict nature conservation goals. Yet many experienced foragers report awareness of biodiversity conservation problems (de Jong & Varley 2017, Landor-Yamagata et al. 2018), and report integrating sustainable harvesting practices in general (McLain et al. 2017). Translating such voluntary "codes of conduct" into greenspace policies could help to include urban foraging as a nature-oriented outdoor activity in people's urban life. In parallel, and supported by local specifications, monitoring of target plant populations could ensure that species are not overharvested and prevent population decline.

Second, **health-related disservices** that arise from collecting edible plants in urban areas must be addressed. On the one hand, greenspace management and maintenance can contribute to limitations for collecting. For example, areas treated with herbicides or synthetic fertilizers are not necessarily suitable for the collection of edible plants. Reducing such treatments and informing on where, how and when urban green spaces are treated with herbicides, fertilizers etc. are feasible options. On the other hand, traffic-related air pollution relates to high contamination in wild edible plants—suggesting that edible plants should be collected best in areas with no or low traffic loads (Säumel et al. 2012, Amato-Lourenco et al. 2020). In addition, soil ingestion should be avoided and all harvest should be rinsed well if plants are collected in urban areas (Stark et al. 2019).

Third, importantly, **administrative guidelines and laws** can limit urban foraging in two ways: regulations that deal with access to urban green spaces such as to vacant land; and administrative guidelines that regulate the collection of (wild) plant material. Limited access can prevent people from foraging, as they are not allowed to enter a certain area (see Hurley et al. 2015). Accessibility may also depend on whether places are privately owned or otherwise physically inaccessible such as fenced off etc. (Unnikrishnan & Nagendra 2015). In this regard, making green spaces physically accessible for collecting edible species, such as allowing people to enter urban vacant lots may help to support legal foraging in such spaces.

In many regions from which urban foraging is reported, the collection of wild plants in urban areas is quite restricted, if it is permitted at all (e.g., in the U.S.; McLain et al. 2014). This is due to regulations and laws that prohibit the collection of edible plant parts. In other regions, such as in Germany, sustainable collection of unprotected plant species for personal use in small quantities is permitted in many places (see German Nature Conservation Act). Thus, especially on public land where the collection of edible species may not compromise nature conservation goals, more openness towards collection of small quantities of edible plants may be one approach to encourage this type of nature interaction in cities.

While urban foraging scholarship has developed over the past few years, the area of inquiry is still relatively young. Being so, there are a number of limitations and gaps within it. For example, most foraging research to date has occurred within the social sciences and data about the relationships between foraging and the urban ecosystems in which it takes place is sparse. For example, in many places it is unclear if and how foraging may influence urban species' abundance or plant community composition, and vice versa. Comparisons between regions are also lacking: while urban foraging has been documented in many areas of the world, so far studies from the U.S. are more abundant than other countries. Understanding how foraging traditions persist, adapt or cease in urban situations is also an area for continued research. Finally, including foraging into the urban agroecology literature would open up opportunities to share and potentially extend insights from both areas to further reveal and enhance food provisioning practices in cities.

5.6 Conclusions

Drawing from the insights of the diverse disciplines that shed light on urban foraging today, it is understandable why the call for appropriate management of urban greenspaces and a more positive attitude towards the collection of edible plants within them is growing louder, especially in places where strong regulations exist. In sum, greenspace management can play a key role to support urban foraging by (i) monitoring edible species populations, (ii) planting more species that are edible within open spaces, (iii) tolerating wild edible plants within greenspaces, and (iv) controlling some pollution and contamination sources. Doing so could strengthen the positive aspects of urban foraging as part of the urban agroecology movement and of the broad realm of edible green infrastructure (Russo et al. 2017)—which in the long run may enhance food provisioning, support people's connection to nature and lead to stronger acceptance of natural elements and their promotion and protection in the city.

REFERENCES

Amato-Lourenco, L. F., Ranieri, G. R., de Oliveira Souza, V. C., Junior, F. B., Saldiva, P. H. N., & Mauad, T. (2020) Edible weeds: Are urban environments fit for foraging? *Sci Total Environ*, 698, 133967.

Antolín, F., Bleicher, N., Brombacher, C., Kühn, M., Steiner, B. L., & Jacomet, S. (2016) Quantitative approximation to large-seeded wild fruit use in a late Neolithic lake dwelling: New results from the case study of layer 13 of Parkhaus Opéra in Zürich (Central Switzerland). *Quaternary International*, 404, 56–68.

Arnold, M., Powell, B., Shanley, P., & Sunderland, T. (2011) EDITORIAL: Forests, biodiversity and food security. *The international Forestry Review*, 13, 259–264.

Arrington, A. B., Diemont, S. A. W., Phillips, C. T., & Welty, E. Z. (2017) Demographic and landscape-level urban foraging trends in the USA derived from web and mobile app usage. *Journal of Urban Ecology*, 3(1), jux006-jux006. doi:10.1093/jue/jux006

Barthel, S., Folke, C., & Colding, J. (2010) Social-ecological memory in urban gardens—Retaining the capacity for management of ecosystem services. *Global Environmental Change-Human and Policy Dimensions*, 20(2), 255–265.

Bharucha, Z., & Pretty, J. (2010) The roles and values of wild foods in agricultural systems. *Philosophical Transactions of the Royal Society B-Biological Sciences*, 365(1554), 2913–2926.

Bunge, A., Diemont, S. A. W., Bunge, J. A., & Harris, S. (2019) Urban foraging for food security and sovereignty: Quantifying edible forest yield in Syracuse, New York using four common fruit- and nut-producing street tree species. *Journal of Urban Ecology*, 5(1). doi:10.1093/jue/juy028

Charnley, S., McLain, R. J., & Poe, M. R. (2018) Natural resource access rights and wrongs: Nontimber forest products gathering in urban environments. *Society & Natural Resources*, 1–17. doi:10.1080/08941920.2017.1413696

Chipeniuk, R. (1995) Childhood foraging as a means of acquiring competent human cognition about biodiversity. *Environment and Behavior*, 27(4):490–512.

Corlett, J. L., Dean, E. A., & Grivetti, L. E. (2003) Hmong gardens: Botanical diversity in an urban setting. *Economic Botany*, 57(3), 365–379.

De Jong, A. & P. Varley (2017) Foraging tourism: Critical moments in sustainable consumption. *Journal of Sustainable Tourism*, 26(4), 685–701.

Fischer, L. K., & Kowarik, I. (2020) Connecting people to biodiversity in cities of tomorrow: Is urban foraging a powerful tool? *Ecological Indicators*, 112, 106087.

Fischer, L. K., D. Brinkmeyer, S. J. Karle, K. Cremer, E. Huttner, M. Seebauer, U. Nowikow, B. Schütze, P. Voigt, S. Völker & I. Kowarik (2019) Biodiverse edible schools: Linking healthy food, school gardens and local urban biodiversity. *Urban Forestry & Urban Greening*, 40:35–43.

Fischer, L. K., J. Honold, A. Botzat, D. Brinkmeyer, R. Cvejic, T. Delshammar, B. Elands, D. Haase, N. Kabisch, S. J. Karle, R. Lafortezza, M. Nastran, A. B. Nielsen, A. P. van der Jagt, K. Vierikko & I. Kowarik (2018) Recreational ecosystem services in European cities: Sociocultural and geographical contexts matter for park use. *Ecosystem Services*, 31:455–467.

Fleischhauer, S. (2003) *Enzyklopädie der essbaren Wildpflanzen* AT Verlag, München.

Frazee, L., Morris-Marano, S., Blake-Mahmud, J., & Struwe, L. (2016) Eat your weeds: Edible and wild plants in urban environmental education and outreach. *Plant Science Bulletin*, 62(2), 72–84.
Garekae, H., & Shackleton, C. (2020) Urban foraging of wild plants in two medium-sized South African towns: People, perceptions and practices. *Urban Forestry & Urban Greening*, 49, 126581.
Glowa, K. M., Egerer, M., & Jones, V. (2019) Agroecologies of displacement: a study of land access, dislocation, and migration in relation to sustainable food production in the beach flats community garden. *Agroecology and Sustainable Food Systems*, 43(1), 92–115.
Guitart, D. A., Pickering, C. M., & Byrne, J. A. (2014) Color me healthy: Food diversity in school community gardens in two rapidly urbanising Australian cities. *Health & Place*, 26, 110–117.
Hartig, T., R. Mitchell, S. d. Vries & H. Frumkin (2014) Nature and health. *Annual Review of Public Health*, 35(1):207–228.
Hummer, K. E. (2013). Manna in winter: Indigenous Americans, Huckleberries, and Blueberries. *Hortscience*, 48(4), 413–417.
Hurley, P. T. & M. R. Emery (2018) Locating provisioning ecosystem services in urban forests: Forageable woody species in New York City, USA. *Landscape and Urban Planning*, 170:266–275.
Hurley, P. T., Emery, M. R., McLain, R., Poe, M., Grabbatin, B., & Goetcheus, C. L. (2015) Whose urban forest? The political ecology of foraging urban nontimber forest products. Sustainability in the global city: myth and practice, Cambridge University Press, New York, 187–212.
Kabisch, N., M. van den Bosch & R. Lafortezza (2017) The health benefits of nature-based solutions to urbanization challenges for children and the elderly – A systematic review. *Environmental Research* 159: 362–373.
Kaoma, H. & C. M. Shackleton (2014) Collection of urban tree products by households in poorer residential areas of three South African towns. *Urban Forestry & Urban Greening*, 13(2):244–252.
Klein, T., Gildhorn, K. & Jurasinski, G. (2019) Der "Mundraub"-Datensatz: Stand und Potenzial für Naturschutz und Forschung. *Journal für Angewandte Geoinformatik*, 2019(5):246–260.
Kowarik, I. & M. von der Lippe (2018) Plant population success across urban ecosystems: A framework to inform biodiversity conservation in cities. *Journal of Applied Ecology*, 55(5):2354–2361.
Kubiak Martens, L. (1999) The plant food component of the diet at the late Mesolithic (Ertebolle) settlement at Tybrind Vig, Denmark. *Vegetation History and Archeobotany*, 8(1–2), 117–127.
Ladio, A. H. & M. Lozada (2004) Patterns of use and knowledge of wild edible plants in distinct ecological environments: a case study of a Mapuche community from northwestern Patagonia. *Biodiversity & Conservation*, 13(6):1153–1173.
Landor-Yamagata, J., I. Kowarik & L. Fischer (2018) Urban foraging in Berlin: People, plants and practices within the metropolitan green infrastructure. *Sustainability*, 10(6):1873.
Larondelle, N. & M. W. Strohbach (2016) A murmur in the trees to note: Urban legacy effects on fruit trees in Berlin, Germany. *Urban Forestry & Urban Greening,* 17: 11–15.
Lockett, C. T., C. C. Calvert & L. E. Grivetti (2000) Energy and micronutrient composition of dietary and medicinal wild plants consumed during drought. Study of rural Fulani, Northeastern Nigeria. *International Journal of Food Sciences and Nutrition*, 51(3):195–208.
Łuczaj, Ł. & A. Pieroni (2016) Nutritional ethnobotany in Europe: From emergency foods to healthy folk cuisines and contemporary foraging trends. *Mediterranean wild edible plants: ethnobotany and food composition tables.* M. d. C. Sánchez-Mata and J. Tardío. New York, NY, Springer New York: 33–56.
Łuczaj, Ł., A. Pieroni, J. Tardío, M. Pardo-de-Santayana, R. Sõukand, I. Svanberg & R. Kalle (2012) Wild food plant use in 21st century Europe: the disappearance of old traditions and the search for new cuisines involving wild edibles. *Acta Societatis Botanicorum Poloniae*, 81(4):359–370.
Maller, C., L. Mumaw & B. Cooke (2019) Health and social benefits of living with 'wild' nature. *Rewilding.* J. T. du Toit, N. Pettorelli and S. M. Durant. Cambridge, Cambridge University Press: 165–181.
McLain, R. J., Hurley, P. T., Emery, M. R., & Poe, M. R. (2012) Gathering in the city: an annotated bibliography and review of the literature about human-plant interactions in urban ecosystems. Portland, OR: United States Department of Agriculture, Forest Service, Pacific Northwest Research Station General Technical Report PNW-GTR-849.
McLain, R. J., M. R. Poe, L. S. Urgenson, D. J. Blahna & L. P. Buttolph (2017) Urban non-timber forest products stewardship practices among foragers in Seattle, Washington (USA). *Urban Forestry & Urban Greening*, 28:36–42.

McLain, R. J., P. T. Hurley, M. R. Emery & M. R. Poe (2014) Gathering wild food in the city: rethinking the role of foraging in urban ecosystem planning and management. *Local Environment,* 19(2):220–240.

Moffett, L. (1991) Pignut tubers from a Bronze age cremation at Barrow hills, Oxfordshire, and the importance of vegetable tubers in the prehistoric period. *Journal of Archaeological Science*, 18(2), 187–191.

Mollee, E., M. Pouliot & M. A. McDonald (2017) Into the urban wild: Collection of wild urban plants for food and medicine in Kampala, Uganda. *Land Use Policy*, 63:67–77.

Molnar, V. A., Nagy, T., Loki, V., Suveges, K., Takacs, A., Bodis, J., & Tokolyi, J. (2017) Turkish graveyards as refuges for orchids against tuber harvest. *Ecology and Evolution*, 7(24), 11257–11264.

Nguyen, M. L. T. (2003) Comparison of food plant knowledge between urban Vietnamese living in Vietnam and in Hawai'i. *Economic Botany*, 57(4):472–480.

Palliwoda, J., I. Kowarik & M. von der Lippe (2017) Human-biodiversity interactions in urban parks: The species level matters. *Landscape and Urban Planning*, 157:394–406.

Pemberton, R. W. & N. S. Lee (1996) Wild food plants in South Korea; market presence, new crops, and exports to the United States. *Economic Botany*, 50(1):57–70.

Petersen, L.M., Moll, E.J., Collins, R. & Hockings, M.T. (2012) Development of a compendium of local, wild-harvested species used in the informal economy trade, Cape Town, South Africa. *Ecology and Society*, 17(2): 26.

Pieroni, A., Nebel, S., Santoro, R. F., & Heinrich, M. (2009) Food for two seasons: Culinary uses of non-cultivated local vegetables and mushrooms in a south Italian village. *International Journal of Food Sciences and Nutrition*, 56(4), 245–272.

Poe, M. R., J. LeCompte, R. McLain & P. Hurley (2014) Urban foraging and the relational ecologies of belonging. *Social & Cultural Geography*, 15(8):901–919.

Poe, M. R., R. J. McLain, M. Emery & P. T. Hurley (2013) Urban forest justice and the rights to wild foods, medicines, and materials in the city. *Human Ecology*, 41(3):409–422.

Pouliot, M. (2011) Relying on nature's pharmacy in rural Burkina Faso: Empirical evidence of the determinants of traditional medicine consumption. *Social Science & Medicine*, 73(10), 1498–1507.

Poulsen, M. N., McNab, P. R., Clayton, M. L., & Neff, R. A. (2015) A systematic review of urban agriculture and food security impacts in low-income countries. *Food Policy*, 55, 131–146.

Redžić, S. (2010) Use of wild and semi-wild edible plants in nutrition and survival of people in 1430 days of siege of Sarajevo during the war in Bosnia and Herzegovina (1992–1995). *Collegium Antropologicum* 34(3):551–570.

Reyes-García, V., G. Menendez-Baceta, L. Aceituno-Mata, R. Acosta-Naranjo, L. Calvet-Mir, P. Domínguez, T. Garnatje, E. Gómez-Baggethun, M. Molina-Bustamante, M. Molina, R. Rodríguez-Franco, G. Serrasolses, J. Vallès & M. Pardo-de-Santayana (2015) From famine foods to delicatessen: Interpreting trends in the use of wild edible plants through cultural ecosystem services. *Ecological Economics*, 120:303–311.

Robbins, P., M. Emery & J. L. Rice (2008) Gathering in Thoreau's backyard: nontimber forest product harvesting as practice. *Area*, 40(2):265–277.

Russo, A., F. J. Escobedo, G. T. Cirella & S. Zerbe (2017) Edible green infrastructure: An approach and review of provisioning ecosystem services and disservices in urban environments. *Agriculture Ecosystems & Environment*, 242:53–66.

Russo, A., & Cirella, G. T. (2019) Edible urbanism 5.0. *Palgrave Communications*, 5(1). doi:10.1057/s41599-019-0377-8

Sachdeva, S., M. R. Emery & P. T. Hurley (2018) Depiction of wild food foraging practices in the media: Impact of the great recession. *Society & Natural Resources*, 31(8):977–993.

Säumel, I., Kotsyuk, I., Holscher, M., Lenkereit, C., Weber, F., & Kowarik, I. (2012) How healthy is urban horticulture in high traffic areas? Trace metal concentrations in vegetable crops from plantings within inner city neighbourhoods in Berlin, Germany. *Environmental Pollution*, 165, 124–132.

Shackleton, C., P. Hurley, A. Dahlberg, M. Emery & H. Nagendra (2017) Urban foraging: A ubiquitous human practice overlooked by urban planners, policy, and research. *Sustainability*, 9(10): 1884.

Shackleton, S., B. Campbell, H. Lotz-Sisitka & C. Shackleton (2008) Links between the local trade in natural products, livelihoods and poverty alleviation in a semi-arid region of South Africa. *World Development*, 36(3):505–526.

Soga, M. & K. J. Gaston (2016) Extinction of experience: the loss of human–nature interactions. *Frontiers in Ecology and the Environment,* 14(2):94–101.

Stark PB, Miller D, Carlson TJ & de Vasquez KR (2019) Open-source food: Nutrition, toxicology, and availability of wild edible greens in the East Bay. *PLoS ONE*, 14(1):e0202450.

Sujarwo, W., I. B. K. Arinasa, G. Caneva & P. M. Guarrera (2016) Traditional knowledge of wild and semi-wild edible plants used in Bali (Indonesia) to maintain biological and cultural diversity. *Plant Biosystems*, 150(5):971–976.

Svanberg, I. (2012) The use of wild plants as food in pre-industrial Sweden. *Acta Societatis Botanicorum Poloniae*, 81(4):317–327.

Synk, C. M., B. F. Kim, C. A. Davis, J. Harding, V. Rogers, P. T. Hurley, M. R. Emery & K. E. Nachman (2017) Gathering Baltimore's bounty: Characterizing behaviors, motivations, and barriers of foragers in an urban ecosystem. *Urban Forestry & Urban Greening*, 28:97–102.

Taylor, J. R., & Lovell, S. T. (2015) Urban home gardens in the Global North: A mixed methods study of ethnic and migrant home gardens in Chicago, IL. *Renewable Agriculture and Food Systems*, 30(1), 22–32.

Unnikrishnan, H. & H. Nagendra (2015) Privatizing the commons: impact on ecosystem services in Bangalore's lakes. *Urban Ecosystems*, 18(2):613–632.

van Andel, T., & Carvalheiro, L. G. (2013) Why urban citizens in developing countries use traditional medicines: the case of Suriname. *Evid Based Complement Alternat Med*, 2013, 687197.

Wang, H. F., S. Qureshi, S. Knapp, C. R. Friedman & K. Hubacek (2015) A basic assessment of residential plant diversity and its ecosystem services and disservices in Beijing, China. *Applied Geography*, 64: 121–131.

Wehi, P. M. & W. L. Wehi (2010) Traditional plant harvesting in contemporary fragmented and urban landscapes. *Conservation Biology*, 24(2):594–604.

Willcox M. L. (1999) A clinical trial of 'AM', a Ugandan herbal remedy for malaria. *J Public Health Med*,: 21(3):318–24.

6

Agroecology as Public Health: The Island Example of Tasmania

Pauline Marsh
[1] *Centre for Rural Health, University of Tasmania*

CONTENTS

KEYWORDS: *Community Gardening*; *public health*; *therapeutic landscapes*; *positive risk*; *health equity*; *wellbeing*

6.1 Introduction

In 1792 French explorer Bruny d'Entrecasteaux sailed into Lyluequonny lands in the South East of what was then Van Dieman's Land, now the Australian island state of Tasmania. With him was a gardener, Felix DeLahaye, who set about establishing a vegetable garden on the shore (Mulvaney and Tyndale-Biscoe 2007). According to the voyage records, the intention of the garden was to provide fresh food to protect future European explorers from scurvy and the French spent three days preparing the soil and sowing seeds for a large range of vegetables (Galipaud et al 2007). When they returned the following year there were little signs of the vegetables—the garden was a horticultural failure. Not surprising, perhaps, given how little the visitors knew of the natural environment. Nevertheless, "DeLahaye's garden" has become highly significant in Tasmania, the extent of being recreated in the Government House grounds. Its importance lies in its successful cultivation of a positive settler–Indigenous contact history narrative that resonates with Tasmanians 230 years later. It forms a part of our contemporary understanding of the relationships between gardening, health and wellbeing.

World over, contact between European colonisers and Indigenous peoples has been characterised almost exclusively by gross misunderstandings, disrespect and violence. Tasmania under British colonial rule was no different. However, the interactions between Aboriginal and French people during DeLahaye's garden experiment were described by the explorers as friendly, passive, and mutually curious. The garden is crucial to these encounters; it is the garden that enables this peaceful moment of history. It is difficult to dispute that the ultimate purpose of gardening is to improve health and wellbeing—of people and the planet. Contemporary Tasmania is now an island full of gardens and gardeners, and a place populated by 'caring cultivators' (Pogue Harrison 2008) of fruits, vegetables and livestock. Tasmania is an excellent location by which we might understand the complex physical, cultural and emotional intersections between acts of gardening and human wellbeing.

To explore this idea further, in this Chapter I present four case studies of gardens established across urban and peri-urban sites in which the concepts and principles of contemporary agroecology and public health coalesce in complementary ways. I focus on one particular type of garden that is an agroecology modality: community gardening. Improved understanding of the nexus between community gardening and the principles and aims of public health allows us to better understand the significance of agroecology and its potential contribution to improvements in human health and wellbeing.

6.1.1 Tasmanian Agriculture, Agroecology, and Community Gardening

Separated by the Australian mainland by a 240-kilometre-wide body of water, Tasmania is the southernmost state of Australia. It comprises 68 500 square kilometres of undulating hills and mountainous ranges, contrasting against the flats of plateaus and plains. At 42 degrees latitude, a cool temperate climate creates lush rainforests, some of which contain the tallest flowering plants in the world—*Eucalyptus regnans*—as well as ideal gardening conditions for the half a million inhabitants. Rainfall rates vary across the island (averaging 1390 mm in 2018) and collects in glacial lakes, tarns, rivers and dams. Land use is a mix of conservation areas, grazing lands, areas of crops, forestry and limited mining.

Up to 42% of the island is set aside for commercial, larger scale agricultural practices, and a statewide irrigation scheme diverts water to select areas (Brown 2010). However, alongside sheep farms, hop fields, vineyards and dairies, small-scale urban agroecology is also booming. Tasmania is the birthplace of permaculture in the 1970s—a movement that applies principles designed to create resilient, ecologically sustainable living and land use (Permaculture Principles). Permaculture is often described as being related to agroecology (Ferguson and Lovell 2014), although this relationship is dynamic (see Chapter 8 of this volume). There are currently many movements and individuals applying ecological and sustainability principles to the way they plant, grow and harvest foods and develop communities. Local organisations such as Urban Farming Tasmania, Backyard Permaculture, and Slow Food Tasmania, for example, are steering this movement.

Community gardens marry agroecology and community development (Egli et al 2016; Glover 2004). Generally established by communities in socially and economically disadvantaged areas, they provide opportunities to grow produce, eat well, practice sustainable land management, and socialise in a free and open space. The extent to which each Garden may implement agroecology practices (that is, those ecologically based practices that increase soil and functional biodiversity and environmental sustainability [Wezel et al 2009]) - they nonetheless have been referred to as sites of "embodied sustainability" (Turner 2011). Importantly, they aim to help address social, economic and health inequities, as sites of socio-cultural-ecological interconnectivity between people, food and food production. The synergies with the community development functions of urban agroecology are evident: "to partially solve problems associated with food quality and affordability, reduce ecological footprints, increase community cohesion, achieve greater community resilience and promote urban sustainability" (Tornaghi 2014).

Since first emerging 30 years ago, Tasmanian community garden movement is now well-established. According to the register of Community Gardens kept by the Australian City Farms and Community Garden Network, the island state is responsible for approximately 16% of the nation's Gardens, even though the island comprises less than 0.9% of Australia's total area and 2% of the population. One third of the Gardens are located alongside Community Houses (also called Neighbourhood Houses). These are community centres established in low socio-economic areas which aim to bring people together to work collaboratively to address various social, economic and health needs (*Neighbourhood Houses Tasmania*).[1] Currently 33 of the state's 35 Neighbourhood Houses have community gardens.

[1] It should be noted that there are many other types of community gardens located in the middle of more affluent suburbs, or communal gardens in schools and aged care facilities; however they are not the focus of this Chapter.

TABLE 6.1

Agroecology and Public Health Coalescence of Principles in the Four Case Studies

Case Study	Community Garden Model	Agroecology Principles	Health Approach
Goodwood	Neighbourhood House Community Garden	Organic gardening, soil and plant biodiversity, environmental sustainability	Community resilience, empowerment
DIGnity Supported Community Gardening	Social and Therapeutic Horticulture Program	Organic gardening, environmental sustainability, community cohesion	Social, cognitive, mental health: prevention and restoration
Royal Tasmanian Botanical Gardens Community Food Garden	Food Production Community Garden with Trauma Support	Organic gardening, biodiversity	Social, physical and mental health: restoration
Edible Precinct	Inner-city Reconciliation Garden	Organic gardening, biodiversity, eco-sustainability, community cohesion	Cultural health, health equity

I was born on the island and have lived here, for the most part, all of my life. My varied career (spanning nursing, gardening and social research) has been enabled by an environment characterised by easy access to wilderness, coastal areas and rich cultivated landscapes of gardens and farms. My local town has a thriving Community Garden. Through this personal and professional lens I explore how gardens and small-scale urban agroecological gardening promotes, remedies and regenerates human health and wellbeing.

Each of the four community gardens chosen as case studies (Table 6.1) for this Chapter are examples of how community gardening can be simultaneously both an agroecology and a public health intervention. The first, Goodwood Community Garden, provides whole-of-community caring through gardening in an area of historical socio-economic disadvantage. Second, the Royal Tasmanian Botanical Gardens Community Food Garden focuses on growing food for low socio-economic communities, and incorporates a program designed to improve mental health for returned veterans with post-traumatic stress disorder. The third example, DIGnity Supported Community Gardening, supports people with various disabilities to garden with others in shared community spaces. The final case study, the Edible Precinct, utilises gardening for its reconciliation capacity. Tasmania's colonial history is a persistent contributor to entrenched cultural health inequities in Tasmania.

Collectively, these four case studies provide insights into the ways that urban and peri-urban agroecology support the aims and outcomes of public health interventions. Before we take an excursion through the gardens, however, I will firstly provide a brief overview of the theoretical context that informs this work and provides the philosophical framework for understanding community gardening as simultaneously agroecology practice, social movement and public health intervention.

6.1.2 Public Health and Therapeutic Landscapes Theory

Public Health aims to reduce health inequities, to ensure that good health is not the privilege of the financially or socially privileged. Its platforms for wellbeing are not only biological and genetic, but importantly also social and ecological: the conditions under which people can maintain and improve their health and wellbeing, or prevent the deterioration of their health (Baum 2016). Although population-level disease eradication might be the most recognisable feature of Public Health interventions (for instance widespread vaccination programs for polio and tuberculosis), in theory it encompasses a much broader entire spectrum of initiatives. Public Health initiatives incorporate salutogens (factors that influence health) as well as pathogens (disease causing factors) and take a whole-of-community approach to improving health. It is poetically defined as "the art and science of preventing disease, prolonging life and promoting health through the organized efforts of society" (World Health Organisation 1988).

The conceptual link between a Public Health approach to health and the perceived health benefits from community gardening is articulated by the theoretical concept of Therapeutic Landscapes. The notion of Therapeutic Landscapes was first suggested by Will Gesler (1992), who examined the healing capacity of places. While the biophilia hypothesis—that humans have an innate desire to connect with non-human nature (Wilson 1986)—underpins this thinking, the Therapeutic Landscape concept is more complex than this. It acknowledges that contact with nature is in itself therapeutic, but also that there are more complex inter-relationships at work. That is, a single intrinsic factor does not create a therapeutic place, but rather a multitude of factors can contribute (Williams 2007). A therapeutic landscape is, therefore, any space, which—through a combination of environmental, social, political, cultural, ecological, and/or emotional factors—has positive impacts on human health.

This dual theoretical lens—Public Health together with Therapeutic Landscapes—is the means by which we currently understand the mechanisms for community gardening as a platform for human wellbeing. Health and wellbeing benefits from community gardening are generally attributed to combinations of physical activity, social connection and healthy eating (Austin et al 2006; Glover 2004; Marsh and Spinaze 2016; Milligan et al 2004; Pitt 2014; Wang and MacMillan 2013). There are clear synergies with agroecology, specifically in its relationship to and dependence on ecosystem services – those ecological factors that are "critical to human health and wellbeing" (McPhearson et al 2015). However, if we are to understand community gardening as a public health intervention we need to move beyond a discussion contained within the established domains of exercise, social engagement and eating. We need to understand the ways in which gardening addresses the range of factors that contribute to health inequities: social, physical, cultural, ecological and emotional.

In Tasmania, there is an urgent imperative for this work. Curiously, the richness of Tasmania's physical landscape is not mirrored by the health and social status of the island's inhabitants. The Index of Relative Socio-economic Disadvantage rates Tasmania's socio-economic disadvantage as high (IRSD 960.8) compared to the Australian average (1000) (Australian Bureau of Statistics 2011). Poor health is currently experienced by 27.2% of those in the most disadvantaged quintile, as compared to 15.8% of those in the least disadvantaged quintile (Department of Health 2016). Across a range of indicators, Tasmanians experience poor health. Consistent with residents of peri-urban and rural areas in Australia and globally, we have high rates of chronic diseases, cancers, and obesity. We smoke more, take greater risks with alcohol, and eat less healthily than we should. The differences between Indigenous and non-Indigenous life expectancy of 10 to 17 years is redolent of longstanding, persistent and preventable health inequities that dog the state and the nation.

6.2 Case Studies: Exploring Four Community Gardens

In earlier work, colleagues and I undertook to ascertain the mechanisms of health improvements particular to the Tasmanian community garden (Marsh et al 2018). We found that community gardens were not delivering healthcare in deliberate or obvious ways, or even practicing Horticultural Therapy, which utilises particular gardening activities to achieve physical or mental health goals (Haller and Kramer 2006). Nor did many gardens incorporate specific therapeutic design components such as enabling gardens (Raske 2010) or dementia-designed 'wander' gardens (Detweiler et al 2002). Rather, in that qualitative study we found evidence of health care that was more covert, which drew on salutogenic functions of participation in community gardening. Through thematic analysis of stakeholder interviews with community gardeners across the island we found numerous positive health benefits from gardening, including exercise and socialising, as well as from sustainable food production and nutrition education. Participants also provided descriptive examples of more surprising health giving benefits that arose from things like the solace, safety and comfort of garden-based end-of-life care, of bereavement support or from the support for people living with dementia. This research provided some insights into the ways in which health was improved and inequities were tackled, but we were keen to know more. Applying case study methodology, and using methods of textual analysis and interviews, this Chapter delves more deeply into the nuances of community gardens and their health impacts.

6.2.1 A Healthy Community around the Goodwood Garden

The Goodwood Community Garden is a small garden, which wraps around a community house located in one of the oldest public housing areas in Hobart. Amongst ageing weatherboard housing, in an area sandwiched by light industrial hubs and major highways, here the gardeners practice sustainable horticulture: seed saving, crop rotation, composting, worm farming, cooking and preserving the harvest. Preserves and seeds are sold to the community in the local shop, and garden produce is freely available to anyone who wanders through the garden, which is open at all times. It is a deliberately intergenerational space, where creativity is welcomed and encouraged. Gardeners and neighbourhood house staff integrate gardening and healthy eating and provide education in relaxed and informal ways.

Although there is a deliberate looseness to the Goodwood garden daily activities, everything that happens there is underpinned by a palpable commitment to improving health outcomes. A combination of biodiversity, horticultural activities and empowerment form the basis for tackling a reduction in population health inequities. The garden coordinator works 10 hours per week, alongside a team of regular volunteers. People with a range of neurological challenges and mental health issues garden alongside carers, children, and other volunteers. Gardeners are guided by the garden coordinator, who in turn is guided by each person's interest and skills. Caring through cultivation is encouraged. People are given the space and opportunity to interact with each other, and to actively care for each other. The garden coordinator describes people 'wrapped in the support of fellow gardeners'. For some, the garden is the only place where they access support – they may not speak with anyone else, nor listen to anyone else in other parts of their daily lives. They are also enabled to feel ownership of the garden and its outputs. All that is produced is the shared work of the community.

Goodwood Garden shares a unique partnership with the Tasmanian Junior Beekeepers Association. What started out as a plan to celebrate a local beekeeper's 100 birthday and 92nd year of beekeeping turned into ongoing school holiday activities, a Bee Book produced by the local children, public presentations, media coverage and the establishment of a Bee Literate Grant Scheme. The timing of this coincided with a surge of interest in small-scale, backyard beekeeping in Tasmania. So enthusiastic had Tasmanians become for beekeeping that at one stage the Tasmanian Beekeepers Association reported demand for starter nuclei (or Nucs) had outstretched supply.

Backyard beekeeping forms a key element of salutogenic gardening. Understanding the role of social bees like the European honeybee (Apis mellifera) helps us to appreciate the importance of biodiversity, and the interconnectedness between plants, pollinators and humans. The vulnerability of bees alerts us to the dangers of inorganic pesticides and the value of organic food production. And as the Goodwood Community Gardening Bee Book reminds us, honey itself has a myriad of health-giving properties (boosting energy and immunity, minor burn treatment, insomnia, skin and digestion). Although not totally aproblematic (concerns about the over-popularity of social bees threatening the function of native or wild bees have been raised (Norfolk, 2018)) saving bees from annihilation is another well-intentioned expression of caring cultivation, and community empowerment.

This is a quintessential public health model of gardening. The individual health benefits come from increased health knowledge, increased confidence, self-worth and empowerment. Importantly, these benefits are also aimed at benefiting the wider community. The gardeners combine agroecology with community development principles to develop a resilient neighbourhood. This is a caring village with biodiversity, beekeeping and communal small-scale agriculture as its beating heart.

6.2.2 Royal Tasmanian Botanical Gardens Community Food Garden with Trauma Support

In a sheltered part of the 200+ years-old, carefully manicured Royal Tasmanian Botanical Gardens (RTBG) in the south of Tasmania, a community food garden boasts 40+ beds of healthy, bountiful produce. The garden is maintained by RTBG staff and a team of regular volunteers. At the time of writing, the produce was being collected by a community organisation, Loaves and Fishes, which distributed the food through community houses and other welfare organisations to supply emergency food relief for people experiencing food insecurities.

This garden-within-the-Gardens was originally established as a food growing area for a national Gardening Australia television program. In 2012 the Tasmanian Food Security Council expanded the space into a community garden, as a means of enacting a Food Security policy. The Policy aimed to try and improve health outcomes by making access to healthy fresh food easier for people on low incomes and living in disadvantaged areas. Plots were set aside for use by various community groups, including plots for veterans and their families. The Department of Veterans Affairs (DVA) were keen to see these garden beds provide "security, control and confidence" for people with Post Traumatic Stress Disorder (PTSD) (Department of Veterans Affairs 2014) as a means of mitigating and helping to overcome this debilitating condition.

Post-traumatic stress disorder (PTSD) is characterised by intrusive memories of the traumatic trigger event(s), severe anxiety, negative thoughts and mood, and a sense of loss of control. These symptoms may manifest in social isolation, depression, inactivity/avoidance, anger or mood numbing. The cures range from "time and good self-care" (Black Dog) to clinical psychological treatments such as trauma-focused Cognitive Behavioural Therapy (CBT) (Watkins et al 2018). One of the garden-related non-pharmaceutical treatments is exercise, which has been found to improve depressive symptoms, sleep quality and sedentary behaviours (Rosenbaum et al 2015). Social and therapeutic horticulture has been shown to have wide-ranging benefits for mental illnesses and disorders (Lorber 2011; Sempik 2010).

The RTBG has a long history of offering therapeutic horticulture and horticultural therapy. I undertook workplace experience there in the 1980s as part of my horticulture study, on placement with the resident occupational therapist. She worked with people with a range of abilities and diversities using various gardening tasks to improve function and capacity. Although that particular program no longer runs, the tradition is continued through the intentional design of the trauma support in the community food garden. The trauma support was designed to improve mental health and wellbeing through an emphasis on peer support and easy conversation, in a space for a non-clinical informal support group. Peter Cundell, war veteran and high-profile Tasmanian gardener, explained at the time of its establishment that gardening provided particularly appropriate support for people experiencing post-traumatic stress disorder (PTSD) because of its predictability: "we know exactly what's going to happen. We know that when we sow seeds that they'll be germinating in two or three weeks' time and that they're going to grow into plants. There's a certainty and a reassurance about gardening, and this is what people need" (Department of Veterans Affairs 2014).

Another of the gardening-based health benefits is thought to come from the opportunity to contribute to the community—to not feel as though you are on the receipt of hand outs or special supports. Self-worth is an important part of returning to civilian life post traumatic events. In the original model, the gardening was combined with education sessions: guest speakers, beekeeping courses, landscaping projects in the space. Large numbers of regular veterans were volunteering and a news story from the time highlights just how therapeutic people found it: "It gives me a positive place to come, to get out of my head space ... we are all from similar backgrounds ... and I can just be myself" (Australian Broadcasting Commission 2014).

Nevertheless, the extent to which returned veterans engage with the community food garden has oscillated over the five years. The coordinator of community programs at the DVA attributes varying interest levels to the passion and availability of the volunteer coordinator at the time. There is no specific funding for a garden coordinator, so when there are too many competing demands, and stress levels rise, the program is put to the side. There have been periods when there has been no coordinator at all, and the program has entered into a hiatus. Moreover, veterans are part of a broader, general volunteering team, and reportedly have felt little sense of ownership of the garden plots.

The RTBG Community Food Garden highlights some of the challenges of running a health program that uses principles of therapeutic horticulture, public health and agroecology located in shared community spaces. As this, and the following example, demonstrates, it can be difficult to create a communal gardening environment where people feel surrounded by and supported by people with shared experiences and needs, and simultaneously an environment that provides opportunities for social engagement with people from the broader community also.

6.2.3 DIGnity Supported Community Gardening

DIGnity Supported Community Gardening (DIGnity) is a model of support that taps into and enhances the salutogenic functions of already established community gardens. Through DIGnity, health workers, artists and researchers work alongside community members, garden coordinators and volunteers onsite at neighbourhood house community gardens. The goal is to extend the reach of the gardens and to welcome people who otherwise could not participate, particularly people who have lost the confidence or their physical or cognitive capacity to garden. The team includes an Occupational Therapist, fibre artist and mental health counsellor who merge into the everyday running of the garden—they are deliberately low-key to avoid creating the feel of a therapy program. Their goal is to enable participation in the garden in any way possible, and in whichever ways people might find meaningful. They liaise closely with local service providers, general practitioners, carer organisations and community groups to get people involved. Community volunteers work alongside people with all sorts of needs and all sorts of abilities.

After three years of operating DIGnity sessions at the Okines community garden, 40 kilometres from Hobart in the South East of Tasmania, between 10–30 people participate regularly each week. Regular visits by groups of people include residents from local aged care facilities. These are people with multiple age-related barriers to gardening: people in wheelchairs, with walking aids, people with dementia and a multiple of physical and cognitive challenges that otherwise keep them indoors. Local mental health care workers also accompany people with severe and debilitating mental illness to the sessions.

This active participation in agroecological urban gardening has a range of health and wellbeing impacts. DIGnity gives people regular opportunities for company—to engage with others from their local community who they might not otherwise see. People eat fresh healthy food, have gentle (or rigorous) exercise, there are opportunities for socialisation and activity, but also for mental rejuvenation. As people garden, weave, have cups of tea and relax, they may be talking with others or be quiet and away from the crowds. One of the regular volunteers articulated his observations of DIGnity's approach to wellbeing. He states:

> *I'm on my own ... a lot of the time. Nice to get out and talk to somebody else and see another face instead of looking at the four walls ... A bit of company, a bit of mixing with people. A lot of people in the same boat as I'm in I suppose are around as you talk to them. It's a way of a bit of company sometimes. [Interviewer: The actual gardening do you like it?] Yeah, I like a bit of that. I should be up doing my own, but I come down here this morning. Talking to the plants by yourself is not much good is it?* (DIGnity 2018)

The aim of this specialised support model is not simply to create a safe and welcoming space for people with particular vulnerabilities, it is also to facilitate positive risk taking. The philosophy underpinning DIGnity is the Dignity of Risk (Perske 1972) which refers to the belief that all of us take risks as part of our everyday life—indeed, taking risks is an important part of being human (Nay 2002). To deny people risks, due to age or physical abilities, is to deny a human right. Taking risks benefits physical and mental health as well as self-esteem (Marsh and Kelly 2018). Gardening, like many outdoor activities, is basically low-risk. DIGnity evaluations suggest that it is possible to enact the Dignity of Risk principle in community gardens provided you have sufficient human support (safety in numbers), a stimulating environment and a range of activities available for people to choose from. As Neil Mapes (2017) points out, in his research on dementia-friendly outdoor activities, "staying indoors all day everyday has its associated risks".

DIGnity is tapping into the wellbeing benefits from urban agroecological activities for people living with the impacts of dementia. Other work has shown how nature-based activities are important for people with dementia (Gigliotti et al 2004; Marsh et al 2018; Myren et al 2017). Dementia refers to a condition characterised by cognitive decline and decreasing capacity to undertake everyday physical and cognitive activities, the most common form of which is Alzheimer's disease. Dementia is the second leading cause of death in Australia, and a growing global challenge for people, families, and the health system. Maintaining quality of life as the disease progresses is the goal of dementia care, and also the way by which DIGnity makes a positive contribution. Evaluative studies of the DIGnity Project showed that participants with dementia were observed to be more socially engaged, physically mobile and psychologically and emotionally relaxed than they were when they were at the Residential Aged Care Facility (Marsh et al 2018). For example, people with dementia are free to walk about the garden, and at no stage did anyone "wander" away—the stimulation of the environment kept people in the space, and all the various attendees kept an unobtrusive eye on people. This finding concurs with that of Myren and colleagues (2017) in their observations of people with dementia participating in care farms in Norway where a combination of outdoor environment, meaningful activities, respectful interactions and low-key professional support facilitates meaningful, satisfying and healthy participation.

6.2.4 Edible Precinct

On the harbourside of Tasmania's capital city, Hobart, is a community garden of a different sort. On approximately an acre of land it comprises a series of varying styles of raised beds, set up on the large slabs of concrete. Concrete typifies the whole of the wharf area—it was until recently used to store containers of freight that were waiting for transport by boat or train, prior to a relocation of the main transport hub to outside of the city. The end of the industrial use of the site gave way to plans for new commercial developments—but these were set to be five years in the planning. In the interim, the Macquarie Point Development Corporation established the Edible Precinct.

In many ways the Edible Precinct is no different from other inner-city community gardens—open to the public, produce donated to community organisations, application of organic farming principles, motivated by greening the urban landscape. However, this one has a deliberate salutogenic strategy that has a particular resonance with DeLehaye's 1872 food garden, that has been so significant in our cultural history. That is, it is an urban agroecology reconciliation garden. The garden beds co-locate European plants with Indigenous edible plants, many of which are unknown and unrecognisable as edible to contemporary Tasmanians. This has been the result of the combined efforts of bush food expert and Indigenous plant specialist Kris Schaffer with a team of local professional horticulturalists. All plants are labelled with details and designed to be deliberately educational.

However, this is more than an exercise in horticultural education; this is an intentional effort to raise awareness of Aboriginal culture, knowledges and continuing culture. The longstanding negative effects of colonisation on Indigenous health outcomes are well documented, and resultant health inequities persist in Tasmania as they do in postcolonial nations globally—the most obvious of which is in the difference in life expectancy in Australia. In a state with a shameful history marked by attempted genocide and subsequent denial of Aboriginal identity, language and culture, this type of gardening seeks to heal the deep ongoing psychological and societal wounds caused by the trauma of colonisation.

It would be naive to suggest that a cross-cultural food garden can fix such an enormous problem. However, the simple concept adopted by the Edible Precinct does work toward improvements. There are three key mechanisms that enable wellbeing outcomes from this project: First, the Edible Precinct facilitates cultural empowerment by reclaiming knowledge and language. Staff work closely with local Aboriginal organisations, the Tasmanian Aboriginal Centre and Nayri Niara. As Pat Anderson, Chairperson of the Lowitja Institute for Aboriginal Health, states "Our medical professionals do a great job of prescribing medicines and devising treatment programs but, to fix the root causes of ill-health, we need something more. As Aboriginal people we need to have a sense of agency in our lives, that we are not stray leaves blowing about in the wind. In a word, we need empowerment" (Whiteside et al 2014). Second, the co-planting of the Indigenous and non-Indigenous food plants is a metaphor for a multicultural reality of postcolonial nations. In the same way as we mythologise the conciliatory power of DeLehaye's garden, this garden is also symbolic of the potential for peaceful intercultural cohabitation. As plant consultant Kris Schaffer said: "We want to use food, including Tasmanian bush foods and introduced species to bring people together in a spirit of collaboration, sharing and reconciliation" (Edible Precinct). Third, culture is a determinant of health and wellbeing, partly due to a pragmatic influence on nutrition and environmental health (Napier et al 2017). Thus, a community garden that pays attention to and incorporates Indigenous knowledges of food sources and their health benefits, and also aims to improve the health of the environment, is also influencing human health and wellbeing.

6.3 Conclusions

Collectively, these case studies illuminate the ways that four communities are applying agroecology and public health principles in practice, in complementary ways, that address the multiple causes and impacts of health inequities in Tasmania. The shared socio-ecological aims such as improving food quality, empowering whole communities, creating sustainable environments and enhancing community cohesion are evident. The case studies illuminate an existing awareness of the broader socio-political, cultural and ecological contexts that impacts communities, and they operate with a mind to these contexts.

There are complex, multifaceted combinations of physical, social, cultural, environmental and psychological factors interacting in these spaces. They coalesce in ways that aim to positively affect the environment, individuals and communities. Indeed, urban agroecology and public health are perhaps most effective because of their complex multi-dimensionality. The interconnectedness of the elements is what enables agroecology interventions in urban and peri-urban areas to be simultaneously ecosystem service, and also therapeutic landscape and public health intervention. Furthermore, these connections allow public health initiatives in community gardens to be beneficial to the physical landscape and health of the planet. This combination in shared outdoor gardening spaces is, I suggest, an underutilised pathway toward a reduction in health inequities.

Nevertheless, despite these best efforts at the local level, we continue to see population health disparities along urbanisation and socio-economic gradients—both in Tasmania and globally. Preventable diseases are not being prevented, and mortality rates are persistently higher amongst people living in disadvantage. Disappointingly, a health inequity qualifier of sorts has started to appear in the Therapeutic Landscapes literature:

> A growing body of epidemiological evidence indicates that greater exposure to, or 'contact with', natural environments (such as parks, woodlands and beaches) is associated with better health and well-being, at least among populations in high income, largely urbanised, societies [my emphasis].
> (White et al 2019)

Why this failure to make significant improvements in measurable health outcomes? One suggestion is that there is a conceptual and pragmatic gap between agroecology, therapeutic gardening and mainstream healthcare services. This may be due to what Bell and colleagues (2018) claim is the struggle for the Therapeutic Landscape concept to find a firm place in health policy. More tangible, simplistic concepts are perhaps accessible to policy makers. Despite evidence that urban gardens and gardening can be "more powerful than medicine" (Sacks 2019) there is little acknowledgement at a policy level of agroecology's capacity for human health improvement.

Future research that unlocks and articulates the role of agroecology in addressing health inequities is vital if these interventions are to be taken seriously as part of the suite of public health measures. Mixed methods research that accommodates the complexity and multi-dimensionality of health determinants and therapeutic functionality will be important. Future research is required that can explore the subjective health benefits such as these case studies demonstrate—e.g. feeling relaxed or more socially confident—as well as social determinants such as cultural and community empowerment. This work should not just stop at gathering research, however. Research findings need to be translated into health policy and practice outcomes that marry agroecology and human health approaches to generate sustainable and meaningful changes.

REFERENCES

Austin, E. N., Y. A. M. Johnston, and L.L. Morgan. 2006. Community gardening in a senior centre: A therapeutic intervention to improve the health of older adults. *Therapeutic Recreation Journal*, 40(1):48–57.

Australian Broadcasting Commission. 2014. A Thriving Patch. https://www.abc.net.au/gardening/factsheets/a-thriving-patch/9435942

Australian Bureau of Statistics. 2011. Community Profiles. *Census 2011.* http://www.abs.gov.au/websitedbs/censushome.nsf/home/communityprofiles

Australian City Farms and Community Garden Network. https://communitygarden.org.au

Baum, Fran. 2016. The New Public Health, Fourth Edition. Oxford: Oxford UP.

Bell, S. L., R. Foley, F. Houghton, A. Maddrell, and A.M. Williams. 2018. From therapeutic landscapes to healthy spaces, places and practices: A scoping review. *Social Science & Medicine*, 196:123–130.

Black Dog Institute. www.blackdoginstitute.org.au

Brown, M. 2010. *Land Use Tasmania: A Technical Report.* Natural Resources Management: Hobart

Department of Health and Human Services, Tasmanian Government. 2016. *Report on the Tasmanian Population Health Survey 2016.* Hobart: Tasmanian Government.

Department of Veterans Affairs. 2014. Gardening helps relieve stress for veterans. https://www.dva.gov.au/about-dva/accountability-and-reporting/annual-reports/annual-reports-2013-14/feature

Detweiler, M. B., D.B. Trinkle, and M.S. Anderson. 2002. Wander gardens: Expanding the dementia treatment environment. *Annals of Long Term Care*, 10(3):68–74.

DIGnity Supported Community Gardening 2018. *Project Report.* www.dignitygardening.com

Edible Precinct. 2019. https://www.macquariepoint.com/edible-precinct

Egli, V., M. Oliver, and E.S. Tautolo. 2016. The development of a model of community garden benefits to wellbeing. *Preventative Medicine Reports*, 3:348–52.

Ferguson, R. S., and S.T. Lovell. 2014. Permaculture for agroecology: Design, movement, practice, and world-view: A review. *Agronomy for Sustainable Development*, 34:251–274.

Galipaud, J.-C., A.D. Biran, G. Jackman, A. Gurnhill, R. Pineda, and A. McGowan 2007. The lost garden of Recherche. In *Rediscovering Recherche Bay.* Eds. J. Mulvaney, and H. Tyndale-Boscoe, 99–118. ACT: The Academy of the Social Sciences in Australia.

Gesler, W. 1992. Therapeutic landscapes: Medical geographic research in light of the new cultural geography. *Social Science & Medicine*, 34(7):735–746.

Gigliotti, C., S. Jarrott, and J. Yorgason. 2004. Harvesting health: Effects of three types of horticultural therapy activities for persons with dementia. *Dementia*, 3(2):161–180.

Glover, T. D. 2004. Social capital in the lived experiences of community gardeners. *Leisure Sciences*, 26(2):143–162.

Haller, R., and C. Kramer Eds. 2006. *Horticultural Therapy Methods.* New York: Haworth Press.

Lorber, H. Z. 2011. The use of horticulture in the treatment of post-traumatic stress disorder in a private practice setting. *Journal of Therapeutic Horticulture*, 21(1):18–29.

Mapes, N. 2017. Think outside: Positive risk-taking with people living with dementia. *Working with Older People: Community Care Policy & Practice*, 21(3):157–166.

Marsh, P., S. Brennan, and M.D. Vandenberg. 2018. 'It's not therapy, it's gardening': Community gardens as sites of comprehensive primary healthcare. *Australian Journal of Primary Health*, 24(4):337–342.

Marsh, P., and L. Kelly. 2018. Dignity of risk in the community: A review of and reflections on the literature. *Health, Risk and Society*, 20(5):297–311.

Marsh, P., H. Courtney-Pratt, and M. Campbell. 2018. The landscape of dementia inclusivity. *Health and Place*, 52:174–179.

Marsh, P., and A. Spinaze. 2016. Community gardens as sites of solace and end-of-life support: A literature review. *International Journal of Palliative Nursing*, 22(5):214–219.

McPhearson, T., E. Andersson, T. Elmqvist, and N. Frantzeskaki. 2015. Resilience of and through urban ecosystem services. *Ecosystem Services*, 12:152–156.

Milligan, C., A. Gatrell, and A. Bingley. 2004. Cultivating health: Therapeutic landscapes and older people in northern england. *Social Science & Medicine*, 58:1781–1793.

Mulvaney, J., and H. Tyndale-Biscoe. Eds. 2007. *Rediscovering Recherche Bay.* ACT: The Academy of Social Sciences in Australia.

Myren, G. E. S., I. Enmarker, O. Hellzen, and E. Saur. 2017. The influence of place on everyday life: Observations of persons with dementia in regular day care and at the green care farm. *Health*, 9:261–278.

Napier, A.D., M. Depledge, M. Knipper, R. Lovell, E. Ponarin, E. Sanabria, and F. Thomas. 2017. Culture matters: Using a cultural contexts of health approach to enhance policy-making. *Cultural Contexts of Health and Well-being Policy Brief No.1.* Geneva: World Health Organisation.

Nay, R. 2002. The dignity of risk. *Australian Nursing Journal*, 9(9):33.

Neighbourhood Houses Tasmania. www.nht.org.au

Norfolk, O. 2018. Keeping Honeybees doesn't save bees - or the environment. *The Conversation* (September 12).
Permaculture Principles. www.permacultureprinciples.com
Perske, R. 1972. The dignity of risk and the mentally retarded. *Mental Retardation* 10(1):24–27.
Pitt, H. 2014. Therapeutic experiences of community gardens: Putting flow in its place. *Health and Place*, 27:84–91.
Pogue Harrison, R. 2008. *Gardens: An essay on the human condition*. Chicago: University of Chicago.
Raske, M. 2010. Nursing home quality of life: Study of an enabling garden. *Journal of Gerontological Social Work*, 53(4):336–351.
Rosenbaum, S., C. Sherrington, and A. Tiedemann. 2015. Exercise augmentation compared with usual care for post-traumatic stress disorder: A randomized controlled trial. *Acta Psychiatr Scand*, 131(5):350–359.
Tasmanian Found Security Council. 2012. *Food for all Tasmanians: A food security strategy*. Hobart.
Sacks, O. 2019. *Everything in its place: First loves and last tales*. New York: Picador.
Sempik, J. 2010. Green Care and mental health: Gardening and farming as health and social care. *Mental Health and Social Inclusion*, 14(3).
Tornaghi, C. 2014. Critical geography of urban agriculture. *Progress in Human Geography*, 38(4): 551–567.
Turner, B. 2011. Embodied connections: Sustainability, food systems and community gardens. *Local Environment*, 16(6):509–522.
Wang D, and T. MacMillan. 2013. The benefits of gardening for older adults: A systematic review of the literature. *Activities, Adaptation and Aging*, 37(2):153–181.
Watkins L., K.R. Sprang, and B.O. Rothbaum. 2018. Treating PTSD: A review of evidence-based psychotherapy interventions. *Frontiers in Behavioural Neuroscience*, 12:258.
Wezel, A., S. Bellon, T. Dore, C. Francis, D. Vallod, and C. David. 2009. Agroecology as a science, a movement and a practice: A review. *Agronomy for Sustainable Development*, 29:503–515.
White, M., A. Alcock, J. Grellier, B. Wheeler, T. Hartig, S. Warber, A. Bone, M. Depledge, and L. Fleming. 2019. Spending at least 120 minutes a week in nature is associated with good health and wellbeing. *Scientific Reports*, 9(7730).
Whiteside, M., K. Tsey, Y. Cadet-James, and J. McCalman. 2014. *Promoting Aboriginal Health: The Family Wellbeing Empowerment Approach*. USA: SpringerBriefs in Public Health.
Williams, A. Ed. 2007. *Therapeutic Landscapes*. Hampshire: Ashgate.
Wilson, E.O. 1986. *Biophilia*. Harvard: Harvard University Press.
World Health Organisation. 1988. *Public Health Services*. http://www.euro.who.int/en/health-topics/Health-systems/public-health-services.

7

From Individual Seeds to Collective Harvests: Urban Agroecology as Political Action

Helda Morales[1], Bárbara Lazcano[2], and Ana García[3]

[1] *Grupo de Agroecología. El Colegio de la Frontera Sur. Carretera Panamericana y Periférico Sur, S-N, Barrio de María Auxiliadora, San Cristóbal de Las Casas, Chiapas, México*

[2] *Solidaridad Internacional Kanda AC (SiKanda). Oaxaca de Juárez. Oaxaca, México*

[3] *Independent Researcher. Alicante, Spain*

CONTENTS

KEY WORDS: *social change, food politics, collective action, networks, agroecology*

7.1 Introduction

Urban agriculture has existed since the foundation of cities (Aben & De Wit 1999, Kois & Morán 2015), but it has been in times of crisis that urban gardens have been promoted by public policies. Famous are the "Relief Gardens" and "Victory Gardens," community gardens that were established in many cities in the United States as a measure to promote access to food during the Great Depression and the Second World War (Lawson 2014). Today, we are experiencing a food crisis due to the invasion of industrialized food in our plates (Holt-Gimenez & Patel 2012), an alarming loss of biodiversity (Thrupp 2000), pollution of soils, water and our bodies (Conway & Pretty 2013), a sedentary life (Powell & Blair 1994)) and the loss of green spaces in our cities (Wolch et al 2014). These contemporary challenges are likely what have spurred the new boom in urban agriculture around the world (Kois & Morán 2015, Tornaghi 2014). Currently, over 800 million people worldwide practice urban farming (FAO 2020).

This new wave of urban agriculture, with both local and global impacts, has mainly emerged from the practice of urban residents themselves, driven by people concerned with producing healthy and culturally appropriate food, beautifying their neighbourhoods, as a family hobby or to contribute to sustainability and biodiversity conservation (McClintock 2010, Partalidou & Anthopoulou 2017). Urban gardens are now germinating on rooftops, as vertical gardens, or as elevated containers in private yards, on sidewalks, and in large community garden spaces on municipal lands and parks. The practice is so important and widespread that some local and federal governments already consider it in their food production statistics (USDA 2018, City of Vancouver 2016), and an entire market of materials and services have emerged for urban gardens.

But, is urban agriculture just a hype? A feel-good practice for a particular sector of society? Or can it lead to wider transformations of the agrifood system?

Although urban agriculture is touted by international organizations and government institutions for its benefits to human and environmental well-being, these proclamations ignore that in many contexts, urban agriculture is practiced as a form of political resistance to the status quo. Examples can be found both in the global north and south. In the US, groups like Phat Beets, a project started in 2007 in Oakland, California, and FoodWhat, an organization that works with youth in Santa Cruz, California, promote social justice and access to healthy food for low-income communities (FoodWhat 2019, PhatBeetsProduce 2009). In Medellin, Colombia, Agroarte has been connecting urban gardening and artistic activities for social transformation for over 17 years (Agroarte Colombia 2019). Practices like "Guerrilla Gardening" have spread between different countries, promoting a direct intervention on public spaces from the UK, to Denmark, to Brazil and Argentina cultivating food and taking over abandoned plots, even if temporarily. Likewise, following the #Occupy movement, #OccupytheGarden invited people from around the world to grow their own food, buy local and refuse chemicals and modified seeds, addressing the issues of "public spaces, vegetable gardening and seed saving as a means of self-sufficiency (Gonzalez in Karim 2014).

Research, international reports, city policies and specialized literature widely recognize the environmental and social benefits of urban agriculture. However, less valued and acknowledged, even by some urban farmers, is urban agriculture's potential to foster wider systemic changes, not only linked to the agrifood system, but also to sociopolitical structures, contributing to the social re-appropriation of nature (Harvey 2008, Larder et al 2014), while demanding just and sustainable relationships between producers and consumers.

However as the number of people taking action to contribute to the environmental and social problems we are facing grows, a relevant critique has emerged, questioning the ability of individual actions and responsible consumption (Shove 2010) to improve our food systems (Bacon et al 2012, McClintock 2010). The critics argue that sociological and economic changes arise only from organized social movements that impact public policies, and argue that putting emphasis on the individual responsibility of citizens is counterproductive. When the individual is burdened with ecological responsibility, it deflects responsibility and attention away from powerful governments and big business, and prevents people from becoming politically involved as a collective (Adger et al 2005, Seyfang 2005). In the case of urban agriculture, Bacon et al (2012) takes this position, arguing that actions to change the food system will only be effective if linked to institutional environments.

In this chapter, we concede that, to achieve food sovereignty (Patel 2009), protect biodiversity, address climate change and advance social justice in cities, we require structural changes brought about by collective actions as a growing urban society. However, these actions are built by individuals who partake in these efforts, and these individuals have different motivations and situations related to their own positionality and experiences. To ignore the power of individual actors may actually *hinder* the urban agroecology movement: human behavior experts examining environmental movements have found that devaluing small-scale, localized and individual practices may lead to demoralization, resignation, demobilization, actor paralysis and even depression (Soliman & Wilson 2017).

As a localized practice, urban gardening narrows the gap between production and consumption that characterizes modern day food systems. Urban agriculture that adopts agroecological practices also threatens the power of big agrifood industry. By our own experience as urban growers and by our observations and conversations with urban farmers in southern Mexico, we hypothesize that caring for a garden often begins as an individual practice that can then turn into a collective action. Even for people who are not politically conscious, growing food can lead to raised awareness about issues like climate change, farmers´ struggles, unjust rural-urban relations, and other problems with the global food system. For some, this is a first step towards becoming political allies for food or social justice movements. Others will become activists themselves, getting involved with groups pushing for change in agricultural and social spheres. In this chapter, we use a case study in Chiapas, Mexico, as well as data from surveys and research conducted in the state of Veracruz, to explore urban farmers' motivations to grow, their political involvement, and what conditions could facilitate the transition from individual practices to wider collective action and change.

7.2 Methodology

We are two agroecologists and a local development specialist who founded "Sembrando Jovel", a network of people who garden in the city of San Cristóbal de Las Casas, in the southern state of Chiapas, Mexico. We also actively participate in the Red Internacional de Huertos Educativos (RIHE, International Network of Educational Gardens, redhuertos.org), a network that brings together educators, academics, promoters of urban agriculture and nutrition from several Latin American countries to exchange experiences on education and agroecology in schools and other educational gardens (Morales & Ferguson 2017), and we have participated in some activities of the Red de Agricultura Urbana y Periurbana de Xalapa (Urban Agricultural Network of Xalapa) in the state of Veracruz.

In the past seven years we have conducted action and participatory research in urban agriculture. Here we present and analyze the results of workshops, interviews and a survey conducted between 2012–2019 with members of the aforementioned networks to: 1) describe why urban farmers grow, 2) describe their political involvement, 3) describe their perspective on individual as a venue for change and 3) how we can all contribute to make deep changes in our food system.

7.2.1 Action Research in Theory and Practice in Urban Agroecology

Action research promotes interplay between scientific theory and local practice (Méndez et al 2013) and has been central to the development of agroecology and our own methodology. In action research, the purpose of inquiry is to contribute to the construction of solutions to a particular problem (Stringer 2014). To do so, action research systematically documents how a group of people implement actions to achieve a desired goal (Stringer 2014). It was born from peasant accompaniment in Colombia and activist documentation of the process (Fals-Borda 1987), but it is now practiced in multiple, different contexts. There are now several definitions and methodologies of action research. Ideally, the initiative to carry out the research to find and implement solutions to a problem comes from the affected community itself. Researchers should work to identify concerns, doubt and pain in a community, decide on a plan of action, take the agreed action or actions and record the results. Together, community and academics make the observations and analyze the data collected, and finally they reflect on the findings and make proposals to take further action if the problem was not resolved. They may redefine their target problem, and so continue with the action research cycle (Ferguson et al 2019).

In practice, many times the initiative to do action research arises from a researcher who invites the community to participate, and involves them as much as possible in some parts or in all the rest of the process (Ferguson et al. 2019). That was the case in our experience. Although we are urban growers and we are members of the networks that we studied, the initiative to conduct research and the data analysis was ours. For many years we have been interested in urban agriculture and we have conducted research to document the growth of urban agriculture in this region. In 2009, the first author (H.M), together with other colleagues, called for a meeting to form the International School Gardens Network (RIHE by its by its Spanish acronym). During the meeting, teachers, researchers, and NGO technicians from the United States, Mexico and Guatemala identified strengths and challenges that school gardens have and decided to work together to find strength and inspiration. Since then, the RIHE has hold nine meetings in Mexico,

Uruguay and Chile, and its members have work together in several projects (redhuertos.org). In 2014, the 3 authors organized a meeting with urban farmers in San Cristóbal de Las Casas, resulting in the foundation of the network *Sembrando Jovel*. More than 50 families that grew food in the city or wanted to learn how to do it participated. We all shared what we could contribute and what we needed from the network. The members meet once or twice a year to share experiences and visit each other gardens, and constantly share concerns and tips in social media. We understand the need to document our actions and processes, and we have come to an agreement with the other participants that we will conduct research and write about it.

In working towards that goal, we interviewed participants to learn about their motivations to grow food in urban settings, their opinions about the implications of their work, and their political agenda. We also conducted a series of workshops to develop a better understanding of the beliefs and action of the groups, and to present the data with them. This allowed us to validate the data and analyses that we conducted. By doing so the debate about individual versus collection action emerged. This research is still on-going and is an iterative process, but we present here our current methods and findings.

7.2.2 Surveys

We have conducted two surveys. In 2016 the third author (A.G.) conducted a household survey in San Cristóbal de Las Casas to develop a broader perspective of urban agriculture in the city. She determined that it was necessary to survey 263 of San Cristóbal's 40,714 households (García-Sempere et al 2019). She then used a two-step cluster method (Scheaffer et al 2006) to select the specific households that would be surveyed. As clusters, she used the 51 postal codes into which San Cristóbal's urban area is divided. In the first step of the cluster method, she selected a simple random sample of 30 clusters (postal codes). In the second step, she selected a sub-sample of households from each of the 30 clusters by numbering the blocks in each cluster and choosing one designated block at random. To reach the required total of 263 households, she selected eight or nine households from each cluster's designated block. The survey inquired about household food acquisition, what food sovereignty means for them, and about perceptions and practices on urban agriculture. Some of the information obtained in that survey is presented here and helped us interpret our interviews and workshops.

In 2019 we conducted a survey using social media (Facebook and WhatsApp), sent to all the members of Sembrando Jovel, the Red Chiapaneca de Huertos Educativos and the Red de Huertos Urbanos y Periurbanos de Xalapa. The purpose of the survey was to document the opinion of urban farmers about the criticism that individual actions do not have social impacts.

Using qualitative methods we analyzed the interviews, the minutes of the workshops and the social media survey. We used descriptive statistics to analyze the data collected during the 2016 survey to have a broader perspective of urban agriculture in San Cristóbal.

In the following sections we will briefly describe the context of our case study, the reasons why people started to plant and their different political involvement. We will also discuss the conditions that may facilitate the transition from individual practices to wider political change and collective action: networking, maintaining hope, an agrocological perspective for urban agriculture, learning to negotiate with governments and an intersectional positionality and intersectional perspective analysis to study urban agriculture.

7.3 Experiences of Resistance and Change from Urban Farmers in Southern Mexico

San Cristóbal de Las Casas (SCLC), is a small city in the southern state of Chiapas, Mexico, inhabited by close to 158,027 people (INEGI 2010) and famous for being home to people belonging to different Indigenous groups, as well as non-Indigenous people from San Cristobal itself and from other parts of Chiapas, Mexico and abroad. The city maintains an active tradition of food production in home gardens dating back to colonial times. In the survey that we conducted in the city in 2016 (Garcia et al 2019), 30% of households surveyed grow edible plants and 16% raise animals for consumption. Currently, around 50 families participate in Sembrando Jovel, an informal network that meets irregularly to visit their gardens, share tips, and celebrate harvest with a meal. Likewise, more than 200 educators have formed a chapter of the International School Gardens Network (RIHE): the Red Chiapaneca de Huertos Educativos (Chiapanecan network of educational gardens), where they meet every two months to visit their school gardens and to attend workshops to learn from each other.

Similarly, Xalapa, is a medium size city in the southern state of Veracruz, inhabited by around 425,000 people. Xalapa is a university city that attracts artists and academics, and it has become the center for many environmental initiatives. The *Red de Agricultura Urbana y Periurbana de Xalapa* congregates more than 2,000 people from diverse backgrounds, meeting every other week. Most of the urban farmers cultivate vegetables in their roofs or backyards, some tend gardens in schools and universities. Despite the nonexistence of a tradition of community gardens in Mexico, in Xalapa, the Network has started several community agriculture initiatives: occupying public land to grow vegetables, producing community compost in a public park, and supporting a community seedling production initiative.

The participants in these networks come from diverse paths, educational, ethnic and socioeconomic backgrounds. Their gardens vary from vertical gardens to agrosilvopastoral systems with fruit trees, hens, pigs or sheep. They have in common that they do not use pesticides, nor synthetic fertilizers, and that their gardens are diversified. The majority of the gardens are private, with the exception of the school gardens, and some community gardens initiatives emerging in Xalapa.

7.3.1 Different People, Different Reasons to Plant a Seed

For some urban farmers in southern Mexico their gardens are a reflection of their own culture. For people with rural or Indigenous backgrounds that moved to the city, their gardens are an expression of resistance to the loss of cultural identity, and help preserve their feeling of connection with the earth. In these cases, urban gardens become lifelines to a cultural identity, guardians of particular types of native crops that have all but disappeared from urban menus. "See this cucumber? It's sweet. They used to sell it everywhere, but now you only find products that come from abroad. My mom got me this plant and I'm taking care of it" (Cristina, personal interview 2014). "Where I come from, we eat these spiders. I managed to reproduced them here in my garden" (María, personal interview 2012).

To people who grew up in San Cristobal, their gardens are places of nostalgia. They are the final frontier in face of rising land costs, and their carers refuse to fragment their plots and sell even when land prices are up. The traditional house layout in San Cristobal included a "sitio" or area destined particularly to food production. Neighborhoods like "Las Delicias" (The Delicious) gained its name from the orchards that were once located on this site and that provided fruit for the production of candy and desserts. "(I remember) the Cuxtitali apple. It is a local variety of a tiny apple, very acid, colored a bit red and green. It is wonderful, but hard to find (…) and the local pear, that was larger and very juicy", remembers Sergio (SCLC, personal interview 2014). Albeit few, some of these orchards remain—thanks to local gardeners, local varieties that would have otherwise disappeared in the area are still grown.

Gardens preserve local practices, and even words, as Pedro states when addressing the differences in which people from different states refer to the same plant: coriander:

> *I plant to defend the way that we speak here. At the market people are now talking like ´chilangos´ (people from Mexico City) and correct us if we don´t. Here, I can call this plant ´culantro´* (as opposed to cilantro), because that is its name
>
> (Pedro, SCLC, personal interview 2012)

For others that migrated to San Cristóbal recently, their gardens are places of memory, in which they maintain practices and family traditions:

> *At first I felt it was a lot like repeating my grandmother's story. (…) She was of Chichimeca origin, no longer a speaker of the language, but culturally. Since I was little she took me to the vegetable garden, to the ranch… it is something that I have present in my life. When she died, to me it was a very difficult loss, very very painful. So I think that planting also reminds me of her. Like "good grandmother, I follow your tradition.*
>
> (Mon, SCLC, personal interview 2014)

Urban gardens might not be as productive as needed to have a large-scale impact on daily nutrition, or to allow their carers to save significantly on food expenses. In fact, none of the participants in this study mentioned "economic reasons" as their main motivation for planting. Participants in our study expressed other motivations, referring mostly to aspects such as nostalgia, health reasons, aesthetics, tradition, the pleasure of performing a direct practical work, or even open political resistance (Lazcano 2014).

Whatever their reasons for starting, people continue to plant. If it is still less expensive, and less time consuming, to buy a tomato in the local market than to grow it, why do people continue to plant? *the point of sowing is precisely that, to sow*, mentioned SCLC urban farmer Rubén (2014) in a personal interview, referring to urban gardening as a valuable practice in itself, one that cannot be measured merely by its utility, but rather by the pleasure, enjoyment, lessons and transformation it provides.

7.3.2 Urban Farmers and Politics

Plating a garden can become a political practice. Some urban farmers are very aware of this particular dimension, and name it as the main reason they began their gardens:

> *On the action day against Monsanto there was a lot of talk about the situation and I had to ask myself, "good, but what are we proposing?" I mean, it's not just about being in opposition, there has to be a proposal. Next to our complaints, there must also be action. And if I'm going to talk about transgenic seeds, it's because I want to share this organic seed to make a counterweight. (...) because life does not have a price, the seeds cannot have a price. Seeds are exchanged, period. They're shared.*
>
> (José, personal interview 2013)

Having a garden is therefore an expression of their political ideas. A form of open resistance, or protest. A way for carers to take direct action in their daily lives and homes. Carmelita, an Urban Farmer from San Cristóbal, plants aromatics in soda bottles mentioning *That is my way to protest against imperialism* (personal interview 2012). Urban Farmers like Sandra, from Xalapa, became aware of the importance of orchards as tools for spreading awareness when she understood "the negative impacts of industrial agriculture" (Facebook survey 2019).

For other urban farmers, it was through this practice that they gained insight on the unequal relations behind rural producers and urban consumers. This consciousness pushes them to pay "more attention" to the products they acquire and the production and distribution process behind them (Rubén, SCLC, personal interview 2014), and facilitates their recognition for rural work:

> *Imagine! I´m so tired just to cultivate this little piece of land. I´m so thankful for peasants who work all day to bring us food.*
>
> (Carolina, SCLC, workshop 2013)

Most farmers are also conscious about how a localized practice, sometimes deemed marginal for structural change, can have a larger political impact:

> *In these times of globalization, not only neoliberal economy and capitalism are globalized, but there is also a globalization of resistance and struggles. (...) There is a very interesting dialogue between these (local) experiences of struggles and resistance and it begins to weave something a little larger, to give a more global dimension to that particular struggle that is taking place....*
>
> *(José, SCLC, personal interview 2013)*

For many farmers in San Cristobal, change does not come through a direct involvement in mainstream politics. This is, perhaps, a reflection of the national trends analyzed in the 2012 National Poll on Political Culture and Citizen Practices. The results of the poll state that 84% of the population have little or no interest at all in politics, 62% of participants mentioned having little interest in the problems of their community, while 71% mentioned never having attended a meeting aiming to resolve problems in their neighborhoods or communities (Inegi 2012).

Likewise, in a series of workshops that we conducted in 2016 in schools, churches and neighborhood organizations, the participants expressed a lack of confidence in the possibility of transforming the food system via institutionalized political channels, reflecting a great detachment towards the political class.

> *I'm not interested in pressuring politicians, we found that it's useless.*
>
> (Carmelita, SCLC, workshop 2016)

> *The government will not solve anything. If we want change, we have to do it ourselves.*
>
> (Yolanda, SCLC, workshop 2016)

Nonetheless, in the survey that we conducted in 2016, gardeners underlined their confidence on the community's power as an agent of change: 87% of the families surveyed believe it is necessary to organize themselves. Frustrated with local and national partisan politics, they value the power of their individual and collective actions:

> *I don't think individual actions are distractions from the real problem, I think they are the first indispensable stage to be carried out. We have to be congruent between what we say and what we do. Grassroots work is fundamental because it is the work that allows real and transcendent changes in people's habits. That is, there may be a law that sanctions you if you do not separate your garbage, but if this law is not accompanied by a real process on the basis of consciousness, then the day another government arrives and changes this law, you will return to your old habits of not separating*
>
> (Laura, Xalapa, Facebook survey 2019)

We recognize that despite the commitment and energy that the urban farmers belonging to these three networks, there is still a lot of work and changes to made to achieve a more just and healthy food system in the region. In San Cristobal, although 30% of the surveyed families grow something, the percentage of families that participate in agroecological initiatives is low (6%), and most of the families that don't have a garden would like to have one but lack the necessary space (48%).

Far from underestimating the diversity of individual agricultural practices in the south of Mexico, we see an opportunity for NGOs, academia, schools, and social movements to promote the spread of agroecology; to raising the visibility of alternatives; and to channeling citizens' discontent toward the undertaking of collective, transformative actions.

But what can we do to not discourage people and at the same time motivate them to get organized and make structural changes? Here we call for recognition of the power of networks, the importance of recognizing individual actions to maintain hope, a better understanding in actor positionality and intersectionality, an agroecological approach to urban farming, and pertinent public policies driven from bellow.

7.4 Future Directions for Urban Agroegocology: Transition from Individual Practices to Wider Political Change and Collective Action in Urban Contexts

It is relevant to emphasize that even if all practices of urban agriculture in southern Mexico could be classified as an act of resistance, whether it is in the face of rising land costs or the loss of cultural identity, not all expressions of urban agriculture are consciously political, or contribute to wider structural change. Moving from an individual practice towards a collective one that contributes to a different agrifood system, requires adopting alternative values and practices to those promoted by large-scale agriculture[1]. To do so, individuals traverse through two main processes: gaining awareness on how their

[1] Large-scale agriculture heavily relies on agrochemicals (or agrotoxics as some urban farmers call them), monocultures, commercial seeds, and increased distance between producers and consumers. These practices cannot be considered sustainable and contribute to soil depletion, pollution, and exploitation.

action connects to, impacts, and contributes to a larger goal; and coming into contact with the idea that it is possible for their own small-scale, individual action, to be transformative.

7.4.1 The Power of Belonging to a Network

This dual movement (from a common practice to a political one) is not linear, and its influenced by the contact with other urban farmers and political movements. What can begin as an individual practice is encouraged to become a collective action. Most of our interviewees belong to a garden network. Most of them were initially attracted to the networks to learn more about agriculture and to share their own knowledge and seeds, but through contact with more experienced or politicized farmers, they recognized the value to be part of a larger group or movement:

> *You immediately need support, advice, exchange ... that contributes to social cohesion. Cultivating is an act of peace, and I don't see now how that can be done individually. I believe that politics can be done with hands and sweat when cultivating, because moving the body not only makes you think, but also reflect, that expands the dimension of action and political positioning.*
>
> (Isabel, SCLC, WhatsApp survey 2019)

Urban farmers also acknowledge that individual practices have limits when contributing to wider change. Thus, collaborating in collective work, facilitating access to knowledge and seeds and participating in networks is key:

> *I think it is super important to recognize the agroecological community that gives us the school garden network. It has sown awareness, see how we share and infect others with these new practices and vision to achieve a transformation in our environment, in our state, in our country.*
>
> (Loreto, SCLC, WhatsApp survey 2019)

As underlined by Santos (2011), resistances are in a constant state of fragility. Therefore, creating links between them allows for their survival, and promotes the possibility to imagine and act on the creation of alternatives.

7.4.2 Maintaining Hope to Take Action

Surveyed urban farmers in 2019 highlighted the need to maintain hope to take action, and they showed concern about the arguments that invalidates the action that each one of us can do.

Monica Soliman and Anne Wilson (2017) are two psychologists who have documented that it is essential for people to believe that they can change society to be sufficiently motivated to engage in collective action. By focusing on the importance of effecting structural changes, individuals often fail to allocate value to small-scale actions in their daily lives (Thomas 2014). This can lead to discouragement and demobilization. For individuals to move from an isolated practice to a collective one, they must be able to recognize their own capability to contribute to large-scale transformations. When facing complex and seemingly unresolvable problematics, taking the first steps towards collective action is often a matter of perception. As Long (2008) stated, "The outcomes and effectiveness of specific forms of resistance or contestation rest, then, not only on the organizing capacities and strategic capabilities of so-called subordinate actors, but also on how rigid or malleable the "dominant" institutional frameworks and discourses are perceived to be".

7.4.3 Understanding Actor Positionality and Intersectionality in Urban Agriculture

Our research highlights that to deepen our understanding on urban agroecology, we could greatly benefit on the introduction of both actor positionality and intersectional perspective analysis to the study, discourse and practice of urban farming.

Positionality is a term influenced by feminist theory that describes "the situated positions from which subjects come to know the world" (Sheppard 2002, p. 318). It refers to the social situatedness in relation to class, gender, ethnicity, sexuality and geography of one actor amongst others. According to Sheppard, positionality is a relational construct that involves power relations and a constant reenactment of its configuration. In this sense, it is important to acknowledge the positionality of urban farmers from a critical perspective to understand and visibilize the diversity within urban agriculture as an individual and collective practice. The particular positionality of urban farmers shapes their practices, changing when an urban farmer is practicing in a city located on the global north, or the global south, or in a capital city, as opposed to a smaller, peripheral one. Power relations experienced by urban farmers will vary, and so will their possibilities of entering into contact with other movements or discourses, accessing land, training or tools.

Likewise, intersectionality is a term that emerged in feminist theory to critique the ethnocentrism and homogeneity of mainstream feminism. Intersectionality names the "multiple, layered, intersecting and compounding forms of discrimination" (Di Chiro 2006, p.99), allowing for a better understanding of a differentiated experience of oppression. Di Chiro underlines how the combination or intersection of specific economic, social and environmental conditions might enable or disable the individual or community's ability to participate in collective actions, activism or civic life. Intersectionality can address how urban agriculture practices can directly benefit some people, while at the same time excluding others. By centering analysis around the actors of urban agriculture themselves, including specific barriers or potentiators faced by people based on their sex, gender, race, geography or social class, and how these interact with each other (Hovorka, 2005; White 2011), we can begin to identify more elements that push individuals towards collective action, or deter them from it.

These concepts can serve as a theoretical approach, but also as a political lens to review, acknowledge and deconstruct our own research and practices of urban agriculture. They can make visible the advancement of more than food security or sovereignty through urban agriculture, but also underline how it can relate to social and economic justice.

7.4.4 The Contribution of Urban Agroecology: Politics Mixed with Practice

Most urban farmers we interacted with for this chapter practice agroecological principles in their urban gardens (they do not use synthetic pesticides nor fertilizers, they use organic soil amendments and their gardens have many species and plant varieties). However, many are not fully aware that agroecology can transcend the technical aspects of sustainable agricultural practices, impacting social dynamics as well.

Agroecology is based on a critical perspective and a clear political position. Urban agroecology is embedded in integrated in a wider network of movements pushing for food sovereignty and justice (Tornaghi & Hoekstra 2017), with the objective of transforming the reality in which the agrifood system is immersed-in, and derives from.

As a science, agroecology emerged in the 1970s as a response to ecological, economical and social problems related to the hegemonic industrialized and Green Revolution agriculture (Gómez et al 2015). Agroecology made a critical approach towards three "epistemological bases" of conventional science, "disciplinarity, epistemological monism and the principle of simplicity" (idem) and translated this critique into an integrated and diversified approach towards agriculture. Although Gomez clearly points towards a need for more robust conceptual construction in agroecology, as a practice itself, Urban agroecology promotes relations between urban farmers, institutions and consumers based on a critical perspective, aiming to pose a sustainable alternative of industrial agriculture through urban agriculture.

Urban agroecology questions fundamental aspects about who participates in caring for a garden and how. It also questions power relationships surrounding the garden and how the benefits from it are accessed or distributed. Some of the questions posed in the practice of urban agroecology are:

- How are the gardens managed and by whom?
- What influence does urban agroecology have on food habits of both producers and consumers?
- How and where are the products and benefits derived from urban agroecology distributed?

TABLE 7.1
Agroecological principles in urban agriculture, classified in the different dimensions of agroecology

Ecological And Techno-Productive Dimension:	Socioeconomic and Cultural Dimension:	Sociopolitical Dimension:
• Optimize and balance the use of nutrients. • Lower use of inputs. • Prioritize the use of renewable energy sources and reduce dependence of fossil fuels and CO2 emissions, contributing to the fight against climate change. • Manage pests through safe prevention and treatment. • Favor the complementarity and synergism of genetic resources and conserve agrodiversity. • Emphasize the conservation of soil, water, energy and biological resources. • Provide healthy and nutritious food, contributing to improved diet and consumption habits.	• Favor short food supply chains between producers and consumers. • Reduce food miles. • Influence the relationships between producers and consumers, transforming them into relationships of trust and mutual support. • Enhance social cohesion. • Generate decent work.	• Allow both producers and consumers to regain control of the food system. • Transform wastelands in productive areas. • Favor culturally appropriate consumption habits. • Convert public spaces into community gardens. • To be accompanied by planning and evaluation processes. • Defend cultivated biodiversity. • Constitute spaces and processes of participation and collective learning. • Contribute to critical thinking and responsible consumption. • Converge with movements that promote the transformation of the food system and other aspects of social justice.

- What type of relationships are established between urban producers and consumers?
- What social groups are partaking in the activity? What is their positionality? Is it just urban elites or underrepresented communities as well?

By integrating the values and principles of agroecology into urban agriculture, we can underline the possibilities of these experiences to push for wider change. Urban agriculture based on agroecological principles presents a series of ecological, productive, socio-economic, cultural and political characteristics that differentiate it from other types of urban agriculture. Table 7.1 shows some of these principles, classified in the different dimensions of agroecology (Sevilla Guzmán 2006; Soler-Montiel & Rivera-Ferre 2010).

7.4.5 Negotiating with Governments

Urban farmers usually begin their practice as individuals. Their positionality can facilitate, or hinder the possibility of these individual actions to become part of larger, organized groups. These collective actions can then shape political movements related to social, environmental and economic justice. Demands made by such movements can have an impact on local formal and non-formal institutions, public policies, or individual practices and representations, shaping the ways in which urban agroecology is signified and promoted, and the way its impacts are distributed amongst local communities.

Urban gardeners who self-organize can foster change by directly negotiating with government institutions and promoting incidence on public policy, or by following more autonomous organization and non-policy paths. Organized urban farmers can choose to participate in more formal, structured groups that demand local and state governments to actively promote and legislate for urban agroecology. To some, public investment and policies are necessary to implement the agroecological model and move towards food sovereignty (Altieri et al 2012). However, to others, public policies by themselves are not enough. For a specific policy to favor a truly agroecological, democratizing transition of the agrifood system, it would have to come from below, from social and community networks (Calle & Casadevente 2015), thus the relevance of localized urban farmers.

Local urban governments and citizens—through food policy councils or other similar spaces for participation—may legislate and act to promote urban transitions toward agroecology and food sovereignty (García et al 2018). The proposed policies should refer to regulating use of urban ecosystem goods and economic activities, marketing and consumption of agroecological products, development of campaigns and public education that are consistent with the objectives of food sovereignty and social justice, protection of regional peasant culture and gastronomy, and cooperation with other municipal governments to promote integrated rural-urban planning (Table 7.2).

TABLE 7.2

Public-policy proposals to promote agroecology and food sovereignty in urban municipalities (García et al 2018)

Thematic Areas	Proposals for Developing Public Policy to Promote Agroecology and Food Sovereignty in Urban Municipalities
Natural goods	• Urban planning to include permanent green spaces and areas for urban agriculture. • Convert municipal land into community gardens over which citizens hold-land use and management rights. • Regulate land access, favoring people who have greater difficulty accessing land (ex. women, migrants, and youth). • Promote agroecological land use, sustainable water and energy use, distribution of organic seeds, and cultivation of autochthonous plant varieties in municipal nurseries.
Urban agriculture	• Carry out binding consultations in which urban residents may make their voices heard and discuss urban agriculture; offer legal, technical, and economic mechanisms to regulate urban and peri-urban farming while favoring public health.
Marketing agroecological products	• Promote and vitalize markets in which regional farmers may sell their products, with sanitary and certification regulations that favor small-scale producers. • Provide public funding for recruiting personnel to promote agroecological markets and avoid relying upon non-remunerated labor. • Purchase food from regional farmers to supply hospitals, schools, prisons, and other municipal institutions, favoring those farmers whose methods promote social and environmental sustainability.
Protecting traditional markets	• Promote local food distribution by regulating supermarkets (e.g., prohibit selling below cost and "a posteriori" payment). Limit supermarket expansion and regulate hours to reduce unfair competition while simultaneously promoting traditional markets and dignified working conditions for small-scale farmers.
Consumption of agroecological products	• Carry out diagnostic studies regarding urban citizens' consumption habits, motivations, and concerns regarding the food system. • Promote consumption of agroecological products through regulations and programs regarding responsible consumption; carry out public purchase of such products, and create measures to counteract social inequality.
Education	• Promote development of a school curriculum that addresses agroecological principles and food sovereignty, and is related to children's experiences. • Establish public-education projects consistent with agroecology and food sovereignty objectives, including school gardens and kitchens, agroecological dining halls supplied via public food purchase, etc. • Promote the inclusion of cultural, biological and gastronomical diversity in school curriculum to foster mutual understanding.
Cultural patrimony	• Fund programs promoting appreciation of regional peasant culture and gastronomy.
Nutrition and health	• Outreach to inform the public regarding healthy diets and health consequences of consuming high levels of processed foods. • Protect vulnerable sectors of the population; for example, by prohibiting distribution of junk food in schools and providing agroecological products through municipal food-assistance programs.
Cooperation	• Develop holistic rural-urban planning through alliances with nearby rural municipalities to develop regional food systems. • Work with other urban municipalities to share ideas and identify successful urban food policies and projects.

Actively collaborating in the creation of public policies is one approach towards turning urban agroecology to a practical response, a concrete proposal of change. Public policies and ordinances can turn individual actions to collective, institutionalized practices that can yield positive benefits for local communities. Examples of this are programs that support the creation of community gardens and local ordinances that recognize the importance of urbanization that considers green common areas that can be harvested, facilitating equal and just access to water and land, and promoting local farmers' markets.

Nonetheless, urban agroecology is as diverse as urban farmers themselves. This means that in certain contexts and for particular individuals and groups, government or formal institutional intervention is not necessary for urban agriculture to grow. For some farmers, there is no need of legislation or specific policies aimed towards urban agroecology. Rather, urban agroecology is an expression of networks of solidarity and open resistance, based on revolutionary or anarchic ideas. Occupying spaces to farm, exchanging and transporting seeds through borders, or taking advantage of horizontal solidarity between farmers are all practices performed by farmers, mostly from the global south, that openly oppose over-regulation and government inherence. These groups are based on alternative ways to understand and make politics, following a tradition built upon autonomous movements. This can translate into important differences on how actors and groups promote urban gardening.

7.5 Conclusion

Although here we focus on our case study in Southern Mexico, the ideas presented could be relevant for cities around the globe:

Individual practices and small-scale initiatives will not single-handedly balance out the negative impacts of industrial agrifood systems. However, they are crucial to promote and channel a sense of collective possibility.

Individual practices of urban agroecology can become part of a larger social movement. For many urban farmers, their first approach towards urban agroecology is not necessarily a political one. It is only through the practice itself, and by having contact with other urban farmers, groups, and activists, that knowledge and experience gained take on a political aspect. The passage between one and the other is not always linear or unidirectional. A person can collaborate in a collective garden and then start one of their own, eventually disengaging from the group, or move from their own individual practice towards a collective one. Moving between one and the other (individual vs. collective, reactionary vs. revolutionary tomatoes) point towards the importance of understanding that individual efforts can become a starting point towards wider, political participation. They can also be a small beacon of hope, an example of the possibilities of change, a refuge not only of biodiversity, seeds and land, but of our capability to produce the world around us, to be active participants of transformation and not just mere observers of the menaces and threats we face as humans in our own communities and in the world.

For urban agriculture to represent a true alternative in defiance of hegemonic practices—not only in the agrifood system but also in terms of social and environmental justice—values, symbols, perspectives and policies related to urban agriculture need to be founded in practices that represent an alternative, such as agroecology. Thus, agroecologists can make great contributions to urban farming and social justice movements by participating actively with grassroots organizations, such as the urban farmers networks that are emerging around the world.

Likewise, by including positionality and intersectionality in our practices and research, we can begin to understand what particular situations can lead an individual towards political action and how to address the barriers that they can face in relation to sex, gender, class or geography, amongst others.

Individual practices can become part of collective actions, either through organized movements aiming for changes in policies, or through more autonomous collectives that exist in parallel to formal institutions. If we underestimate the relevance and potential of individual actions and their possibilities to connect with wider movements at a local and global scale, we deny the possibility for actor-driven structural change.

Food gardens are rooted on relationships. The diversity of the individuals that care for them and the communities around them, nurture their complexity and richness. There is much soil within them to welcome practitioners, researchers, scientists and community members alike. Whether individually or as part of a larger group, it is always a good moment to begin planting in the city.

REFERENCES

Aben, R., & De Wit, S. 1999. *The Enclosed Garden: History and development of the Hortus Conclusus and its reintroduction into the present-day urban landscape.* 010 Publishers. Rotterdam.

Agroarte Colombia. 2019. Sobre nosotros. Retreived from: https://www.agroartecolombia.co/

Altieri, M. A., Koohafkan, P., & Gimenez, E. H. 2012. Agricultura verde: fundamentos agroecológicos para diseñar sistemas agrícolas biodiversos, resilientes y productivos. *Agroecología*, 7(1), 7–18.

Bacon, C. M., Getz, C., Kraus, S., Montenegro, M., & Holland, K. 2012. The social dimensions of sustainability and change in diversified farming systems. *Ecology and Society*, 17(4).

Calle-Collado, A., Casadevente, J. 2015. Economías sociales y economías para los bienes comunes. *Otra Economía: Revista Latinoamericanas de Economía Social y Solidaria* 9(16):44–68.

City of Vancouver. 2016. Policy report development and building. Retreived from: https://council.vancouver.ca/20160223/documents/p1.pdf

Conway, G. R., & Pretty, J. N. 2013. *Unwelcome harvest: agriculture and pollution.* Routledge. London, 676 p.

Delgado, R. 2007. Los marcos de acción colectiva y sus implicaciones culturales en la creación de ciudadanía. *Universitas humanística* 64:45.

Di Chiro, G. 2006. Teaching urban ecology: environmental studies and the pedagogy of intersectionality. *Feminist Teacher* 16(2):98–109.

Dunford, R. 2015. Autonomous peasant struggles and left arts of government. *Third World Quarterly* 36(8):1453–1471. https://doi.org/10.1080/01436597.2015.1037388

FAO. 2018. Five ways to make cities healthier and more sustainable. Retreived from: http://www.fao.org/fao-stories/article/en/c/1260457/

Fals-Borda, O. 1987. The application of participatory action research in Latin America. *International Sociology* 2(4):329–347.

Ferguson, B., Morales, H., Hernández, C., López, L. 2019. *Alimentación, comunidad y aprendizaje: recursos para docentes.* Chetumal, México: El Colegio de la Frontera Sur, 208 p.

Food What. 2019. About us. Retreived from: http://www.foodwhat.org/our-story

García-Sempere, A. 2018. La Transición Urbana Hacia la Soberanía Alimentaria: Un Marco Teórico-Conceptual y Metodológico Para su Análisis. Estudio de Caso en San Cristóbal de Las Casas, Chiapas (México). *Tesis de doctorado.* Chiapas, México: El Colegio de la Frontera Sur, San Cristóbal de Las Casas.

García-Sempere, A., Hidalgo, M., Morales, H., Ferguson, B., Nazar-Beutelspacher, A., & Rosset, P. 2018. Urban transition toward food sovereignty. *Globalizations.* DOI: 10.1080/14747731.2018.

García-Sempere, A., Morales, H., Hidalgo, M., Ferguson, B., Rosset, P., & Nazar-Beutelspacher, A. 2019. Food Sovereignty in the city?: A methodological proposal for evaluating food sovereignty in urban settings. *Agroecology and Sustainable Food Systems.* DOI: 10.1080/21683565.2019.1578719

Giacchè, G. 2014. L'agriculture urbaine révélatrice de formes de résistance. *ESO, Travaux & Documents* 37:17–25.

Gómez, F., Ríos-Osorio, L., Eschenhagen, M. 2015. Las bases epistemológicas de la agroecología. *Agrociencia* 49(6):679–688.

Harvey, D. 2008. The right to the city. *The City Reader*, *6*(1), 23–40.

Holt-Gimenez, E., & Patel, R. (eds) 2012. Food Rebellions: Crisis and the Hunger for Justice. *Food First Books.* 267 p.

Hovorka, A. J. 2005. The (Re) production of gendered positionality in Botswana's Commercial Urban Agriculture Sector. *Annals of the Association of American Geographers* 95(2):294–313.

INEGI. 2010. Censo de Población y Vivienda.

INEGI. 2012. Encuesta Nacional de Participación Política y Prácticas Ciudadanas ENCUP.

Jarosz, L. 2008. The city in the country: growing alternative food networks in Metropolitan areas. *Journal of Rural Studies* 24(3):231–244. https://doi.org/10.1016/j.jrurstud.2007.10.002

Karim, A. 2014. "Occupy gardens? A case study of the people's peas garden in Toronto, Canadá. Dalhousie University. Retrieved from: https://dalspace.library.dal.ca/bitstream/handle/10222/56040/Karim-Alia-MES-SRES-Nov-2014.doc.pdf?sequence=1&isAllowed=y

Kois, J. L. F. C., & Morán, N. 2015. *Raíces en el asfalto: pasado, presente y futuro de la agricultura urbana.* Madrid: Libros en Acción.

Larder, N., Lyons, K., & Woolcock, G. 2014. Enacting food sovereignty: values and meanings in the act of domestic food production in urban Australia. *Local Environment, 19*(1), 56–76.

Lawson L. J. 2014. Garden for Victory! The American Victory Garden Campaign of World War II. In: Tidball K., Krasny M. (eds) *Greening in the Red Zone.* Dordrecht: Springer, 181–195 pp.

Lazcano, B. 2014. *¿Acción conformista o acción rebelde? La práctica de la Horticultura Urbana en San Cristóbal de las Casas y sus aportes a la sustentabilidad local. Tesis. Maestría en Desarrollo Local.* Universidad Autónoma de Chiapas.

Lutz, J., & Schachinger, J. 2013. Do local food networks foster socio-ecological transitions towards food sovereignty? Learning from real place experiences. *Sustainability* 5(11):4778–4796. https://doi.org/10.3390/

Long, N. 2008. Resistance, Agency and Counterwork. In *The fight over food. Producers, consumers and activists challenge the global food system.* Pennsylvania: Pennsylvania State University Press, 69–89 pp.

Lyson, T. A. 2000. Moving toward Civic Agriculture. *Choices*, (Third Quarter), 42–45. Retrieved from http://ageconsearch.umn.edu/bitstream/132154/2/ThomasLyon.pdf

Méndez, V. E., Bacon, C. M., & Cohen, R. 2013. Agroecology as a transdisciplinary, participatory, and action-oriented approach. *Agroecology and Sustainable Food Systems* 37:3–18. doi:10.1080/10440046.2012.736926.

Morales, H., & Ferguson, B. 2017. La red internacional de huertos escolares. *Decisio* 46:8–10.

McClintock, N. 2010. Why farm the city? Theorizing urban agriculture through a lens of metabolic rift. *Cambridge Journal of Regions, Economy and Society* 3(2):191–207.

McClintock, N. 2014. Radical, reformist, and garden-variety neoliberal: coming to terms with urban agriculture's contradictions. *Local Environment: The International Journal of Justice and Sustainability* 19(2):147–171. DOI: 10.1080/13549839.2012.752797

Partalidou, M., & Anthopoulou, T. 2017. Urban allotment gardens during precarious times: from motives to lived experiences. *Sociologia Ruralis* 57(2):211–228.

Patel, R. 2009. Food sovereignty. *Journal of Peasant Studies* 36(3):663–706.

Phat Beets Produce. 2009. Our vision. Retreived from: http://phatbeetsproduce.org/our-vision/

Powell, K. E., & Blair, S. N. 1994. The public health burdens of sedentary living habits: theoretical but realistic estimates. *Medicine and Science in Sports and Exercise* 26(7): 851–856.

RUAF, 2017. Urban agroecology. *Urban Agriculture Magazine* 33(November):1–75.

Santos, B. de S. 2011. *Producir para Vivir: Los Caminos de la Producción No Capitalista.* Fondo De Cultura Economica, Mexico.

Sevilla Guzmán, E. 2006. *De la sociología rural a la agroecología.* Barcelona: Icaria Editorial.

Seyfang, G. 2005. Shopping for sustainability: can sustainable consumption promote ecological citizenship?. *Environmental politics*, 14(2), 290–306.

Shove, E. 2010. Beyond the ABC: climate change policy and theories of social change. *Environment and planning A*, 42(6), 1273–1285.

Sheppard, E. 2002. The spaces and times of globalization: place, scale, networks, and positionality. *Economic Geography* 78(3):307–330.

Soler Montiel, M. M., & Rivera Ferre, M. G. 2010. Agricultura Urbana, Sostenibilidad y Soberania Alimentaria: Hacia una propuesta de indicadores desde la Agroecología. X Congreso Español de Sociología, Pamplona.

Soliman, M., & Wilson, A. E. 2017. Seeing change and being change in the world: The relationship between lay theories about the world and environmental intentions. *Journal of Environmental Psychology*, 50, 104–111.

Stringer, E. 2014. *Action Research.* Sage: Los Angeles, 68 p.

Tornaghi, C., & Hoekstra, F. 2017. Urban Agroecology. *Urban Agriculture Magazine* 33:3–4.

Tornaghi, C. 2014. Critical geography of urban agriculture. *Progress in Human Geography* 38(4): 551–567.

Thomas, M. 2014. Climate depression is for real. Retreived from: https://grist.org/climate-energy/climate-depression-is-for-real-just-ask-a-scientist/

Thrupp, L. A. 2000. Linking agricultural biodiversity and food security: the valuable role of agrobiodiversity for sustainable agriculture. *International affairs* 76(2):265–281.

USDA. 2018. Vertical farming for the future. Retreived from: https://www.usda.gov/media/blog/2018/08/14/vertical-farming-future

White, M. M. 2011. Sisters of the soil: urban gardening as resistance in Detroit. *Race/Ethnicity: Multidisciplinary Global Conext* 5(1):13–28.

Wolch, J. R., Byrne, J., & Newell, J. P. 2014. Urban green space, public health, and environmental justice: the challenge of making cities 'just green enough'. *Landscape and Urban Planning* 125:234–244.

8

Surveying the Landscape of Urban Agriculture's Land Politics: Civic, Ecological, Heritage-Based, Justice-Driven, and Market-Oriented Fields

K. Michelle Glowa[1] and Antonio Roman-Alcalá[2]

[1] *Anthropology and Social Change, California Institute of Integral Studies, San Francisco, CA, USA*

[2] *International Institute of Social Studies (ISS), Erasmus University, The Hague, Netherlands*

CONTENTS

KEY WORDS: *land politics, land sovereignty, urban gardens, tenure security, property*

8.1 Introduction

Urban agriculture (UA) has been continuously present in the United States (US), taking many forms including market gardens, home gardens, school gardens, community gardens, job-training gardens, and horticultural therapy gardens. Unlike European gardens, which have generally been more institutionalized and supported by the state throughout the 20th century, US UA has occurred with state support only in waves (Basset 1981). At the end of such waves, UA has often receded from the urban landscape as gardeners lost access to the land. Both with state support and without it, UA practitioners and advocates have utilized various strategies in order to access land to pursue their gardening goals. This chapter explores the politics that inform land access strategies and dynamics, presenting five heuristic categories—Civic, Ecological, Heritage-based, Justice-driven, and Market-oriented. These categories help us see how lineages of (urban) agroecologies intersect to generate existing UA's land politics and agroecological dynamics, and their interrelations.

We define land politics as processes of defining norms of land use via political claims and actions. We are especially interested in how utopian desires are enacted on the land and on pre-existing property relations. Through land politics gardeners recreate old or develop new socio-spatial relations, setting

BOX 8.1

La Via Campesina's political positions about land are rooted strongly in its origin as a movement of peasants and Indigenous people mostly from the 'developing' world. In other words, its land politics reflect the legacies of colonial-era land and resource dispossession, as well as historical struggles for land reform through the twentieth century, more recently expanded into the notions of 'integral agrarian reform' (Rosset 2013). This idea that without land access and tenure there is no potential for agroecology or 'food sovereignty' (Rosset and Torres 2013; Mendez et al. 2016), though useful and generative, seems less taken up by many sectors of UA in the US (Roman-Alcalá 2015). UA practitioners *do* discuss and seek to address land access, as do their rural counterparts in sustainable agriculture movements, but generally there have been less transformative proposals for gaining that access, such as interim use agreements, temporary access, public-private partnerships, and calls for access to public lands for UA uses (Glowa and Roman-Alcalá 2020). Rarely do UA movements argue for land redistribution from those who currently hold private title to those who do not; even rarer are calls to decommodify land culturally, economically, or politically (as some have interpreted LVC's land politics). Still, LVC-type global agrarian anti-corporate movement politics find some resonance and reflection in the US, and in UA specifically. In some UA, the 'food justice' frame (see below) has evolved into 'food sovereignty' and to some degree has begun to emphasize 'land justice' and 'land sovereignty' (Williams and Holt-Giménez 2017).

direction, and foreclosing on other possibilities if only for the moment. In this sense a land politics focus asks us to consider practical questions of how gardeners access and use space, and to ask symbolic and historical questions of how relations to property and land are cultivated through and shaped by tenure arrangements. Urban agriculturalists in the US have been influenced by multiple traditions and politics of agroecology, with historical roots from inside and out of the country. One international influence has been peasant-based and explicitly agroecological food producer movements like the transnational network La Via Campesina (LVC), whose radical concepts of "food sovereignty" and "land sovereignty" point to the necessity of democratized access to land for just agricultural systems. Yet this is not a widely-mobilized framework in the land politics of US urban agroecologies (see Box 8.1 for details on LVC's land politics). We are interested in the relation of land politics of US urban agriculture to LVC's radical proposals, but knowing that influences on UA do not stop at LVC, we look at UA at large, seeing how *varying* land politics come about, intersect, and evolve.

Historically, the specific natures of urban agroecologies (for instance, whether or not urban gardeners use ecological growing techniques) have been shaped by cultural legacies, sociopolitical orientations, and larger political-economic contexts, including dynamics of land access. For instance, during periods of economic downturn or reduced access to resources gardeners frequently produce crops using as little outside inputs as possible, emphasizing agroecological techniques. Similarly, wider upticks in environmental consciousness have articulated urban gardening as part of an ecological ethic. In this chapter, we will explore the various projects and instances of UA, and their histories of land access, land politics, and agroecology, drawing out different lineages of practice. In so doing we hope to provide a fuller understanding of how UA's land politics in the US have evolved and are evolving, and to help integrate land politics into an urban agroecology framework.

8.2 Urban Agriculture Land Politics as Urban Political Agroecology

How we define and understand agroecology creates the basis for seeing how diverse sets of land politics relate to agroecological projects in both urban and rural areas. Giraldo and Rosset (2017) warn that there is a risk in debates over its meaning that "agroecology will be co-opted, institutionalized, colonized and stripped of its political content" (1, 2017). In writing this chapter we reassert the need to understand agroecologies as tied to and embedded within relationships to land, and thus as inherently raising important questions of property, land access, and sovereignty. Although not limited to urban contexts, these issues are fruitfully addressed through a combination of urban political ecology and political agroecology, focused on issues of urban agriculture.

While agroecology can be and is often treated as merely ecological approaches to agriculture, we find this approach inadequate. It is important to note that an increasing emphasis has been placed on agroecology as engaged with questions of food systems, politics, and social movements, not just individual behaviors of farmers (Wezel et al. 2009). Steve Gliessman, Miguel Altieri, John Vandermeer, and Ivette Perfecto, along with many other agroecological scholars, have led this charge since the 1970s. The field of agroecology as it has developed has had close connections to peasant-based movements associated with food sovereignty. Smallholder, traditional agriculture—particularly in Latin America—has provided both the socio-cultural and ecological basis of study for the field (Altieri 1995; Gliessman 2006; Holt-Giménez and Altieri 2012). Agroecology is "a social movement with a strong ecological grounding that fosters justice, relationship, access, resilience, resistance, and sustainability." (Gliessman 2012, 19); its goals are greater than simply developing more environmentally sustainable agricultural production.

Urban agriculture can likewise be an important area of study in understanding how the social and ecological are intertwined. In gardens, cultivation unfolds with cultural knowledge being passed down generations, creativity and expressiveness coming through in design and planning, food security needs being met, gardeners acting with soil microbes and cultivars to meet various goals. Biodiversity, nutrient flows, and water cycles: all play important roles in urban gardening processes just as much as meeting human needs. All these processes take place on land. Yet agroecology studies do not always directly address land, and even with social science literature on agroecology developing rapidly, agroecological science has not sufficiently developed its understanding of land politics. Meanwhile, urban political ecology is increasingly being leveraged in geography, environmental studies, sociology, and other social sciences, and offers well developed theories on land politics, but is rarely specifically applied to food and farming issues. Here we attempt to combine the two.

Bolstering our case for a land-focused agroecological lens is Gonzales de Molina's (2017) call for 'political agroecology' in the reworking of agroecological approaches. He identifies that the connection between politics and agroecology is not novel to the field, stating that "many authors have demanded the need for socioeconomic structural reforms in order to be able to achieve sustainable agrarian systems (Buttel 1997, 2003; Altieri and Toledo 2011)". For Gonzales de Molina, 'political agroecology' can provide both the analysis of the industrial food system, "the precise diagnosis of the crisis", and ideas to move forward "a theory of collective action and action by the political institutions to progress toward agroecological transition" (67). In this sense political agroecology becomes both an approach to analysis and a program that advises which are the "most suitable ways to participate" in moments of socioecological change (60). We see political agroecology as a key tool in understanding how agricultural production changes through time, in relation to institutions, discourses, and ways of knowing. The continuing commodification of land is one such change that has shaped modes of land and resource access and use in both rural and urban contexts. In response to marginalization and dispossession, grassroots social movements to maintain agrarian peasant livelihoods have demanded "land sovereignty". Land sovereignty entails community-based control over productive and political resources, as essential requisites to ensure the right to food (Borras and Franco 2012, 7). In cities in both the developed and developing world, urban movements against displacement have converged to address injustices in housing affordability, fresh food and green space access, and livelihood maintenance. Yet recent US-based food and urban movements, while increasingly engaging with some form of land politics, have made relatively "few references to global historical trends and movements around land grabbing, land concentration, and land reform" (Kerssen and Brent 2017). Scholars such as Roman-Alcalá, Borras and Franco argue that we must directly articulate the connection between land sovereignty—the right of working people to control land and benefit from the its use—and agroecology in both rural and urban contexts (Borras and Franco 2012; Roman-Alcalá 2013; Borras, Franco, and Suárez 2015).

Despite calls for more critical framings of the underlying land politics that have limited who and where people can garden within cities, some gardeners have worked to develop land access strategies for short-term tenure. That is, some gardeners have opted for a land politics devoid of the normative position described by land sovereignty. The approach to inquiry we take here helps us understand why and how this is so by using the tools of a political agroecological approach. In the sections that follow we explore how urban land access strategies have connected to different 'fields' of urban agricultural ideologies and practices, asking how urban agroecologies relate to the development of power vis-a-vis land politics. As we will see, diverse UA projects and their politics of land are tied to specific histories (e.g. migration, previous lineages of social struggle) and continued processes of racialization, colonization, and state-building.

8.3 Fields of Land Politics

The "fields" of land politics seen within US UA movements represent a broad spectrum of experiences of (and beliefs about) the past, fantasies about the future, and actions in the present which link the two. Although these fields/categories—Civic, Ecological, Justice-driven, Heritage-related, and Market-oriented—may not encompass all of the various threads that weave together in creating possibilities for UA land politics in the US, they contain within them many of the threads that we believe are most important for understanding the current context and future potentials (Table 8.1). Of course, the names and descriptions we offer for various fields are merely heuristic devices, as these fields are non-homogenous, and not mutually exclusive. Overlaps between them and contradictions within them are to be expected—and should be expected to have some of the most interesting dynamics for analysis. Similarly, utilizing these fields in defining how UA manifests its politics, and specifically its land politics, we should keep in mind the graduated and contradictory rather than dichotomous nature of politics. Jan Douwe van der Ploeg's reminder that peasants are characterized by "degrees of commoditisation" (2010, 10) helps us recall that UA practitioners look, act, and evolve in dynamic ways. That is, UA practitioners do not fit snugly into one or another category. Similarly, land politics can be considered to be characterized by "degrees of commoditisation" as well as enmeshed in other influences that we will discuss in the chapter. We feel that these categories may find resonance in other geographies of study beyond the US, though our particular discussion and findings are linked to the US context, and even more specifically to the San Francisco Bay Area where the authors are most familiar with UA through personal experience and research.

8.3.1 Civic

The Civic field of UA is defined by the attitude and normativity of civic virtue. Instances of this field—gardens and institutions for food production and the (re)development of food culture—are associated with state institutions, and the projects of state-building and citizen-subject creation that are embedded within them (Pudup 2008). 'Civic' land politics are the result. Examples we explore in this section include 'Victory Gardens' developed during war times (but also seen in resurging during times of economic crisis); home gardens and school gardens (and the ways in which they have created 'proper'

TABLE 8.1

Fields of land politics in US urban agriculture

"Fields" of Land Politics	Where Do These Occur?	Who Is Involved?
Civic	• Primarily public land • Also in expanded areas during times of crisis	• Governmental agencies and the publics they address
Ecological	• Across public, private, and hybrid properties	• Environmental advocates
Heritage/Home	• Back and front yards of privately owned and rented homes (single family or otherwise)	• Home gardeners, home makers, and garden advocates
Justice-driven	• Across public, private, and hybrid properties	• Advocates addressing racial and economic justice, community sovereignty, and other concerns
Market-oriented	• On private property or (increasingly) public land slated for development	• Garden advocates, municipalities and developers

citizens while also supporting the state's need for self-provisioning among the masses); and various institutions related to the 'land grant' university system established by the US federal government in 1863, such as its 'extension' programs meant to engage the public. By connecting itself to the political projects of the state and cultivation of state subjects into 'productive' members of unified society, the Civic field tends to align with land politics of modernism: private property rights, a dichotomy of land as public (i.e. state) or private with little space for collective self-determined land management, and deference to states as the rightful arbiter of land and property relations. In examining land access in the context of Civic UA, it is essential to recognize the dominance of what Singer (2000) calls the "ownership model" of property. This model reifies private property as an essential tool for productivity and social stability. Its other is public or state-owned land. In much of UA, and particularly its Civic fields, this public/private dichotomy is normalized, taken as "pre-political, obvious, and unproblematic" (Blomley 2003, 6). Although Civic UA has tended to abide this normalization, all garden projects have the potential to destabilize the meaning of land and its use for agriculture; the Civic field is hegemonic regarding land politics but not purely so.

During both World War I (WWI) and World War II (WWII) large, federally supported gardening programs enrolled civilians in supporting war efforts to improve national diets and habits while making resources available for the war efforts. Gardens became essential tools in campaigns to advance patriotism and encourage public participation in war efforts (Lawson 2005). As John Brucato, San Francisco Victory Garden leader, observed "food was considered one of the most important weapons of war" (Brucato 1993, 142). During WWI when it was necessary to export portions of the domestic food supply, 'liberty' gardens were also used to supplement food during shortages through a federal strategy of asking citizens to voluntarily substitute food purchases with garden produce. During WWII, citizens were encouraged to grow nutrient dense vegetables and were also mandated to comply with national rationing and price controls. During both wars, gardens were a key strategy to both produce food for nutritional needs and encourage at home participation in the war efforts, and as such significant federal and state government support assisted the rapid development of extensive garden networks. Liberty gardens in 1918 numbered 5,285,000 and produced $525 million worth of food (Lawson 2005). By 1944, victory gardens provided 40% of American's domestic food (McClintock 2010). In both wars, 'unused' property was mobilized from public areas and vacant lots, but home and school gardens also played roles in generating these great statistics. Victory Gardens show most clearly the Civic logic of some UA: gardens are about food production, but justified as necessary in order to exhibit patriotic support for a nation-state at war. Tangibly supported by that state, they produce an improved citizen: state subjects who are more likely to hold affinity for its values and order, including its central arbiter role in land-property regimes.

Similarly to civic improvement ideologies, school gardening had by the early 1900s become a popular avenue to promote agrarian ethics, entrepreneurial skills and work ethic, and provide opportunities for developing connections to nature. Gardens were frequently initiated through the work of women's clubs, mother's associations, and horticultural clubs, with advocates urging that gardens should hold a permanent place in public education. University extension offices became advocates for UA as an integral piece of public schooling. The University of California (an original 'land grant' college) developed a project entitled the Garden City, which promoted agrarian ideals through gardening and agricultural activities for a wide age range of students. By 1911, over 200 students were allocated plots in a one-acre site on the UC Berkeley campus, where they worked individually to produce and sell vegetables and flowers (Lawson 2005). Communal plots were used to demonstrate agricultural technologies and best practices, as well as to do team building activities. This combination of individual and communal gardening became a common strategy used across the nation to both encourage individual ownership and work ethic while engaging students in collective learning (Lawson 2005). University students were sent throughout the state to replicate the extension service's model by establishing clubs and Garden Cities. Not all school gardens operated on plots of land on school property, unlike many school gardens today. Many garden programs focused on teaching school children skills they were expected to use in backyard home gardens (Lawson 2005). Home gardens were used to cultivate children's sense of ownership and pride in their homes, frequently predicated on valuing private property, while also teaching children how to reduce household expenses on food. Teachers visited student homes and interacted with parents, which was a common strategy of turn of the century social reform charities interested in improving the moral and physical health of youth at home (Taylor 2009). Though individual private property plays a big role in the development of US land use and capitalism writ large, such examples show that Civic UA does not always simply promote a purely individual vision of property; still, such efforts reinforce the public/private dichotomy in land that renders invisible and undermines alternative legal and structural relations of property, such as commons.

Another way in which such civic improvement was institutionalized was via the 'land grant' university system established by the Federal Morrill Act of 1862 to serve the vanguard of settlement along the expanding colonial border. These universities followed and implemented the 'manifest destiny' doctrine to develop virtuous communities of self-actualized pioneers along the ever-expanding frontier of the late nineteenth century US (Stein 2017). Though seemingly different and distant from this initial purpose, programs aimed at adults like Cooperative Extension and the Master Gardeners program of the University of California Cooperative Extension agency continued, well beyond the initial stages of western expansion and the initial mandate of generating and improving rural industry, to bring forward the state's central role in civilizing its own people and offering an agriculturally-based vision by which citizens can see themselves as a part of a larger 'productive' political project. In this case, today the Master Gardeners program is more likely to serve retirees and aspiring hobbyists than pioneering agriculturists. Yet, the ethics of a hard-working hobbyist that aims for greater self-sufficiency through garden cultivation be it at home or on more collective sites (such as community gardens which are still popular at Cooperative Extension offices), has continued to promote a land politics that is tied to a sense of settler-agrarian mandate to cultivate the land productively.

8.3.2 Ecological

In the Ecological field, we consider ecological motivations for UA. These are overtly ecological—outwardly concerned with problems of the environment both large and small—and the role of (food) gardening as a means to address such issues. Ecological concerns motivate (among others) the permaculture movement, biodynamic agriculture, and Organic-certified agriculture in both its movement history and its commercialized contemporary reality. All of these tendencies appear in contemporary UA in the US. These movements and orientations have often been supporters of gardening projects in various contexts, helping elevate the importance of eating as a political act, connecting food and farming to their ecological impacts. The ecological field is important to address, as it is a very dominant part of the discourse in UA as well as the larger food movement in the US. It is often tied to particular kinds of people, namely white-identified people and people of relative economic and social status privilege. This is not to say, of course, that non-white people or the poor are not interested in ecological ideas or values, but rather that the participation in such spaces has been dominated by white-identified people. This has begun to change, with pressure from marginalized communities both within these movements and outside of them, as well as from the larger popular culture wherein a shift to prioritizing people of color, women, and queer populations has influenced much of what are known as 'left', 'liberal', and 'progressive' politics.

Permaculture originated as the brainchild of two Western-trained naturalists, who subsequently taught its framework around the world. Permaculture is controversial, as critics claim it appropriates existing land based knowledges, mostly of Indigenous peoples, and its origins and continuing spread reflect a 'colonizing' approach to making change. However, permaculturists are increasingly aware of these problems (Ferguson 2014, Watson 2016). In the US, it has been associated with the agrarian imaginary of back-to-the-land self-sufficiency. The ecological ethos of its practitioners might be presented shorthand as: "I will go get my own acreage and be self-sufficient while the world burns"—an attitude that reflects a particular political vision and context. Organic agriculture has in some cases been based in a similar imaginary, harkening to its origins in the hippie back-to-the-land movement, when many wanted to drop out of the mainstream of society and fight it by retreating into food production. However, in contrast to permaculture, Organic became a regulated marketing category, and those who succeeded in pioneering early Organic businesses have gone on to dominate the industry, as well as pave the road for later corporate intrusion into the market (Guthman 2004; Fromartz 2006). The converse argument in favor of Organic, of course, is that this very commercial success has translated its originally niche ecological ethic into wider-scale, mainstream forms, therefore helping to 'democratize' ecological consciousness and 'defetishize' commodities (Allen & Kovaks 2000).

Permaculture and organic agriculture overlap in many of the practices, but tend to differ in terms of land politics. Organic agriculture has no overt land politics, as it constitutes such a diversity of interests, from ideological to commercial. It has not substantially formed part of any movements to disrupt existing private property relations since at least the 1960s. Permaculture, on the other hand, has at its pedagogical/canonical core ethical values around land management that make clear the mandate to not over-accumulate wealth, including land, and that prioritize the use of land for biodiversity and human needs. In sum, permaculture as it has been developed as a thought system has clear anti-capitalist implications, but in practice it also has been adopted by doomsday preppers and others with less explicit anti-capitalist values. Some permaculture promoters have even begun discussing and pursuing the economic viability of permaculture production within capitalist markets, but these are largely on the margins. For both permaculture and organics in rural and urban settings alike, questions of who can access land for cultivation and who has historically and systematically been excluded from access are increasingly being raised, paralleling increased food movement concern for social injustices.

Biodynamic agriculture also has influenced UA. Though it is more commonly associated with vineyards and rural farms that are attempting to achieve a whole systems biological approach to commercial production, home gardeners (some of whom are urban) make up 40% of existing members of the US Biodynamic Association, according to the director of the association (personal comm. 2019). Biodynamics also has a certification ('Demeter Certified') which, like Organic, creates a marketing status for practicing farmers. Biodynamic has a more overt spiritual dimension with cosmovisions rooted in connections with the land, and engaging with the purported powers of celestial objects to influence biological dynamics. These beliefs overlap with the spiritual imperative to get beyond raw economic logics and Western modalities. This is ironic enough given that it was developed by a very Western philosopher, Rudolf Steiner. Yet it is meaningful to any discussion of land politics that Biodynamics promotes an approach that focuses on the spiritual dimensions of human interactions with place-based life—seeing land not as a generic 'factor of production' but as a living thing of its own, deserving of respect and reverence—and in this way denies capitalist/modernist traditions of treating land as alienable 'property'. It might be noted that this approach to land as relation is not dissimilar to the ways in which many Indigenous people conceive of land and their place with regards to it. In practice, Biodynamic practitioners abide by property laws, 'own' farms, and so on; however, as an influence on the at-large composition of UA practices vis-a-vis land, biodynamic is likely to have a counter-hegemonic influence.

The fundamental take-home from examining this field for land politics dynamics is that ecological interests do not in themselves have a land politics. Yet in UA contexts, ecological concerns can push those interested in growing food *as ecological intervention* into situations where the essentially political nature of making ecological change possible becomes unmistakable: evictions from gardens, the prioritization of higher-income development by governments, and increasing discussions of justice among UA groups all can lead to questioning of the land politics normalized in Civic fields.

8.3.3 Heritage and Home

In this section we take a closer look at the relationships between heritage and differing forms of land politics that urban agriculturalists articulate. In the US, a country composed of many people of immigrant descent (with the exception of course of Africans who were brought to the landmass as enslaved people and the Indigenous who were there to begin with), there are many lineages of food production practices that have been brought to these soils. These food production practices, brought to both rural and urban areas, have been rooted in historical experiences in relation to land access and land politics. These experiences shape how many communities develop and articulate contemporary land politics through UA, even as their practices and values change through assimilation and shifting racial categories. One key difference where various lineages of heritage intersect with land politics is in how social groups have differentially related to home ownership and therefore access to the potential of a home garden.

Following WWII, while some advocated for the importance of permanent public gardening, vacant-lot and community gardening largely disappeared from the US urban landscape. Home gardens however did not. Land use discourses that claimed gardening was contrary to "modern" development played a key role in the post-war retreat of gardens from more visible and public spaces. Post-war planners increasingly saw agriculture as a threat to urban health and safety (and to continued urbanization) and used zoning to move this threat out of the city (Hodgson, Caton Campbell, and Bailkey 2010). In addition, the still dominant discourse of 'gardening as a response to crisis' helped to normalize their erasure once crisis had passed and other urban development schemes surged. Backyard gardening was promoted as a hobby by magazines like *Better Homes and Gardens*, but for those without access to backyards it was unclear where, if at all, *their* gardens had a place in the city. The Washington D.C. Victory Garden commission went as far as to state, "[victory] gardening has no place as a 'proper peacetime municipal function'" (quoted in Lawson 2005, 202).

While gardens became less visible in the public sphere, private home gardens grew in popularity. Garden clubs and social reforms advocated that a well-maintained, orderly home garden indicated a responsible homeowner or tenant who valued health, frugality, nutrition, and positive occupation of one's personal time regardless of a person's economic status or cultural heritage (Lawson 2005). In some major cities, women's garden clubs played a significant role in promoting conservation and civic improvement campaigns, including the development of home gardens (Walker 2009). These efforts also trace back to the early decades of the twentieth century, when landscape architects, city planners, and social reformers alike dreamed of the possibilities of improvement of urban civic life through better order of the physical environment. An improved physical environment was believed to lead to improved behavior, health, and society by conservative social reformers and radical material feminists alike (Hayden 1982). This belief in environmental determinism was promoted by various plans and movements including the garden city and city beautiful movements (Fishman 1982; Daniels 2009). For both of these movements gardens were part of designs intended to address a multitude of social problems, including disease and lack of physical health, crime, and social unrest. Well tended gardens—whether in the home, on a vacant lot, or at a school—would go far to improve society according to certain social norms of 'improvement' (which are typically if not overtly classed, gendered, and racialized). The 'chaotic' intercropping of many traditional agroecologies might not pass muster, in contrast to the British-style orderly rows held as ideal. While the land politics here are implicit and perhaps subtle, they indicate a view of land use as rightfully constrained to certain kinds of people and ways of use, and wherein the dominant notion of 'common good' dictates a hierarchy of accepted land uses and outcomes.

During this period, racialized home lending and restricted sales practices (which mainly prevented non-whites from buying homes) combined with government benefits, particularly those directed at returning war veterans, making homeownership a reality for many white families. Home gardening with secure tenure was a possibility for these families but was not for many African American, Chinese American, Japanese American, Mexican American, and other racially excluded communities. Many white family units with a male head of household now had access to spaces to develop home food production gardens, and were encouraged through institutional means, such as Extension Master Gardener programs, and cultural materials, such as home improvement magazines. Whiteness, as a consolidating category of particular European ethnicities, thus has historically developed in relationship to home ownership, with implications of this access for the land politics of UA.

Still, gardening has persisted in many non-white communities. For example, African American families fled the Jim Crow south to move north and west and brought agricultural practices with them, just as Mexican Americans continued practicing agriculture even after having lost land access to the advancing US border. Gardening in Chinese communities in Southeast San Francisco and Oakland was a common practice in the first half of the twentieth century and provided significant amounts of produce to local markets (Brahinsky 2012), yet Chinese gardeners are notably missing from Brucato's account of WWII gardener and truck farmer efforts. Chinese and Japanese gardeners were denied rights to own land and most gardeners had lost access to their gardens by the 1940s through urban development, or the expansion of Italian and Portuguese gardens. Post WWII homeownership became a demobilizing force working against public garden efforts in white communities, as increasing numbers of individuals had access to backyard gardens. At the same time, in racially marginalized communities, where homeownership was suppressed (and in the case of the interned Japanese, forcefully taken away), collective garden projects grew in importance during subsequent moments of marginalization and resistance.

As additional waves of immigration occurred various cultural and ethnic groups moved to urban areas in the US and recreated cultural landscapes within urban gardens. Numerous articles and books document the ethno-botanical diversity of gardens across the US (Corlett et al. 2003; Baker 2004; Schupp and Sharp 2012; Carter et al. 2013; Gray et al. 2014; Taylor and Lovell 2014). Each group brings specific values around food production and land use, environmentalism, culinary traditions, and most pertinent to this discussion, land politics. For example, in studying Mexican immigrant run gardens in Los Angeles and Seattle, Mares and Peña (2010) argue that UA provides an avenue for immigrant gardeners to assert their identity onto a new location, creating a transnational sense of place. Glowa et al. (2019) document how immigration and displacement open opportunities for both the migration and transfiguration of agroecological traditions, making human movement an important factor in how agroecological knowledges and practices change. For the Mexican and Salvadoran immigrants of their case example, economic and war-driven migration meant that Latin American farming traditions were brought to and shared with many on the Central Coast of California, while the risk of imminent displacement through gentrification impacted choices of how to continue traditions—for example choosing not to invest in perennial crops. Here we see how histories and contemporary relationships of land access impact the kinds of gardens we see develop in the landscape.

In another example, an organization, La Mesa Verde, that promoted home gardening—helping community members set up home gardens with access to seeds and starts—demonstrates that in a predominantly Latino immigrant community with low housing security and thus precarious home garden tenure, resistance to disempowering property regimes can be linked to cultural histories of land struggle. The greatest challenge for many of their participants was gaining permission from their landlords to have a garden (Glowa 2017). In interviews, program participants noted that those who work the land in order to produce food are those who deserve to have access to it and decision-making power over its use, with one gardener citing the Mexican revolutionary Emiliano Zapata directly, stating "*La tierra es de quien la trabaja con sus manos*/the land belongs to those that work it with their own hands." The remembered cultural history of a social revolution largely focused on land redistribution impacted how the gardeners viewed the question of access to home gardening as tenants. The work of home gardeners who are tenants provide a counter-narrative and counter-practices to those land politics of home gardens that uphold the middle class private property owner as the ideal backyard producer, instead alluding to property relations that prioritize use over ownership. Additionally, work by advocates like the Sustainable Economies Law Center to legislatively prevent individual landlords from prohibiting tenants from having food gardens at home shows that home gardens (in which heritage is a common component) are a contested and rich area for evolving urban agroecologies.

In considering influences of migrant lineages, there are also elements of conflict, especially relevant in how diverse land politics converge through UA. Collective strategies for survival—for instance producer cooperatives, or collectively managed community gardens—are common and sensible to many Mexican immigrants from Indigenous areas of Mexico where such collective action forms a long and culturally relevant tradition. However, such strategies are not only alien, but can actually appear as offensive or scary to other migrant groups, such as the Hmong, whose South Asian homeland was subject to authoritarian communist governments. During fieldwork in California's urbanized Central Valley, one author heard a local Hmong leader discuss (in response to a suggestion from Mexican American attendees to collaborate on collectivization efforts for producers of color, both urban and rural) how the South Asian-heritage farmers he worked with commonly viewed the practice of collectivization as an unmitigated evil. Considering that an entire generation of South Asian migrants gained entry to the US through collaboration in the country's anti-communist war on Vietnam, many such communities, even if they are not directly of the generation that migrated, maintain this sort of cultural skepticism. Therefore, thinking through how land politics develop, we must keep in mind the many and diverse heritages, and how specificities of one group, family, or even individual, articulate with another.

8.3.4 Justice-driven

In recent years there has been an upsurge in the use of the term "food justice" within food movements across the US, and with this, an increased interest in "land justice" (see the edited volume by Williams and Holt-Giménez, 2017). The food justice movement has come together from several points of origin including environmental justice activism, organizing against hunger and disease in communities of color, struggles against institutionalized racism that view food activism as an entry into making change, critiques of racism in the food system, and critiques of racism in the food movement (Holt-Giménez and Wang 2011). This work "contextualizes disparate access to healthy food within a historicized framework of institutional racism" (Alkon and Norgaard 2009). Several scholars (taking the lead of activists like New York UA organizer Karen Washington) have suggested the term "food apartheid" to refer to racially exclusionary practices that result in unequal food access (Bradley and Galt 2013). Food justice as an analytical framework, in its emphases on histories and geographies of racism, classism, and gender oppression in the food system, recognizes space as consequential, and as socially created.

Racial justice forms a central component to contemporary food justice movements (Alkon and Agyeman 2011). Among antecedents to contemporary racial justice thought are major thinkers like Malcolm X who directly described the basis of justice as access to land. This access to land has also been linked by movements to community self-determination, and indeed self-determination is a key principle of *some* food justice UA projects. Justice-driven projects in urban contexts often arise from long-term patterns of marginalization, and operate on common principles based on lived experiences and analyses of injustice. Many food justice movements come out of long-term struggles, often tied to heritage; often those movements formed out of displacement from land, and desires for land redistribution. We might point to African American and Native American experiences of generational trauma and marginalization that have inspired the adoption or continuation of various strategies of resistance (White 2018; Estes 2019).[1] Because of intergenerational trauma caused by histories of racism and colonialism, healing (broadly considered) is a main *means* towards justice, and a desired *outcome* of justice for many projects. This healing—personal, spiritual, interpersonal, communal—happens often through (re)connection to land-based experiences, thus creating a tie between food production and land justice on multiple levels (White 2018; Penniman 2019). Yet, the origins and usages of the term "justice" can vary depending on who is using it—importantly, not all projects that adopt the title of 'food justice' are rooted in communities or leadership of those most marginalized from land access and land tenure security. Some urban projects self-defined as food justice are more oriented towards service and more so overlap with the orientations of the 'Civic' field. Thus, land politics towards justice are multifaceted. Therefore it is important to understand how UA projects have developed particular orientations towards justice, and how these influence their relationships to land politics.

The local food production focus of US food movement projects, including UA, have been critiqued as isolationist, individualistic, conservative, inadequately strategic, and in tension with social justice desires (Dupuis & Goodman 2005). However for many in the food justice movement localist politics are not an unexamined commitment to localization ideals. The impulse of communities marginalized around common identities to create self-defined food production or exchange systems has been a common theme of new social movements since the 1960s. In such communities, mistrust of dominant power structures and understanding of systemic, historical and contemporary institutionalized discrimination and oppression have influenced some organizers to emphasize strategies that promote local community self-reliance and community self-determination (Alkon and Noorgard 2009; Gottlieb and Joshi 2010). The Her Lands, women and lesbian separatist communes of the 1970s, frequently integrated gardening and agriculture into their work as a means to empower members and distance themselves from what were seen as inherently unjust and patriarchal agricultural systems (Lee 2003).

Similarly, self-determination projects have been a central theme of black liberation organizing for the last half century (Hilliard 2008). In Detroit, the D-Town Farm enacts a contemporary politics of community self-determination in the African American community, focusing specifically on the need for land and collective initiatives based on it. The farm emphasizes that UA, local policy development, and co-operative food buying should be directed by the black community in Detroit. D-Town and its leaders have also critiqued land deals promoted in mainstream local politics, linking their positive vision of land

[1] Of course, there are many other groups who have experienced marginalization and have struggled for justice within the US; we focus on these two groups for purposes of illustration, considering length limitations.

justice to a struggle against racialized capitalist development and the inequality it causes (Wallace 2011). Malik Yakini, D-Town farm leader, and others have noted that long-term control over land resources is an important issue for their goal of self-determination. Activists have been putting pressure on the City of Detroit for long-term leases and to develop mechanisms to release control of parks and vacant land to communities. Yakini states:

> "We're black self-determinationists, frankly. And Detroit is at least an 85-90% African American city, so any project must benefit the majority population... We're concerned with control and ownership. So this is an example of an agricultural project which is controlled by African Americans and where we're able to control the revenue generated from this farm and we want to model not only the growing technique but model the social and political and economic dynamics that are appropriate for a city like Detroit." (Goodman 2010)

Scholars have also noted the use of food activism in efforts for Indigenous self-determination and justice. For many Indigenous people, connection to the land and gaining sustenance from it have played key roles in intergenerational teaching and processes of cultural identity articulation. For instance, the Karuk tribe of northern California has used the framing of food justice and decolonization to articulate their right to traditional foods and ecosystem management appropriate to maintain these food sources. In working to remove dams that have harmed the salmon runs on which they have traditionally relied, the tribe has

> "locate[d] their current food needs in the history of genocide, lack of land rights, and forced assimilation that have so devastated this and other Native American communities. These processes have prevented tribal members from carrying out land management techniques necessary to food attainment." (Alkon and Norgaard 2009, 297)

Many food and farming-related projects led by Indigenous organizations and leaders have similarly framed their efforts in centuries-long histories of genocide, forced removal from lands, and attempts at systematic erasure. While much of this Indigenous food justice work has taken place in rural places, and this work has been increasingly addressed in scholarship (Grey and Patel 2015; Jäger et al 2019), there has not been as much attention to the efforts of urban Indigenous food movements. This is an even more concerning gap considering that Indigenous people have lived in cities for generations, and over half of the Indigenous population since the 1960s is urban (Fixico 2000).

One recent project, the Sogorea Te' Land Trust of California's East Bay, provides an example of intertribal work aimed toward the healing from legacies of genocide and colonialism by reclaiming heritage, practicing self-determination, accessing land, and healing towards justice. This Indigenous women-led organization works to return Chocohenyo and Karkin Ohlone lands to Indigenous stewardship through the acquisition of urban parcels. For the organizers this work is "guided by the belief that land is the foundation that can bring us together ... as Indigenous people from many tribes, working together to create healing on this land" (Sogorea Te' n.d.). As a part of this work, the Sogorea Te' Land Trust has worked with urban food justice organizations to create access to space for reconnection with land and community, and to reinvigorate traditional food ways. The physical healing from illnesses such as diabetes and heart disease is deeply intertwined with the healing from continued processes of colonization—marking active *de*colonization as resistance. Leader Johnella LaRose again brings this back to the question of land access: "In the Indian community we're in a crisis around food, and we have no place to grow this food. There are many community gardens in the Bay Area, but the native community does not have one. We *do not* have one" (ibid). The need for land is centered in a politics that also addresses food justice, cultural sovereignty, and community healing.

Food justice activists construct a notion of justice that is focused on historical stories of who is included and excluded from economic and social structures of power. Activists argue that in the US communities of color and low-income communities have had less access not only to physical food resources, but also less access to the institutions that provide opportunity to feed one's self and community. This is centrally tied to the questions of property ownership, land access and land tenure security. Freedom from the injustices of discrimination in land access based on racial and other differences is key to this movement, offering a land politics grounded in understandings of the histories of property that continue to impact projects concerned with racial and economic justice. In this way, Justice-driven UA pushes a rethinking and restructuring of property, bringing its practitioners closer to the LVC-like land politics of food sovereignty. Indeed, many food justice UA projects are now aligning overtly with food sovereignty, for instance by joining the US Food Sovereignty Alliance.

8.3.5 Market-oriented

As we have seen in the various fields covered above, urban gardeners in various historical periods and social situations have produced not just for self and community, but also for various markets and selling opportunities. They have sought land to produce food to sell and exchange, though the purpose of that exchange has not always been the same: it has overlapped with inherited practices in food and agriculture; with the desire to advance ecological goals (as we see with parallels to the movement for Organic production); with civic values of supporting one's nation; and with efforts to achieve justice for marginalized communities—as in the burgeoning field of 'social entrepreneurial' gardens through the turn of the 21st century. In a heavily market-driven, capitalist society it makes sense that the sale of garden produce would be an element in many garden projects. In addition to understanding how gardeners have oriented their production towards sales, our land politics lens focuses attention to shifting orientations of gardeners toward the question of land markets as well.

Commodified land, as one of the most profitable markets in some areas of the US and world, is an obstacle to gardeners who may want to cultivate land for a purpose not aligned with the most profitable use for a parcel. The commodification of land is also tied to underlying logics of land politics that are picked up in UA projects. Municipal agencies and private landowners typically hold other priorities for long-term use of land, and with the increase in neoliberal urbanization those purposes are generally aligned with putting the land to its 'highest and best use', i.e. its most profitable use. Gardening almost never falls into this category and thus UA projects are continually at risk. As gardener and planning professor Sam Bass Warner explained, "Control of land has always been the rock that smashed American urban garden projects" (Warner 1987).

Since the 1980s when the UN World Commission on Environment and Development released the report *Our Common Future*, UA has been held up as a tool within the sustainable development toolbox. Through entrepreneurial UA projects, gardeners could increase food security, home income, and create environmental benefits in the city. As an example, in 1991 a study in Tanzania, it was found that small urban farmers in Dar es Salaam made approximately 1.6 times the annual minimum salary through farm profits (Sawio 1998). In the 1990s funding increased both internationally and in the United States for entrepreneurial urban agricultural programming and in this context of UA as sustainable development, many advocates suggested municipal and state governments should support gardening as an interim, or temporary, land-use. Scholar and advocate Luc Mougeot asserted that aid workers and some municipal officials began to see that "urban agriculture is neither the short-lived remnant of rural culture nor a nasty symptom of arrested urban development" and that instead it was coming to be seen as compatible with the goals of contemporary urban development (2006, xiv).

In the United States today there are tensions between approaches that highlight the need to create amicable relationships with developers by embracing short-term land tenure, and approaches that seek longer tenure security. One such example of the former can be seen in the San Jose, California based organization Garden to Table which developed a partnership with landowner Berry Swenson to develop a one-acre urban farm in downtown on land owned by Swenson (Personal communication 2014). Swenson is Chairman of the Board of Berry Swenson Builder, a prominent real estate and construction company. To make a partnership attractive to Swenson, Zach Lewis, Director of Garden to Table, agreed to start paying property taxes in the second year of the project (2015) with the hope that they could work together to convince San Jose to implement AB 551, California legislation that incentivizes UA by decreasing property taxes on garden parcels. Swenson and Lewis agreed this would be a temporary project as the land would eventually be developed. Predicting the need to eventually move, Lewis worked to develop portable beds and mechanisms to move the farm when they relocated. For Lewis portability is at the heart of their approach:

> People need to be clever... We build mobile beds so you can move - everything is above ground so it can be moved with so much volunteer support... we build it into the concept. We created a pleasant relationship with the owner/developer - with the community – we're lucky to have this. You have to go into it with the mind frame that it's temporary. There's no incentive right now for landowners to do it. We need to make owners/developers feel comfortable. (Personal communication 2014)

Garden to Table is not alone in taking this approach to the development of portable beds and temporary use agreements in order to gain access to land (see Glowa and Roman-Alcalá 2020). Yet other urban gardeners have expressed concern about the potential negative impacts that short-term projects could have. Concerns range from a relationship to the soil and land that does not take long-term stewardship as a central tenet, to developing gardens without deep engagement or involvement of the surrounding community. One garden advocate working with a temporary, interim-use project that served houseless people described that being a temporary project was often interpreted by their gardeners to mean the city did not think their needs were a priority. When the parcel was slated to be sold for development the advocate stated, "Frankly, I'm not interested in interim use. I mean it's not useless of course, but it's not really extremely beneficial in the long run by any means, except for developers who really do have a lot to gain from it." (Personal communication 2014). For many, this explicitly raises the concern that gardens may unintentionally be part of processes of ecological gentrification where green projects can be used to raise the property values in a neighborhood, thus increasing profit potentials for an owner or developer and contributing to the gentrification/transformation of a neighborhood and displacement of historical residents.

In terms of the market orientations at play, to adopt a land politics that centers the ease at which a parcel can be put to its 'highest and best use' engages a central logic of contemporary capitalism. Land use change, construction, and redevelopment can all serve to facilitate investment in land as a commodity. Since we have seen the rise of real estate financialization over the last 40 years, we have also seen gardeners take on some of these priorities for profitable investment in land and a politics that can bolster the authority of landlords as the ultimate decision-makers for land management and access (Glowa 2017).

8.4 Conclusions

This chapter has sought to combine disciplinary traditions of urban political ecology and political agroecology to deepen our understanding of contemporary land politics in UA. Combining the latter's attention to sociopolitical dimensions of agriculture and the former's more developed theories of land politics enables research in urban agroecology to better represent the truly diverse nature of urban agricultures. UA practitioner and promoter Wayne Roberts (2017) makes the obvious but needed point that analysts should address UA as diverse in practices, motivations and outcomes. As has been made clear above, this plurality is paralleled in UA's diverse land politics. We can see that diversity across the various examples examined in this chapter (See Table 8.2).

Based on our heuristic categories, future research might focus on one or another field, or the interactions between them, helping understand the development of land politics: in specific locations, in the larger UA movement, among state officials and civil society groups who influence UA, and in relation to the agroecological production methods UA efforts utilize. Using such heuristics, researchers can unpack why, how, and to what extent UA projects subvert or reproduce existing norms and structures of property relations. Some, as we have seen, in particular in the Civic and Market-oriented fields, tend towards reproduction. Yet there is also an increasing focus on questions of land access and control within some UA, as demonstrated in some Justice, Ecological, and Heritage-based projects. Most gardeners and garden advocates recognize that their work is bounded—or to some degree shaped—by contemporary property relations. Their assessment that property relations are a determining dynamic for the future of

TABLE 8.2

Summary of land politic examples with heuristic category and land politics approach*

Example/Project	Heuristic Category	Central Land Politic Approach
Liberty/Victory Gardens	Civic	State-mandated, temporary access to 'unused' property for patriotic self-provisioning
Early 1900s school gardens	Civic	Public school led garden education on site at schools (public property) to promote agrarian ethics, entrepreneurial skills and work ethic, and provide opportunities for developing connections to nature
Early 1900s home gardening education at schools	Civic	Public education support of learning in student home gardens used to cultivate sense a of ownership and pride in their homes, reduce household expenses on food, and improve the moral and physical health of youth at home
Permaculture	Ecological	Critiqued for appropriation of Indigenous land based knowledges. Core ethical values mandate to not over-accumulate wealth, including land, and prioritize the use of land for biodiversity and human needs
Biodynamics	Ecological	Focuses on the spiritual dimensions of human interactions with place-based life—seeing land not as a generic 'factor of production' or alienable 'property' but as a living thing of its own, deserving of respect and reverence
Post-WWII home garden clubs	Heritage	Home gardening promoted as a healthy and responsible part of single family home ownership model, linked to the predominantly white and male new home owning class
Chinese market gardens (first half of the 20th century)	Heritage	A common practice in mainly rented yards and open lands in San Francisco Bay Area to produce for local markets until communities were stripped of land access
La Mesa Verde	Heritage	Mixed approaches to property, including resistance to landlords limiting food production, citing cultural history of a social revolution largely focused on land redistribution
D-Town Farm	Justice-driven	Use of public park land to create community self-determination for Detroit's black residents
The Sogorea Te' Land Trust	Justice-driven	Cross-tribal work to reclaim urban land, which is to be held in common, to create opportunities for healing from continued colonization
Garden to Table	Market-orientation	Built temporary garden to facilitate access land through amicable relationship with a developer, facilitating easy return and profit making on property

***Note:* The summary of land politics examples discussed in this chapter, along their associated heuristic category and central land politic approach, demonstrates the complexity of land and property relations even within heuristic categories.**

their gardens can be acute, especially since garden (and housing) tenure insecurity and displacement for working people have only worsened in recent decades across the urban United States. In conclusion, gardeners can contribute to destabilizing conceptions of property as apolitical 'things', when they participate in movements demanding access to space through various frameworks including land sovereignty, 'right to the city', and decolonization.

REFERENCES

Airriess, C.A. and Clawson, D.L. 1994. "Vietnamese Market Gardens in New Orleans." *Geographical Review* 84(1): 16.

Alkon, A.H. and Agyeman, J. 2011. *Cultivating Food Justice: Race, Class, and Sustainability.* The MIT Press.

Alkon, A., and Norgaard, K. 2009. "Breaking the Food Chains: An Investigation of Food Justice Activism*." *Sociological Inquiry* 79(3): 289–305.
Allen, P. &Kovaks, M. 2000. "The Capitalist Composition of Organic: The Potential of Markets in Fulfilling the Promise of Organic Agriculture." *Agriculture and Human Values* 17: 221–232.
Altieri, Miguel A. 1995. *Agroecology: The Science Of Sustainable Agriculture, Second Edition*. Boulder, Colo. : London: Westview Press.
Altieri, M.A., and Toledo, V.M. 2011. "The Agroecological Revolution in Latin America: Rescuing Nature, Ensuring Food Sovereignty and Empowering Peasants." *Journal of Peasant Studies*. 38(3): 587–612.
Baker, L.E. 2004. "Tending Cultural Landscapes and Food Citizenship in Toronto's Community Gardens." *The Geographical Review* 94(3): 305–325.
Basset, T. 1981. "Reaping on the Margins: A Century of Community Gardening in America.". *Landscape* 25(2): 1–8.
Benford, R. &Snow, D. 2000. "Framing Processes And Social Movements: An Overview and Assessment." *Annual Review of Sociology* 26: 611–639.
Blomley, N. 2003. *Unsettling the City: Urban Land and the Politics of Property*. New York: Routledge.
Blomley, N. 2005. "Flowers in the Bathtub: Boundary Crossings at the Public–private Divide.". *Geoforum* 36(3): 281–296.
Borras, S. and Franco, J. 2012. *A 'Land Sovereignty' Alternative? Towards a Peoples' Counter-Enclosure*. Amsterdam: Transnational Institute Agrarian Justice Programme Discussion Paper.
Borras, S.M., and Franco, J.C. 2012. "Global Land Grabbing and Trajectories of Agrarian Change: A Preliminary Analysis." *Journal of Agrarian Change* 12: 34–59.
Borras, S.M. Franco, J.C., and Suárez, S. Monsalve 2015. "Land and Food Sovereignty." *Third World Quarterly* 36(3): 600–617.
Buttel, F. 1997. "The Politics and Policies of Sustainable Agriculture: Some Concluding Remarks." *Society and Natural Resources* 10: 341–344.
Buttel, F. H. 2003. "Environmental Sociology and the Explanation of Environmental Reform." *Organization & Environment* 16(3): 306–344.
Bradley, Katharine, and Galt, Ryan E. 2013. "Practicing Food Justice at Dig Deep Farms & Produce, East Bay Area, California: Self-Determination as a Guiding Value and Intersections with Foodie Logics.". *Local Environment* 19(2): 172–186.
Brahinsky, R. 2012. *The Making and Unmaking of Southeast San Francisco*. Berkeley, CA: University of California.
Brucato, J. 1993. *Sicilian in America*. Millbrae, CA: Green Hills Pub. 142
Carter, E.D., Silva, B., and Guzman, G. 2013. Migration, acculturation, and environmental values: the case of Mexican immigrants in central Iowa. *Annals of the Association of American Geographers* 103: 129–147.
Carlson, J. and Chappell, M.J. 2015. "Deepening Food Democracy: The tools to create a sustainable, food secure and food sovereign future are already here—deep democratic approaches can show us how". Institute For Agriculture And Trade Policy. https://www.google.com/url?q=https://www.iatp.org/sites/default/files/2015_01_06_Agrodemocracy_JC_JC_f_0.pdf&sa=D&ust=1580333607620000&usg=AFQjCNGlM4U4alHt1-isllo3euQj-HW1-g
Corlett, J.L., Dean, E.A., and Grivetti, L.E. 2003. "Hmong Gardens: Botanical Diversity in an Urban Setting." *Economic Botany* 57(3): 365–379.
Daniels, T. 2009. "A Trail Across Time: American Environmental Planning From City Beautiful to Sustainability." *Journal of the American Planning Association* 75(2): 178–192.
DuPuis, E. Melanie, and Goodman, D. 2005. "Should We Go "home" to Eat?: Toward a Reflexive Politics of Localism." *Journal of Rural Studies* 21(3): 359–371.
<bookref>DuPuis, E. Melanie, Harrison, J., and Goodman, D. 2011. "Just Food? Cultivating Food Justice: Race, Class, and Sustainability." *Food, Health, and the Environment*. Cambridge, MA.: Massachusetts Institute of Technology.
Estes, N. 2019. *Our History Is the Future: Standing Rock Versus the Dakota Access Pipeline, and the Long Tradition of Indigenous Resistance*. Berkeley, CA: Verso.
Fishman, R. 1982. *Urban Utopias in the Twentieth Century: Ebenezer Howard, Frank Lloyd Wright, Le Corbusier*. Cambridge, Mass: The MIT Press.
Fixico, Donald. 2000. *The Urban Indian Experience in America*. Albuquerque: University of New Mexico Press.

Gliessman, Stephen R. 2006. *Agroecology: The Ecology of Sustainable Food Systems, Second Edition*. Boca Raton: CRC Press.

Gliessman, S. 2012. "Agroecology: Growing the Roots of Resistance." *Agroecology and Sustainable Food Systems* 37(1): 19–31.

Gliessman, S.R. 2016. "Agroecology: Roots of Resistance to Industrialized Food Systems." In *Agroecology: A Transdisciplinary, Participatory and Action-oriented Approach*, edited by V. Ernesto Méndez, Christopher M. Bacon, Roseann Cohen, and Stephen R. Gliessman, 23–36. Boca Raton, FL: CRC Press

Giraldo, O.F., and Rosset, P.M. 2017. "Agroecology as a Territory in Dispute: Between Institutionality and Social Movements." *The Journal of Peasant Studies*: 1–20.

Glowa, M. and Roman-Alcalá, A. 2020. "Diverse Politics, Difficult Contradictions: Gentrification and the San Francisco Urban Agriculture Alliance," Chapter 8 in *A Recipe for Gentrification: Food, Power, and Resistance in the City*, edited by Yuki Kato, Josh Sbicca, and Alison H. Alkon. New York: NYU Press.

Glowa, M., Egerer, M. and V. Jones 2019. Agroecologies of Displacement: A Study of Land Access, Dislocation, and Migration in Relation to Sustainable Food Production in the Beach Flats Community Garden. *Agroecology and Sustainable Food Systems*, 43(1): 92–115. DOI: 10.1080/21683565.2018.1515143

Glowa, M. 2017. "Urban Agriculture, Food Justice, and Neoliberal Urbanization: Rebuilding the Institution of Property" In *The New Food Activism: Opposition, Cooperation and Collective Action*. Eds. Alison Hope Alkon and Julie Guthman. University of California Press

Gonzalez de Molina, M. 2013. "Agroecology and Politics. How To Get Sustainability? About the Necessity for a Political Agroecology." *Agroecology and Sustainable Food Systems* 37(1): 45–59.

González de Molina, M. 2017. "Political Agroecology: An Essential Tool to Promote Agrarian Sustainability." In *Agroecology: A Transdisciplinary, Participatory and Action-oriented Approach*, edited by V. Ernesto Méndez, Christopher M. Bacon, Roseann Cohen, and Stephen R. Gliessman, 55–72. Boca Raton, FL: CRC Press.

Goodman, A. 2010. "Detroit Urban Agriculture Movement Looks to Reclaim Motor City" Democracy Now segment. https://www.democracynow.org/2010/6/24/detroit_urban_agriculture_movement_looks_to

Gottlieb, R., and Fisher, A. 1996. "Community Food Security and Environmental Justice: Searching for a Common Discourse". *Agriculture and Human Values* 13(3): 23–32.

Gottlieb, R., and Joshi, A. 2010. *Food Justice. Reprint edition*. Cambridge, MA: The MIT Press.

Estes, Nick. 2019. *Our History Is the Future: Standing Rock Versus the Dakota Access Pipeline, and the Long Tradition of Indigenous Resistance*. London:Verso.

Ferguson, R.S. 2014. "Critical Questions, Early Answers," *Permaculture Activist 93* http://liberationecology.org/critical-questions-early-answers/

Fromartz, S. 2006. *Organic, Inc.: Natural Foods and How They Grew*. Harcourt Books: Orlando Florida.

Gray, L., Guzman, P., Glowa, K.M., and Drevno, A.G. 2014. "Can home gardens scale up into movements for social change? The role of home gardens in providing food security and community change in San Jose, California." *Local Environment* 19(2):187–203.

Grey, S. and Patel, R. 2015. "Food sovereignty as decolonization: some contributions from Indigenous movements to food system and development politics." *Agriculture and Human Values* 32(3): 431–444

Guthman, J. 2004. *Agrarian Dreams: The Paradox of Organic Farming in California*. Oakland, CA: University of California Press.

Hayden, D. 1982. *The Grand Domestic Revolution: A History of Feminist Designs for American Homes, Neighborhoods and Cities*. Cambridge, Mass.: The MIT Press.

Hilliard, D. ed. 2008. *The Black Panther Party: Service to the People Programs*. First Printing edition. Albuquerque: University of New Mexico Press.

Hodgson, K., Caton Campbell, M. and Bailkey, M. 2010 *Urban Agriculture: Growing Healthy, Sustainable Communities*. Chicago, Ill: APA Planners Press.

Holt-Giménez, E., and Altieri, M. 2012. "Agroecology, Food Sovereignty, and the New Green Revolution." *Agroecology and Sustainable Food Systems* 37(1): 90–102.

Holt Giménez, E., and Shattuck, A. 2011. "Food Crises, Food Regimes and Food Movements: Rumblings of Reform or Tides of Transformation?" *Journal of Peasant Studies* 38(1): 109–144.

Holt-Giménez, E., and Wang, Y. 2011. "Reform or Transformation?: The Pivotal Role of Food Justice in the U.S. Food Movement.". *Race/Ethnicity: Multidisciplinary Global Contexts* 5(1): 83–102.

Jäger, M.B., Ferguson, D., Huntington, O., Johnson, M., Johnson, N., Juan, A., Larson, S., Pulsifer, P., Reader, T., Strawhacker, C., Walker, A., Whiting, D., Wilson, J., Yazzie, J., Carroll, S., & Foods Knowledges Network, I. 2019. "Building an Indigenous Foods Knowledges Network Through Relational Accountability." *Journal of Agriculture, Food Systems, and Community Development* 9(B): 45–51.

Kerssen, T.M., and Z.W. Brent. 2017. "Grounding the US Food Movement." In *The New Food Activism: Opposition, Cooperation, and Collective Action*, edited by A. Alkon, and J. Guthman, 284–315. Oakland, CA: University of California Press.

Lawson, L. 2005. *City Bountiful: A Century of Community Gardening in America.* Oakland, CA: University of California Press.

Lee, P. 2003. "Setting Up Women's Land in the 1970s: Could We Do It?" *Off Our Backs*, April: 43–47.

Mares, T., and Peña, D. 2010. "Urban Agriculture in the Making of Insurgent Spaces in Los Angeles and Seattle". In *Insurgent Public Space: Guerrilla Urbanism and the Remaking of the Contemporary City.* London: Routledge.

Mayer, M. 2010. "The 'Right to the City' in the context of shifting mottos of urban social movements." *City* 13(2-3): 362–374.

McClintock, N. 2010. "Why Farm the City? Theorizing Urban Agriculture through a Lens of Metabolic Rift". *Cambridge Journal of Regions Economy and Society* 3(2): 191–207.

McClintock, N. C. 2011. "Cultivation, Capital, and Contamination: Urban Agriculture in Oakland, California." Ph.D Dissertation, University of California, Berkeley. http://search.proquest.com.oca.ucsc.edu/docview/929198964/abstract/ABEB10C04F744DC6PQ/4?accountid=14523, accessed December 3, 2014.

Méndez, E.V., Bacon, C.M. and Cohen, R. 2016. "Introduction: Agroecology as a Transdisciplinary, Participatory, and Action-Oriented Approach." In *Agroecology: A Transdisciplinary, Participatory and Action-oriented Approach*, edited by V. Ernesto Méndez, Christopher M. Bacon, Roseann Cohen, and Stephen R. Gliessman, 1–21. Boca Raton, FL: CRC Press.

Merchant, C. 1980. *The Death of Nature: Women, Ecology, and the Scientific Revolution.* New York: Harper and Roe.

Mougcot, L.J.A. 2006. *Growing Better Cities: Urban Agriculture for Sustainable Development.* Ottawa: IDRC Books.

Moragues Faus, A. 2017. "Urban Food Policy Alliances as Paths to Food Sovereignty? Insights from Sustainable Food Cities in the UK." In: Annette Aurélie Desmarais, Priscilla Claeys and Amy Trauger (Eds) *Public Policies For Food Sovereignty: Social Movements And The State.* London: Earthscan, Routledge)

Penniman, L. 2019. *Farming While Black: Soul Fire's Farm Practical Guide to Liberation on the Land.* Chelsea, Vermont: Chelsea Green Publishing.

Pudup, M.B. 2008. "It Takes a Garden: Cultivating Citizen-Subjects in Organized Garden Projects." *Geoforum* 39(3): 1228–1240.

Roberts, W. 2017. "Urban Agriculture Is Dead! Cities Have Two, Three .... Many Urban Agricultures!!" *Medium Blog* https://medium.com/@wayneroberts/the-ways-and-the-way-to-urban-agriculture-digging-into-three-new-books-29d43b84c585

Roman-Alcalá, A. 2013. "Occupy the Farm: A Study of Civil Society Tactics to Cultivate Commons and Construct Food Sovereignty in the United States", paper presented at Yale University 'Food Sovereignty: A Critical Dialogue' Conference (September 2013).

Roman-Alcalá, A. 2015. "Broadening The Land Question In Food Sovereignty: A Case Study of Occupy the Farm." *Globalizations* 12(4):

Roman-Alcalá, A. 2018. "(Relative) Autonomism, Policy Currents, and the Politics of Mobilization for Food Sovereignty in the United States: The Case of Occupy the Farm." *Local Environment: The International Journal of Justice and Sustainability* 23(6): 619–634. https://doi.org/10.1080/13549839.2018.1456516

Roth, M, 2011. Coming Together: The Communal Option. *In Ten Years That Shook the City: San Francisco 1969-1978.* San FranciscoL City Lights Foundation Books. 192–208.

Rosset, P. 2003. Food Sovereignty:Global Rallying Cry of Farmer Movements. *Food First Backgrounder* 9(4): 1–4. http://foodfirst.org/wp-content/uploads/2013/12/BK9_4-Fall-2003-Vol-9-4-Food-Sovereignty.pdf.

Rosset, P., and Martinez-Torres, M.E. 2012. "Rural Social Movements and Agroecology: Context, Theory, and Process." *Ecology and Society* 17(3): 17.

Rosset, P., and Martinez-Torres, M.E. 2013. "Alternative Agrifood Movements: Patterns of Convergence and Divergence". In *Alternative Agrifood Movements: Patterns of Convergence and Divergence*. Published online: 03 Dec 2014; 137–157.

Peter Rosset, P. 2013. "Re-thinking Agrarian Reform, Land and Territory in La Via Campesina", *The Journal of Peasant Studies* 40(4): 721–775.

Sagorea Te' n.d. "Our vision" accessed 1/20/2020 at https://sogoreate-landtrust.com/our-vision/

Sawio, C.J. 1998. "Managing Urban Agriculture in Dar Es Salaam." Cities Feeding People 20. International Development Research Cenre (IDRC).

Schupp, J.L. and Sharp, J.S. 2012. "Exploring the social bases of home gardening." *Agriculture and Human Values* 29(1): 93–105.

Singer, J.W. 2000. *Entitlement: The Paradoxes of Property*. First Edition edition. New Haven Conn.: Yale University Press.

Stein, S. 2017. "A Colonial History of the Higher Education Present: Rethinking Land-Grant Institutions Through Processes of Accumulation and Relations of Conquest." *Critical Studies in Education* 61 (2): 212–228. https://www.tandfonline.com/doi/abs/10.1080/17508487.2017.1409646

Taylor, D. E. 2009. *The Environment and the People in American Cities, 1600s-1900s: Disorder, Inequality, and Social Change*. Durham: Duke University Press Books.

Taylor, J.R. and Lovell, S.T. 2014. "Urban Home Food Gardens in the Global North: Research Traditions and Future Directions". *Agriculture and Human Values* 31(2): 285–305.

van der Ploeg, J.D., Jingzhong, Y., and Schneider, S. 2012. "Rural development through the construction of new, nested, markets: comparative perspectives from China, Brazil and the European Union", *the Journal of Peasant Studies* 39(1): 133–173.

Walker, R. 2009. *The Country in the City: The Greening of the San Francisco Bay Area*. University of Washington Press.

Wallace, H. 2011. "Malik Yakini of Detroit's Black Community Food Security Network" *Civil Eats:* accessed at https://civileats.com/2011/12/19/tft-interview-malik-yakini-of-detroits-black-community-food-security-network/

Warner, S.B. 1987. *To Dwell Is To Garden: A History Of Boston's Community Gardens. First Edition*. Boston: Northeastern.

Watson, J. 2016. "Decolonizing Permaculture" *Permaculture Activist*, reprinted at: https://www.resilience.org/stories/2016-02-19/decolonizing-permaculture/

Wezel, A., Bellon, S., Doré,T. et al. 2009. "Agroecology as a Science, a Movement and a Practice." *In Sustainable Agriculture Volume 2*. Lichtfouse, E., Hamelin, M., Navarrete, M., and Debaeke, P. eds. Springer: Netherlands. 27–43. http://link.springer.com.oca.ucsc.edu/chapter/10.1007/978-94-007-0394-0_3, accessed December 8, 2014.

White, M. 2018. *Freedom Farmers: Agricultural Resistance and the Black Freedom Movement*. Chapel Hill, NC.: University of North Carolina Press.

Williams, J. and Holt-Giménez, E. 2017. *Land Justice: Re-imagining Land, Food, and the Commons*. Oakland, CA: Food First Books.

Zigas, Eli. 2014a. "San Francisco Establishes California's First Urban Agriculture Incentive Zone" - Urban Agriculture - ANR Blogs. http://ucanr.edu/blogs/blogcore/postdetail.cfm?postnum=15017, accessed December 6, 2014.

9

Co-Producing Agro-Food Policies for Urban Environments: Toward Agroecology-Based Local Agri-food Systems

Daniel López-García[1] and Manuel González de Molina[2]
[1] Fundación Entretantos, Valladolid, Spain
[2] Universidad Pablo de Olavide, Sevilla, Spain
[*] Corresponding author– Email ID: daniel@entretantos.org, Address: Fundación Entretantos, Valladolid, Spain

CONTENTS

KEY WORDS: *political agroecology, urban food policies, city-region food systems, agroecology-based local agri-food systems*

9.1 Introduction

The great challenge facing agroecology today is to become a widely adopted, real alternative to the dominant food regime. This challenge is especially relevant in urban and metropolitan contexts, which are strongly dependent on the imports of diverse basic goods (water, energy, etc.). Beside the practical disappearance of primary food production in the main urban areas of the world, large distribution and agricultural input companies are located and concentrated in urban areas. Paradoxically, large cities have also become spaces of innovation for the reorganization of agri-food systems towards sustainability, especially since the signing, in 2015, of the Milan Pact on Urban Food Policies (Calori and Magarini 2015; EIP-Agri 2016; DeCunto et al. 2017). The urban sphere is under-represented in the debate on the scaling up of agroecology, but it should be at the focus.

The discussion of bringing agroecology to scale has recently focused on the tension between *up-scaling and out-scaling. Up-scaling* is commonly understood to follow a vertical approach as a way to change power relations within the food system, including market rules, institutions and policies. *Out-scaling* focuses on the horizontal spread of the agroecological transitions, in geographical and social terms, to more people and communities, and is often associated with grassroots movements (Altieri and Nicholls 2008; Rosset and Martínez-Torres 2012; Parmentier 2014; Giraldo and Rosset 2017; Mier y Terán et al. 2018). Proposals for out-scaling agroecological transitions have so far mostly been oriented towards rural contexts, where peasant movements are strong; they focus on the dialectic between the State and social movements or peasants; and they tend to give less importance to urban settings and to social groups not directly involved in agri-food production. Some of these proposals for out-scaling also reject or questions the importance of public policies and the ability of the State to push forward the agroecological transition (Giraldo and Rosset 2017; Mier y Teran et al. 2018; Giraldo and McCune 2019).

Across both positions there is a wide consensus on the centrality of the food movements and peasantries in order to develop deeply transformative programs in urban environments, both regarding to agreocological (Levidow et al. 2014; López-García et al. 2019; Tornaghi and Dehaene 2019) and to food systems (Holt-Giménez and Shattuck 2011; DeCunto et al. 2017; Moragues and Sonnino 2018) approaches. Yet, despite the importance of public policies to the agroecological transition, there has been little discussion about how do the dialectics between public administrations and food movements work and how such policies should be promoted, specifically in urban environments (González de Molina 2013; López-García et al. 2017; González de Molina et al. 2019; Giraldo and McCune 2019). The objective of this chapter is to discuss the role that urban food policies can play in up-scaling agroecological transitions to the level of food systems in urban contexts. For this purpose we will analyse the urban agroecology movement in Europe using Spain as a case study, and highlight its strengths and weaknesses in relation to up- and out-scaling agroecology through urban food policy co-production between local administrations, peri-urban farmers, grassroots movements and NGOs, and scholars.

9.1.1 Urban Agroecology in Spain

In 2015, European cities from all over the world came together to sign the Milan Urban Food Policy Pact, which had an important impact in Spain (López-García et al. 2018c). Over the previous years the agroecological movement has significantly grown and developed (Guzmán et al. 2013; López-García et al. 2017). In Spain and other European countries, the global financial crisis, political corruption, and the degradation of welfare programs called into question the legitimacy of the traditional party system and of the neo-liberal model of economic development. The strong social mobilizations that took place throughout Spain starting in 2011—known as the "15-M" Movement, or the "*Indignados*", as it was referred to by the international press—led to the beginning of numerous constituent processes at the local level with a municipalist vision. In 2015 municipal elections, many municipalist parties accessed local governments in important cities[1] and also in rural municipalities, on what was called "the assault on institutions" from social movements (Observatorio Metropolitano 2014; Calle-Collado and Vilaregut 2015). Such social mobilization created a favourable situation for scaling up agroecology. Although the food issues were not among the explicit priorities in the municipalist programmes, the social movement for agroecology and food sovereignty worked hard to place the food system within the municipality agenda (López-García et al. 2017; Morán et al. 2018). These organizations developed political advocacy

[1] For example Madrid, Barcelona, Zaragoza, Palma de Mallorca, Cádiz or Santiago de Compostela; and allied with other left wing parties in Sevilla, València, Las Palmas de Gran Canaria, Córdoba and others.

actions from outside the institutions, and in numerous cases prominent agroecological activists occupied positions of responsibility in local governments, or joined them as technical assistants.

In this sense, the period 2015–2019 in Spain can be understood as a massive experiment of up-scaling agroecology through the development of urban food policies, and doing so in a representative sample of big and middle size cities of a country located in the Global North. Following a comparative approach, in this chapter we analyse the urban food policies developed in several cities, initially promoted by grassroots movements related to agroecology. In a second step, we focus on two case studies—Zaragoza and Valladolid—in order to understand the way such policies are co-produced. Rather than questioning if the policies developed are transformative or not, we would like to discuss in which ways the urban public policies developed are favourable (or not) to agroecological transitions. The discussion builds on three elements: the central role of urban food policy co-production on processes for up-scaling agroecology; the proposal of "Agroecology-based Local Agri-food Systems," (ALAS) as a preliminary framework to set up rural–urban linkages to move from urban agroecology experiences to understanding the food system as an agroecosystem; and the need of a diverse and fluid social ecosystem of food system change towards an urban agroecology.

9.2 The Role of Cities in Bringing Agroecological Experiences to Scale

The academic literature reflects a broad debate concerning the sustainability and transformative potential of alternative food systems. That has been taking place for more than two decades, and whose epicentre is in urban spaces (Blay-Palmer et al. 2018). In some cases there have been experiments with agroecological policies at the national level (Sabourin et al. 2017; Ajates et al. 2018), the greatest example of which may be Brazil in the years 2003–2016 (Petersen 2017). In other cases work has been carried out at the regional level, as occurred in Ontario, Canada (Friedmann 2007) and Brussels, Belgium (Van Dyck et al. 2018). But the most creative and potent developments in food public policy are to be found, worldwide, at the local (municipal) level. This is due, perhaps, to a greater proximity to citizens and their demands; but it is also because at this administrative level competencies in agriculture are limited, which means lighter pressure from agribusiness lobbies (López-García et al. 2018a). Cities have worked to promote alternative, sustainable distribution and consumption ways, rather than on production itself, since the role of urban and peri-urban soils is still seen as a reserve for capitalist space production (Tornaghi and Dehaene 2019). In addition, the discussion on the incorporation of agroecological criteria into urban food policies remains scarce, both from a theoretical and practical perspective (Vaarst et al. 2017), and the analysis of the movement for agroecological transitions in urban or metropolitan environments lacks empirical evidence on their role (Mier y Terán et al. 2018; López-García et al. 2019).

9.2.1 Political Agroecology and the Debate on Scaling

While some argue for the growth of agroecology in the policy realm, others highlight the risks involved when agroecology is institutionalized, such as its conceptual co-optation and the associated loss of its transformative features (González de Molina 2013; Levidow et al. 2014; Giraldo and Rosset 2018; Mier y Terán et al. 2018; Rivera-Ferre 2018). In particular, there are concerns that the translation of agroecology into public policy (*up-scaling*) might lead to hybridizations between agroecological proposals and conventional (market-oriented) agricultural policies. There is a risk that up-scaling will exclude the diversity of stakeholders involved in the food system, and that up-scaling will select for policies that still uphold hegemonic interests (Ajates et al. 2018; Rivera-Ferre 2018). We argue that the expansion or up-scaling of agroecology may potentially result in radical challenges to the current food system, because, when properly enacted, up-scaling reinforces the control of food systems by local communities (Parmentier 2014; Giraldo and McCune 2019). We propose a political approach to agroecology, related to food systems, called Political Agroecology (González de Molina 2013; Rosset and Altieri 2017; González de Molina et al. 2019). The two main objectives of Political Agroecology are to design institutions that favour the pursuit of agricultural sustainability, and to organize agroecological movements in such a way that ensures that policies can be implemented according to the interests of grassroots stakeholders (González de Molina et al. 2019).

9.2.2 Agroecology, Cities and Food Policies

The role of cities in promoting sustainability and health in relation to food systems is increasingly recognized in the policy realm by both local governments and global organizations such as the United Nations (e.g. Food for Cities, UN 2001). However, cities of the Global South and North differ in their approaches to food policy. In the global South, emphasis is placed on access to adequate food for large impoverished social groups, usually through the promotion of self-production of food in urban settings, or through the provision of community food equipment, often from a welfare perspective. On the other hand, Global North policies are founded on issues ranging from the mitigation of climate change; sustainability (regarding food waste, for example); the economic activation of metropolitan environments; providing impoverished social groups for access to good food; or the fight against obesity and other diet-related non-infectious diseases (Calori and Magarini 2015; DeCunto et al. 2017).

The most common actions developed by urban food policy programs relate to access to fresh and good quality food for marginalized social groups; community activation; and education and awareness-raising among different stakeholders in the food chain, especially final consumers. Also included are actions aimed at generating bottom-up, multi-stakeholder and multi-level participatory food governance processes. A smaller but growing number of projects develop specific actions aimed at increasing professional agricultural production and rural–urban linkages. Notwithstanding, the nature of cities as centres of consumption provides an important opportunity for them to support agroecological transitions outside their administrative boundaries, as an act of co-responsibility with the territory on which they depend (Moragues et al. 2013; *Milan Urban Food Policy Pact 2015*; Calori y Magarini 2015; EIP Agri 2016; DeCunto et al. 2017; Renting et al. 2017).

One of the most significant features of urban food policy development today is the emphasis on linking the city with the countryside. The concept of City-Region Food Systems (CRFS) was first proposed by a group of global entities and researchers, concerned with the critical role of food in a globalized and urbanized world, and the growing dependency of mega-cities on physical supplies (such as energy, water or food) provided by their surrounding territories, in a context of global change (Jennings et al. 2015). According to Blay-Palmer et al. (2018:2), it is a theoretical framework and an operational approach that "integrates (biophysical) flows across sectors and resources (related to food), [...] offers an integrative method with which to consider and develop policies and programs across scales, including urban, peri-urban, and rural". In addition, some authors propose CRFS as a suitable approach for up-scaling agroecological experiences in the urban environment (Vaarst et al. 2017).

Another significant feature of urban food policy development is the configuration of multi-stakeholder governance. Some key factors include the involvement of diverse stakeholders to foster policy innovation and depth; governance structures that clarify and amplify the terms in which these diverse stakeholders can cooperate; and cooperation with research bodies (DeCunto et al. 2017; Moragues-Faus and Sonnino 2018). The capacity to build hybrid forums that bring together the agroecological experiences with local administration and other conventional actors, is a key element in the construction of transitions towards sustainability in local agri-food systems (López-García et al. 2018a). In addition, hybrid marketing strategies—between conventional and alternative marketing strategies and food networks—have been identified as the only way to achieve economic viability within a globalized food system, both for small conventional actors (family farms, food retailer shops or others) and for small alternative experiences (Ilbery and Maye 2005; Goodman et al. 2012; Darnhofer 2014; López-García et al. 2018b). New approaches to food system transformation require new alliances, which has to be recognised by the food policies.

Because the urban environment lacks the presence of rural farmers, urban food policies are limited in their capacity to generate transformations in productive and territorial systems, and often struggle to bring together subjects with an agroecological vision (Giraldo and Rosset 2017; Mier y Terán et al. 2018). For this reason, combining CRFS proposals with agroecological programmes is key to the success of the up-scaling agroecology process (Vaarst et al. 2017).

The urban food policies developed so far show numerous deficiencies from an agroecological point of view. For example, the lack of a biophysical or material approach to adequately assess the sustainability of urban agroecological experiences, as well as of the spatial designs needed for more sustainable food systems (González de Molina et al. 2019; Simón-Rojo 2019). Although most of the urban food policies developed around the world highlight sustainability as a central action area, the proposed actions are generally limited to reducing food waste, protecting the headwaters of river basins whose waters are collected for urban use, or protecting to some extent the environment of peri-urban agricultural spaces. The processes that have the greatest environmental impact along the food chain, such as fertilisation, irrigation and pest control are intimately related to the stages of input production or crop production, which for the most part are not covered by these policies. Such deficiencies in urban food policies have led to talk on 'agroecological urbanism', which bases urban agroecological transitions on the redirection of urban nutrient flows, restoration of peri-urban agricultural land, 'community food pedagogies', and empowering infrastructures for localised food networks (Tornaghi and Dehaene 2019).

The objective of the following section is to analyse these elements using empirical data obtained from selected case studies. Throughout these case studies, we will address the discussion on how to promote a transformative agenda for scaling agroecology through urban food policies; and which multi-stakeholder alliances are to be made, in order to deploy all its transformative potential.

9.3 Four Years of Urban Food Policies in Spain from an Agroecological Perspective

As has been mentioned before, cities today define the shape and functioning of food systems because they are political, economic and consumption centres. In this section, we provide case studies of policy co-production for the promotion of urban agroecology in Spain. The analysis of such experiences will then allow us, in the following section (number 4), to discuss to what extent have been the agroecological approach developed into food policies, which are the main weaknesses and strengths, and which are the main challenges to face when up-scaling urban agroecology to food systems scale. To do so, we first we conducted a comparative study of food policies implemented during the 2015–2019 period in 13 Spanish cities belonging to the Spanish Red de Ciudades por la Agroecología (Network of Cities for Agroecology). We further focus in two case studies, with the aim to analyse the multi-stakeholder configurations and processes undertaken during the setting up of municipal food strategies in two medium-sized cities included in the previous analysis: Zaragoza and Valladolid. This comparison highlights how specific multi-stakeholder configurations play a role in shaping policy.

All data used in this section is based on empirical experience from processes of urban food policy co-production with an agroecological approach, in which the authors have participated, respectively, as research director and advisor.

9.3.1 Agroecology-Oriented Urban Food Policies in 13 Spanish Cities: A Critical Assessment[2]

As explained above, the last years have shown in Spain an interesting laboratory for the experimentation of bottom-up municipalist proposals (Morán et al. 2019). This momentum from below, together with the example of some pioneering cities (especially Zaragoza, Vitoria and Lleida) that already had food policies in place, and the emergence of the *Milan Urban Food Policy Pact* (2015), led to the development of important examples of urban food policies, based explicitly on agroecology. Table 9.1 shows the main action areas in which some of the most prominent cities have developed their activity. These cities are members of the Spanish Red de Ciudades por la Agroecología, and almost all are signatories to the Milan Pact.

Table 9.1 summarizes the results of a survey to city officers in charge of the food policies in 13 selected Spanish cities, complemented by personal interviews and official documentation review. The typology of "action areas of food policies" lies on a meta-analysis of typologies used in previous research comparing

[2] This section is based on an adaptation and update of the report "Políticas alimentarias urbanas para la sostenibilidad: análisis de experiencias en España, en un contexto internacional" (López-García et al. 2018c).

TABLE 9.1

Actions and Main Action Areas of Food Policies Developed in Selected Spanish Cities from 2015

City (Inhabitants)	Year in Which Food Policies Began	Governance and Social Activation	Research, Communication, Education and Behavioural Change	Access and Equity	Sustainability	Production and Urban–Rural Linkages	Supply and Distribution
Barcelona (1,6M)	2015	-Food strategy	-Cooperation with University -Awareness campaigns	-Cooperation with social policies and services -Public procurement -Community gardens -School gardening	-Organic farming promotion -Legal protection of agricultural land -Composting	-Agricultural park -CRFS approach for public procurement -Food system perspective in metropolitan urban planning	-Farmers markets -Food hub -Small retailers support -Agroecological entrepreneurship promotion
Córdoba (326.000)	2015	-Food Council -Food strategy	-Cooperation with University -Awareness campaigns -Research on food affairs	-Cooperation with social policies and services -Public procurement -Community gardens -School gardening	-Organic farming promotion -Traditional irrigation systems restoration	-Coordination with province administration -Collective structures promotion	-Farmers markets -Food hub -Small retailers support -Agroecological entrepreneurship promotion
El Prat de Llobregat (63.000)	1998			-Social enterprises promotion -Community gardens	-Organic farming promotion -Legal protection of agricultural land	-Agricultural park	-Farmers markets -Agroecological entrepreneurship promotion
Fuenlabrada (195.000)	2012		-Cooperation with University -Awareness campaigns	-Social enterprises promotion -Public procurement -Community gardens -School gardening	-Agricultural biodiversity promotion	-Agricultural park -Land bank -Collective structures promotion	-Farmers markets -Small retailers support -Agroecological entrepreneurship promotion
Granollers (60.000)	2007		-Cooperation with University -Awareness campaigns	-Cooperation with social policies and services -Community gardens	-Organic farming promotion -Legal protection of agricultural land	-Agricultural park -Collective structures promotion	-Farmers markets -Small retailers support -Agroecological entrepreneurship promotion

Lleida (138.000)	2011	-Agriculture Council	-Cooperation with University -Awareness campaigns	-Community gardens -School gardening	-Organic farming promotion -Traditional irrigation systems restoration	-Land bank -Agroecological incubator	-Farmers markets -Small retailers support -Agroecological entrepreneurship promotion
Madrid (3,2M)	2015	-Food Council -Food strategy	-Cooperation with University -Awareness campaigns -Research on food affairs	-Cooperation with social policies and services -Public procurement -Social enterprises promotion -Community gardens -School gardening	-Organic farming promotion -Organic waste composting -Food loss reduction	-CRFS approach for public procurement -Collective structures promotion	-Farmers markets -Food hub -Small retailers support -Agroecological entrepreneurship promotion
Manresa (74.000)	2012			-Social enterprises promotion -Community gardens	-Organic farming promotion -Legal protection of agricultural land	-Coordination with County administration -Collective structures promotion	-Food hub -Small retailers support -Agroecological entrepreneurship promotion
Pamplona (195.000)	2016	-Food Council	-Cooperation with University	-Public procurement -Social enterprises promotion -Community gardens -School gardening	-Organic farming promotion	-CRFS approach for public procurement -Collective structures promotion	-Farmers markets -Food hub -Small retailers support
Rivas-Vaciamadrid (82.000)	2008		-Awareness campaigns	-Community gardens -School gardening	-Organic farming promotion -Organic waste composting	-Agricultural park -Land bank -Agroecological incubator -Collective structures promotion	-Farmers markets -Food hub -Small retailers support -Agroecological entrepreneurship promotion

(Continued)

TABLE 9.1 (Continued)

City (Inhabitants)	Year in Which Food Policies Began	Governance and Social Activation	Research, Communication, Education and Behavioural Change	Access and Equity	Sustainability	Production and Urban–Rural Linkages	Supply and Distribution
València (790.000)	2015	-Food Council -Food strategy	-Cooperation with University -Awareness campaigns -Research on food affairs	-Cooperation with social policies and services -Public procurement -Social enterprises promotion -Community gardens	-Organic farming promotion -Traditional irrigation systems restoration -Agricultural biodiversity promotion -Legal protection of agricultural land -Organic waste composting -Food loss reduction	-Agricultural park -CRFS approach for public procurement -Food perspective in metropolitan urban planning -Land bank -Agroecological incubator -Collective structures promotion	-Farmers markets -Food hub -Small retailers support -Agroecological entrepreneurship promotion -Collective processing centre
Valladolid (300.000)	2017	-Food Council -Food strategy	-Cooperation with University -Research on food affairs Food system diagnosis	-Public procurement -Community gardens	-Organic farming promotion -Organic waste composting -Food loss reduction	-CRFS approach for public procurement -Collective structures promotion -Food system perspective in metropolitan urban planning	-Farmers markets -Food hub -Small retailers support -Agroecological entrepreneurship promotion -Collective processing centre
Zaragoza (661.000)	2009	-Food Council -Food strategy	-Cooperation with University -Awareness campaigns -Research on food affairs Food system diagnosis	-Public procurement -Social enterprises promotion -Community gardens -School gardening	-Organic farming promotion -Traditional irrigation systems restoration -Agricultural biodiversity promotion -Organic waste composting -Food loss reduction	-Agricultural park -CRFS approach for public procurement -Land bank -Agroecological incubator -Collective structures promotion	-Farmers markets -Food hub -Small retailers support -Agroecological entrepreneurship promotion -Collective processing centre

Source: Authors' Own Work.

urban food policies (Calori and Magarini 2015; DeCunto et al. 2017; EIP-Agri 2016; Renting et al. 2018). The table shows fairly comprehensive agendas in the cities analysed. However, the detailed analysis of the actions carried out shows important differences between action areas. The action areas that related more closely to economic and productive activities ('production and rural–urban linkages', and 'supply and distribution') concentrated a bigger amount of actions in most of the cities analysed. On the other hand, action areas with greater social and ecological relevance ('governance and community activation', 'equity and access' and 'sustainability') were rather neglected. The agroecological approach is very present in the discourse of cities, perhaps due to the strong push that social movements have given to food policies.

However, we find important weaknesses in the development of an agroecological praxis oriented to the development of sustainable food systems. These include a general lack of agroecosystems perspective in regard to the *upstream* phases (prior to agricultural cultivation) of the production cycle being considered – taking into account factors such as water supply, organic matter cycling or traditional agricultural landraces; which implies a strong focus on integrated territorial planning. In addition, lack of a clear multi-stakeholder process, which fails when articulating within food policies co-production some local actors expelled from global food markets (small farmers and food retailers) or marginalised (low income groups, racialized minorities).

The promotion of organic agricultural production, both urban and peri-urban (through agricultural parks, training, local quality schemes, etc.), as well as the establishment of specific distribution networks for organic food (in special farmers markets and public procurement in school canteens), were the action areas that were developed the most. However, regarding an agroecological perspective that encompasses the ecological food cycle in a holistic manner, the most common actions at the *upstream* phases of the production cycle were reducing and reusing food waste, in some cases linked to organic waste composting programmes. Agroecology-oriented territorial planning, green and blue infrastructure planning according to urban food production, or other interventions regarding soil fertility or landscape-level biodiversity management were absent.

The food systems approach never reached, for Spanish municipalist projects, a core place in the model of sustainable city they imagined -such as, for example, mobility or housing policies (Morán et al. 2018). Therefore the most ambitious features of what should be Agroecology-based Local Agri-food Systems (in advance, ALAS) have stayed underdeveloped. Top-down policies have never fulfilled the expectations of the food movements, and have also weakened its strength by instrumentalizing it in some ways. In addition, limiting local food policies to the municipal level, especially in urban or metropolitan contexts, this results in scarce attention being given to *upstream* processes (input production and distribution, agroecosystem design, etc.) as opposed to *downstream* processes (distribution and marketing). The lack of multi-stakeholder perspectives likely resulted in the one-sided focus on *downstream* agricultural processes, as the farming sector (mainly located outside of city boundaries) and its interests showed a scant presence in the urban food governance processes. Lastly, the difficulty in developing community activation dynamics around food, especially in impoverished neighbourhoods where agroecological experiences are scarce and very incipient, and economic access to good food is neglected (Simón-Rojo 2019), clearly illustrates an undesirable but clear focus of food policies on urban middle classes.

Three main issues emerge from this analysis, as tools to strengthen an agroecological approach to urban food policies. First, the agroecological perspective requires the implementation of an agroecosystemic approach to City-Region Food Systems, which implies territorial planning processes that focus on food and local communities (Tornaghi and Dehaene 2019). Second, there is a need to develop specific tools and processes in order to give space in urban food policy production for certain stakeholders which are marginalized within the agri-food system – especially impoverished social groups, but also small conventional farms in the metropolitan area and small, conventional food retailers (López-García et al. 2018b), with special attention to groups of urban and peri-urban rural farmers (both conventional and organic). Third, there is a need for narratives linking local, urban agroecological experiences with emerging socio-ecological issues both in the local and the global agenda, such as food-related health issues or climate change, in order to overwhelm activist and privileged social groups, and develop wider alliances for food systems transformation.

9.3.2 Multi-Stakeholder Processes in the Construction of Two Urban Food Strategies in Spain

The objective of this section is to take a deeper look into the relational and processual dimensions of urban food policies. Concretely, two case studies are discussed in which local food strategies were developed with a participative and multi-stakeholder approach in two medium-sized Spanish cities: Zaragoza (661,000 inhabitants) and Valladolid (300,000 inhabitants), both being capitals of large and very unpopulated regions. Both processes gave as output a document—the 'local food strategy' itself—which were later approved by the respective City Councils, as the document which defines and drives the official, local food policies in the medium term (Ayuntamiento de Valladolid 2019; Ayuntamiento de Zaragoza 2019). The two cases were similar in terms of the make-up of the teams guiding each participatory process, and in terms of their methodological design[3]. In both cases the institutional justification and the structure of action areas corresponded to the Milan Pact on Urban Food Policies, with explicit mention to agroecology. However, differences in both the starting contexts and the political and administrative framework of each territory determined differences in their outcomes.

In the case of Valladolid, more than 500 individuals and 134 organizations participated in the process, including neighbourhood associations, NGOs, ecological consumer groups, small commerce, production, distribution, parents' associations and federations, 19 public organizations, and other agents involved in the food system. Some stakeholders participated in the group leading the process with greater stability and involvement than others. This was particularly the case with provincial organic producers, the federation of neighbourhood associations, (non-professionalized) grassroots environmental groups, the University of Valladolid (especially the department of Geography), and the regional academy of dietitians/nutritionists. To a lesser extent, representatives of parents' associations and small commerce specialized in organic food have also participated. The absence of hotel, restaurant and small food retailers sectoral associations has been notable, perhaps due to their ideological position, which was opposed to that of the municipal government. This diverse "hard core" of participants evolved into a shared identity and commitment to the objectives of the process and of the local food strategy itself (Méndez et al. 2017). In parallel to this, a specific process was developed to accompany the structuring of the provincial organic production sector around a farmer market, which resulted in the establishment of an organic farmers' association strongly involved in the implementation of actions set out by the local food strategy. At the same time, a broad political support from the Environment Councelor gave rise to an important involvement of various departments in the development of the Action Plan that derived from the local food strategy.

In the case of Zaragoza, 72 people participated in the drafting of the local food strategy, representing 56 organizations – social and economic organizations, and branches of the local and regional administrations. The most involved actors were the Municipal Federation of Schoolchildren Parents' Associations and the provincial organic production sector. In this process, a Steering Group was not formed, neither formal nor informally, and the stakeholders that participated in the few meetings held were rather belligerent and demanding in the face of the local administration. In the case of the organic production sector, testimonies were gathered that delegitimized both the participatory process and the draft local food strategy, expressing mistrust towards the role of the administration and its ability to listen to and address the needs and demands of the local agroecological fabric. In fact, deliberative workshops were scarce and had little attendance; and some actors also expressed unrest regarding the ability to participate in the co-production of food policies in previous experiences of food policies deliberative processes.

Based on the differences observed between the two case studies, we can identify some elements that facilitate or hinder the development of multi-actor engagement dynamics. Among these is the need to allow for the "natural time" it takes for a participatory process to adapt to the rhythms and processes of local communities (López-García et al. 2019); the need to involve local stakeholders from the very beginning of the process, and to be honest in the management of communication and information (Méndez et al. 2017); or the importance of a clear commitment to the participatory process and its results, both from the administration's technical staff and from elected officials (Mier y Terán et al. 2018; Giraldo

[3] The technical coordinator of both processes is the first author of the present chapter.

and McCune 2019). Farmers were present in both processes, but they showed to be more engaged in the process (Valladolid) in which special attention was posed to attract them to the workshops, and regarding its time availability. Economic, racial and gender equity, despite often being mentioned in scientific debates on agroecology and even in Zaragoza's local food strategy, showed to be absent topics in the deliberative workshops in both processes, and thus neglected to be consistently included in the processes unless there is a clear and explicit will to do so (including methodological designs) (Zuluaga Sánchez et al. 2018; Van Dick et al. 2019; Tornaghi and Dehaene 2019).

9.4 Agroecological Transitions in Urban Contexts: From Agroecological Experiences to Agroecology-based Local Agri-food Systems

In this section we highlight the main learnings from the previous analysis, to create a general framework of food policy co-production, favourable to agroecological transitions in urban contexts of the Global North. The discussion builds on three elements: the central role of urban food policy co-production on processes for up-scaling agroecology; the proposal of Agroecology-based Local Agri-food Systems, as a preliminary framework to set up rural–urban linkages in order to move from urban agroecology experiences to understanding the food system as an agroecosystem; and the need of a diverse and fluid social ecosystem of food system change towards an urban agroecology.

9.4.1 Up-Scaling Agroecology in Urban Environments?

The scientific debate focused on the scaling of agroecology emphasizes the leading role played by peasants and peasant movements in true processes of large-scale agroecological transitions (Giraldo and Rosset 2017; Mier y Terán et al. 2018; Giraldo and McCune 2019). However, these positions rest almost exclusively on rural experiences in the Global South. In fact, they point out the need to study further this type of processes in other contexts, especially in the Global North (Mier y Terán et al. 2018)—and in urban environments. The urban or metropolitan realities analysed in this chapter involve different processes than those in some rural areas of the Global South. These experiences problematize, from an empirical point of view, the theoretical and methodological tools applied in the debate on scaling agroecology particularly in the urban realm, as well as the proposals for action that derive from this debate.

In urban spaces and in the Global North, peasant movements are weak, if not already incorporated into the dynamics of the Corporate Food Regime (McMichael 2016). Both urban food movements and local authorities have shown to be more linked to consumers' issues or urban 'free-time activities'—such as community gardens—rather than to rural issues. Thus, it has been necessary to push local authorities (in a *bottom-up* policy co-production process) to set up links among rural and urban movements; to develop policies to expand demand of local, organic foods among urban dwellers and public procurement services; and to create infrastructures in order to facilitate the delivery of such food and improve the access to it of different urban, social groups (*top-down* policies). In the cities analysed there were no (strong) collective subject demanding agroecological transitions at a food system scale, but rather support to small scale urban experiences such as food-coops, community gardens and some small and isolated production initiatives. Thus, there were some organizations which assisted local authorities in the co-creation of agroecology oriented public policies at a food system scale, bringing rural—or food system scale—approaches to urban food policies. And, at the same time, supported the creation of a collective subject to receive such policies and to develop activities favoured by such policies (e.g. metropolitan and regional organic farmers groups, and regional platforms of NGOs, farmers and grassroots groups to push for deepening in the transition), trying to engage both urban experiences—such as community gardens or consumer groups- with rural experiences—organic farmers' groups, rural grassroots initiatives, etc.

Actions to strengthen the agroecological social fabric have been incorporated into various local food strategies as a specific action area, as is the case in Zaragoza and Valladolid. Nevertheless, the assimilation of activism by local administrations reduces the strength and responsiveness of grassroots organizations when facing higher levels of administration (regional, national or European)—the levels at which most of the competences in agriculture and food reside. The incapacity to formulate a true multi-level food policy, given the global political context, represents a ceiling for food policies with an agro-ecological approach. This generates frustration within the local agroecological fabric, and undermines the transformative potential of such policies.

The cases analysed suggest that, in situations where agroecological movements are not strong (such as urban contexts in the Global North), there is probably no other way to scale up agroecology than to seek the support of public policies through social and peasant protagonism (González de Molina et al. 2013; Giraldo and McCune 2019). This represents a combined action on bottom-up and top-down food policy approaches, which have shown to be complementary on promoting up-scaling (as an introduction of an agroecological approach within local administration) through bottom-up action; and promoting out-scaling (facilitating the development and articulation of new agroecological experiences) through top-down policies. Public administrations would have to develop decisive actions with a top-down approach; at the same time as they strengthen, from a bottom-up approach, the agroecological social fabric – which is in many cases fragmented and fragile in urban and peri-urban contexts (Paül 2007; López-García et al. 2019). In this sense, out-scaling and up-scaling are not that isolated from each other or contradictory, but interdependent and synergistic processes (see Figure 9.1) (Fergusson et al. 2019). At a global scale, such perspective could fit also in widespread situations in the Global South, as far as the urbanization of the population, as well as the rapid loss of the active agrarian population, keeps on growing in many countries of Latin America and Asia.

Regarding the impact of the cases that have been presented here, four years of work only allowed for developing a few concrete and often unrelated actions. The actions carried, actually, out only fit weakly with the narratives and positions of a transformative and strong approach to agroecology (Levidow et al. 2014). Regarding to this weak performance of agroecology; a door has been opened for its co-optation, and for the co-optation of the processes that were initiated (Giraldo and Rosset 2017; Rivera-Ferre 2018). But, despite the important deficiencies that have been pointed out, the cases presented describe some first steps in the scaling of agroecology. At the end, they leave the agroecological transition in a position significantly more advanced than where it was before. To develop its potentialities, bottom-up and top-down approaches should be coordinated and articulated by multi-stakeholder and multi-level structures of food

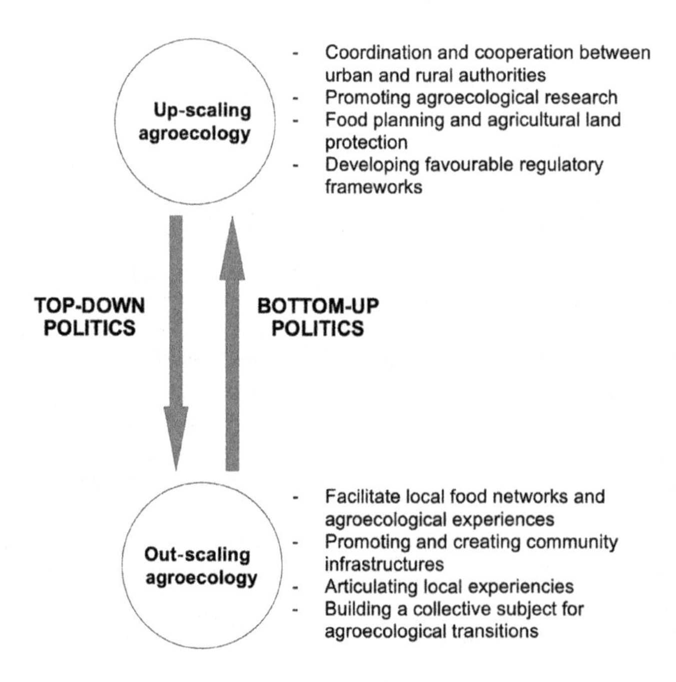

FIGURE 9.1 Synergies among up-scaling and out-scaling agroecology, through bottom-up and top-down politics.

governance that currently do not exist. The dependence on the political orientation of the governments of the day to activate processes of escalation of agroecology reinforces this need to combine top-down and bottom-up approaches. For this reason, the building of a multiple and strong subject for the agroecological transition capable of imposing these governance structures takes on central importance.

In the following sections, we will try to deepen the debate around the scaling of agroecology, trying to address some operational aspects in urban contexts. A look at the methodological tools for its development has been developed in other papers (*i.e.* López-García et al. 2019), and in other chapters of this book (e.g. see chapter 14 by Caswell et al.). These all could be the central elements for a debate on the scaling of agroecology from an urban perspective.

9.4.2 Setting Up Agroecology-based Local Agri-food Systems and the Rural–Urban Alliance

One of the main problems facing urban agroecology is the need to consider metropolitan areas as incorporating agroecosystems. The inadequate connection between urban agroecology and the rural context, where the bulk of the food that feeds cities is produced, is a major strategic problem. It therefore seems essential to re-link urban agroecological experiences with agricultural production, which is the first link in the chain, through alternative food networks and community infrastructures (Tornaghi and Dehaene 2019); and through this, to foster links with the rural world in general. This re-linking must also be based on the application of agroecological principles of agroecosystem design to spatial planning on a CRFS scale—a proposal that has not yet been formulated and that undoubtedly implies a high level of political and technical complexity.

We propose Agroecology-based Local Agri-food Systems (ALAS) as an articulation of agroecological experiences, linking territories and actors within a City Region Food System, and oriented to gain greater degrees of social-ecological sustainability in a comprehensive approach. Such ALAS are to be sustained by the strength of social movements and local food networks, but also by its own socio-economic viability (González de Molina et al. 2017; González de Molina et al. 2019). It is a question of seeking the synergies that result from rural–urban cooperation so as to produce, distribute and consume among various agroecological experiences in different locations, and to facilitate the organized incorporation of new projects. And such aims need to be achieved within favourable environment of nested food policies at different territorial scales.

The ALAS should pursue a dual strategy of cooperation (see Figure 9.2). On the one hand, the *horizontal strategy* would develop a local, territorialized agri-food systems approach, which links the potential for social and ecological sustainability with its ability to adapt to the territory from a multi-stakeholder perspective, including both agrarian and non-agrarian stakeholders (Marsden et al. 2000; Ventura et al. 2008; Bowen 2010; Bowen and de Master 2011; Goodman et al. 2012; López-García et al. 2018a, 2019). On the other hand, the *vertical strategy* would bring together the different stakeholders involved in the local food chain into a common project of social and ecological sustainability, based on cooperation and within their own territory (Marsden and Sonnino 2008; Darnhofer 2014; Bui et al. 2016; López-García et al. 2018c). The so-called vertical strategy can be subdivided into two perspectives (upstream and downstream), according to the different processes and stakeholders involved in each phase of the food chain.

Within the vertical strategy, the *upstream* perspective of an ALAS should promote connections between local production and producers, so that nutrient cycles are closed and direct energy consumption throughout the food chain is reduced. It is not by chance that the most important energy expenditure in the agricultural sector is associated to the importation of chemical fertilizers (especially the nitrogenous kind) and large quantities of animal feed (Infante and González de Molina 2013). By its side, the objective of the *downstream* perspective of ALAS is the expansion and consolidation of shorter and more sustainable channels of distribution and commercialization. The territorial adaptation of production would favour locating agro-industrial activities close to areas of agricultural production, and grouping producers in order to sell jointly, organize production, regulate supply and secure provisioning, while developing specific infrastructure for these purposes. The *horizontal or territorial strategy* of the ALAS would also play a key role in effectively linking production with consumption, based on alliances with

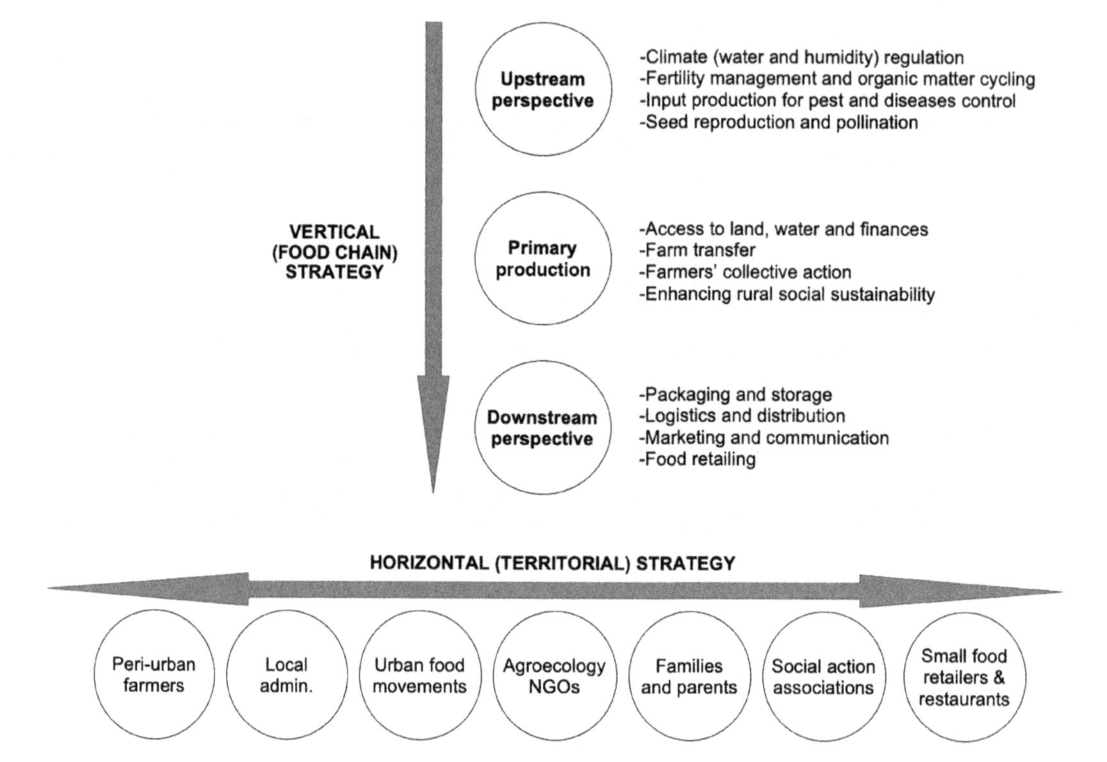

FIGURE 9.2 Combination of vertical (food chain) and horizontal (territorial) strategies for promoting to Agroecology-based Local Agri-food Systems (ALAS).

local stakeholders from outside the food chain. It would also facilitate changing the food consumption behaviour that sustains the current corporate food regime (McMichael 2016).

This territorialized approach to the organization of the food chain follows the same criteria applied in the design of agroecosystems that aim for maximum productivity, stability over time and resilience. These are the expression of processes of social self-organization, that is, of processes that link stakeholders with territorial resources that are sometimes hidden or sequestered by the hegemonic actors (Petersen 2015). The social mobilization process implicated in setting up ALAS must also involve public administrations, especially local ones.

9.4.3 A Plural Vision of the Social Ecosystem of Food Change

It is critical to get the farming sector—and specially, where it exists, the so-called 'agroecological peasantries'—on board if we want to enact agroecology-oriented urban food policies. Currently family farms (including peasants) constitute over 98% of all farms, and work on 53% of agricultural land, producing 53% of the world's food (Graeub et al. 2016, 1). They constitute the basis of the world's food and the maintenance of global agroecosystems. It would seem natural that any strategy to advance an agroecological transition, scaling agroecological experiences and building an alternative food regime, should be based around the social protagonism of peasants. But, the corporate food regime not only harms peasants, it "affects marginalized races, nationalities, genders, and classes of people, [so that] natures must be restored with the consent, participation, and design of those so affected", (Garvey 2016, cited in Cadieux et al. 2019). Women, young new farmers and urban low-income dwellers suffer the impacts particularly harshly. Without them, it will be difficult to achieve change in large parts of the world, a world that is already more urban than rural.

However, the marginal presence of the agricultural sector in metropolitan territories (Paül 2007; López-García et al. 2019), coupled with its high level of disconnection and the remoteness of farmers unions with respect to urban food policies, makes it difficult to integrate rural and agricultural views into urban policies. The dispersion and disorganization of the organic production fabrics in the areas of influence of cities is generating significant tensions in cases where urban demand increases, because it cannot be satisfied by the incipient alternative and local food networks. On the other hand, it is markedly difficult to find business models that are in line with agroecology and the social and solidarity economy, and at the same time are economically viable – especially for new entrants into farming (López-García 2018c). New entrants into organic farming and conventional farmers have profiles of a very different nature, and therefore they require different approaches for becoming involved in ALAS construction processes, which adds complexity to the processes (López-García et al. 2019).

The weak presence of the agrarian sector—and even more so of organic agriculture—in urban and metropolitan areas requires that we speak of a multiple subject for promoting the agroecological transition, along the lines of alternative food networks (Goodman et al. 2012). As we have seen, it is often non-agricultural stakeholders (neighbourhood associations, federations of school children's parents' associations, research groups) that are most involved in the promotion of urban food policies, ahead of the agricultural sector itself. Two key stakeholders that could complete this multiple subject for agroecological transitions in metropolitan environments may also be found among social actors that are in crisis or that have been expelled from the globalized food market. On the one hand, marginalized or impoverished social groups, which were not involved in the Spanish case studies analysed, meanwhile experiences in this regard exist in other territories (Holt-Giménez and Shattuck 2011; Nelson et al. 2013). On the other hand, the world of feminism is one with which agroecology finds convergent views, activist involvement, and practices, especially through ecofeminism (Shiva and Mies 2014; Zuluaga et al. 2018; Tornaghi and Dehaene 2019). Yet so far weak outreach towards feminism has taken place, both in theory and through activism. The strengthening of a multiple subject—both through the nature of its actors and by bringing together rural and urban territories—to push for agroecological transitions appears to be the only way to overcome the debate between transforming or conforming to the corporate food regime.

Scaling agroecological experiences to ALAS can only be done by mobilizing a social majority, led by peasants—whether traditional or "new"—in a global struggle for food sovereignty. Merely adding up claims, which are already fragmented and even contradictory, will not be sufficient to cement such a heterogeneous social alliance. A holistic political proposal is necessary, capable of promoting changes in production as well as in distribution and consumption[4]. This proposal has been called *food populism* (González de Molina et al. 2019; Cadieux et al. 2019), and transcends the rural–urban dichotomy, as a way in which cooperative and solidary exchange between the two extremes of the food chain, the basis of a sustainable food system, will be possible. Here, populism should not be understood as an ideology but as a political language of communication for the "people" oppressed by a privileged minority. The people are thus not considered to be a truly really existing social category, but a plural category seeking to reduce social complexity. Populism can be understood as a grammar that produces the people, a historical subject of change (Retamozo 2017:170).

9.5 Conclusions: Food Systems Agroecological Transitions Will not Be Possible without Involving Urban Environments

The generalization of agroecological practices requires, more than in any other aspect of agroecology, the engagement of politics (see, among others, González de Molina 2013; Levidow et al. 2014; Giraldo and Rosset 2017; Mier y Terán et al. 2018; Giraldo and McCune 2019). That is, it requires the design and implementation of institutions—both outside and inside the State—that make it possible to change the hegemony of the corporate food regime (McMichael 2016), and the development of an alternative regime based on sustainability. The urban realm concentrates a growing number of the population, holds

[4] In some Spanish territories, social multi-actor platforms with this aim -such as 'Madrid Agroecológico' or 'Alliance for Food Sovereignity of Córdoba' have been created, with an important impact on bottom-up food policies when they found a receptive political environment.

economic and political power, and steers cultural interest. Cities have also become a central space for the development of experiences of urban agroecology. Thus, cities are critical to food systems' agroecological transitions.

Based on the experiences analysed, we propose transition paths to Agroecology-based Local Agri-food Systems performed through approaches that transcend urban–rural segregation. We recommend the application of the CRFS to wide-scale agroecosystems design, combined with multi-stakeholder and multi-level approaches in the co-production of public policy, and paying special attention to bio-physical flows and sustainability. The conception of ALAS focus on the redesign of production processes, going beyond food production to consider the phases of input production and provisioning (upstream), and food distribution and consumption (downstream). It articulates both the different actors in a territory (horizontal strategy) and along the food chain (vertical strategy). The aim is to look for the synergies produced by the urban–rural cooperation, in order to expand and supply local/regional consumption with healthy products, grown sustainably in the territory itself, with fair remuneration for work and accessible to consumption for all social groups, both in price and physical location. The Spanish cases analysed in the previous pages seem to be significant to other parts of the World suffering deagrarianization processes, which includes many countries and regions. The different paths and situations of such deagrarianization in each territorial context should drive local adaptations of the concept of ALAS.

The change in the dynamics of agroecosystems and of the food system as a whole can only come from the momentum of social agents, with the support of favourable institutional mediations, coordinating top-down and bottom-up approaches to food policies. The subjects promoting this change should bring together different local stakeholders, from both inside and outside the food chain, and urban and rural realms, in a shared project of social and ecological sustainability that crosses the rural and urban categories to draw a food system approach. This is because urban food flows transcend urban geographical limits, and also because cities have a clear responsibility on their impact—social, economic and ecological—on the surrounding territories. The proposal of food populism (González de Molina et al. 2019; Cadieux et al. 2019), as a grammar capable of converting domination into cooperation among oppressed social actors, can be useful in the construction of the alliances that will make up this multiple subject, transcending rural and urban boundaries. In any case, the leading role that should be played by the agroecological agricultural sector in this new social subject must be a central theme of agroecological transition building—even in urban and metropolitan environments. Issues that we still need to address in depth include: (1) How to place the voice of the agroecological peasantries at the core center of urban food policies, and (2) how to create networks of cooperation between the diversity of urban social groups—including 'urban agroecology' groups (Tornaghi and Dehaene 2019)—and the diversity of peasantries around the world. In the end, it is the question of who is the subject of the agroecological transitions at food system levels that needs to be more precisely addressed.

Thus far, experiences of ALAS are still scarce and weak, and are usually developed in parallel to and away from urban agroecology experiences. For example, those in community gardens, community kitchens, food co-operatives, etc. Therefore a greater accumulation of empirical knowledge will be necessary to refine the proposal of urban political agroecology; and to better know the contradictions it generates inside and outside the boundaries of cities. All these considerations, together with the social technologies used to set them in action through social and peasant protagonism, must be subjected to further empirical research.

Acknowledgements

We would like to thank the food activists and the staff of the cities involved on the Spanish Red de Ciudades por la Agroecología their help and support as providers of the data used in this chapter, and specially the staff of the cities of Valladolid and Zaragoza, and the Technical Secretariat. And we would really like to thank the editors Monika Egerer and Hamutahl Cohen for their sustained, invaluable editing work and comments. This research has been done thanks to the support of the Fund for the Third Sector Grants from the Spanish Ministry of Ecological Transition (2019), and has used data from projects supported by Fundación Daniel y Nina Carasso, Ayuntamiento de Valladolid and Ayuntamiento de Zaragoza.

REFERENCES

Ajates Gonzalez, R., J. Thomas and M. Chang. 2018. Translating Agroecology into Policy: The Case of France and the United Kingdom. *Sustainability* 10, 8. DOI: 10.3390/su10082930

Ayuntamiento de Valladolid. 2019. Estrategia Alimentaria de Valladolid. Ayuntamiento de Valladolid. http://www.alimentavalladolid.info/wp-content/uploads/2019/05/EstrategiaAlimentariaValladolid.pdf (accessed December 11, 2019)

Ayuntamiento de Zaragoza. 2019. Estrategia de Alimentación Sostenible y Saludable de Zaragoza. Ayuntamiento de Zaragoza. https://www.zaragoza.es/contenidos/medioambiente/alimentacion/EASS_Zaragoza_Mayo2019_definitivo.pdf (accessed December 11, 2019)

Blay-Palmer, A., G. Santini, M. Dubbeling, H. Renting, M. Taguchi, and T. Giordano. 2018. Validating the City Region Food System Approach: Enacting Inclusive, Transformational City Region Food Systems. *Sustainability* 10:1680. DOI: 10.3390/su10051680

Bowen, S. 2010. Embedding Local Places in Global Spaces: Geographical Indications as a Territorial Development Strategy. *Rural Sociology* 75, 2:209–243.

Bui, S., S. Bui, A. Cardona, C. Lamine and M. Cerf. 2016. Sustainability transitions: insights on processes of niche-regime interaction and regime reconfiguration in agri-food systems. *Journal of Rural Studies* 48:92–103.

Bowen, S., and K. De Master. 2011. New rural livelihoods or museums of production? Quality food initiatives in practice. *Journal of Rural Studies* 27:73–82.

Cadieux, K.V., S. Carpenter, A. Liebman, R. Blumberg and B. Upadhyay. 2019. Reparation Ecologies: Regimes of Repair in Populist Agroecology. *Annals of the American Association of Geographers* 109, 2:644–660. DOI: 10.1080/24694452.2018.1527680

Calle-Collado, A. and R. Vilaregut-Sáez. 2015. *Territorios en democracia. El municipalismo a debate.* Barcelona: Icaria.

Calori, A. and A. Magarini. 2015. *Food and the cities. Food policies for sustainable cities.* Milán: Edizioni Ambiente.

Darnhofer, I., 2014. Contributing to a Transition to Sustainability of Agri-Food Systems: Potentials and Pitfalls for Organic Farming. In *Organic farming, prototype for sustainable agricultures*, eds. S. Bellon and S. Penvern, 439–452. Dordrecht: Springer. DOI: 10.1007/978-94-007-7927-3_24

DeCunto, A., C. Tegoni, R. Sonnino, and C. Michel. 2017. *Food in cities: Study on innovation for a sustainable and healthy production, delivery and consumption of food in cities.* Brussels: European Commission/Eurocities/Cardiff University/Comune di Milano. Available at: https://ec.europa.eu/research/openvision/pdf/rise/food_in_cities.pdf

EIP-Agri. 2016. *Cities and food. Connecting producers and consumers. Final report. EIP-agri workshop.* Brussels: European Commission.

European Commission. 2018. Sustainable Use of Land and Nature-Based Solutions Partnership. Draft Action Plan. European Commission. <https://ec.europa.eu/futurium/en/system/files/ged/final_draft_action_plan_27-07-2018v3.pdf (Accessed July 14, 2019)

Ferguson, B.G., M. Aldasoro Maya, O.F. Giraldo, M. Mier y Terán Giménez Cacho, H. Morales, P. Rosset. 2019. Special issue editorial: What do we mean by agroecological scaling? *Agroecology and Sustainable Food Systems* 43, 7–8:722–723, DOI: 10.1080/21683565.2019.1630908

Food and Agriculture Organization of the United Nations (FAO). 2005. *Voluntary guidelines to support the progressive realization of the right to adequate food in the context of national food security.* Rome: FAO.

Friedmann, H. 2007. Scaling up: Bringing public institutions and food service corporations into the project for a local, sustainable food system in Ontario. *Agriculture and Human Values* 24:389–398.

Giraldo, O.F. and N. McCune. 2019. Can the state take agroecology to scale? Public policy experiences in agroecological territorialization from Latin American. *Agroecology and Sustainable Food Systems* DOI: 10.1080/21683565.2019.1585402

Giraldo, O.F. and P.M. Rosset. 2017. Agroecology as a territory in dispute: between institutionality and social movements. *Journal of Peasant Studies* 45, 3:545–564. DOI:10.1080/03066150.2017.1353496

González de Molina, M. 2013. Agroecology and Politics. How To Get Sustainability? About the Necessity for a Political Agroecology. *Agroecology and Sustainable Food Systems* 37, 1:45–59

González de Molina, M., D. López García and G.I. Guzmán. 2017. Politizando el consumo alimentario: estrategias para avanzar en la transición agroecológica. *Redes* 22, 2. DOI: 10.17058/redes.v22i2.9430

González de Molina, M., P.F. Petersen, F. Garrido Peña and F.R. Caporal. 2019. *Political agroecology: Advancing the transition to sustainable food systems.* Dordrecht: Springer.

Goodman, D., DuPuis and M.K. Goodman. 2012. *Alternative food networks: Knowledge, practices and politics.* London: Routledge.

Graeub, B.E., M.J. Chappell, H. Wittmand, S. Ledermann, R. Bezner Kerr and B. Gemmill-Herren. 2016. The State of Family Farms in the World. *World Development* 87:1–15. DOI: 10.1016/j.worlddev.2015.05.012

Guzmán Casado, G.I., D. López García, L. Román Bermejo and A.M. Alonso Mielgo. 2013. Participatory Action-Research in Agroecology: building organic food networks in Spain. *Journal of Sustainable Agriculture* 37, 1. DOI: 10.1080/10440046.2012.718997

Holt-Giménez, E. and A. Shattuck. 2011. Food crises, food regimes and food movements: rumblings of reform or tides of transformation? *Journal of Peasant Studies* 38, 1:109–144. DOI: 10.1080/03066150.2010.538578

Ilbery, B. and D. Maye. 2005. Alternative (shorter) food supply chains and specialist livestock products in the Scottish-English borders. *Environment and Planning A*, 37:823–844.

Infante-Amate, J. and M. González De Molina. 2013. Sustainable de-growth'in agriculture and food: an agro-ecological perspective on Spain's agri-food system (year 2000). *Journal of Cleaner Production* 38:27–35.

Jennings, S., J. Cottee, T. Curtis and S. Miller. 2015. Food in an urbanised world. The role of City Region Food Systems in Resilience and Sustainable Development. FAO & International Sustainability Unit. http://www.fao.org/fileadmin/templates/agphome/documents/horticulture/crfs/foodurbanized.pdf (accessed June 30, 2019)

Levidow, L., M. Pimbert and G. Vanloqueren. 2014. Agroecological Research: Conforming or Transforming the Dominant Agro-Food Regime? *Agroecology and Sustainable Food Systems* 38:1127–1155.

López-García, D., J.L. Fernández-Casadevante (*Kois*), N. Morán and E. Oteros-Rozas (eds.). 2017. *Arraigar las instituciones. Propuestas de políticas agroecológicas desde los movimientos sociales.* Madrid: Libros en Acción

López-García, D., L. Calvet-Mir, M. Di Masso and J. Espluga, 2018a. Multi-actor networks and innovation niches: university training for local Agroecological Dynamization. *Agriculture and Human Values* 36:567–579. DOI: 10.1007/s10460-018-9863-7

López García, D., B. Pontijas, M. González de Molina, M. Delgado, G.I. Guzmán-Casado and J. Infante-Amate. 2018b. Saltando de escala…¿ hacia dónde? El papel de los actores convencionales en los sistemas alimentarios alternativos. *Ager* 25:99–127.

López-García, D., N. Alonso and P.M. Herrera. 2018c. *Políticas alimentarias urbanas para la sostenibilidad. Análisis de experiencias en el estado español, en un contexto internacional.* Valladolid: Fundación Entretantos.

López-García, D., V. García-García, Y. Sampedro-Ortega, A. Pomar-León, G. Tendero-Acín, A. Sastre-Morató and A. Correro-Humanes. 2019. Exploring the Contradictions of Scaling: Action Plans for Agroecological Transition in Metropolitan Environments. *Agroecology and Sustainable Food Systems* 44, 4:467–489. DOI: 10.1080/21683565.2019.1649783

Marsden, T., J. Banks and G. Bristow. 2000. Food Supply Chain Approaches: Exploring their Role in Rural Development. *Sociologia Ruralis* 40, 4:424–438.

Marsden, T. and R. Sonnino. 2008: Rural development and the regional state: Denying multifunctional agriculture in the UK. *Journal of Rural Studies* 24:422–431.

McMichael, P. 2016. *Regímenes alimentarios y cuestiones agrarias.* Barcelona: Icaria.

Méndez, V.E., M. Caswell, S.R. Gliessman and R. Cohen. 2017. Integrating Agroecology and Participatory Action Research (PAR): Lessons from Central America. *Sustainability* 9, 5:705. DOI: 10.3390/su9050705

Mier y Terán Giménez Cacho, M., O.F Giraldo, M. Aldasoro, H. Morales, B.G. Ferguson, P. Rosset, A. Khadse, and C. Campos. 2018. Bringing agroecology to scale: key drivers and emblematic cases. *Agroecology and Sustainable Food Systems* 42,6:637–665. DOI: 10.1080/21683565.2018.1443313

Milan Urban Food Policy Pact, 2015. Milan Urban Food Policy Pact. http://www.foodpolicymilano.org/wp-content/uploads/2015/10/Milan-Urban-Food-Policy-Pact-_SPA.pdf (accessed august 11, 2020)

Moragues-Faus, A., K. Morgan, H. Moschitz, I. Neimane, H. Nilsson, M. Pinto, H. Rohracher, M. Thuswald, T. Tisenkopfs, and J. Halliday. 2013. Urban Food Strategies: the rough guide to sustainable food systems. http://www.foodlinkscommunity.net/fileadmin/documents_organicresearch/foodlinks/publications/Urban_food_strategies.pdf (accessed July 11, 2019)

Moragues-Faus, A., R. Sonnino, and T. Marsden. 2017. Exploring European food system vulnerabilities: Towards integrated food security governance. *Environmental Science & Policy* 75:184–215.

Moragues-Faus, A. and R. Sonnino. 2018. Re-assembling sustainable food cities: An exploration of translocal governance and its multiple agencies. *Urban studies* 56, 4:778–794.

Morán, N., J.L. Fernández Casadevante (Kois) and F. Prats. 2018. *Ciudades en movimiento. Avances y contradicciones de las políticas municipalistas ante las transiciones ecosociales.* Madrid: FUHEM

Nelson, E., I. Knezevic and K. Landman. 2013. The uneven geographies of community food initiatives in southwestern Ontario. *Local Environment* 18, 5:567–577. DOI: 10.1080/13549839.2013.788489

Observatorio Metropolitano. 2014. *La apuesta municipalista. La democracia empieza por lo cercano.* Madrid: Traficantes de Sueños.

Parmentier, S. 2014. *Scaling-up agroecological approaches: what, why and how?* Brusels: OXFAM.

Paül, V. 2007. Agricultural Marginality and Marginal Agriculture in Metropolitan Areas A proposal for systematisation based on some Spanish case studies. In *Issues on geographical marginality*, eds. G. Jones, W. Leimgruber and E. Nel, 38–46. Grahamstown: Rhodes University.

Petersen, P. 2015. Hidden treasures. Reconnecting Culture and Nature in Rural Development Dynamics. *Research in Rural Sociology and Development* 22:157–194.

Petersen, P. 2017. Avances y límites de una política de agroecológía frente a la hegemonía de la agroindustria: reflexiones a partir de Brasil. In *Arraigar las instituciones. Propuestas de políticas agroecológicas desde los movimientos sociales*, eds. D. López-García, J.L. Fernández-Casadevante (*Kois*), N. Morán and E. Oteros-Rozas, 292–302. Madrid: Libros en Acción.

Renting, H., R. van Veenhuizen, M. Dubbeling and F. Hoekstra (Eds.). 2017. European case studies on governance of territorial food systems – Project GouTer. International Urban Food Network. https://www.ruaf.org/sites/default/files/European%20case%20studies%20on%20governance%20of%20territorial%20food%20systems%20Gouter-RUAF%20final.pdf (accessed July 11, 2019)

Retamozo, M. 2017. La teoría política del populismo: usos y controversias en América Latina en la perspectiva posfundacional. *Revista de estudios latinoamericanos* 64:125–151

Rivera-Ferre M.G. 2018. The resignification process of Agroecology: Competing narratives from governments, civil society and intergovernmental organizations. *Agroecology and Sustainable Food Systems* 42(6): 666–685. DOI: 10.1080/21683565.2018.1437498

Rosset, P. and M.A. Altieri. 2017. *Agroecology: science and politics. Agrarian change & peasant studies.* Rugby: Practical Action Publishing.

Sabourin, E., M.M. Patrouilleau, J.F. Le Coq, L. Vásquez and P. Niederle (Orgs.). 2017. *Políticas públicas a favor de la agroecología en América Latina y El Caribe.* Porto Alegre: Red PP-AL/FAO.

Shiva, V. and M. Mies. 2014. *Ecofeminismo: teoría, crítica y perspectivas.* Barcelona: Icaria.

Simón-Rojo, M., I. Morales, and J. Sanz-Landaluze. 2018. Food Movements oscillating between autonomy and co-production of Public Policies in the City of Madrid. *Nature and Culture* 13, (1):47–68. DOI: 10.3167/nc.2018.130103

Simón-Rojo, M. 2019. Agroecology to fight food poverty in Madrid's deprived neighbourhoods. *Urban Design International* 24, 2:94–107.

Tornaghi, C. and M. Dehaene. 2019. The prefigurative power of urban political agroecology: rethinking the urbanisms of agroecological transitions for food system transformation. *Agroecology and Sustainable Food Systems* 44, 5:594–610. DOI: 10.1080/21683565.2019.1680593

Vaarst, M., A. Getz Escudero, M.J. Chappell, C. Brinkley, R. Nijbroek, N.A.M. Arraes, L. Andreasen, A. Gattinger, G. Fonseca De Almeida, D. Bossio and N. Halberg. 2017. Exploring the concept of agroecological food systems in a city-region context. *Agroecology and Sustainable Food Systems* 42, 6:686–711. DOI: 10.1080/21683565.2017.1365321

Van Dyck, B., A. Vankeerberghen, E. Massart, N. Maughan and M. Visser. 2018. Institutionalization of Participatory Food System Research: encouraging reflexivity and collective relational learning. *Agroecología* 13, 1:21–32

Ventura, F., G. Brunori, P. Milone, G. Berti. 2008. The Rural Web: A Synthesis. In *Unfolding webs. The dynamics of regional rural development*, eds. J.D. Van der Ploeg and T. Marsden, 149–174. Assen: Van Gorcum.

Zuluaga Sánchez, G.P., G. Catacora-Vargas and E. Siliprandi (Coords.). 2018. *Agroecología en femenino. Reflexiones a partir de nuestras experiencias*. La Paz: SOCLA-CLACSO.

10

Holistic Pedagogies for Social Change: Reflections from an Urban Agroecology Farmer Training

Ana Galvis Martinez,[1] Brooke Porter,[2] Paul Rogé,[3] Leah Atwood,[4], and Natalia Pinzón Jiménez[5]

[1] *Independent Consultant, Founder of www.cafepanamericano.org and www.holisticsustainabilities.org*

[2] *Graduate Student, MSc in Agroecology, Norwegian University of Life Sciences*

[3] *Faculty, Environmental Management & Technology, Merritt College*

[4] *Wild and Radish Ecovillage, LLC and Agroecology Commons*

[5] *Graduate Student, PhD in Geography, Department of Human Ecology, University of California, Davis*

CONTENTS

KEY WORDS: *agroecology*; *constructivist pedagogies*; *food sovereignty*; *humanistic education*; *popular education*; *urban agroecology*

Land Acknowledgement

This chapter was written on occupied, unceded Chochenyo and Karkin Ohlone land. We would like to offer our deepest gratitude and respect for the Indigenous people of this planet and this place, honoring that their stewardship from time immemorial has created the foundation for what we understand to be agroecological principles today.

10.1 Introduction

> Agroecology is actually a peasant Indigenous movement, it's a social movement for liberation, undoing all the harm that has been caused by industrial agriculture, the harm that white supremacy and patriarchy has caused with the pursuit of capitalism and the commodification of resources. In my role of being someone who identifies as an educator or even as a mentor, after BAFT I feel like I'm more prepared and better equipped to have those types of conversations with people and that I'm able to help to engage folks with those types of thinking, helping them become system thinkers themselves, this is how BAFT has been helpful to me.
>
> Samuel Madrigal, BAFT participant, November 12, 2019

Samuel Madrigal shared this reflection approximately one year after graduating from the Bay Area Farmer Training (BAFT), a program implemented from 2015–2019 that sought to meet the growing demand for agroecological training in urban settings of California.[1] In less than a century, global urban populations have rapidly expanded from 15% to 55% of the total (UN DESA 2018). A complex matrix of power dominates urban geographies, forming a landscape highlighted by its inequalities (Deelstra and Girardet 2000). In this context, agroecological education has an important role to play in scaling up, or massifying, the ability of urban people to meet their own basic needs for healthy food while simultaneously building community and defending territories.

Agroecology has long been practiced and protected by Indigenous and peasant farmers across the globe. Rooted in Indigenous traditions of reciprocity with the land (Kimmerer 2018), agroecology arose as a response to the Green Revolution, which promoted ecological destructive, chemical intensive, maximum yield breeding strategies, and monoculture specialization (Wezel et al. 2009). Agroecology is often defined as the "application of ecological principles to the study, design and management of agroecosystems that are both productive natural resource conserving, culturally sensitive, socially just and economically viable" (Miguel Altieri 1995; Gliessman 2015; Tornaghi 2017; A. B. Siegner, Acey, and Sowerwine 2020). While it is very much a science, it is also a practice and movement (Wezel et al. 2009). As a social movement, it has a strong ecological foundation and is backed by peasants, farmers, and activists seeking to ensure global food sovereignty (Steve Gliessman 2013; "Forum for Agroecology, Nyéléni" 2015).

In recognizing that the extractive industrial agriculture model doesn't serve people or the planet (IPES-Food 2016; Stephen Gliessman, social movements such as La Via Campesina,[2] the Landless Rural Workers Movement,[3] and the farmer-to-farmer movement have massified agroecology through popular education (Holt-Giménez 2006; McCune, Reardon, and Rosset 2014; Meek and Tarlau 2016. The horizontal nature of popular education and farmer-to-farmer exchange have helped facilitate the preservation and proliferation of agroecology in Indigenous and peasant communities around the world (Holt-Giménez 2006; Wilson 2011). Borrowing from these movements, urban agroecology education has

[1] BAFT was designed and implemented by staff at two nonprofits: the Multinational Exchange for Sustainable Agriculture (MESA) and Planting Justice.

[2] Founded in 1993, La Via Campesina is an international movement bringing together in solidarity small and medium sized farmers, landless people, rural women and youth, Indigenous communities, migrants and agricultural workers to defend a fight for agroecology, food sovereignty and gender equality around the globe (La Via Campesina 2020).

[3] The Landless Workers' Movement—"Movimento dos Trabalhadores Sem Terra" (MST)—is a Brazilian social movement which actively fights for agrarian reform by occupying unproductive lands, a constitutional right as outlined by Brazil's post dictatorship constitution of 1988 (MST 2020).

the capacity to stand as the protagonist in the transition to create resilient urban communities by encouraging food and farming models that center equity, cooperation, and solidarity.

Transforming how humans relate with each other and to the ecosystems of which they are a part of is a central challenge to urban agroecology education. As the world becomes increasingly urbanized, it is vital to maintain and reclaim land-based relationships and wisdom rooted in agroecological principles, which have the potential to serve as valuable tools to mitigate climate change, biodiversity loss, fresh water depletion, land and ocean degradation among other major global environmental problems. While agroecology has a strong focus on production, it also seeks to address a larger paradigm shift within food and farming systems through social equity, one in which many urban communities play a central role.

Transitioning towards agroecology within urban geographies is multifaceted and manifests within the ecological, political, economic, and social realms of society (Dehaene, Tornaghi, and Sage 2016; Tornaghi 2017; M Altieri and Nichols 2019). Providing consumers, particularly urban populations with direct supply networks not only decreases the geographical distance in which food travels—addressing its ecological footprint—but simultaneously builds relationships between producers and consumers, oftentimes strengthening urban and rural relations (Dumont et al. 2016). Following socioeconomic principles of agroecology, cooperative models present opportunities to strengthen urban communities by increasing agency, collaboration, and profit-sharing. Agroecology also has the potential to serve as a bridge between a wide array of social movements and platforms: ecofeminism, racial justice, LGBTQIA+, Indigenous sovereignty, agrarian reform, land reparations, and more. Public policies that increase urban farms have a wide range of benefits such as: interception of solar radiation, waste and nutrient recycling, increased soil fertility, filtration of atmospheric pollution, microclimate improvement and overall community wellness (Deelstra and Girardet 2000). Urban agriculture is a vital aspect of city infrastructure to promote health, peace and interdependence by creating places for residents to connect to food, nature, and each other (Reynolds and Cohen 2016).

This chapter discusses the challenges and opportunities for applying agroecology to the interwoven environmental and social issues of urban places. Urban agroecology education occurs in different contexts, within academic institutions, grassroots organizing of social movements, and non-profits and community-based organizations. In the case of BAFT, it emerged from the context of the non-profit sector in the United States. Its educators had experience in both social movements and traditional academic settings. BAFT took the shape of a community-based farmer training program focused on social justice.

In this context, BAFT provides insights into contra-hegemonic pedagogies with a focus on critical, constructivist, humanistic approaches emerging from non-academic spaces. This case study highlights some of the challenges in creating these types of learning environments. Many of the authors of this chapter formed the BAFT educator and program team. We weave together our own perspectives with interviews of former BAFT participants and program evaluations. The following analysis of BAFT illustrates one way to design and implement urban farmer training programs rooted in agroecology and supported by humanistic values and decolonial frameworks.

10.2 BAFT: A Case Study in Politicized Urban Agroecology Education

Through funding from the United States Department of Agriculture's (USDA) Beginning Farmer and Rancher Development Program (BFRDP), BAFT trained 122 aspiring urban farmers in agroecology and food sovereignty between 2015 to 2019. It was specifically crafted for underserved aspiring and beginning urban farmers with a focus on people of color, women, immigrants, formerly incarcerated people, and LGBTQIA+ individuals. The majority of those who participated in BAFT lived in urban places. Most were landless, facing severe challenges in accessing farmland and lacking secure housing. Systematic disparities have barred some of the communities participating in BAFT from accessing institutional and academic opportunities. BAFT attempted to offer a high-quality educational experience for an unconventional student demographic that faced ongoing challenges in entering the farming sector.

The BAFT program consisted of two main components: a three-month course and a follow-up mobilization phase. The BAFT course introduced agroecology and food justice theories and practices to 122 participants. Each BAFT course spanned three months, with eight hours of classes per week.

The curriculum used didactic tools such as field trips, participatory presentations, on-farm practice, anti-oppression training, project-based learning, online resources, and mentorship support to create an environment that celebrated different learning styles. Field trips included visits to farms, food preparation facilities, aquaponic systems and nurseries. The online course contained a learning network with multimedia lessons and readings that supported the in-person classes. Each BAFT course concluded with a celebratory graduation ceremony, where students reflected on their learning and presented their visions for future businesses, projects, and other endeavors.

In recognition of societal inequalities affecting many BAFT participants, the program was designed to reduce barriers to participation and meet some of their basic resource needs. To accommodate working students, the course took place in the evenings and on weekends. Eight hours per week of in-person meetings were supplemented with optional 3 hours per week of online course materials for the week's topic. The BAFT course was offered at a sliding scale rate with scholarships and participation stipends available for low-income applicants. Over 50% of graduates received a stipend between $350 and $800—in addition to a fee waiver—to support their participation. Participants put the stipends toward transportation, childcare, and/or meals. This greatly facilitated their involvement in the classes. Participants were allowed to bring their children to class, where they frequently received childcare support from both staff and fellow classmates. Laptop computers were also available on loan, which allowed some participants to engage with the online materials and prepare their applications for the mobilization phase.

Graduates of the course could continue in the BAFT mobilization phase, which provided guidance on the development of participants' farming, food business, and education projects. Out of the mobilization program formed projects such as the East Bay Farmers Collective, which was founded by a group of BAFT graduates seeking to cultivate agroecological produce and medicinal herbs (Paxton 2019). The collective focuses on distributing nourishing food and medicine to predominantly people of color, Indigenous communities, women, trans, and fem residents of the Bay Area.

The BAFT mobilization phase included mentorship for participants from specialists in their field of interest, on-farm apprenticeships, and mini-grants to support their mobilization projects. The application for all forms of support required a basic project proposal or business plan. The hours of mentorship during the incubation phase varied in length, depending on the needs of each project. The BAFT provided matchmaking services and $15/hour stipends for on-farm apprenticeships in the Bay Area and surrounding rural regions. BAFT graduates also applied for competitive mini-grants toward material costs of their projects. BAFT educators strove to foster a culture of transparency, inclusivity, and engagement through participatory budgeting and an emphasis on the formation of worker cooperative farms and projects.

At its core, BAFT sought to address the structural inequalities that shape the current hegemonic food system. The program provided tools to overcome imminent challenges that participants would likely encounter—difficulties in accessing land, financial and social capital, and technical support—with alternatives such as cooperativism, local markets, connections to locally available resources, relationship-based networks, and mentorship support. Rather than seeing these inequalities as personal shortcomings to be overcome, the program sought to understand the origins of these structural inequalities, which are produced by a society plagued by colonization, white supremacy, extractive capitalism, and patriarchy. This radical vision of agroecology from the perspective of social justice set BAFT apart from many other farmer training programs funded by the USDA BFRDP.

10.3 Guiding Pedagogies and Didactics

BAFT educators approached agroecology as a multidimensional means to achieve food sovereignty, and as a living concept that evolves as it is adapted to diverse contexts. Until recently in the United States, agroecology research, education, and practice has emphasized the natural science components to the detriment of a holistic understanding of sustainable food systems. However, a politically aware agroecology is common in many parts of the world. "Agroecologists recognize a wider sense of agricultural purpose that goes beyond mere production of commodities, and includes issues of environment, community, and justice. This wider understanding of the agricultural context requires the study of relations between agriculture, the global environment, and society" (David and Bell 2018). For this reason agroecologists must grapple with a structural analysis of inequality within the food and farming system. Toward this end, the BAFT course borrowed from an array of pedagogies and didactics. The focus was primarily on constructivist, critical, and humanizing pedagogies. We review each of these dimensions of the curriculum before describing the key curricular aspects of BAFT.

10.3.1 Critical Pedagogy

Critical pedagogy is a teaching philosophy that encourages participants to examine power structures and patterns of inequality within society (McGuire 2016). Through facilitating discussions around power structures and patterns, participants can critically evaluate opinions they may have inherited or absorbed, and feel a greater sense of agency in their own learning process (McGuire 2016). This is essential because agroecology must challenge the ideological system that protects the corporate food regime and it must take issue with the concentration of power and the unequal distribution of wealth that lie at the heart of the way the food system operates (Gliessman 2015; Chohan 2017). It is through this lens of questioning power structures that Brazilian philosopher Paulo Freire (2014) critiqued the Western "banking method" of education that treats students as empty vessels to be filled with the values of the dominant class. "[This] practice dehumanizes and disempowers students, whose culture, experience, language, and ideas are subjugated in order to indoctrinate the students with the ideology of those in power. It fails to teach critical thinking skills, and it doesn't teach the value of dialogue" (Mink 2019). BAFT actively sought to incorporate these philosophies by questioning and critiquing the ways the global industrial food system displaces small farmers, colonizes traditional diets, exploits labor, and monopolizes the market.

BAFT's critical pedagogy borrowed from traditions of popular education, which is a people-oriented and people-guided approach to education (Freire 2014). It encourages participatory activities and learning methods that value participants' life experiences resulting in the development of critical consciousness (Intergroup 2012; Freire 2014. This approach strives for horizontal relationships between teachers and students, rather than the more traditional, static, and vertical transfer of knowledge from teacher to student. By implementing popular education principles, BAFT educators sought to incorporate participants' recommendations and feedback on an ongoing basis. "Many political and educational plans have failed because their authors designed them according to their own personal views of reality, never once taking into account (except as mere objects of their actions) the men-in-a-situation to whom their program was ostensibly directed" (Freire 2014). The process of continuously integrating feedback allowed for the course to be collectively constructed and honor the diverse realities of the students.

Acknowledging that learning is not a purely individual process (McCune and Sánchez 2018), dialogue is a key component that assures that the production of knowledge is formed through a collective process within the classroom. BAFT emphasized the importance of horizontal facilitation that allowed for students to have agency in exploring and developing their critical voices. "We must continually remind students in the classroom that expression of different opinions and dissenting ideas affirms the intellectual process. We should forcefully explain that our role is not to teach them to think as we do but rather to teach them, by example, the importance of taking a stance that is rooted in rigorous engagement with the full range of ideas about a topic" (hooks 2014). The course's emphasis on dialogue encouraged cross-cultural understanding and movement building. "If it is in speaking their word that people, by naming the world, transform it, dialogue imposes itself as the way by which they achieve significance as human beings" (Freire 2014). Fostering dialogue resulted in a strong sense of interconnectedness and community, reinforcing a sense of shared territoriality that brought together a diverse group of people from the Bay Area around visions of agroecology and food sovereignty.

10.3.2 Humanizing Pedagogy

Humanization strengthens a person's capacity to recognize our commonality, rather than furthering divisions and othering based on distinct human identities and social constructs. BAFT emphasized the importance of humanizing classroom environments, as implemented in various social movements such as La Via Campesina. "Education as the practice of freedom—as opposed to education as the practice of domination—denies that man is abstract, isolated, independent, and unattached to the world; it also denies that the world exists as a reality apart from people" (Freire 2014). In pointing to this freedom, Paulo Freire outlines how humanization enhances our collective capacity to oppose isolation toward liberation. "One does not liberate people by alienating them. Authentic liberation—the process of humanization—is not another deposit to be made in men. Liberation is a praxis: the action and reflection of men and women upon their world in order to transform it" (Freire 2014). This speaks to the broader ecological realm of "humanization" that Paulo Freire emphasized in 1970, stating that at "the center of education is no longer only a transformation of the relations among people, but also between people and all other forms of life" (Meek and Lloro-Bidart 2017). The very act of farming the city has the potential to mend the alienation from nature felt by many urban people in modern capitalist societies (McClintock 2010). BAFT educators sought to address farmers' feelings of isolation within urban food justice spaces by reestablishing meaningful connections with each other and the land.

A humanistic pedagogy in BAFT was realized through a values-centered curriculum—discussed in greater detail in a section that follows—which invited discussions on systemic oppression and societal traumas connected to racism, sexism, and classism in the food system. In practice, this approach brought various challenges with multiple site visits and guest speakers from diverse backgrounds (Landzettel 2018). Some of these challenges were important learning opportunities for participants on how to engage people with different viewpoints and awareness of systemic oppression, or the lack thereof. These incidences also provided some important lessons for the educator and program team on how to better structure the class and engage with guest speakers.

For example, on a site visit to a farm owned and operated by an immigrant and farmer of color, the BAFT class was joined by another tour of farmers from the Midwestern United States. At the end of the tour the farmer shared his story as an immigrant starting out with limited financial resources to now owning his own thriving operation as a testament there was no systemic barrier preventing someone's success, and if you put your mind to it and worked hard, you could thrive, regardless of your background, financial status, or race. The educators noticed that many participants from the Midwest group were nodding, while many participants from BAFT were not expressing agreement. On the return trip, the class engaged in a thoughtful discussion about the "self-made and pull yourself up by your bootstraps" mentality. While at times inspiring or motivating, it can also invisibilize the struggle of people who have been discriminated against for their race, culture, gender, or sexual orientation. The class discussed how a person who has experienced discrimination can still internalize oppression by either adopting, normalizing, or ignoring a pervasive discriminatory mentality, sometimes as a means of assimilation or protection.

In another example, a speaker, in sharing his experience of working in seed saving with Indigenous communities, described the communities as "having no culture." Although the intention was to share about the loss of food culture and seed saving practices in these communities, by using this specific phrasing, the speaker participated in the erasure of the violent history of displacement and genocide experienced by Indigenous peoples, which is the reason why so many cultural practices have been lost. Deep historical awareness is vital in order to recognize the global impact and normalization of colonization and white supremacy.

In both of these cases, a humanistic approach supported the class to share feedback with the speakers about the impact of their rhetoric. It also supported lessons in compassionate engagement and restorative justice to see the speakers as human beings who have been conditioned to perpetuate these patterns over time but are open to adopting new behaviors and values, as opposed to recreating trauma through shaming and silencing. "Shaming is one of the deepest tools of imperialist, white supremacist, capitalist patriarchy because shame produces trauma and trauma often produces paralysis" (hooks 2013) These experiences helped clarify the need for restorative justice training for facilitators, as well as the need for deeper communication with guest speakers about the decolonial course framework, and to inform course participants in advance about the background and perspectives of guest speakers.

Regular community-building exercises were critical to fostering dialogue on challenging issues. First, each class began with a round of check-ins, allowing students to talk about what was alive for them and inviting them to bring their whole humanity to the classroom space. Although this sometimes took more time than expected, participants said this space for open sharing helped them feel valued and some expressed that it was the best part of their week. Educators also provided food during class, which later led to students preparing and sharing their own dishes accompanied by family recipes and stories. Finally, it was encouraged to substitute or re-frame highly academic and alienating rhetoric with more commonly used vocabulary based on lived experiences.

10.3.3 Constructivist Pedagogy

Constructivism posits that every individual constructs their own understanding and knowledge of the world based on their own unique experiences (Bada 2015). Constructivist pedagogies specifically recognize the learner's innate knowledge prior to entering spaces of learning. In embodying constructivist education in BAFT, educators placed great value in the rich social and biocultural knowledge that each individual brought to class, and they developed trust with the students through honest and open dialogue about their own backgrounds, lived experiences, and social positionality.

One of the co-lead educators of BAFT, Ana Galvis, presented her background as an immigrant and single mother, which allowed her to connect with many of the students who shared similar backgrounds. The challenges she faced to become an agroecology educator with two Masters degrees in the United States resonated with many participants and fostered trust in the class. In contrast, co-lead educator Paul Rogé consciously stepped out of certain roles in recognition of his social privileges as a cis-white male with a PhD in Agroecology so that Ana and BAFT participants could cultivate leadership in the classroom. His personal dedication to service manifested in simple actions—driving to field visits, providing technical assistance, and meeting with students outside of class time—and he was invaluable in presenting complex agroecology concepts in a very accessible way, all of which led to the formation of deep, meaningful, and lasting connections with BAFT students over time. The end result was that no one individual dominated the discourse, and both educators shared teaching responsibilities and class time conscientiously, knowing when to step in when their expertise was needed and when to step back to allow others to be heard.

10.4 Key Curriculum Concepts

BAFT overall addressed the key integrated approaches involving agroecological education outlined by David and Bell (2018): bringing agroecological practitioners and activists into the classroom as instructors and sources of knowledge; developing and expanding an active and experiential learning program; diversifying the origins of agroecology students and instructors, including diversity of gender, sexuality, cultural heritage, and national origin; and creating a sense of agroecology as a publicly-oriented endeavor with important policy implications. In the sections that follow, we discuss some of the key concepts embedded within the BAFT curriculum.

10.4.1 Values-Centered Curriculum

In designing courses, an approach that urban agricultural educators can use to engage with humanizing pedagogy is to create a value-centered curriculum such as the one developed for BAFT (Table 10.1).[4] From this list one can reflect on the ways in which different values can become unbalanced, especially within an individualistic and capitalist society. For example, care and compassion are contrasted with neglect and cruelty. Each value is then placed within the social and technical contexts of agroecology, and practical, experiential learning activities are identified that provide students with opportunities to actively embody those values. Eventually, this framework permeated the chosen topics within four broad categories: regenerative agriculture and agroecology, social movements, models of agroecological production, and business incubation (Table 10.2).

[4] Co-lead educator Ana Cecilia Galvis created the values-centered curriculum for BAFT.

TABLE 10.1

The Values-Centered Curriculum of the Bay Area Farmer Training Course[1]

Core Value Balanced	Core Value Unbalanced	Domain	Topics
Care and compassion	Neglect and cruelty	Social Theory	Social movements and their practices; Risk management planning; Business planning and enterprise budgeting
		Technical Theory	Food safety, post-harvest handling, and food distribution; Cover crops and soil-plant health; Irrigation and evapotranspiration
		Practical Activity	Building terraces and contour ditches with the "A" tool; Soil tillage and cultivation
Diversity	Hegemony	Social Theory	Diversifying income streams; Management of economic risks
		Technical Theory	Polycultures; Functional biodiversity to enhance fertility, control pests and diseases, and attract pollinators; sexual/asexual plant propagation
		Practical Activity	Transplanting and direct seeding; Vegetative propagation through cuttings and divisions
Harmony	Hatred	Social Theory	Food empires, regimes, and injustice
		Technical Theory	Ecological management of soil; Ecological management of pests, disease, and weeds
		Practical Activity	Bed design for ecological management of pests, diseases, and weeds
Fairness	Unfairness	Social Theory	Gender and agroecology; Traditional agriculture and Indigenous agroecological knowledge; Sustainability; Agriculture and nature
		Technical Theory	Economic thresholds of pest damage
		Practical Activity	Farm design based on the biointensive model
Autonomy	Dependency	Social Theory	Decolonization of diets and medicines; Cooperative Businesses and the Sharing Economy; Access to land; Marketing plans; Community Supported Agriculture and other direct marketing outlets; Access to land and capital through community support and local governance
		Technical Theory	How to prepare herbal medicine, make compost, harvest water, save seed, conserve food through pickling and preserves etc.; Analysis of market conditions; Building community with social media and events; Dynamic cash flow planning, bookkeeping, farm taxes, etc.
		Practical Activity	Compost production and use; Seed saving and selection; Soil evaluations
Integrity	Dishonesty	Social Theory	Kinds of product certification
		Technical Theory	Farm record keeping; Managing on-farm food safety risks; Assessments of sustainability and resilience
		Practical Activity	Business and market plan
Renewal and cycling	stagnancy or lack of flow	Social Theory	Animal health and well-being
		Technical Theory	Season extension; Small farm equipment; Aquaponics; rangeland management; raising small animals; Whole farm design and management; Crop planning software
		Practical Activity	Greenhouse propagation; Milking animals

[1] The values-centered curriculum for BAFT was created by co-lead educator Ana Cecilia Galvis.

TABLE 10.2

The Four Curriculum Categories of the Bay Area Farmer Training Course

Regenerative Agriculture and Agroecology	Social Movements
• Agroecology and Permaculture Ethics • Garden Design • Vegetable Production • Ecological Pest Management • Irrigation and Water Management	• Intersectionality and Social Movements • Decolonization of Diets • Gender and Agroecology • Seed Sovereignty
Models of Agroecological Production	Business Incubation
• Evaluation of Agroecosystems • No-Till Vegetable Production • Rooftop Gardening • Nursery Production • Organic Farming • Herbal Medicine and Food Preparation	• Product Certification • Business Planning and Marketing • Financial Planning and Fundraising • Democratic Workplaces • Business Incubation

10.4.2 The Relational over the Technical

In contrast to many agroecology training programs within the United States that offer curricula emphasizing the technical aspect of sustainable agriculture—a depolitized agroecology—BAFT intentionally balanced both the technical and social components, establishing an organic link between education as a training process for political action and practical skills.

One of the many learning objectives for the course was to create conscious political subjects while simultaneously building community within the classroom for solidarity, collaboration, and mobilization in urban farming movements. Acknowledging that many participants were already politically conscious, BAFT educators sought to encourage collective agency. The curriculum emphasized the importance of political education in recognition of the historic socio-political forces that have shaped contemporary food and farming systems.

The fragmented and reductionist ways of thinking that have been historically promoted by Western science has left out the social and cultural components of agriculture. It is vital to comprehend how relations, both human and ecological, function alongside the technical and agronomic aspects of agroecology. "Some scientists (and among them agroecologists) are proposing that a paradigm shift is needed, a transformation toward ways of knowing and doing that are contextual and relational and can address sensitively the complexity that is at the heart of living systems" (Ferdowsi 2013). In practice, BAFT sought to strengthen participants' ability to be systems thinkers and to see the many different factors that make up the whole in order to understand both the ecological and social components of the food system.

10.4.3 Decolonial Framework

Just as agroecology principles embrace biodiversity within ecosystems, there is also great value to be found in a diverse classroom. The wide range of diversity within the BAFT program allowed for deep conversations around identity, systemic oppression, privilege, and intersectionality. In teaching agroecology, it is essential to talk about race within the food system in the context of the United States, where racism and xenophobia are deeply woven into the threads of society. The food system has a long history of reinforcing violence against people of color, from the expropriation of Indigenous lands, to slavery, to the exploitation of farm laborers. One program participant stated the following when reflecting on their own identity and desire to participate in BAFT.

"I think when a lot of people think of Black peoples' relationship to the land, they think of folks that were enslaved and I knew there was more and I wanted to look for that, and I wanted to feel that rather than searching for it in a book. I knew that that information was inside me already and I wanted to wake it up" –Shelley Hawkins, BAFT participant

Participant experiences such as the one above influenced the course's decolonial framework that critically examined the colonial legacy behind food and farming in the United States. The objective of a decolonial classroom is to make power visible, to understand the ways in which settler mentalities have formed racial hierarchies, and to map the ways in which global politics affect the distribution of resources (Avalos 2019).

While there are many dimensions of colonization, including the power it holds over the production of knowledge, one of the key colonial forces within farming and food systems is the emphasis on the domination of the land and extractivist production as part of the larger colonial project. This concept is deeply linked to the notion that humans dominate nature and that nature exists to serve humanity. This perception was born out of the Renaissance period of the 16th century and advanced with the development of the reductionist sciences in the historical period known as the modern era (Moura 2015). Historically, this development was concurrent to the colonization of the Americas and many other parts of the world by European forces (Moura 2015). This objectification of the Earth laid the groundwork for contemporary industrial and extractive agriculture. The philosophy of control is deeply rooted in the concept of modernity fostered by a colonial-imperialist mentality of the West. Agroecology seeks to break away from these agricultural practices that attempt to control the environment, and in doing so implements practices that harmoniously grow food alongside natural ecologies (Miguel Altieri and Nicholls 2002).

In many urban places within the US, the colonial history of the food system has greatly played a role in who has access to nutritious and fresh food, hence the growing need and demand for a food justice movement (A. Siegner, Sowerwine, and Acey 2018). It is a movement that goes beyond consumers' individual food choices by addressing the systemic inequalities that bar certain communities from accessing nutritious food. "Food justice thus pursues a liberatory principle focusing on the right of historically disenfranchised communities to have healthy, culturally appropriate food, which is also justly and sustainably grown" (Sbicca 2012).

Food justice within urban communities is multifaceted, ranging from issues around environmental racism, access to land, and labor in terms how food is grown and processed. The Bay Area in particular has a long history of grassroots organizing around food justice. For example, West Oakland was the birthplace of the Black Panthers Free Breakfast for Children program, which sought to provide nutritious food for youth living in the highly industrialized neighborhood (Sbicca 2012). The area's long history of discriminatory redlining played a central role in the neighborhood's lack of grocery markets. The Black Panther Parties efforts to address issues around equity and access to fresh food can be seen in the emancipatory spirit of the neighborhood today, with entities such as the Mandela Grocery Cooperative serving as a thriving local community hub that is a worker-owned and Black-owned business (Figueroa and Alkon 2017). These historical and contemporary examples were woven into the BAFT course through group discussions and guest speakers.

The decolonial framework within BAFT spanned from a historical analysis of how colonization has affected land tenure in the United States, to practical tools to decolonize one's diet from production to consumption, all while challenging colonial ideologies and assumptions. One BAFT participant Samuel Madrigal reflects, "the course taught me to be thinking about the cultural significance of food, that it isn't just that it is representative of our culture but that it also holds our lineage." Within the classroom an array of conversations were sparked, ranging from the ways in which the forced enslavement of people to work the land has resulted in deep-seated historical trauma and internalized colonialism, and how these narratives must be centered when discussing farming and access to land, to how to reclaim traditional dietary practices and the implications of eating sugary and highly processed foods produced by industrial agriculture.

Agroecology in theory seeks to decolonize knowledge by reclaiming traditional Indigenous and peasant based agricultural knowledge, acknowledging that there is a multiplicity of epistemologies. In order to integrate humanizing pedagogies into a decolonial framework, it was essential to allow students the time to reflect on their own people's histories around land and land tenure in this country. This diverse group of students shared narratives that are often absent from popular dialogues around the history of land and farming. These conversations in turn created space to discuss food sovereignty and environmental justice, as well as to recognize how resistance movements—frequently led by people of color and impoverished communities—have long counteracted extractive capitalist models of farming.

BAFT not only sought to uphold a decolonial framework within the classroom, but also to decolonize federal resources from the USDA. BAFT diffused institutional power by reallocating funds and resources to underserved beginning farmers. Recognizing that the USDA has a long history of unlawfully barring the distribution of federal resources to farmers of color (Williams and Holt-Giménez 2017), BAFT directly allocated funds to beginning farmers from similar backgrounds.

10.4.4 Gender and Sexuality

When discussing agroecology, it is essential to critically look at gender to understand how patriarchy has influenced land tenure, labor, and resource distribution within the food system. According to a UN report, "Gender issues are incorporated into less than ten percent of development assistance in agriculture, and women farmers receive only five percent of agricultural extension services worldwide" (De Schutter 2010). Agroecological models strive for self-sufficiency so that farmers are not dependent on high inputs. In this regard, low-input agroecological practices may have the potential to benefit women and fems who frequently struggle to find access to capital, external inputs, and/or subsidies. Agroecological practices seek to address gender inequalities and help pave the way for resilient, regenerative, self-sufficient, and empowered farmers.

The following data highlights the disparities faced by female-identified farmers in the United States and female-headed households seeking to address food security. According to the USDA Agriculture Census of 2017 only 36 percent of all farms have a woman as the principal operator. Women farm operators as a whole receive 61 cents on the dollar made by men, resulting in one of the largest wage gaps of any industry (Kruzic and Hazard 2017). Women own only 2 percent of all titled land worldwide (Milgroom et al. 2015). Structural gender based oppression is also visible within households resulting in female-headed homes being 30.3 percent food insecure, in comparison to a mere 22.4 percent of male-headed homes (Kruzic and Hazard 2017). Within the course, BAFT participants were asked to reflect statistics such as these, the implications of patriarchy within agriculture, and the ways in which agroecology has the potential to dismantle these inequalities.

The BAFT course provided students with interactive opportunities to reflect on their own identity within the broader intersection of gender and sexual orientation in agroecology. Students were invited to express their gender through creating a visual or written art piece and then place it on a spectrum of masculine to feminine, exploring gender as both a spectrum and social construct, rather than a binary biological determination. By starting the dialogue from a personal reflection rather than an abstract theoretical approach, students were able to contextualize their own narratives in theoretical ideas.

Even more broadly, gender and sexuality were examined in relation to other social "isms" within a context of structural oppression. A module focused on intersectionality was used to explore the impact of patriarchy, white supremacy, neoliberalism, and colonialism and to promote dialogue and understanding between participants from diverse backgrounds. "A central tenet of modern feminist thought has been the assertion that 'all women are oppressed.' This assertion implies that women share a common lot, that factors like class, race, religion, sexual preference, etc. do not create a diversity of experience that determines the extent to which sexism will be an oppressive force in the lives of individual women" (hooks 2014). In an intersectionality exercise led by guest educator, Shakirah Simley, descriptive signs were placed around the classroom such as race, gender, class, sexual orientation, appearance, physical ability, mental ability, language, religion, citizenship, and academic education. Participants were then asked the various questions for reflection, including:

- "What posed the greatest challenge for you growing up?"
- "What has had the biggest influence on your life?"
- "What do you feel has given you the most privilege?"
- "What is something that you have or currently struggle with that other people may not know?"

After each question was read, participants would stand near the category they felt most impacted by. Everyone was invited to share about why they selected a category, often resulting in informative, eye-opening, and sometimes challenging dialogue.

These examples reflect a feminist pedagogy that not only recalls oppression but also resistance, privileges testimonies over written text, and thus forms a collective testimony capable of questioning and affirming identities (Korol 2007). BAFT's feminist pedagogy allowed students to explore the ways in which agroecology challenges hegemonic relationships of power and domination perpetuated by patriarchy, while simultaneously opening a space for dialogue on how to observe and dismantle patriarchal and heteronormative behaviors within students own lives.

10.4.5 Honoring Queer Identities within the Classroom

One of the many realms of diversity represented in the BAFT classroom was participants' gender identities. Both participants and educators were challenged to expand their perceptions of gender. As an educator, if you mistake a person's pronoun, it is important to acknowledge the mistake and apologize. A standard classroom introductory practice is for educators to share their preferred gender pronoun and to invite participants to share their preferred pronouns. This fosters inclusivity within the classroom and normalizes the importance of checking assumptions. Further, we suggest educators incorporate a module around gender and agriculture that expands on the concepts discussed in this chapter and defines important terms such as sex, gender, and sexual orientation. We invited a guest speaker who developed a module called "The Garden is Queer" which showed students how some plants can be "male" and "female" at the same time. The course then looked at ways in which heteronormative scientific botanical terms can be redefined to label plants as "pollen-producing" and "pollen-receiving" rather than "male" or "female", directly disrupting common institutional articulations of "normative" plant biology.

10.4.6 Spirituality and Mysticism

Inspired by global peasant movements such as La Via Campesina and MST, BAFT also greatly emphasized the spiritual component that comes with stewarding land, honoring the many different traditions of people across the globe and recognizing the somatic healing experiences that form when reconnecting with the Earth. Welcoming the spiritual aspects of agroecology and different cosmologies linked to land stewardship is another form of decolonizing the classroom. "Revolutionary theorist Frantz Fanon noted that colonization estranges the colonized from their own metaphysical worlds, their epistemologies, knowledges, and ways of being. Multiple forces of power (institutional, epistemological, religious) collude over time to produce this estrangement" (Avalos 2019).

Spiritual components brought into the classroom included "místicas" offered by guest speakers of the MST in Brazil, sharing circles, and other rituals. While BAFT had various mystic and/or spiritual components interwoven into its curriculum, the BAFT educators recognized the importance of expanding these elements of the curriculum, especially in the context of the Global North where much of Earth-based spiritual practices have been forcefully erased or banned as part of the colonial legacy. Urbanization and colonial legacies present in the United States have a long history of attempting to sever communities' connections to land. BAFT educators incorporated these elements into the curriculum, seeing them as important opportunities to honor a wide range of ancestral traditions and recenter the spiritual component of farming, especially with urban communities that have limited access to the land.

10.5 Beyond the Course

10.5.1 Mobilization Phase Design and Implementation

A significant part of BAFT is the mobilization phase that created pathways for participants to take the next step as urban farmers and agroecologists. This process started during the BAFT course with participants engaging in a visionary process of designing a business or project plan. This component of the course assisted in bridging the relationship between theory and practice to form praxis. Participants received mentorship and paid internships with other farmers and community entrepreneurs working within a similar realm, as well as seed grants to help participants get their land access or their food and farming business up and running.

While recognizing that the concept of "business incubation" is still functioning within the capitalist paradigm, educators tried to introduce concepts such as exchange, sharing, and cooperativism.

Leah Atwood, BAFT program co-director reflects, "Extractive capitalism creates inequality in the way that it is designed and unregulated. With the interconnected influences of settler colonialism, patriarchy, and white supremacy, there is no way unfettered capitalism can promote social equity. It's exactly the opposite." The mobilization phase allocates public funds via mini-grants, paid internships and mentorships to serve individuals from structurally oppressed communities to cultivate agroecological food and farming businesses. "The aim is to promote economic viability for farmer livelihoods, and to increase food sovereignty in the communities most impacted and most deserving," states Leah. She envisions the potential of public subsidies, not for cash crops, but to recognize and support farming as an essential public service, as a part of food security and social and ecological resilience. The BAFT program, through the mobilization phase, aspired to increase food sovereignty and to create networks of solidarity between consumers and producers toward cooperative economic models.

BAFT aimed to reduce barriers commonly present within on-the-job training opportunities, while simultaneously building resilient and equitable community food systems with experienced food and farming leaders fighting for agroecology and food sovereignty. Together with like-minded organizations and experienced farmers, BAFT directly connected individuals to build relationships and learn new skills and perspectives. By directing grant funds to support leaders and learners to cultivate community and provide paid skills training through mini-grants, paid internships, and paid one-on-one consultation with mentors of their choice, BAFT provided an alternative to the unsustainable model of unpaid internships.

A key part of BAFT was not only to provide education but to strengthen a movement within the Bay Area. "The more relevant measure of formative processes may not be in the quality of the thinkers they produce, but in the territoriality of the movement they reproduce" (McCune and Sánchez 2018). Reinforcing territoriality and movement building within the bioregion of California's Bay Area is an essential component of forming the local food sovereignty mobilization network.

10.5.2 Challenges and Future Directions for BAFT

While the concept of humanization was foundational is designing the BAFT pedagogy, trying to embody humanization in the classroom did not come without challenges. Due to the background of the BAFT educators and program designers, many of the pedagogical influences came from Latin America. When teaching agroecology in the US, it is important to incorporate the US legacy of land-based oppression and its historical and ongoing impact on people of color. Specifically, the centuries of genocide, mass enslavement, pillaging, internment, and exploitation has impacted Indigenous peoples, and people of the African, Asian, and Central and South America diasporas, in unique and explicit ways. By building awareness around racism in the food system and the historic and present-day impact upon diverse racial and cultural identities, educators can shed light on the trauma of white supremacy and examine what is needed for healing and collective liberation. For future directions of BAFT, it was recommended to have a more specific focus on racial justice and agroecology as it's own curriculum topic, as well as be embedded throughout the course. Educators were challenged to discuss race in a multiracial classroom, to discuss class in a space where a wide range of privileges were represented, and to discuss gender when there is a spectrum of gender identities. Historical traumas arise, white fragility is confronted, and accountability is demanded.

A challenge facing many farmer training programs is how to measure the impact of their programs. Depending on the source of funding, measures can be predominantly reductionistic, focusing on the number of people impacted, businesses started, and the number of people being served from socially disadvantaged backgrounds, etc. These gross standards of success can motivate organizations to increase participant counts without focusing on the depth of impact. California farmer-educator networks have discussed the need to create shared metrics that better reflect our educational programs and to recommend those changes to funding agencies. We encourage funders to expand their impact vocabulary when it comes to program evaluation, and, rather than create metrics from the top down, to give organizations autonomy to set and design their own course metrics. The goal is to create metrics that can help both the organization and their funders gauge the overall success of a program. Measuring early and frequently is key, as well as incorporating measurements that reflect a holistic and deep impact. The pressing challenge of improving metrics of success for farmer training programs mirrors the need to expand the ways in which we understand and measure ecological productivity on farms. The current standard is to measure on-farm productivity by quantitative yield or profit, rather than the quality of ecosystem services provided, of human physical-emotional health maintained and of socio-ecological sustainability promoted.

As BAFT created a larger and larger network of graduates who are working within the farming and food systems, a process of inviting graduates back to come teach different cohorts was established. These active teaching methods encourage previous graduates to reflect on their experience and continue to engage with the course content, formulating a collaborative process and deepening in their skillset. This methodology assists in the construction of knowledge originating from peers, who are able to contextualize knowledge for underserved beginning urban farmers.

10.6 Recommendations for Urban Agroecology Educators

10.6.1 The Need for Politicized Urban Agroecology Education

A core objective of BAFT was to prepare urban farmers as agroecologists to dismantle the structural inequalities that shape the current hegemonic food system. Programs such as these require educators who are highly competent at facilitating dialogue among diverse students and communicating social justice principles. The diversity that exists within urban areas presents an important opportunity to build solidarity between diverse communities. However, this requires a nuanced unpacking of the impact of systems of oppression. Fostering cross-cultural dialogue in agroecology training programs can help build bridges between the diversity of urban communities that are often historically and intentionally fragmented.

Politicized urban agroecology education—as presented in this chapter—provides opportunities to strengthen food sovereignty, contribute to cultural healing, and achieve personal transformation. Educators must consciously avoid indoctrinating students to accept or adapt to the conditions of the global food system. Agroecology is a life-honoring philosophy that is linked to a long history of resistance. Its transformative potential is rooted in shifting paradigms. Our recommendations follow for integrating social justice and politicalized agroecology into urban farmer training programs.

10.6.2 Educational Tactics to Support Politicized Urban Agroecology Education

Train educators: We recommend that core educators and facilitators receive training in restorative justice, anti-oppression, conflict engagement. This will allow the program team to maintain awareness of power dynamics in the classroom while also increasing their capacity to address with care and compassion the triggering of traumas that arise when discussing challenging issues. The quality of facilitation makes the difference between deepening divisions and causing harm versus deepening cross-cultural connection and a sense of healing.

Cultivate cross-cultural competency: Educators can cultivate cross-cultural dialogue in classrooms by avoiding alienating rhetoric or theoretical abstraction. One useful way to avoid alienation is by contextualizing discourse in lived experiences, and to remain receptive to student feedback. Examples include: student reflections on their own people's histories around land and land tenure; honoring queer identities students self-identifying their gender pro-nouns.

Build trust: Trust is built based on an awareness of social positionality and privileges, either by identification with the struggle and overcoming obstacles, or through solidarity and humble service. This is particularly important in programs aiming to serve structurally oppressed, overburdened, or multicultural communities. Representation by the communities served in the educator and leadership team is critical for building trust.

Humanize your class: Encourage your students to be their whole selves in class and share their own knowledge and stories. To do this, you must cultivate respectful and sincere relationships with students, truly listen to them, and value the wisdom they bring. A humanistic pedagogy allows for more authentic learning and exchange. Strategies for creating spaces for dialogue and community-building include:

personal check-ins, shared meals, artistic and creative self-expression, and spiritual connections with land and territory.

Prepare guest speakers: Challenges can arise when field visits and guest speakers have limited familiarity with social justice and anti-oppression frameworks. Prepare guests in advance by sharing overview documents of program goals, core values, and class agreements. Follow up with conversations to address any questions they may have.

Create a values-centered curriculum: Mapping the curriculum to values—as was done for BAFT (Table 10.1)—can help ground farming activities and classroom learning in relatable terms.

10.6.3 Programmatic Tactics to Support Politicized Urban Agroecology Education

Address participation barriers: Addressing barriers to participation and structural inequalities is imperative to long term success of these programs. Underserved beginning farmers are likely to confront challenges in access to land, resources, markets, and institutional support throughout their careers. Think deeply about how to meet as many of the basic needs of participants as possible in recognition of societal inequalities and barriers to access. For example, BAFT provided fee waivers and scholarships to students with financial need, hourly wages for on-farm apprenticeships, and guest speaker stipends. In addition, carefully track and measure the allocation of resources to understand their use and impact.

Prioritize organizational health: The organization needs to be financially healthy as these programs can require significant financial, mental, and often emotional resources. We recommend prioritizing emotional, mental, and physical wellness. In our experience, a weekly check in to acknowledge one another's personal lives and explore ways to cultivate health, both individually and collectively, was tremendously powerful. We recommend prioritizing paid time to engage in wellness practices, such as acupuncture, massages, potlucks, and hikes as a team. This can strengthen relationships and improve individual and collective wellness.

Cultivate representative staffing and leadership: Consciously make space for and hire educators and staff who reflect the racial, sexual and/or cultural diversity of the people your organization serves. If possible, train students or participants to transition into staffing, educator, and leadership roles. For example, BAFT hired two graduates of its program.

Integrate metrics: Create metrics that can help both the organization and your funders gauge the overall success of your program. Both your team and the people your organization serves need to understand the value of the metrics. Measuring early and frequently through various modalities can help collect comprehensive and quality data.

Adopt participatory planning for the future: The long-term impact and success of any farmer training program is contingent on the future opportunities that follow an educational program. Most farmer training programs for new or beginning farmers see a small percentage of graduates continue on farming career paths. When serving structurally oppressed communities, this percentage can be even smaller. For this reason, the BAFT mobilization phase was developed through a participatory process as a follow-up to the BAFT course and offered these kinds of next steps for graduates:

- **Mentorship:** Funding covered one-on-one mentorship and consultation support to work with a mentor of their choice. Mentors should be offered compensation, and can also donate their fees to the scholarship programs for participants.
- **Paid internships:** Funding for graduates continued experiential training and relationship development. Offering living-wage internships is critical for low-income graduates who often cannot afford to participate in unpaid volunteer internships.
- **Participatory mini-grant allocation:** Graduates were invited to lead the process of directing and allocated funds based on an individual's need. These strategies empower participants to decide how funds are pooled or distributed to achieve the greatest impact and best meet participants' needs.
- **Land access:** We recommend participants have the opportunity to actively steward land, either through an incubator site or in a mentorship capacity where they can receive guidance, resources, and build community with other urban farmers.

It is essential to consistently listen to the students, to value the deep inner-knowledge present in both the students and facilitators identities, privileges, and traumas. This allows for authentic learning, stemming from human humility. Love sincerely the students, truly see them as human beings with stories. When implementing educational programs, have the courage to implement the knowledge and techniques in their entirety. Be sure to pause and listen to the group, taking in the pulse and energy of the classroom, and remember to stay true to the original learning objectives of the course. Agroecology is linked to a long history of popular culture. It stands as a protagonist in the resistance against industrial agriculture that tries to erase those histories. Agroecology is a philosophy, a way of life, that honors life and should also uphold these values when the focus of educational spaces. As a transformative process, the praxis of agroecology aims for paradigm change—not only for the individuals within the classroom—but for the students' broader communities of which they are a part.

10.7 Conclusions

Agroecology education can serve as a driver for radical change, one that is rooted in systems transformation, behavioral change, and paradigm shifts for culture and society. As a concept, agroecology is often interpreted as a blending of agriculture and ecology. Indeed, much of the sustainability and organic farming discourse has focused on environmental conservation in agriculture, with social justice as an afterthought. However, a core ethos of agroecology is centered around humans as an inseparable part of nature. Therefore, without human rights and social justice, there is no ecological resilience. In an urban context, agroecology is a nexus for people from diverse backgrounds, including a confluence of communities who have been structurally oppressed.

The Bay Area Farmer Training sought to strengthen a community of agroecological practitioners who collectively have the tools to dismantle the extractive industrial agricultural model. It offered a unique pedagogy, weaving together popular education, critical, constructivist and humanizing pedagogies within decolonial and feminist frameworks. Such an approach to agroecology education, set it apart from many beginning farmer training programs in the United States. At its core, BAFT not only focused on agroecology as the foundation for regenerative farming practices, but actively reimagined what it looks like to live in urban communities. The program created a vibrant group of 122 participants who continue to actively care for and protect the territory in which they live. Many have gone on to create cooperatives, businesses, and projects that fight for food sovereignty and agroecology.

By centering the humanizing and critical pedagogies in urban agroecology education, we have the opportunity to build bridges and promote social healing to massify agroecology as a movement. In order to do this, an intersectional educational approach that actively builds solidarity is critical. This requires a process in which individuals see themselves as protagonists confronting environmental and social problems while also in service of something bigger than themselves. Urban agroecology education can empower, unite, and mobilize individuals to build resilient community food systems that treat the Earth and one another not as ecological or human resources to be exploited, but as an interconnected living ecosystem for which we are all responsible.

Acknowledgments

This chapter honors those who have been deeply dedicated to BAFT's work in creating a grassroots global and local food sovereignty network. Thank you to those that strive to live in reciprocity with the land and are at the forefront of the struggle to shift our global food system in order to protect agroecological farming practices. We recognize the students in BAFT for their camaraderie, wisdom, and guidance that contributed to a vibrant farmer training program.

We also recognize our many institutional partners and guest speakers, including The Multinational Exchange for Sustainable Agriculture (MESA), Planting Justice, Spiral Garden, Singing Frogs Farm, Soul Farm, Wild and Radish, Catalan Family Farm, Oakland Sol Collective, La Cocina, Farmlink, Dark Heart Nursery, Organic Seed Alliance, Gill Tract, Don Bugito, the Latin American Scientific Society of

Agroecology, BlueDog Consulting and many more supporters that kindly shared their time and knowledge with the program.

All authors contributed to the writing of this chapter. Leah Atwood, Natalia Pinzón Jiménez, and Gavin Raders co-authored the proposal to the USDA Beginning Farmer and Rancher Development program that designed and funded the BAFT concept and framework. Ana Galvis and Paul Rogé co-designed the BAFT curriculum. Ana Galvis served as the lead educator for most BAFT cohorts. Paul Rogé and Leah Atwood supported BAFT classes. Paul Rogé, Natalia Pinzón, and Leah Atwood designed an online learning platform and course to support BAFT's in-person learning. Leah Atwood served as a co-director for BAFT and became more directly involved as a facilitator in the last two BAFT cohorts. Leah Atwood and Paul Rogé led the design and implementation of the mobilization phase of the program. Brooke Porter conducted collaborative research with MESA in 2019 that included interviewing BAFT participants and educators a like for her MSc in agroecology.

BIBLIOGRAPHY

Altieri, M, and C Nichols. 2019. "Urban Agroecology: Designing Biodiverse, Productive and Resilient City Farms." *AgroSur*, no. 46: 49–60. https://doi.org/10.4206/agrosur.2018.v46n2-07.

Altieri, Miguel. 1995. *Agroecology: The Science of Sustainable Agriculture.* Second Edition. Boulder Colorado: Westview Press.

Altieri, Miguel, and Clara Nicholls. 2002. "Ecologically Based Pest Management: A Key Pathway to Achieving Agroecosystem Health." *Managing for Healthy Ecosystems.* https://www.researchgate.net/publication/328717813_Ecologically_based_pest_management_A_key_pathway_to_achieving_agroecosystem_health.

Avalos, Natalie. 2019. "Insurgent Pedagogies: Decolonization Is for All of Us." In *Teaching Resistance: Radicals, Revolutionaries, and Cultural Subversives in the Classroom*, edited by John Mink. Oakland, CA: PM Press. http://search.ebscohost.com/login.aspx?direct=true&scope=site&db=nlebk&db=nlabk&AN=2221367.

Bada, Steve Olusgun. 2015. "Constructivism Learning Theory: A Paradigm for Teaching and Learning." *Journal of Research & Method in Education*, 1, 5, no. 6: 66–70.

Chohan, Jaskiran Kaur. 2017. "Reclaiming the Food System: Agroecological Pedagogy and the IALA María Cano." *Alternautas 4 (2) 13–26.* http://www.alternautas.net/blog/2017/8/17/reclaiming-the-food-system-agroecological-pedagogy-and-the-iala-mara-cano

David, Christophe, and Michael M. Bell. 2018. "New Challenges for Education in Agroecology." *Agroecology and Sustainable Food Systems* 42 (6): 612–19. https://doi.org/10.1080/21683565.2018.1426670.

De Schutter, Olivier. 2010. "Promotion and Protection of All Human Rights, Civil, Political, Economic, Social and Cultural Rights, Including the Right to Development." New York City. https://www.un.org/development/desa/disabilities/promotion-and-protection-of-all-human-rights-civil-political-economic-social-and-cultural-rights-including-the-right-to-development.html.

Deelstra, Tjeerd, and Herbert Girardet. 2000. "Urban Agriculture and Sustainable Cities." In *Growing Cities, Growing Food: Urban Agriculture on the Policy Agenda*, edited by N Bakker, M Dubbeling, S Gundel, U Sabel-Koschela, and H de Zeeuw, 43–65. Feldafing: German Foundation for International Development.

Dehaene, Michiel, Chiara Tornaghi, and Colin Sage. 2016. "Mending the Metabolic Rift – Placing the 'Urban' in Urban Agriculture." In *Urban Agriculture Europe*, edited by Frank Lohrberg, Lilli Lička, and Axel Timpe, 174–77. Berlin: Jovis. https://www.jovis.de/en/books/product/urban-agriculture-europe.html.

Dumont, Antoinette M., Gaëtan Vanloqueren, Pierre M. Stassart, and Philippe V. Baret. 2016. "Clarifying the Socioeconomic Dimensions of Agroecology: Between Principles and Practices." *Agroecology and Sustainable Food Systems* 40 (1): 24–47. https://doi.org/10.1080/21683565.2015.1089967.

Ferdowsi, Mir A. 2013. "UNCTAD – United Nations Conference On Trade And Development." In *A Concise Encyclopedia of the United Nations*, edited by Helmut Volger. Brill | Nijhoff. https://doi.org/10.1163/ej.9789004180048.i-962.602.

Figueroa, Melieza, and Alison Hope Alkon. 2017. "Cooperative Social Practices, Self-Determination, and the Struggle for Food Justice in Oakland and Chicago." In *New Food Activism*, edited by Alison Hope Alkon and Julie Guthman. University of California Press.

"Forum for Agroecology, Nyéléni." 2015. Nyéléni Center, Sélingué, Mali. http://www.foodsovereignty.org/forum-agroecology-nyeleni-2015/.

Freire, Paulo. 2014. *Pedagogy of the Oppressed.* Translated by Myra Bergman Ramos. New York ; London : Bloomsbury, 2014.

Gliessman, Steve. 2015. *Agroecology: The Ecology of Sustainable Food Systems.* Boca Raton: London: New York: CRC Press/Taylor & Francis Group.

———. 2018. "Breaking Away from Industrial Food and Farming Systems: Seven Case Studies of Agroecological Transition." IPES-Food. www.ipes-food.org.

———. 2013. "Agroecology: Growing the Roots of Resistance." *Agroecology and Sustainable Food Systems* 37 (1): 19–31. https://doi.org/10.1080/10440046.2012.736927.

Holt-Giménez, Eric. 2006. *Campesino a Campesino: Voices from Latin America's Farmer to Farmer Movement for Sustainable Agriculture.* Oakland, Calif. : New York: Food First Books ; Distributed by Client Distribution Services (CDS).

hooks, bell. 2013. *Black Female Voices: Who Is Listening - A Public Dialogue between Bell Hooks + Melissa Harris-Perry.* https://www.youtube.com/watch?v=5OmgqXao1ng.

———. 2014. *Teaching to Transgress: Education as the Practice of Freedom.* New York : Routledge.

Intergroup. 2012. "Popular Education – Intergroup Resources." *Popular Education.* 2012. https://www.intergroupresources.com/popular-education/.

IPES-Food. 2016. "From Uniformity to Diversity: A Paradigm Shift from Industrial Agriculture to Diversified Agroecological Systems." Brussels. http://www.ipes-food.org/_img/upload/files/UniformityToDiversity_FULL.pdf.

Kimmerer, Robin. 2018. "Mishkos Kenomagwen, the Lessons of Grass: Restoring Reciprocity with the Good Green Earth." In *Traditional Ecological Knowledge: Learning from Indigenous Practices for Environmental Sustainability,* edited by Melissa K. Nelson and Daniel Shilling, 1st ed., 27–56. Cambridge University Press. https://doi.org/10.1017/9781108552998.

Korol, Claudia, ed. 2007. Hacia una pedagogía feminista: géneros y educación popular: Pañuelos en Rebeldía. 1a. ed. *Colección Cuadernos de educación popular.* Buenos Aires: Editorial El Colectivo.

Kruzic, Ahna, and Erik Hazard. 2017. "How Female Farmers Are Fighting Big Ag's Gender Injustice by Taking Control of Their Food Systems – Alternet.Org." *AlterNet.* May 10, 2017. https://www.alternet.org/2017/05/how-female-farmers-are-fighting-big-ags-gender-injustice-taking-control-their-food-systems/.

La Via Campesina. 2020. "Via Campesina - Globalizing Hope, Globalizing the Struggle !" 2020. https://viacampesina.org/en/.

Landzettel, Marianne. 2018. "Growing Food, Leadership, and Social Justice." *Oregon Tilth.* November 21, 2018. https://tilth.org/stories/growing-food-leadership-and-social-justice/.

McClintock, Nathan. 2010. "Why Farm the City? Theorizing Urban Agriculture through a Lens of Metabolic Rift." *Cambridge Journal of Regions, Economy and Society* 3 (2): 191–207. https://doi.org/10.1093/cjres/rsq005.

McCune, Nils, Juan Reardon, and Peter Rosset. 2014. "Agroecological Formación in Rural Social Movements." *Radical Teacher* 98 (February): 31–37. https://doi.org/10.5195/rt.2014.71.

McCune, Nils, and Marlen Sánchez. 2018. "Teaching the Territory: Agroecological Pedagogy and Popular Movements -." *Agriculture and Human Values.*

McGuire, Della. 2016. "Critical Pedagogy in the Classroom." September 1 2016. https://study.com/academy/lesson/critical-pedagogy-in-the-classroom.html.

Meek, David, and Teresa Lloro-Bidart. 2017. "Introduction: Synthesizing a Political Ecology of Education." *The Journal of Environmental Education* 48 (4): 213–25. https://doi.org/10.1080/00958964.2017.1340054.

Meek, David, and Rebecca Tarlau. 2016. "Critical Food Systems Education (CFSE): Educating for Food Sovereignty." *Agroecology and Sustainable Food Systems* 40 (3): 237–60. https://doi.org/10.1080/21683565.2015.1130764.

Milgroom, Jessica, Henkjan Laats, Janneke Bruil, Edith Van Walsum, Leonardo Van Den Berg, and Danielle Peterson. 2015. "Farming Matters - Women Forging Change with Agroecology | FAO." *ILEIA, Center for Learning on Sustainable Agriculture* 31 (4). http://www.fao.org/family-farming/detail/en/c/385789/.

Mink, John, ed. 2019. *Teaching Resistance: Radicals, Revolutionaries, and Cultural Subversives in the Classroom.* 1 online resource vols. Oakland, CA: PM Press. http://search.ebscohost.com/login.aspx?direct=true&scope=site&db=nlebk&db=nlabk&AN=2221367.

Moura, Abdalaziz de. 2015. *Uma filosofia da educação do campo que faz a diferença para o campo.* Vol. 1.

MST. 2020. "Nossa História - MST." *MST Movimento Dos Trabalhadores Rurais Sem Terra*. 2020. https://mst.org.br/.

Paxton, Maura. 2019. "Circle of Change: Food Sovereignty through Community Support." *Oregon Tilth (blog)*. November 14, 2019. https://tilth.org/stories/circle-of-change/.

Reynolds, Kristin, and Nevin Cohen. 2016. Beyond the Kale: Urban Agriculture and Social Justice Activism in New York City. *Geographies of Justice and Social Transformation 28*. Athens: The University of Georgia Press.

Sbicca, Joshua. 2012. "Growing Food Justice by Planting an Anti-Oppression Foundation: Opportunities and Obstacles for a Budding Social Movement." *Agriculture and Human Values*.

Siegner, Alana Bowen, Charisma Acey, and Jennifer Sowerwine. 2020. "Producing Urban Agroecology in the East Bay: From Soil Health to Community Empowerment." *Agroecology and Sustainable Food Systems* 44 (5): 566–93. https://doi.org/10.1080/21683565.2019.1690615.

Siegner, Alana, Jennifer Sowerwine, and Charisma Acey. 2018. "Does Urban Agriculture Improve Food Security? Examining the Nexus of Food Access and Distribution of Urban Produced Foods in the United States: A Systematic Review." *Sustainability* 10 (9): 2988. https://doi.org/10.3390/su10092988.

Tornaghi, Chiara. 2017. "Urban Agriculture in the Food-Disabling City: (Re)Defining Urban Food Justice, Reimagining a Politics of Empowerment." *Antipode* 49 (3): 781–801. https://doi.org/10.1111/anti.12291.

UN DESA. 2018. "68% of the World Population Projected to Live in Urban Areas by 2050, Says UN." United Nations Department of Economic and Social Affairs. May 16, 2018. https://www.un.org/development/desa/en/news/population/2018-revision-of-world-urbanization-prospects.html.

Wezel, A., S. Bellon, T. Doré, C. Francis, D. Vallod, and C. David. 2009. "Agroecology as a Science, a Movement and a Practice. A Review." *Agronomy for Sustainable Development* 29 (4): 503–15. https://doi.org/10.1051/agro/2009004.

Williams, Justine M, and Eric Holt-Giménez, eds. 2017. *Land Justice: Re-Imagining Land, Food, and the Commons in the United States*. Food First Books.

Wilson, J. 2011. "Irrepressibly toward Food Sovereignty." In *Food Movements Unite! Strategies to Transform Our Food Systems*, edited by Eric Holt-Gimenez, 71–92. Food First Books.

11

Growing Together: Participatory Approaches in Urban Agriculture Extension

Lucy O. Diekmann[1*] **and Marcia R. Ostrom**[2]
[1] *University of California Cooperative Extension*
[2] *Washington State University*
[*] *Corresponding author Email: Lodiekmann@ucanr.edu*

CONTENTS

KEY WORDS: *Community engagement, Extension, Land-Grant Universities, urban agriculture*

11.1 Introduction

In the United States, the Land-Grant University (LGU) Extension System originated to serve a primarily rural and agricultural population (Hayden-Smith and Surls 2014). In the intervening 100 years, however, Extension's programming has expanded in keeping with key demographic shifts and now includes suburban and urban audiences as well as a broad array of food system-related topics. In response to a renewed interest in urban farming and gardening, which picked up in the early 2000s, some Extension systems have developed programs and dedicated staff time specifically to urban agriculture (e.g., Diekmann et al. 2017; Surls et al. 2015). While urban agriculture Extension is relatively new terminology, it is important to recognize that urban agriculture itself is centuries old (Schupp and Sharp 2012) and Extension support for urban agriculture and community food systems also has deep roots (Smith 1949/2013). Urban agriculture is popular in part because it is perceived as a strategy for addressing complex urban issues by offering an interconnected set of health, community, environmental, and economic benefits to urban residents (e.g., Daftary-Steel, Herrera and Porter 2015; Lovell 2010). It also faces unique social and ecological challenges because of its urban setting, such as high land and water costs, urban soil contamination, and city zoning restrictions (Hendrickson and Porth 2012; Oberholtzer, Dimitri and Pressman 2014). Expanding Extension resources for urban agriculture has been identified as one strategy for supporting the benefits it offers while helping to address the challenges entailed (Reynolds 2011). However, to support urban agriculture, Extension cannot only develop information tailored to the unique circumstances of this group. To be an effective actor in the urban food system, Extension must also adapt its approach to community engagement to recognize the knowledge and strengths that exist within urban communities, engage in reciprocal partnerships and participatory processes, and be open to engaging with the full suite of social, cultural, political, and economic issues that that both challenge and motivate many urban agriculture initiatives.

Urban agroecology provides a useful framework for conceptualizing a more participatory form of Extension. First, agroecology values and incorporates farmer-generated and other forms of local knowledge (Méndez et al. 2013). When Extension professionals engage in urban agriculture programming, they may sometimes create or disseminate new information, but they can also learn from farmers, gardeners, and other community members and conduct inquiry in partnership with them to co-create urban agroecological knowledge. Second, Getz and Warner (2006) observe that systems-oriented agroecological knowledge is not likely to be adopted through one-way delivery of information from Extension to farmers. Instead, they suggest that an Extension model that involves co-learning and facilitation is more effective for sharing agroecological knowledge. Finally, urban agroecology speaks to an important role for Extension and public universities, as the very topics that private food and agriculture research do not address—nutrition, environment, and social and community development (Anderson 2019)—are the topics most related to the mission-driven work done by many urban agriculture organizations and embodied by an urban agroecological framework (Siegner et al. 2019).

This chapter uses a survey of Extension practitioners and case studies of Extension urban agriculture programming to provide an example of the role that publicly-supported LGUs can play in the creation and dissemination of urban agroecological knowledge when engaged at the community level. This chapter proceeds in three parts. First, we first introduce the US LGU Extension System and discuss how urban agriculture fits within it. Second, we outline key characteristics of urban agriculture extension. Finally, we illustrate how urban agriculture extension is being put into practice through case studies of education and research carried out by Extension in Indiana, Pennsylvania, Michigan, Washington, and California, U.S.

11.2 Overview of Extension and Its Involvement in Urban Agriculture

A series of legislative acts between 1862 and 1914 established American Land-Grant Universities and created their tripartite system of teaching, research, and extension (Buttel 2005; National Research Council 1995). First, the *Morrill Act* of 1862 granted federal lands to each state to fund the establishment of a college that would teach practical subjects, including agriculture, and expand access to higher education (Hayden-Smith and Surls 2014).[1] In many southern states, African Americans were not able to attend land-grant universities until the passage of the second *Morrill Act* of 1890, which conditioned federal funding for land-grant universities on states *either* prohibiting racial discrimination in admissions *or* establishing a separate land-grant institution for Black students (Lee Jr. and Keys 2013; National Research Council 1995). With the passage of the *Hatch Act* of 1887, the federal government subsequently expanded the research function of LGUs by providing funding for agricultural experiment stations that emphasized locally adapted, applied agricultural research (Buttel 2005). The final addition to this three-part system, Cooperative Extension was envisioned as a cooperative funding effort by county, state, and

[1] Stein (2017) and Nash (2019) argue that it is important to place the federal land grants that lay the foundation for the LGU system in the broader context of Indian dispossession. They note that although the federal government did not acquire Indian lands for the purpose of creating LGUs, the Morrill Act was possible only because of Indian lands purchased or taken by the federal government.

federal government and was intended to extend land-grant universities' reach beyond their campuses and research stations, bringing education and applied research into rural communities and homes. President Woodrow Wilson, who signed the *Smith-Lever Act* of 1914 into law, called it "one of the most significant and far-reaching measures for the education of adults ever adopted by government" (Wilson quoted in Hayden-Smith and Surls 2014).

As the statewide arm of LGUs, Extension functions as a bridge between the university and the community, creating "a unique space that connects academic knowledge with practical purposes" (Collins 2015, 48). Extension does this work by engaging communities in learning and research activities, creating access to existing research-based information, developing local leadership, and facilitating collaborative applied research and the application of research (Henning et al. 2014; Smith 1949/2013). All of these activities are intended to help community members conduct their own analyses of their situations, solve local problems and to share knowledge so that people can improve their work, homes, lives, and communities (Collins 2015; Peters 2014). Extension employs a variety of formal and informal educational approaches to inspire learning and create access for a broad audience (Henning et al. 2014). For instance, as an early form of extension education, the Tuskegee Institute Moveable School brought an agricultural agent, a home demonstration agent, and a registered nurse to people throughout Alabama to conduct on-site demonstrations in partnership with farm families (Mayberry 1989; Zabawa 2008). Other early examples include locally-led reading circles for farm women and men, correspondence courses, farm and home institutes, short courses and conferences, agricultural fairs, nature study programs, and youth clubs (Smith 1949/2013). Although constantly evolving, many of these educational approaches continue into the present. As part of its model for adapting programs to local needs and building community relationships, Extension has staff located in nearly all of America's roughly 3000 counties (Clark et al. 2016).

Each state administers its own Extension system, so the terminology used to describe Extension personnel varies from state to state (e.g., specialist, faculty, professor, agent, advisor, educator). Despite differences in terminology, positions within Extension are structured similarly across states: the largest two groups of employees are Extension specialists with a statewide appointment and Extension agents or educators with a county or multi-county appointment. Extension specialists typically have disciplinary expertise and doctoral degrees and are located on university campuses or at research stations. At the county or regional level, Extension professionals work directly with communities and key stakeholder groups, where they address locally identified issues through research and education. Some county-level positions are responsible for conducting research and education, while others focus primarily on implementing educational programs. County level assignments may include 4-H youth-focused organizations for youth development, agriculture, family and consumer sciences, health and nutrition, community development, water and natural resources, forestry, and gardening (ECOP 2019). As of 2010, roughly 30% of Extension positions were specialists and 60% were county extension agents (Wang 2014). By working together, county- and campus-based Extension personnel strive to integrate research and academic expertise with community expertise and priorities.

While Extension's mission has remained more or less the same over the institution's first 100 years, the context in which it operates has changed profoundly. For instance, the agricultural knowledge landscape has shifted. Contemporary farmers and food system stakeholders have access to many sources of information and Extension is just one node in agricultural knowledge networks (Lubell et al. 2014; National Research Council 1995). The country has also experienced a large demographic transition. In the early twentieth century when the Extension system was established, about half of the US population lived in rural areas and about 30% of the workforce was involved in farming (NIFA n.d.). One hundred years later, about 85% of the population is urban and less than 2% of workers are engaged in farming (NIFA n.d.). To stay relevant to the American public, Extension must engage with socioeconomically and racially diverse rural, urban and suburban audiences, while navigating a crowded field of other service providers. Some recent studies suggest that supporting efforts to develop healthy, just, and sustainable community-based food and farming systems could offer a strong foundation upon which to build more inclusive and relevant urban Extension programming (Henning et al. 2014; Oberholtzer et al. 2014; Reynolds 2011; Smith et al. 2019; Surls et al. 2015; Ventura and Bailkey 2017).

Over time, urban agriculture in American cities has ebbed and flowed, often emerging in response to periods of social or economic crisis (Basset 1981). Examples include Detroit Mayor's Hazen S. Pingree's potato patches in the 1890s and the contemporaneous Vacant Lots Cultivation Program in Philadelphia (Smithsonian Gardens 2020), Liberty Gardens during World War I, and World War II's Victory Gardens. Seen as an important way to cultivate a scientific spirit and a sympathy with nature among urban and rural grade school students, school gardens were a focus of Extension work in the early 1900s (Smith 1949/2013). At that time, school gardens and schoolyard education programs in cities and in the country were supported by Extension as an important way of teaching observational skills and engaging the public with nature (Smith 1949/2013). From the early days of Extension, improving food production and preservation methods were considered critical to improving home life (Mayberry 1989; Smith 1949/2013). In the 1970s, the number of community garden programs expanded rapidly, driven by growing environmental awareness and by an economic downturn (Bassett 1981). It was during this time that Extension agents in Washington State created the Master Gardener program, designed specifically to offer horticultural education to urban audiences by training local volunteers. The Master Gardener program has since spread throughout the US and Canada (Gibby et al. 2008). Extension agents also participated in the USDA-funded Urban Gardening Program, teaching gardening, providing nutrition assistance, and engaging in 4-H type work for low-income families in more than 20 cities between 1976 and 1994 (Reynolds 2011).

Despite a long history of offering successful gardening and food preservation education through county-based extension offices, other forms of urban agriculture have sometimes fallen into institutional cracks within Extension systems. Due to decreases in the proportion of local, state, and federal funding available for Extension and the subsequent increasing reliance on grants and contracts (ECOP 2019; Wang 2014), many County Extension offices are no longer fully staffed with home food and nutrition experts or agricultural professionals. Work that does not fit comfortably into the scope of either the Master Gardener program, staffed primarily by volunteers, or the work of traditional agricultural educators may be difficult to sustain (Diekmann et al. 2017; Reynolds 2011). Further, some Master Gardener programs may be more focused on ornamental horticulture than on home food production, depending on the interests of local volunteers. In many regions, agricultural extension has become narrowly focused on the production of the most commercially valuable crops rather than on diversified cropping systems intended for local consumption, distribution, and marketing. A study of University of California Cooperative Extension found that Extension farm advisors did not differentiate their target clientele based on location (urban vs. nonurban sites of production), but on whether they were commercial or noncommercial operations, referring "noncommercial operators to the [Master Gardener Program] for assistance" (Reynolds 2011, 15). Yet, as mission-driven organizations, many urban agriculture operations do not fit neatly into these categories—being neither fully commercial nor purely home horticulture (Reynolds 2011). Even fully commercial urban agriculture operations may require assistance with non-traditional cropping systems, new financing and business models, and alternative marketing systems and therefore not be well suited for existing areas of Extension expertise and responsibility. A new set of Extension positions and programs specifically addressing urban agriculture are attempting to remedy this situation and respond to the needs of urban growers, urban consumers, and urban food and agriculture organizations and movements. Extension's engagement with local, regional, and community food systems which encompasses urban agriculture—through a variety of food and agriculture-based projects appears to be a new and evolving area of Extension practice (Clark et al. 2016).

11.3 Context of Urban Agriculture Extension

In the past decade, Extension personnel and other researchers have conducted assessments of urban farmers' needs and Extension's involvement with urban agriculture to help identify key themes for Extension's engagement with urban food systems.[2] On the one hand, these assessments have concluded that Extension has the capacity to address many urban agriculture practitioners' informational and technical needs, which are often similar to those of other small, diversified farms for production, business planning, and marketing (Oberholtzer, Dimitri and Pressman 2014; Reynolds 2011). On the other hand, they highlight

[2] Needs assessments are a standard part of program planning within Extension. Extension personnel use the results of needs assessments to set priorities and guide their extension and research efforts.

issues that are unique to urban agriculture because of either the project's setting or purpose, and may necessitate new kinds of research and educational programs. Because of the urban context in which they operate, for instance, urban farmers may need help addressing issues such as city zoning, access to water and water management, and urban soil quality (Brown and Carter 2003; Oberholtzer et al. 2014; Reynolds 2011; Surls et al. 2015). Because of their missions and organizational structure, funding and developing sustainable funding streams are also a challenge (e.g., Hendrickson and Porth 2012).

Urban farms and gardens are not just part of a city's natural environment through the use and management of land, water, and other resources; they are also part of its social fabric. Many urban agriculture operations and organizations have clear social goals that extend beyond food production to include increasing access to healthy food and culturally relevant foods, food systems education, providing employment opportunities, building community power, and community revitalization (Daftary-Steel et al. 2015; Ostrom and Donovan 2018; Smith and Ostrom 2019; Ventura and Bailkey 2017; Vitiello and Wolf-Powers 2014). In addition, urban farms differ from their rural counterparts because they are surrounded by non-farming land uses and neighbors and they often intentionally offer multiple services to their neighborhoods in addition to food production (Opitz et al. 2016; Poulsen, Spiker, and Winch 2014). Consequently, as both researchers and practitioners have observed, "urban farming projects are more likely to survive and thrive if they have local support" (Poulsen, Spiker, and Winch 2014, 1). Given its community-oriented goals as well as the importance of community buy-in for the viability of urban agriculture, it is critical for Extension's urban agriculture research and education to address the social, cultural, and political aspects of urban agriculture (Reynolds 2011; Surls et al. 2015).

Many urban communities "seek to create sustainable and just food systems that provide healthy food" (Ventura and Bailkey 2017, 6). In some cities, like Detroit, social justice is the central focus of community food system work (Ventura and Bailkey 2017). Often urban agriculture groups engaged in these efforts aim specifically to address social inequities manifested in the food system and the urban landscape (Reynolds 2011; Surls et al. 2015). Some of the disparities within the urban food system include food insecurity, lack of access to healthy foods, and low wages for food system workers, all of which disproportionately affect people of color (Giancatarino and Noor 2014). Urban agriculture is one potential strategy for improving these conditions, by improving access to healthy foods, saving families money on food, supplementing incomes, providing job training, and building community connections and engagement (Hagey et al. 2012; Horst et al. 2017; Vitiello and Wolf-Powers 2014). Yet urban agriculture, and other alternative food initiatives, have also been implicated in the perpetuation of social inequities "by benefiting already privileged communities, contributing to the ongoing marginalization and even displacement of disadvantaged groups" (Horst et al. 2017, 278) and because of "disparities in representation, leadership and funding" within urban agriculture organizations themselves (Horst et al. 2017, 283; Reynolds and Cohen 2016). Given the structural inequities in the food system, the LGU system itself (Lee Jr. and Keys 2013), and the explicit food justice and food sovereignty orientation of some urban agriculture organizations, it is important for Extension personnel to have an understanding of racism, implicit bias, and the history of land and resource ownership, as well as the ability to assess and/or teach about existing disparities and their causes.[3]

Finally, community engagement is at the heart of Extension's work because it is essential for ensuring that the resources Extension professionals bring to the table are relevant to and useful for their collaborators and the communities they serve. Urban food and agriculture present an opportunity for Extension to engage communities through reciprocal partnerships that respect the strengths of the university, local organizations, and community members and allow for joint definition of "problems, solutions, and success" (Kellogg Commission 1999, 29; Peters 2002). As University of Wisconsin faculty and Extension personnel observed through a multi-year, multi-city project on community food system initiatives to address urban food insecurity, their community partners "did not want to be studied" (Ventura and Bailkey 2017, 3). Instead, they expected an equal role in project development, based on discussions with

[3] Iowa State University's Core Competencies and Curriculum Project—conducted under a cooperative agreement with the USDA Agricultural Marketing Service–has identified equity as one of the core competencies for extension professionals working on local food systems. Although still under development at the time of this writing, this project identifies nine categories, including equity and community capacity, which each contain core competencies and learning objectives.

community leaders about which issues were important and in what ways the research partners could contribute (Ventura and Bailkey 2017). This type of Extension-community partnership benefits from an asset-based approach to assessing communities, rather than an approach to understanding community issues and potential solutions that is deficit-centered. It also values local knowledge, avoiding Extension "educators' exclusive claim to knowledge" (Colasanti et al. 2009, 6). These types of collaborative or participatory action research and education approaches, in which researchers and stakeholders collaborate throughout the research process to generate information that can be the basis for taking action, is a useful tool for urban agroecological work (Campbell, Carlisle-Cummins, and Feenstra 2013; Méndez et al. 2013; Surls et al. 2015). Another role that has been suggested for Extension in local food systems work is that of a network weaver or coordinator, facilitating information exchange within agricultural networks and creating networks of actors across the food system as a basis for collective action (Clark et al. 2016; Drake and Lawson 2015; Dunning et al. 2012, 104; Oberholtzer et al. 2014, Reynolds 2011). As outlined by Raison (2010), Extension personnel working on urban agriculture could ideally act both as educators, providing access to research-based information, and as facilitators, who engage in collaborative approaches to solving community-identified problems by acting as resource coordinators and network builders. Described as educational organizing (Peters 2002), a local-leader model (Smith 1949/2013), or networked learning (Lubell et al. 2014), such approaches, while not always recognized or celebrated, have long been a part of the fabric of community-based Extension work (Ostrom 2019).

11.4 Urban Agriculture Extension in Practice

To better understand what urban agriculture extension looks like in practice, the authors, along with other members of the Community, Local & Regional Food Systems Community of Practice (CLRFS), surveyed 147 Extension professionals about their involvement in urban agriculture. Survey respondents represented the full spectrum of roles in the Extension system, including administrators and specialists/faculty, but most were Extension educators and agents (50% of respondents). Survey respondents came from 33 states, with the greatest portion of respondents in the Midwest and Southeast. The survey, conducted in 2015, is representative of the CLRFS membership, which received the survey, rather than the Cooperative Extension System as a whole. (A more detailed explanation of the survey methods can be found in Diekmann et al. 2016.) The survey data provides an overview of Extension's urban agriculture activities and clientele, and responses to open-ended questions suggest types of Extension urban agriculture programs to explore in greater depth. In addition, attendance at the National Urban Extension Conference in 2019 provided a snapshot of the current state of urban agriculture extension programming. We have expanded on five examples of Extension urban agriculture practice through conversations with Extension staff. It is important to note that the urban agriculture extension examples described in this chapter are not exhaustive. Instead, it is our intent to illustrate some current programs and practices, recognizing that there is much innovative urban agriculture research and extension work not captured here.

11.4.1 Education

Education forms the basis of Extension work (Adams et al. 2005). When asked to describe their recent urban agriculture work, respondents to the urban agriculture survey most frequently described educational activities (Diekmann et al. 2016). Their responses show that Extension's urban agriculture educational programming aims to deliver a wide range of content to a diverse audience. Some of the examples of educational programming provided include:

- teaching K–12 students about gardening and nutrition at school gardens;
- teaching local residents about bees, backyard poultry, and small livestock production;
- cooking demonstrations for shoppers at farmers markets and mobile produce markets;

- classes for community gardeners on composting, record-keeping in the garden, and garden donation programs;
- teaching urban farmers and gardeners about production, marketing, and business planning;
- food safety training for urban farmers, gardeners, and school teachers and youth gardeners; and
- educating planners and city managers about urban agriculture and healthy food access.

Additionally, survey respondents reported organizing urban agriculture and other local-food-related educational events for a broad audience. These included an urban food systems symposium, local food summits, urban agriculture conferences, farmers' market conferences, and urban farm tours. These types of events are valuable ways for reaching a diverse set of urban stakeholders and building connections among them for the exchange of agroecological knowledge and other resources.

11.4.1.1 Purdue Extension – Urban Agriculture Certificate

Urban farmer or urban agriculture training was one of the educational programming themes. The survey identified several urban farmer training programs run by Extension as of 2015.

One example of this type of programming is the Urban Agriculture Certificate offered by Purdue University Extension. This particular program began in 2016 and is currently offered in two cities: Indianapolis and Fort Wayne, Indiana (Kilbane 2017). The course is tailored to several common features of urban agriculture projects that arise because of their placement within the urban landscape, in particular agriculture that 1) occurs on a very small scale (often one acre or less), 2) is adjacent to many non-farming neighbors or land-uses (Opitz et al. 2016), and 3) relies on partnerships in the community (Poulsen et al. 2014). The intended audience for the Purdue Extension course consists of urban farmers, community garden organizers, school garden leaders, market gardeners, urban homesteaders, and other urban agriculture project leaders.

The yearlong program combines classwork in the winter with monthly field trips to established urban agriculture sites during the summer. The course uses a "flipped classroom model," an educational approach that introduces students to new material before class, usually through videos or readings. In the certificate program, students watch online videos developed by Purdue Extension, complete homework assignments, and have access to a library of urban agriculture resources. Class time is then devoted to a variety of interactive activities in which students apply and assimilate new information (Brame 2013; OMERAD n.d.). These activities include workshopping homework assignments (e.g., peer review of participants' farm/garden mission statements); hands-on activities (e.g., reading fertilizer labels, transplanting); and demonstrations (e.g., pruning). As the culmination of the program's lessons and assignments, students create their own comprehensive farm plan, which can help them to move forward with their plans, seek funding, and more.

The primary goal of the urban agriculture certificate program is to increase participants' knowledge of key aspects of urban farming and gardening. The curriculum focuses on four main topics: 1) building strong roots––which covers creating a mission and vision statement, building and managing teams, and communication; 2) site assessment; 3) organic crop production; and 4) business planning and marketing. The units on building strong roots and site assessment are significant contributions to an urban agricultural programming model because they address challenges specific to urban agriculture endeavors.

One such challenge is soil contamination. There are many pollutants in the urban environment that contribute lead and other heavy metals to the soil, posing a human health risk (Surls et al. 2015). Consequently, the certificate program includes a unit on how to evaluate a property's soil quality and safety. The class also emphasizes the social aspects of having an urban farm or garden with modules on how to build relationships in the neighborhood, understanding the social context of a site (e.g., how community members perceive a site and what they would like to see there), and asset-based community development. Programming around asset-based community development represents an important shift in Extension approaches to community engagement: one in which community strengths are just as important as perceived needs. At home, students create a list of the social resources, or assets, in their community, as well as existing or potential partnerships. In class, they use this information to develop a reciprocity map, identifying what they are able to give to potential partners, alongside what they hope to gain from those partners. The social aspects of urban agriculture are an important theme because so many urban agriculture projects have social goals beyond food production and because, in an urban environment, connections to the broader community are essential for farm success (Poulsen et al. 2014).

While the content of the class prepares students to effectively engage with their community, the course itself also helps to foster connections among students and between students and other urban agriculture projects in the host cities, ultimately strengthening each city's urban agriculture network. During class time, students have an opportunity to connect with one another. The summer farm tours also serve a networking function, connecting students to established urban farmers and gardeners in the area. After tours, some of the students have gone on to volunteer at the featured operations. In these instances, farmers gain skilled, committed volunteers, while students can continue to develop their urban agriculture skills and knowledge. Classes also provide opportunities for people from the community to come in and meet with participants. Guests have included funders, zoning officials, and representatives from other county departments. One of the outcomes of creating this space for interaction is that, in general, urban farmers have become more open to communication with regulators. For instance, course leaders report that the local health department now sees fewer violations because people are willing to come speak to them before a problem arises.

Because of the success of the urban agriculture certificate program, the Allen County Community Development Corporation is providing vouchers for a free parcel of land to help people who complete the course access land (Green 2019)—helping to overcome one of the primary challenges for new entry farmers in the US (Ackoff et al. 2017). The only requirements are that the recipient has completed the urban agriculture certificate program and grows something on the site for a year. The vacant land is owned by the county, which will reduce its maintenance costs if the land is transferred to and managed by a farmer. In this instance, through partnerships in the community, Purdue Extension is not just providing urban agricultural knowledge, but also connecting participants to other assets (i.e., land) that can help them overcome some of the barriers to establishing or maintaining a new farm (Carlisle et al. 2019).

11.4.1.2 Penn State Extension – Growers' Series

Pennsylvania State University Extension operates an urban agriculture program in Philadelphia that provides education, technical support, and engages in community-based research. A city of 1.5 million, Philadelphia is home to more than 35 farms and 470 gardens. Often these nonprofit farms and gardens are located in parts of the city that have been affected by disinvestment, redlining, and deindustrialization. Urban agriculture in Philadelphia is diverse and includes Southeast Asian refugee, African American, and Puerto Rican communities.

Extension urban agriculture programs have found that it is important to offer educational programming for both beginning and experienced growers, since these audiences often have different needs (Smith et al. 2019). Penn State Extension's Growers' Series is an example of continuing education that fits the needs of a variety of local producers and highlights growers' knowledge. This workshop series, one of Penn State Extension's main urban agriculture educational programs in Philadelphia, is led by experienced urban growers with support from Extension. Community-Extension partnership is the foundation for this series, which 1) draws on the strengths of local organizations and Extension; 2) values local knowledge; and 3) facilitates information exchange within an agricultural network.

The Growers' Series is designed to highlight "Philadelphia-based growers implementing innovative, sustainable agricultural practices in their production spaces and their communities" (Penn State Extension 2019.). To identify workshop presenters and hosts, Extension solicits applications from interested farmers and gardeners who want to showcase practices at their sites. Topics have included low-till farming, fertility management for vegetable growing, designing a rain garden, and high tunnel soil health and maintenance. This series recognizes the deep, historically rooted agricultural knowledge that exists within Philadelphia's diverse communities. Many members of the city's migrant and immigrant communities, for example, come from agricultural backgrounds and bring their own growing practices and traditions to the city. The grower series provides a forum for disseminating local, experiential knowledge in partnership with Extension.

As a partner, Extension helps to support the development of the educational program and can either incorporate existing Extension resources or help to develop new ones as appropriate. These resources help to institutionalize agricultural knowledge over the long-term, even as individual staff members come and go—and urban agriculture can be a field with high turnover. Extension is also responsible for the logistics, promotion and outreach, and evaluation of these grower-led workshops. Extension staff are available to help develop or co-lead the program as needed. As part of their commitment to making sure that Extension values the knowledge, expertise, and time that community members contribute to these partnerships, Penn State Extension supports these classes with honoraria for speakers and a materials budget. They also provide culturally relevant snacks when appropriate or possible. These monthly grower-led workshops are held in the off-season (roughly October-May), a time of year when growers are better able to participate. The classes are typically held at a farm or garden, rather than the Extension office, and at different times, depending on growers' schedules. The Growers' Series is one example of Penn State Extension Philadelphia's commitment to collaboration, shared expertise, and co-production of knowledge with community partners.

Partnerships and community engagement can be multifaceted. In Philadelphia, Penn State Extension also coordinates a summer internship program for students from Penn State's main campus. Interns are matched with urban agriculture organizations based on their interests and skills and community partners' needs and vision (Penn State Extension 2018). In response to partner organizations' feedback, Extension has begun to offer regular anti-racism and anti-oppression training to interns, helping to build core competencies in racial equity. These are important skills for work with urban agriculture organizations, many of which serve and/or are led by people of color and which engage in food justice and food sovereignty work (e.g., Reynolds and Cohen 2016; Ventura and Bailkey 2017). The trainings are done in partnership with Philadelphia-based facilitators of color and supported with supplemental funding from the main campus. In addition to building relationships locally, the internship program also deepens partnerships within the Penn State system. During some summers, faculty members from Penn State College of Agricultural Sciences have given presentations to interns and staff on the latest information in fields ranging from soil science to community assessment and youth engagement. At the end of the internship, interns present what they have learned and any resources they have created (e.g., decision trees, curriculum, etc.). Some student internship projects have informed the Growers' Series in the fall. In these instances, Extension has provided support to community-driven projects with an intern and other resources; and subsequently, the lead organization has shared the completed project with the wider community through the Growers' Series.

The sustained community engagement that occurs through the internship program has a broader impact on Penn State Extension Philadelphia's urban agriculture work. As interns and community partners share knowledge, resources, and experiences, they also strengthen Extension's community partnerships and build trust with individuals, communities, and organizations, who may be distrustful of Extension and other institutions of higher education (Penn State Extension 2018). Working with local partners creates avenues and opportunities for Extension staff to develop programming that is more relevant to the community and to share projects more widely.

11.4.2 Research

The renewed public interest in the health, economic development, social justice, and environmental aspects of community food systems has inspired a growing body of academic research and literature. Campbell et al. (2013) found that this body of work shares a focus on collaborative efforts to enhance food system benefits for a particular geographical place. In their analysis of the literature, they call for a greater integration of research and practice and ask "how academics and practitioners might join forces to puzzle through persistent challenges facing the field" (122). As an integral component of academic institutions, Extension engages in research and can serve as a bridge between formal and informal kinds of knowledge. What may complicate these efforts, however, are hierarchical distinctions about what constitutes research and who is considered a researcher. As elaborated by Robert Chambers (1995) and other advocates of Participatory Action Research (PAR) approaches (including authors in chapters of this volume), inquiry with a goal of promoting equitable community development must recognize the inherent power imbalances that result from privileging academic knowledge over people's everyday lived realities and experiences. Instead, in its critique of dominant positivist research paradigms, PAR posits that both practitioners and academic experts should be recognized as legitimate researchers and co-producers of knowledge and that those who are directly affected by a research problem have a right to be involved in the research process (Gaventa and Cornwall 2008, 178). Doing so may require role reversals such that instead of controlling the terms of research and being the gatekeepers who transfer scientific knowledge and technologies from the university to the public, extension professionals serve as, in Chambers' words, "facilitators of learning" who "sit down" and "listen" (1995, 34).

Interestingly, our analysis of open-ended responses revealed that urban agriculture research was the activity Extension personnel mentioned least frequently in the 2015 survey. This may reflect the limited portion of time Extension personnel devoted to urban agriculture – nearly three quarters of respondents spent less than half their time on urban agriculture. It may also have to do with the fact that many respondents identified themselves as educators—positions that may have varied expectations for research depending on the university setting. It is also possible that the highly applied, participatory and community-led types of research engaged in by many Extension professionals may not be recognized or rewarded by their administrators as "research" since these efforts may not fit typical positivist research models or quantitative evaluation metrics for agricultural or economic research or journal publications (Archibald 2019; Collins 2015; Ventura and Bailkey 2017). Finally, as Ventura and Bailkey point out (2017, 3), those involved in community food systems projects may simply not want to be studied.

Despite the challenges, it is worth noting that Extension has a long history of publicly-engaged, community-directed research related to health, nutrition, home improvement, gardening, and farming in both rural and urban setting that can be traced from its early-nineteenth-century roots (Boyte 1999; Mayberry 1989; Peters 2002; Smith 1949/2013) to the modern era (Campbell et al. 2013; Dunning et al. 2012; Peters et al. 2005; Reardon 2019). It seems likely that as Extension urban agriculture programs have developed over the last five years, that urban agriculture research–and with it the co-creation of urban agroecological knowledge by Extension personnel working alongside and in support of urban agriculture practitioners–has expanded as well. Indeed, the National Urban Extension Leaders (NUEL), a group representing Extension personnel working in metropolitan regions, has encouraged urban Extension practitioners to embrace collaborative, applied research and engaged scholarship as one of the public services Extension is able to provide in metropolitan areas (NUEL 2015). To illustrate the many shapes and forms this work could take, we provide a few recent examples that we learned about as a follow-up from our survey.

11.4.2.1 Michigan State University – Detroit Partnership for Food Learning and Innovation (DPFLI)

Internationally known for its urban agriculture, the city of Detroit has more than 1,600 gardens and farms (Edwards 2019, personal communication). Urban agriculture there is a practice with deep roots, stretching at least as far back as Mayor Hazen Pingree's potato patches in the 1890s. Recently, Michigan State University (MSU) Extension has invested in urban agriculture research, establishing the Detroit Partnership for Food Learning and Innovation (more commonly referred to as the DPFLI) as an urban agriculture and forestry research center on the west side of the city. While MSU operates 13 other research and extension centers around the state, this is its first devoted to urban issues, particularly urban agriculture. The DPFLI sits on the site of a former elementary school, which was demolished in 2016. MSU began leasing the property in 2017, hired a director in 2018, and spent much of 2018 and 2019 getting permits and site plans approved. Construction of the first building on the site should be completed by the end of 2019.

Community engagement is integral to MSU Extension's approach to operating the DPFLI. One of the goals for the research center is to build strong, mutual relationships between Extension and Detroit communities, where an exchange of agroecological knowledge can take place. As the Director of MSU Extension explained, the role of the DPFLI director is "to be the point person in connecting MSU's vast research and outreach community to Detroit's urban agriculture stakeholders, neighborhoods and community organizations" (Corp 2018). In the process of designing the site's first research project, the director of DPFLI held monthly meetings with both urban farmers and academic staff. The site director also held roundtable discussions with farmers to see what questions they had. Soil management was one of the issues that emerged and is the subject of the site's first research project. As this initial design process shows, research conducted at the site is supposed to be informed by the local community. In addition to research, MSU Extension has made a commitment to offer ongoing extension programming at the site.

Efforts to rehabilitate the property have also spurred the new Center's first research project. Because of the school demolition, soil quality at the site is poor. However, these conditions are common for urban agriculture in Detroit; gardeners and farmers are often working on a property where a building was demolished. Thus, the poor soil quality at the DPFLI site presents an opportunity for MSU to investigate ways to improve soil that could be broadly applicable to other sites in the city. DPFLI's first research project is documenting the costs and benefits of rehabilitating soil through three tillage techniques (tilling with a tractor, a rototiller, or a broad fork) in combination with cover cropping.

Some of this research project's early results illustrate the ways in which urban environmental and social issues are interconnected, and the value of undertaking interdisciplinary research when investigating urban agroecological issues. While the goal of the project is soil improvement, early results show that some of the tillage and cover crop techniques may also reduce weed pressure and increase water infiltration. Techniques for reducing weed pressure on vacant lots could be helpful in Detroit, which has the highest asthma rates in the state of Michigan and some of the highest asthma rates in the county (Joel 2019; Nelson 2017). Tillage techniques that increase water infiltration could help reduce sewage overflow, which causes flooding, water damage, and contaminates Detroit's rivers (Bienkowski 2013).

11.4.2.2 Washington State University – Bilingual Organic Farming Incubator Partnership

In response to citizen organizing, Washington State University (WSU) initiated a statewide research and extension program to serve urban agriculture, small farms, and community food systems in 2000. As part of this effort, the WSU Cultivating Success Sustainable Farming Education Program was launched in 2002 to link the region's urban farmers with each other through farmer-to-farmer mentoring and with the faculty and research programs at WSU Research and Extension Centers (RECs) and County Extension. To meet the interest of immigrant and refugee farmers and farm workers in new farm entry and business entrepreneurship, this program expanded to offer multilingual education through County Extension offices in 2004 (Ostrom, Cha, and Flores 2010). By employing bilingual staff skilled at offering simultaneous interpreting, this program has been able to adapt and expand access to existing extension and research programs of interest to urban agriculturists, as well as build new connections with university agricultural specialists to influence their research and outreach directions. Eventually because of the challenges encountered by low-income and limited English-speaking households in obtaining farmland in the rapidly urbanizing Puget Sound region north of Seattle, Skagit County Extension and Cultivating Success began partnering with a local non-profit, Grow Food, to build a bilingual Spanish-English farmer incubator program called Viva Farms that could facilitate access to land, farming infrastructure, and urban markets for graduates of the Cultivating Success program.

Through this program, Extension has been able to contribute to a broad-based, and inclusive community coalition to support new entry farmers by 1) supporting a local non-profit through creating welcoming multilingual spaces; and 2) bringing together innovative local farmer mentors, business experts, philanthropists, government agencies, university social and agronomic scientists, and community food system organizers. The local county Extension office has also facilitated the participation of other university specialists in soils, food safety, and evaluation methods. As needed, these specialists have partnered with the project's non-profit leaders and farmer participants to conduct research on production, postharvest handling, and marketing and business management questions, as well as to explore the most effective pedagogical approaches for building both interest and skills in agroecologically-oriented farming practices. Connections with undergraduate and graduate research projects have been critical for allowing the program participants to tell their own stories in their preferred language and for the project partners to assess how well the project has been working from the point of view of the participants. An M.S. research project was structured in accordance with the non-profit's need to design an ongoing participatory evaluation process for the incubator program and resulted in a published article that could support the non-profit, as well as meet Extension research expectations (see Smith et al. 2019; https://wsuenglish.wixsite.com/vivafarms; https://vivafarms.org/impact-and-results/).

11.4.2.3 University of California, Berkeley – Sustainable Urban Farming for Resilience and Food Security

Survey results indicated that campus-based Extension specialists are playing a leading role in researching urban agriculture topics. One such example is the work of the University of California, Berkeley's metropolitan agriculture and food systems Extension specialist. Along with other University of California Berkeley faculty members, the metropolitan agriculture and food systems specialist specialist is leading a multi-year research project on Sustainable Urban Farming for Resilience and Food Security that focuses on the San Francisco Bay Area.

The University of California's Sustainable Urban Farming project intentionally uses an urban agroecology lens to frame their research because it is a concept that addresses the whole food system and captures "the multiple ecological, social, economic, and political dimensions of urban farming" (Siegner et al. 2019, 22). Consequently, researchers from multiple disciplines are working together to consider policy and practical solutions for building more equitable and resilient urban agroecosystems systems (Berkeley Food Institute 2019). To address the ecological sustainability of urban farming systems, researchers are focusing on techniques to build soil health, reduce soil contamination, and promote beneficial insects and water conservation. Other project researchers are evaluating the economic viability of urban farms, as well as both formal and informal distribution networks for urban-produced food and ways to make this food more accessible and affordable for urban populations (Siegner et al. 2018). As a scientific mode of inquiry, one of the ten key elements of agroecology stresses the co-creation and sharing of knowledge (Méndez et al. 2013; Siegner et al. 2019). In keeping with this approach, the Sustainable Urban Farming project uses participatory and collaborative research methods to engage students, farmers, community members and stakeholder institutions throughout the research process. For instance, researchers working on evaluating food distribution and food access organized two stakeholder input sessions for urban farmers and food system advocates to guide the study design, interpret results, and identify community-led solutions to identified challenges (Siegner et al. 2019).

Survey results of urban farmers in the San Francisco East Bay (Siegner et al. 2019) affirm the unique characteristics of urban agricultural operations and their missions as well as threats to their long-term viability including insecure land tenure, gentrification, and lack of funding. These farms were typically small (mean size 1.8 acres), frequently located on public land or land belonging an educational institution, and were primarily non-profit operations (Siegner et al. 2019). When urban farmers were asked to describe their missions, they ranked food sovereignty and food justice, education, and environmental sustainability the highest, while job creation and profit were ranked the lowest (Siegner et al. 2019). Siegner et al. (2018, 2019) make recommendations for researchers, urban farmers, and policymakers to help strengthen the sustainability of urban farming through innovative partnerships, shared learning opportunities, improving food distribution methods to enhance food access, and, more broadly, revaluing urban agriculture as a public good meriting public investment. Their recommendations reiterate the importance of urban agroecology as a framework for evaluating and communicating the many cultural, social, and ecological services urban agriculture provides and stress the need for applying an equity lens to urban agriculture initiatives to ensure they benefit low-income communities and communities of color.

11.5 Conclusions

The urban agriculture extension programs outlined here illustrate some of the different ways in which various Extension systems are providing education and conducting research on urban agroecological topics. As all five case studies demonstrate, each urban agriculture extension program needs to be tailored to the particular conditions in its city, region, and state as well as the diverse goals and organizational structures of the urban agriculture operations there (Diekmann et al. 2017; Drake and Lawson 2015; Hendrickson and Porth 2012). Nevertheless, some common themes emerge.

Various Extension offices across the country are adapting existing curricula on production and farm economics to the situations of urban growers, whose operations are often small-scale and run as non-profit, educational, or public entities. Extension is also developing new materials tailored to common features of urban production sites and shared interests of urban agriculturalists. Some of these include urban environmental conditions, such as the possibility of soil contamination at urban sites, which is part of Purdue Extension's urban agriculture certificate program's curriculum and the focus of initial research at MSU's Detroit Partnership for Food Learning and Innovation. They also include an emphasis on the social context and purpose of urban agriculture, such as community engagement, inclusivity, and partnership building, and social and racial justice within the urban food system–practices and foci that run through each of the five case studies described above. Furthermore, as the Penn State Extension and WSU examples demonstrate, some Extension offices are making strides in lifting up local agroecological knowledge, cultivating local urban agroecological leadership, and building reciprocal partnerships with local partners. Many Extension educational and research activities also contribute to building a network of urban farmers and gardeners, creating connections across race, culture, and class, and between urban agriculture practitioners and other members of the urban food system that enhance the generation and sharing of knowledge and resources.

Working in partnership with diverse community members will remain an important aspect of urban agriculture extension. Extension can develop its internal capacity for this type of work by offering training and resources to its staff on building equitable, reciprocal partnerships and on conducting community-based participatory research. It will be important for Extension academics to continue to conduct research that is action-oriented and responsive to community priorities, such as community building, social justice, and communicating urban agriculture's diverse impacts (Diekmann et al. 2017). Because of the multifunctionality of urban agriculture and the diverse social and environmental factors that influence the context in which it operates, much future urban agriculture research will require interdisciplinary approaches and collaboration. There is growing interest in training on diversity, inclusion, and racial equity, both within Extension and provided by Extension (Ostrom 2019). Extension systems can increase their support for doing this work internally and can partner with community partners to offer racial equity trainings or foster conversations about diversity and inclusion within urban agricultural spaces (Ammons et al. 2018).

While Extension urban agriculture programs are taking root in cities around the country, Extension professionals encounter institutional barriers when pursuing this type of work. The biggest challenge is resources. In recent decades, Extension systems across the country have had their budgets and their staffing levels cut. From 1980 to 2010 the number of full-time Extension specialist and county positions declined by roughly 21% nationwide (Wang 2014). In our survey of CRLFS members, the most commonly cited challenge was inadequate funding to address urban agriculture (52% of respondents). To maintain or expand its urban agriculture programming, Extension will need additional funding, either from internal or external sources (Diekmann et al. 2016). Depending on the institutional culture of various state Extension and LGU systems, it can be difficult for Extension professionals to get support or credit for participatory, community-based programming and research, particularly when they are playing support roles to other organizations and developing local leadership networks outside of Extension (Adams et al. 2005; Archibald 2019; Collins 2015). The logic models employed by many Extension Systems for program evaluation are most suited for unidirectional linear measurements of quantitative inputs and outcomes and presume that university-based Extension professionals are the primary actors and agents of change rather than community members. Relationship building is essential, time-consuming, and a long-term effort required by urban agriculture extension professionals but, because this work is difficult to quantify, Extension professionals may not get credit for it when they are evaluated by their institutions. Another institutional challenge is the distribution of Extension professionals working on urban agriculture; they are often geographically dispersed and isolated, both within their state Extension systems and nationally. Creating a network of urban agriculture Extension educators could facilitate the dissemination of successful program models, materials, and research across the country and help people in these roles to learn from one another.

Going forward, Extension can continue to improve its urban agriculture offerings by carrying out and sharing more urban agroecological research and educational approaches for urban and culturally and racially diverse settings (Reynolds 2011). Recognizing the different levels of expertise among urban agriculture community members, Extension should continue to offer education for both beginning and more advanced farmers, gardeners, and food entrepreneurs. It is also helpful, as in the Purdue Extension and Penn State Extension examples, to create opportunities for both new entry and established growers to come together to learn from one another. Extension professionals can continue to support diverse urban agriculture practitioners and new entry urban farmers in building and sustaining resilient social and knowledge networks, which can be integral to their long-term success and efforts to reform urban food systems (Smith et al. 2019). Finally, it will be critical for Extension administrators to improve the long-term sustainability of such efforts by supporting, recognizing and rewarding community-engaged Extension scholarship.

Acknowledgments

We wish to thank the CLRFS members who participated in the Urban Agriculture Working Group and helped to develop and disseminate the survey. They are Anne Randle, Brian Raison, Christine Coker, Connie Fisk, Courtney Long, Cynthia Nazario-Leary, Douglas Gucker, Garrett Ziegler, Jacqueline Kowalski, Jennifer Sowerwine, Julie Dawson, Lynn Heuss, Michelle Spain, Mike Hogan, Rob Bennaton, Shosha Capps, Stephen Haddock, and Tasha Hargrove. Thanks to James Wolf, Teri Theissen, Naim Edwards, Dorothy Cross, and Jennifer Sowerwine for sharing information about their urban agriculture extension programs. We also appreciate the thoughtful suggestions from Heather Hyden, Monika Egerer, and Hamutahl Cohen. This work was supported in part by the National Institute of Food and Agriculture, U.S. Department of Agriculture, 2014-67012-22270 and 2014-67012-27506.

REFERENCES

Ackoff, S., Bahrenburg, A., & Shute, L. L. (2017). Building a future with farmers II: Results and recommendations from the national young farmer survey. *National Young Farmer Coalition. November*, 1–86.

Adams, R., Harrell, R., Maddy, D., & Weigel, D. 2005. A diversified portfolio of scholarship: The making of a successful extension educator. *Journal of Extension*, *43*(4), 4COM2. https://www.joe.org/joe/2005august/comm2.php

Ammons, S., Creamer, N., Thompson, P. B., Francis, H., Friesner, J., Hoy, C., & Tomich, T. P. (2018). A deeper Challenge of Change: The role of land-grant universities in assessing and ending structural racism in the US food system. *Inter-Institutional Network for Food, Agriculture, and Sustainability.* Available online: https://asi.ucdavis.edu/programs/infas/a-deeper-challenge-of-change-the-role-of-land-grant-universities-in-assessing-and-ending-structural-racism-in-the-us-food-system.

Anderson, M. (2019) Scientific Knowledge of Food and Agriculture in Public Institutions in Vivero-Pol, J. L., Ferrando, T., De Schutter, O., & Mattei, U. (eds.), *Routledge handbook of food as a commons*, London and New York: Routledge.

Archibald, T. (2019). Whose extension counts? A plurality of extensions and their implications for credible evidence debates. *Journal of Human Sciences and Extension*, *7*(2).

Bassett, T. J. (1981). Reaping on the margins: A century of community gardening in America. *Landscape*, *25*(2), 1–8.

Berkeley Food Institute (2019). *Sustainable Urban Farming for Resilience and Food Security.* Retrieved from https://food.berkeley.edu/sustainable-urban-farming-for-resilience-and-food-security/

Bienkowski, B. (2013, Aug. 27). "Sewage overflow adds to detroit's woes." Scientific American. Retrieved from: https://www.scientificamerican.com/article/sewage-overflow-adds-to-detroits-woes/

Boyte, H. (1999). Building the Commonwealth: Citizenship as Public Work in Elkin, S. & K. Soltan (eds), *Citizen competence and democratic institutions*, pp 259–278. Pennsylvania State University Press: University Park, PA.

Brame, C. (2013). "Flipping the classroom." Vanderbilt University Center for Teaching. Accessed March 1, 2020. https://cft.vanderbilt.edu/guides-sub-pages/flipping-the-classroom/

Brown, K., & Carter, A. (2003). *Urban agriculture and community food security in the United States: Farming from the inner-city to the urban fringe.* Community Food Security Coalition: Venice, CA. Retrieved from http://foodsecurity.org/PrimerCFSCUAC_pdf.pdf

Buttel, F. H. (2005). Ever since hightower: The politics of agricultural research activism in the molecular age. *Agriculture and Human Values, 22*(3), 275–283.

Campbell, D. C., Carlisle-Cummins, I., & Feenstra, G. (2013). Community food systems: Strengthening the research-to-practice continuum. *Journal of Agriculture, Food Systems, and Community Development, 3*(3), 121–138. doi:10.5304/jafscd.2013.033.008

Carlisle, L., de Wit, M. M., DeLonge, M. S., Calo, A., Getz, C., Ory, J., … &Iles, A. (2019). Securing the future of US agriculture: The case for investing in new entry sustainable farmers. *Elem Sci Anth, 7*(1).

Chambers, R. (1995). Paradigm shifts and the practice of participatory research and development, *in* N. Nelson and S. Wright (eds), *Power and Participatory Development: Theory and Practice*, pp. 30–42. Intermediate Technology Publications: London.

Clark, J. K., Bean, M., Raja, S., Loveridge, S., Freedgood, J., & Hodgson, K. (2016). Cooperative Extension and food system change: Goals, strategies and resources. *Agriculture and Human Values*, 1–16. doi:10.1007/s10460-016-9715-2

Colasanti, K., Wright, W., & Reau, B. (2009). Extension, the land-grant mission, and civic agriculture: Cultivating change. *Journal of Extension, 47*(4), Article 4FEA1. Retrieved from https://www.joe.org/joe/2009august/a1.php

Collins, C. S. (2015). Land-Grant Extension: Defining public good and identifying pitfalls in evaluation. *Journal of Higher Education Outreach and Engagement, 19*(2), 37–64.

Corp, S. (2018). "MSU names director for Detroit's first urban agriculture research center." Retrieved from https://www.canr.msu.edu/news/msu-names-director-for-detroit-s-first-urban-agriculture-research-center

Daftary-Steel, S., Herrera, H., & Porter, C. M. (2015). The unattainable trifecta of urban agriculture. *Journal of Agriculture, Food Systems, and Community Development, 6*(1), 19–32.

Diekmann, L., Dawson, J., Kowalski, J., Raison, B., Ostrom, M., Bennaton, R., and Fisk, C. (2016). Preliminary results: Survey of Extension's Role in Urban Agriculture.

Diekmann, L., Bennaton, R., Schweiger, J., & Smith, C. (2017). Involving extension in urban food systems: An example from California. *Journal of Human Sciences and Extension, 5*(2).

Drake, L., & Lawson, L. (2015). Best practices in community garden management to address participation, water access, and outreach. *Journal of Extension, 53*(6), Article 6FEA3. Retrieved from https://www.joe.org/joe/2015december/a3.php

Dunning, R., Creamer, N., Lelekacs, J. M., O'Sullivan, J., Thraves, T., & Wymore, T. (2012). Educator and institutional entrepreneur: Cooperative extension and the building of localized food systems. *Journal of Agriculture, Food Systems, and Community Development, 3*(1), 99–112. doi:10.5304/jafscd.2012.031.010

Edward, N. Director, Michigan State University Detroit Partnership in Food, Learning and Innovation. (Personal communication, 19 December 2019).

Extension Committee on Organization and Policy (ECOP) (2019) "How Does Cooperative Extension Work? Accessed December 31, 2019: https://www.aplu.org/members/commissions/food-environment-and-renewable-resources/board-on-agriculture-assembly/cooperative-extension-section/index.html

Gaventa, J. &A. Cornwall (2008). *Power and knowledge, in the SAGE Handbook of Action Research: Participative inquiry and practice*, P. Reason &H. Bradbury (eds). pp 172–189. Second Edition. Sage Publications Ltd.: London.

Getz, C. & K. D. Warner (2006). Integrated Farming Systems and Pollution Prevention Initiatives Stimulate Co-Learning Extension Strategies. *Journal of Extension*, 44(5), Article 5FEA4. Retrieved from: https://www.joe.org/joe/2006october/a4.php

Giancatarino, A., & Noor, S. (2014). Building the case for racial equity in the food system. *Center for Social Inclusion*, 1–35.

Gibby, D., Scheer, W., Collmen, S., and Pinyuh. G. 2008. The Master Gardener Program: A WSU Extension Success Story, downloaded 11/1/2016: http://mastergardener.wsu.edu/wp-content/uploads/2012/12/MasterGardenerProgramHistoryrev2009.8.pdf

Green, R. (2019, Aug 27). Urban Farming Gets Voucher Boost. WBOI. Retrieved from https://www.wboi.org/post/urban-farming-gets-voucher-boost#stream/0

Hagey, A., Rice, S., & Flournoy, R. (2012). Growing urban agriculture: Equitable strategies and policies for improving access to healthy food and revitalizing communities. Oakland: PolicyLink.

Hayden-Smith, R., and Surls, R. (2014). "A century of science and service," *California Agriculture*, *68*(1): 8–15

Hendrickson, M. K., & Porth, M. (2012). Urban agriculture—best practices and possibilities. *University of Missouri*, 1–52.

Henning, J., Buchholz, D., Steele, D., & Ramaswamy, S. (2014). Milestones and the future for Cooperative Extension. *Journal of Extension*, *52*(6).

Horst, M., McClintock, N., & Hoey, L. (2017). The intersection of planning, urban agriculture, and food justice: a review of the literature. *Journal of the American Planning Association*, *83*(3), 277–295.

Joel, L. (2019, Aug. 18). "In Detroit, 'revolutionary' mapping could help residents with asthma." Grist. Retrieved from https://grist.org/article/in-detroit-revolutionary-mapping-could-help-residents-with-asthma/

Kellogg Commission on the Future of State and Land-Grant Universities. (1999). Returning to our roots: The engaged institution (Vol. 3). *National Association of State Universities and Land-Grant Colleges.* Retrieved from: https://www.aplu.org/library/returning-to-our-roots-the-engaged-institution/file.

Kilbane, K. (2017, Nov. 1). "Urban farm certificate program trains people for operating an urban farm or community garden in the Fort Wayne area." News-Sentinel. Retrieved from https://www.news-sentinel.com/news/local-news/2017/11/01/urban-farm-certificate-program-trains-people-for-operating-an-urban-farm-or-community-garden-in-the-fort-wayne-area/

Lee, Jr. J.M., & S.W., Keys. (2013). *Land-Grant but Unequal: State one-to-one match funding for 1890 Land-Grant Universities. Policy Brief.* Association of Public and Land-Grant Universities. Report number: 3000-PB1.

Lovell, S. T. (2010). Multifunctional urban agriculture for sustainable land use planning in the United States. *Sustainability*, *2*(8), 2499–2522.

Lubell, M., Niles, M., & Hoffman, M. (2014). Extension 3.0: Managing agricultural knowledge systems in the network age. *Society & Natural Resources*, *27*(10), 1089–1103.

Mayberry, B. D. (1989). *Role of Tuskegee University in the origin, growth and development of the Negro Cooperative Extension System, 1881-1990.* Brown Printing Company: Montgomery.

Méndez, V. E., Bacon, C. M., & Cohen, R. (2013). Agroecology as a transdisciplinary, participatory, and action-oriented approach. *Agroecology and Sustainable Food Systems*, *37*(1), 3–18.

Nash, M. A. (2019). Entangled Pasts: Land-Grant Colleges and American Indian Dispossession. *History of Education Quarterly*, *59*(4), 437–467.

National Institute of Food and Agriculture (NIFA) (no date). Cooperative Extension History. Accessed at nifa.usda.gov/cooperative-extension-history on January 1, 2020.

National Research Council. (1995). Colleges of agriculture at the land grant universities: A profile. National Academies Press: Washington, D.C. Retrieved from: https://www.nap.edu/read/4980/chapter/6

National Urban Extension Leaders (NUEL): De Ciantis, D., Fox, J., Gaolach, B., Jacobsen, J., Obropta, C., Proden, P., … Young, J. (2015). A national framework for urban Extension: A report from the national urban Extension leaders. Retrieved from http://media.wix.com/ugd/c34867_668cd0780daf4ea18cb1dad dad557c72.pdf

Nelson, R. (2017, Apr. 3). "Plans to reduce the high asthma burden in Detroit." *The Lancet*, *5*(5): 388–389. Retrieved from https://www.thelancet.com/journals/lanres/article/PIIS2213-2600(17)30130-3/fulltext

Oberholtzer, L., Dimitri, C., & Pressman, A. (2014). Urban agriculture in the United States: Characteristics, challenges, and technical assistance needs. *Journal of Extension*, *52*(6), Article 6FEA1. Retrieved from https://www.joe.org/joe/2014december/a1.php

OMERAD, n.d. "What, why, and how to implement a flipped classroom model." Accessed March 1, 2020. https://omerad.msu.edu/teaching/teaching-strategies/27-teaching/162-what-why-and-how-to-implement-a-flipped-classroom-model

Opitz, I., Berges, R., Piorr, A., & Krikser, T. (2016). Contributing to food security in urban areas: Differences between urban agriculture and peri-urban agriculture in the global north. *Agriculture and Human Values*, *33*(2), 341–358. doi:10.1007/s10460-015-9610-2

Ostrom, M. (2019). "Radical Roots and 21st Century Realities: Rediscovering the Egalitarian Aspirations of Land Grant University Extension, Presidential Address, Agricultural, Food, and Human Values Society Annual Meeting, Anchorage, AK, June 29.

Ostrom, M. and C. Donovan (2018). *Organizational Dimensions of Farmers Markets in Washington State. WSU Extension Publication #TB48E,* https://pubs.extension.wsu.edu/organizational-dimensions-of-farmers-markets-in-washington

Ostrom, M., Cha, B., & Flores, M. (2010). Creating access to land grant resources for multicultural and disadvantaged farmers. *Journal of Agriculture, Food Systems, and Community Development*, *1*(1), 89–105.

Penn State Extension (2019). "Growers' Series: Fertility Strategies for Vegetable Growing." Retrieved from: https://www.facebook.com/events/295841127788750/

Penn State Extension (2018). Penn State Extension Philadelphia Community Engagement Internship.

Peters, S.J. (2002.) Rousing the people on the land: The roots of the educational organizing tradition in extension work. *Journal of Extension*, *40*(3), June.

Peters, Scott J. (2014.) Extension Reconsidered. *Choices. Agriculture & Applied Economics Association*, *29*(1): 1–6.

Peters, S., Jordan, N., Adamek, M. and T. Alter. (2005). *Engaging Campus and Community: The practice of public scholarship in the State and Land-Grant University System*. Kettering Foundation Press: Dayton.

Poulsen, M. N., Spiker, M. L., & Winch, P. J. (2014). Conceptualizing community buy-in and its application to urban farming. *Journal of agriculture, food systems, and community development*, *5*(1), 161–178.

Raison, B. (2010). Educators or facilitators? Clarifying Extension's role in the emerging local food systems movement. *Journal of Extension*, *48*(3), Article 3COM1. Retrieved from https://www.joe.org/joe/2010june/comm1.php

Reardon, K. 2019. *Building bridges: Community and University Partnerships in East St Louis*. Social Policy Press: New Orleans.

Reynolds, K. (2011). Expanding technical assistance for urban agriculture: Best practices for extension services in California and beyond. *Journal of Agriculture, Food Systems, and Community Development*, *1*(3), 197–216. doi:10.5304/jafscd.2011.013.013

Reynolds, K., & Cohen, N. (2016). *Beyond the kale: Urban agriculture and social justice activism in New York City* (Vol. 28). University of Georgia Press.

Schupp, J. L., & Sharp, J. S. (2012). Exploring the social bases of home gardening. *Agriculture and Human Values*, *29*(1), 93–105.

Siegner, A. B., Acey, C., & Sowerwine, J. (2019). Producing urban agroecology in the East Bay: From soil health to community empowerment. *Agroecology and Sustainable Food Systems*, 1–28.

Siegner, A., Sowerwine, J., & Acey, C. (2018). Does urban agriculture improve food security? Examining the nexus of food access and distribution of urban produced foods in the United States: A systematic review. *Sustainability*, *10*(9), 2988.

Smith, R. G. (1949/2013). *The People's Colleges: A history of the New York State Extension Service in Cornell University and the State, 1876-1948*. Cornell University Press, Ithaca, N.Y. and Fall Creek Books (2013 edition).

Smith, K., Ostrom, M., McMoran, D. and L. Carpenter Boggs. (2019). Connecting new farmers to place, agroecology, and community through a bilingual organic farm incubator. *Journal of Agriculture, Food Systems, and Community Development*. *9*(1).

Smithsonian Gardens. (2020). "Vacant Lot Gardens." Accessed March 1, 2020. https://communityofgardens.si.edu/exhibits/show/historycommunitygardens/vacantlot

Stein, S. (2017). A colonial history of the higher education present: Rethinking land-grant institutions through processes of accumulation and relations of conquest. *Critical Studies in Education*, 1–17.

Surls, R., Feenstra, G., Golden, S., Galt, R., Hardesty, S., Napawan, C., & Wilen, C. (2015). Gearing up to support urban farming in California: Preliminary results of a needs assessment. *Renewable Agriculture and Food Systems*, *30*(1), 33–42. doi:10.1017/S1742170514000052

Ventura, S. and M. Bailkey (Eds.). (2017). *Good food, strong communities: Promoting social justice through local and regional food systems*. University of Iowa Press: Iowa City, IA.

Vitiello, D., & Wolf-Powers, L. (2014). Growing food to grow cities? The potential of agriculture for economic and community development in the urban United States. *Community Development Journal*, *49*(4), 508–523.

Wang, S. L. (2014). Cooperative extension system: Trends and economic impacts on US agriculture. *Choices*, *29*(316-2016-7709), 1–8.

Zabawa, R. (2008). Tuskegee Institute Moveable School. *Encyclopedia of Alabama*. Retrieved from http://www.encyclopediaofalabama.org/article/h-1870

12

How to Study the Ecology of Food in the City: An Overview of Natural Science Methodologies

Theresa Wei Ying Ong[1*] and Gordon Fitch[2]
[1] Program in Environmental Studies, Dartmouth College, Hanover, NH USA
[2] Department of Ecology and Evolutionary Biology, University of Michigan, Ann Arbor USA
[*] Corresponding author Email: theresa.w.ong@dartmouth.edu.

CONTENTS

KEY WORDS: *Urban agriculture, natural science methods, review, urban agroecology, urban ecology*

12.1 A Focus on Ecology

Research in urban agriculture (UA) tends to focus on its very important social aspects (Guitart et al. 2012, Lin et al. 2015). This is likely because UA is clearly a socio-ecological system, which is defined as a system composed of bio-geo physical and social actors and institutions that interact and adapt to one another (Ostrom 1990, Glaser et al. 2008). Agroecology itself is an interdisciplinary field that positions itself squarely in the middle of the social and ecological disciplines, though many argue that the field moves too far in one direction or the other (Gliessman 2014, Rosset and Altieri 2017). An agroecology for urban systems may find itself pulled even further in the direction of society given the central role of the human-dominated landscape. As such, the ecological component of agroecological research, particularly in UA, remains nascent (Guitart et al. 2012, Epstein et al. 2013). Though there has been a recent explosion of interest in UA, a quick search on web of science of "urban+gardens" reveals only 62 articles in the "Ecology" category, the majority of which were published in 2014–2019 and do not necessarily include natural science methods. A similar search for "urban+agriculture" reveals only 91 ecology related articles. In a review of 87 studies on community gardens, Guitart et al. authors found only a single study focused on ecology (Guitart et al. 2012). This finding may be exacerbated by the review's focus on community gardens rather than UA more broadly, yet socio-ecological systems research in general is undeniably skewed in favor of the social sciences (Epstein et al. 2013). The ecological work that has been conducted since Guitart's review focuses primarily on cataloguing and comparing the biodiversity of UA to peri-urban agriculture (UPA) and rural sites (Ricketts and Imhoff 2003, Goddard et al. 2010, Beninde et al. 2015, Lin et al. 2015, Speak et al. 2015). These studies have demonstrated that an impressive array of species inhabit urban gardens, but cohesive theories to explain the mechanism behind this and divergent trends is less developed (Egerer, Li, et al. 2018). At the same time, there is a pressing need for ecological research in UA to forge closer integration with social science research on the topic, in short, to create an urban agroecology discipline. To encourage advances on these fronts, this chapter overviews some ecological concepts and theories applicable to UA, as well as the overall characteristics of UA itself.

12.2 Understanding the Ecosystem – What is Urban Agriculture?

Designing an ecological study of UA necessarily begins with its definition (Mougeot 2010, Zezza and Tasciotti 2010, Orsini et al. 2013, Lin et al. 2017). Agriculture implies some form of food production, yet most UA sites are composed of food, ornamental and weed species (Mougeot 1999, WinklerPrins 2002, Guitart et al. 2012, Lin et al. 2017). Indeed, many ornamentals are planted as companions to food crops, providing benefits including pollination and pest control services (Baker 1989, WinklerPrins 2002, Colding et al. 2006, Lin et al. 2015). Focusing purely on food production would ignore these synergies and obscure many of UA's unique characteristics, including its high levels of planned and associated biodiversity (Daniels and Kirkpatrick 2006a, 2006b, Goddard et al. 2010, Lin et al. 2015, Speak et al. 2015). Yet, UA comes in many different sizes and forms, from large-scale community gardens, rooftop gardens on top of businesses and restaurants, indoor hydroponics operations, edible city trees and landscapes, potted apartment windowsill plants to guerrilla gardens hiding in city sidewalks or on display in vacant lots (Mougeot 1999, Zezza and Tasciotti 2010, Lin et al. 2017).

In order to provide an appropriate context for study and comparison, ecological studies of UA should define and possibly limit what type of UA they are considering. Some sites may be more appropriate than others, depending on the questions and organisms of interest. For example, larger UA plots including community gardens or urban farms may provide the spatial scale necessary for sampling larger or more mobile organisms such as mammals, lizards and birds (Preston 1962, Bowman et al. 2002). In contrast, rooftop gardens may provide better insight on the effects of novel soil infrastructure and nutrient flows in cities (Buehler and Junge 2016, Pennisi et al. 2016, Harada et al. 2018, Harada et al. 2018, Harada et al. 2019). Potted plants and windowsills can elucidate impacts of space and light, while guerrilla gardens can be particularly useful for studying temporally-dependent dynamics (Orwell et al. 2006, Adams and

Hardman 2013). Comprehensive understanding of UA as an ecosystem necessarily encompasses all of these forms. However, including all types of UA in a single study makes replication, without confounding effects, difficult to achieve, particularly at the scale of a city (Goddard et al. 2010, Egerer, Li, et al. 2018). Thus, the choice of which kind(s) of sites to utilize and at what spatial and temporal scales will depend on the questions at hand.

Conceptually, progress in ecological studies of UA is impeded by inconsistent definitions not only of 'agriculture', but also of 'urban'. Human population density and the percentage of impervious surface are often used to categorize cities. But there is no widely agreed-upon threshold of either population density or impervious surface cover that qualifies as 'urban'. This is due in part to differences in patterns of urban development and settlement in different places. But the lack of consistency makes it difficult to generalize across studies, and is likely one reason for the frequently conflicting narratives about the effect of urbanization on ecosystems and communities (Raciti et al. 2012). Researchers should make explicit how they define 'urban', and perhaps provide a justification for their definition, in order to facilitate comparison across studies.

12.3 Reviewing Ecological Concepts – What Kinds of Questions to Ask

12.3.1 Succession-Legacy Effects, Path Dependence

Ecological studies are often limited for practical reasons to single time snapshots or very short periods (Callahan 1984). However, the history of land-use in and surrounding UA is an undeniably important aspect of its ecology (Lane 2015, Crumley 2018, Isendahl 2018, Isendahl et al. 2018). The extensive use of impervious surface (e.g. concrete, asphalt) in urban areas is hypothesized to have strong, perhaps long-term effects on the space, form and functioning of UA (Rudd et al. 2002, Goddard et al. 2010, Martellozzo et al. 2014). Many urban soils are compacted, and contain contaminants dangerous to human health (De Kimpe and Morel 2000, Li et al. 2001). Brownfields represent a particularly extreme example of contaminated urban soils. These are sites that have a history of poor management practices leading to environmental contamination, which require significant remediation before usable as a city resource for human habitation or food production if converted to UA (Groffman and Tiedje 1988, Reddy Krishna R. et al. 1999, Brown and Jameton 2000, Pouyat et al. 2010, 2015). The study of brownfield sites can therefore be particularly useful for understanding legacy effects in urban soils more generally. Understanding how urban soils and their associated communities change over time and whether they can be managed to create safe, productive agricultural products are questions directly related to the concept of ecological succession.

Ecological succession considers the change in species composition of a community over time. First used by Cowles to describe the changes in vegetation structure and communities he observed along transects running inland from the shores of Lake Michigan, the concept of succession has since been used to explore the role of deterministic and stochastic processes in shaping communities (Cowles 1899, Clements 1916, Gleason 1926, Tansley 1935). Current research on ecological succession stresses the importance of stochasticity in conjunction with legacy effects, which can create a **path dependence** in the community structures that develop (Connell and Slatyer 1977, Bazzaz 1996). This means that the structure of a community is dependent on past processes including for example, the order of colonization events or how soils developed in the area (Fukami 2015). This notion is particularly relevant for UA where the history of urbanization is integral to its form (Isendahl 2018). However, few studies have explicitly looked at succession in UA systems to date (Table 12.1).

There is a rich literature on urban soils and the communities of microorganisms that inhabit them, though these studies are not typically conducted in UA (Craul 1999, De Kimpe and Morel 2000). Studies involving bio-remediation, the use of microorganisms to clean up contaminated sites, are highly relevant to UA but rarely discussed in conjunction with food production (Adhikari et al. 2004, Malik 2004, Iram et al. 2011). This is unfortunate since many UA sites occupy abandoned land (Taylor and Lovell 2012, Gardiner et al. 2014). The danger to public health in UA is widely acknowledged, yet some growers may not test soils before cultivation, or testing may not continue long-term (Brown and Jameton 2000,

TABLE 12.1

Summary of Urban Agriculture Natural Science Methodologies

Target Taxon or Abiotic Factor	Sampling Method	Types of Questions	Pros	Cons	Special Considerations	Alternatives	References
Invertebrates							
General	Sweep netting	Abundance and diversity	Efficient method for sampling across taxonomic groups; can be non-lethal	Often destructive to vegetation;not selective in type of invertebrate collected			Vollhardt et al. 2008†
	Targeted visual survey	Abundance and diversity	Low impact; good for quantifying abundance of uncommon species or species not attracted to traps	Only feasible for relatively large, obvious species (e.g. lady beetles, caterpillars, ants); high potential for observer bias			Edwards 2016
	Sticky traps	Abundance and diversity	Low effort; efficient for sampling range of flying arthropods ; especially useful for sampling adult parasitoids	Lethal; difficult to identify specimens to high taxonomic resolution; traps may be disturbed by garden/ farm users			Egerer et al. 2018b; Lowenstein & Minor 2018
Herbivores	Sentinel plants	Abundance and diversity; impacts of herbivory	Effective way to assess herbivory pressure while controlling for plant quality	Plants may require care	Requires permission to place plants in garden/farm;		Egerer et al. 2018a
Predators and Parasitoids	Sentinel pests	Abundance and diversity; biological control	Most accurate way to quantify predation pressure		must rear or buy pest species; requires placing food plant as well; requires permission		Lowenstein et al. 2017; Philpott & Bichier 2017
Ants (Formicidae)	Tuna baits	Diversity; competitive relationships	Low effort / Allows for non-lethal sampling / Easily adapted to small spaces	Baits may attract unwanted organisms or may be viewed as undesirable by gardeners/ farmers; sampling biased towards certain species			Edwards 2016; Uno et al. 2010
	Pitfall traps	Diversity and activity level	Samples broad taxonomic and functional group range	Traps may be disturbed by garden/farm users; requires disturbing soil			Edwards 2016

	Mini-Winklers	Abundance and diversity	Samples broad taxonomic and functional group range	High effort; involves collecting soil and/or litter			Savage et al. 2015*
	Tanglefoot exclusion	Effectiveness of biocontrol	Can clearly demonstrate effect of ants on pests	Sticky substance may be mildly annoying to garden/farm users			Reimer et al 1993†; Morris et al. 2015†;
Pollinators (Hymenoptera, Diptera, Lepidoptera)	Pan/bowl traps	Abundance and diversity; diet and pollination potential (from pollen collected from bodies)	Low effort / Samples species often missed in netting	Lethal / Traps may be disturbed by garden/farm users; standard protocol requires access to large area			Matteson et al. 2008; Quistberg et al. 2016; Glaum et al. 2017
	Aerial netting	Abundance and diversity; diet and pollination potential (from capture location and/or pollen)	Complements pan/bowl traps; can be non-lethal ; can catch flying bees; adaptable to any space	Can damage vegetation; biased towards larger-bodied bees		Ziplock bag 'net'	Quistberg et al. 2016
	Ziplock bag 'net'	Abundance and diversity; diet and pollination potential (from capture location and/or pollen)	Complements pan/bowl traps; can be non-lethal; not destructive of plants ; adaptable to any space	Generally biased towards large-bodied bees ; can only catch bees when landed		Aerial netting	Fitch et al. 2019
	Sentinel plants	Pollination effectiveness; pollinator foraging behavior; competitive interactions	Allows estimation of pollinator visitation to particular species	Time intensive; only feasible for small number of plant species	Requires permission to place plants in garden/farm		Fitch 2017; Lowenstein et al. 2014
	Sentinel colonies	Pollinator diet; pollinator population dynamics; pollinator pathogen and parasite prevalence	Allows researcher to address a range of questions that are otherwise difficult to answer	Relatively expensive and difficult to maintain			Williams et al. 2012†; Vaidya et al. 2018
	Trap nests	Abundance and diversity of trap-nesting bees; population dynamics; diet; pathogen and parasite prevalence	Allows researcher to address a range of questions that are otherwise difficult to answer	Often low rate of colonization			MacIvor et al. 2014; MacIvor and Packer 2015;
Soil invertebrates	Pitfall traps [see under Ants (Formicidae)]						Gardiner et al. 2014; Burkman & Gardiner 2015
	Mini-Winklers [see under Ants (Formicidae)]						

(Continued)

TABLE 12.1 (Continued)

Target Taxon or Abiotic Factor	Sampling Method	Types of Questions	Pros	Cons	Special Considerations	Alternatives	References
	Baermann funnels	Abundance and diversity	Useful for sampling nematodes	High effort; requires removing soil from site			Sharma et al. 2015
	Sentinel pests [see under Predators and Parasitoids]						Yadav et al. 2012
Vertebrates							
General	Mesh exclosures / fencing	Effectiveness of vertebrate biocontrol; vertebrate herbivore effects	Can clearly demonstrate effect of particular guilds of organisms	High effort; often requires large amount of space and may interfere with use; may be disturbed by garden/farm users			Karp et al. 2013†
Birds	Point counts	Abundance and diversity	Well-established protocol facilitates comparison across studies	Time-intensive; sensitive to researcher bias	Ability to identify birds by sight and sound needed		Rottenborn 1999*; Paker et al. 2014*
	Mist-netting	Abundance and diversity; diet (from excrement); nutritional status and health; population genetics	Not sensitive to researcher bias; allows for collection of tissue samples	Requires expensive materials and training; highly obvious and potentially disruptive to other use of sites	Requires special permits		Evans et al. 2009*; Hamer et al. 2012*
Mammals	Live-trapping (Sherman and/or Longworth traps)	Abundance and diversity; diet (from excrement); nutritional status and health; population genetics			Requires special permits		Munshi-South and Kharchenko 2010*; Wilson et al. 2016*
	Camera traps	Abundance and diversity	Allows assessment of occurence/abundance of uncommon and secretive species	Equipment expensive and likely to be tampered with			Widdows et al. 2015*; Gallo et al. 2019*
Plants							
General	Transect surveys	Abundance and diversity; spatial arrangement				Quadrat surveys	
	Quadrat surveys	Abundance and diversity; spatial arrangement				Transect surveys	Ahrne et al. 2009;
Crop yield	Sentinel plants	Yield of particular crop type; response to treatment	High degree of control by researcher	Limited to a small number of plants and species			Potter & LeBuhn 2015; Bennett & Lovell 2019

	Gardener/user data	Larger-scale yield data for one to many crops	Direct estimate of plot- or site-scale yield	Difficult to assess accuracy of data; requires training of and coordination with site users		
Microbes	Soil microbe DNA extraction	Microbial diversity	Allows for characterization of soil microbial community	Sample processing can be costly		Yan et al. 2016*; Wang et al. 2017*
Abiotic factors						
Climate and temperature	Data logger	Site-level temperature, humidity, insolation	Low-effort; fine-scale data	Can be expensive; data can be lost if loggers are tampered with or removed	Publically-available weather station data	Glaum et al. 2017
	Publically-available weather station data	Local temperature, humidity, precipitation	Low effort; free	Resolution poorer than data logger, esp. for sites in close proximity	Data logger	
Soil characteristics	Soil physical analysis	Soil texture; soil aggregate structure; soil moisture content				Beniston et al. 2016; Pennisi et al. 2016
	Soil chemical analysis	Soil nutrient content (e.g. C, N, P); soil heavy metal content; soil pH	Accurate, precise estimates of soil properties; many labs will accept samples for processing	Requires specialized equipment or ability to pay for sample processing		Beniston et al. 2016; Harada et al. 2019
	Litterbags	Decomposition rates; role of soil biota in decomposition; nutrient fluxes	Relatively low effort; may not require expensive equipment	Requires leaving litterbags for an extended time		Pavao-Zuckerman & Coleman 2005*; Tresch et al. 2019
	Ion-exchange resin bags	Leaching rates for nutrients and heavy metals	Accurate estimates of leaching rates	Requires leaving resin bags for an extended time; requires some processing		Harada et al. 2018a; Harada et al. 2019
Atmosphere characteristics	Atmospheric particulate collectors	Deposition rates for atmospheric particulate matter (e.g. heavy metals)	Accurate estimates of deposition rates	Requires leaving (often large) collectors for an extended time; requires some processing		Harada et al. 2018a; Harada et al. 2019
Vegetation biochemistry	Vegetation biochemical analysis	Nutrient content of crop plants; heavy metal content of crop plants	Accurate estimates of nutrient and/or contaminant concentrations in crops	Requires specialized equipment or ability to pay for sample processing		Harada et al. 2018a; Pennisi et al. 2016

* Study conducted in urban area but not UA.
† Study conducted in agricultural area but not UA.

De Kimpe and Morel 2000, Minca et al. 2013, Beniston et al. 2016). Ecological study of succession in UA could pull techniques from these fields to study the change in soil communities over time and still more, the potentially irreversible effects of past land-use choices on these communities (Huang et al. 1998, Fukami 2015, Ong and Vandermeer 2018, Vandermeer and Perfecto 2019). The idea of historical legacy is tied to the physical concept of **hysteresis**, where the path forwards is different from the path back (Mayergoyz 2012). All cities were built on top of some natural ecosystem at some point in history, and that land-use history is likely to have a strong deterministic effect on agricultural productivity (Bellemare et al. 2002). UA is unique however, in the extensive translocation of soils (Craul 1992). Many gardens are built from scratch, transporting organic material from a variety of distance sources including bogs, city composting facilities or other farming operations (Beniston and Lal 2012, Beniston et al. 2016). This process is expected to have important effects on the communities that assimilate in UA yet extracting the effects of soil and land-use histories on community composition requires more than observations of above and below ground biodiversity. Observational studies need to be combined with more experiments and theory in order to tease apart mechanisms of **community assembly** and its impact on ecosystem services in UA.

One potential avenue for exploring these questions utilizes UA's use of both perennial and annual systems and techniques (Bernholt et al. 2009, Lin et al. 2017). UA can be managed to include both perennial (multi-season) and annual (single season) plants, though some sites or plots in allotment-style gardens may specialize on one or the other, depending on the use and extent of mechanical tilling at the site. Perennial sites or plots offer opportunities to study long-term impacts of UA on ecological processes especially when compared to annual sites or plots. Annual sites and plots are tilled every season to aerate soils and minimize weeds whereas perennial sites and plots refrain from disturbing soils. Mechanical tilling has dramatic effects on soil communities in rural agriculture systems, though the effects on UA are less well known (Giller et al. 1997, Höflich et al. 1999). Perennial crops are woody, which commonly include fruit trees and bushes, while annual crops are herbaceous, presenting potential confounding issues for comparative studies. However, some perennial crops, including many herbs, are managed as annual crops and could be used to tease apart effects of soil disturbance on above and below ground community composition.

12.3.2 Ecosystem Ecology – Soil, Water and Nutrients

Though the city clearly had and has effects on UA, UA can also impose significant effects on the city. Management of UA usually involves a large amount of inputs including soil, water and nutrients, which introduce significant chemical loads to the urban environment (Gregory et al. 2016, Harada et al. 2018, Harada et al. 2018, Harada et al. 2019). The infrastructure of most cities is not designed to deal with the **spillover** of agricultural nutrients in human water streams (Harada et al. 2018, Harada et al. 2018, Harada et al. 2019). For example, the very common practice of composting in UA can have dramatic effects on urban waste streams, sometimes leaching significant levels of phosphorus, a common driver of freshwater eutrophication events, into adjacent water bodies (Carpenter et al. 1999, Riskin et al. 2013, Small et al. 2018). Yet as managed systems, UA has great potential to marry ecological processes with food production and environmental stewardship goals. More studies assessing the effects of garden inputs on **nutrient retention and decomposition**, as well as downstream freshwater communities and water quality, are necessary.

Ecological research in UA is beginning to consider the broader system connected to urban soils and water (Wortman and Lovell 2013). Future work in this area may require creative collaborations between landscape architects and ecologists. Designing urban systems to reduce environmental impact has been an important and well-researched component of landscape architecture and should not be ignored by ecologists (Ahern 2013). For UA in particular, researching potentially negative consequences of management on the environment and human health has not been a popular approach (Guitart et al. 2012). For example, proponents of UA may be surprised by the over-use of water by UA in drought-stricken areas (Egerer, Li, et al. 2018, Egerer, Lin, et al. 2018, Lin et al. 2018), or that the presence of a diverse weed community in vacant lots and UA increases the asthma rates of urban youth (Katz et al. 2014). How does UA management impact these flows and is there room for ecology to improve **public health**? Does UA require the equivalent of riparian buffers in rural agriculture, and how would they impact the ecology of UA? When do design systems fail from an ecological perspective and what are the drivers of these failures from the point of view of architects? These are a just a few suggestions in a wide range of possibilities still left for ecologists in UA to explore.

12.3.3 Population and Community Ecology

The diversity of forms that make up UA also create **novel environments** and ecological conditions for the populations and communities that inhabit them (Kowarik 2011). Novel food sources attract organisms including garbage cans and compost heaps (Cofie et al. 2006). Irrigation or structures designed to collect and retain water alter hydrology and create water sources and green infrastructure that may otherwise be absent (Smit and Nasr 1992). Buildings disrupt airflow and present barriers to movement for many organisms (Specht et al. 2014, Quistberg et al. 2016, Glaum et al. 2017). This may pose significant challenges in UA; for example, UA practitioners may find it challenging to successfully rear beehives on rooftop gardens (Colla et al. 2009). The utilization of vertical space in cities to optimize solar radiation can help reduce heat island effects and cool buildings but will likely also have consequences on the growth patterns of plants in vertical gardens (Perini et al. 2013). Despite the relative wealth of research on biodiversity in UA, we know little about how populations and communities adapt to these novel conditions (Table 12.1). How do crop species and associated biodiversity adapt to novel UA environments? Are some species more suited to novel infrastructure and what are the consequences of planting these species in environments where they may not be native?

UA is designed in ways that force novel species interactions. Though green city infrastructure often touts the use of native plant species, in agriculture, particularly UA, native species may be insufficient to satisfy the diverse diets and resource requirements of urban populations (Simberloff 2003, Brzuszek et al. 2007, Pardee and Philpott 2014, Lin et al. 2017). Diverse diets are largely responsible for the overwhelming diversity of UA systems, creating new ecological communities of plants and organisms that subsequently colonize them. Strong **species interactions** are common in agricultural systems, including mutualism (ie. nitrogen fixing bacteria and legumes, pollination, plant-mycorrhizae symbioses), predator-prey and parasitism (ie. biocontrol systems) (Vandermeer et al. 2010). These interactions can be very specialized, yet UA communities are composed of a diversity of non-native species. Thus, there is potential for UA to harbor very specific interactions between otherwise generalist species.

UA systems are just beginning to accumulate the biodiversity data necessary to create and interpret the network structure of UA communities (Akinnifesi et al. 2009, Goddard et al. 2010, Beninde et al. 2015, Otoshi et al. 2015, Speak et al. 2015, Egerer et al. 2017, Philpott and Bichier 2017). Community ecology is already a very active area of research in agroecology, but the extreme levels of disturbance and migration from the global transportation of food and wildlife products through urban cores are likely to have dramatic consequences on the structure and function of UA communities (Phillips 2006, Gliessman 2014). Are species evolving to adapt to the UA environment? At what point do translocated urban species become native or specialists? Do frequent disturbance events exacerbate or disrupt selection and does this favor food production or other ecosystem services of UA? Understanding how these communities assemble, assimilate, and interact to influence food production has not been well studied, yet this gap in knowledge provides ample opportunity for new discoveries.

12.3.4 Habitat Fragmentation and Connectivity – What's Inside and Outside the Garden

A key finding in ecology is that populations on islands are prone to extinction when there is insufficient migration across the landscape (Gotelli 1991). Some islands are source populations that continuously send migrants out to sink habitats, which do not persist long-term. Whether populations inhabiting UA are viable in the long-term is unknown. As islands of habitat in putatively harsh urban space, UA is reminiscent of the islands in classic **meta-population theory** (Levins 1969, Hanski and Gilpin 1991, Gibb and Hochuli 2002, Egerer et al. 2017, Philpott and Bichier 2017). A fundamental question in UA is the relationship between patches of UA in cities and also external, surrounding agriculture and natural areas that may represent sources. Do some gardens act as sources, and others as sinks? What characteristics or management practices in UA produce source or sink habitats for different organisms, and is there some optimal mix to maintain ecosystem function in the long-term?

The reason why many of these questions have few answers is that most studies of UA focus on habitat quality of the garden itself. However, questions related to metapopulation theory require a larger spatial and temporal scope (Table 12.1). As opposed to rural systems where agriculture is often represented as fragmenting forest or other natural habitat, in UA, gardens are often represented as the natural habitat. As such, management and urban design is considered key to conservation efforts (Edwards 2016). UA are sites of high water and nutrient input. They also represent cooler microclimates in otherwise super-heated urban landscapes. This often creates an oasis type habitat, which dispersing organisms are attracted to (Egerer, Li, et al. 2018, Lin et al. 2018). Yet we know little about what is attractive about UA and how these drivers change from species to species. For example, the spatial and temporal scale necessary to maintain populations of species inhabiting UA is unknown, although work comparing local versus landscape level effects in UA tends to favor smaller, local scales (Bennett and Gratton 2012, Otoshi et al. 2015, Egerer et al. 2017, Philpott and Bichier 2017).

If local factors are more important for retention of biodiversity, the question remains, how do ladybird beetles and bees find gardens in cities to begin with and how long do they remain there? A large gap in ecological knowledge of UA concerns the rate and pattern of species spread. Ecologists have studied the spread of fires in forest habitats and disease in populations at length (Pascual and Guichard 2005). From these studies we know that critical densities of trees or people can tip dynamics towards thresholds where fires or epidemics emerge and do not easily subside (Achlioptas et al. 2009). The same types of questions concerning the spread of pests from site to site and zoonotic disease through human-wildlife contact exist in UA.

To answer these spatially-dependent questions, the network structure of the community may be important to consider. Effects of small world, random and highly modular networks on **ecosystem stability** are a very active area of research in ecology (Barabási and Albert 1999, Newman 2001, Pascual and Dunne 2005, Achlioptas et al. 2009). Hunter and Brown found that UA can spread in a neighborhood as a **social contagion** (2012). Yet we know little about how connected UA sites are to one another in a city and to what extent species move between these sites, let alone the ecological consequences of these movements. There is also no obvious reason why these effects will be consistent across species in a community. Connectivity may be advantageous for the spread of pollinators, but not pests. Thus, understanding different percolation thresholds for different organisms and city layouts would be illuminating.

12.3.5 Biodiversity and Ecosystem Services

Most studies of UA focus on the incredible array of biodiversity inhabiting urban gardens and the potential for this biodiversity to improve the diets and well-being of urban residents (Akinnifesi et al. 2009, Goddard et al. 2010, Guitart et al. 2012, Lin et al. 2015, 2017). However, questions remain as to what biodiversity has to offer cities and what kind of biodiversity can most effectively provision valued **ecosystem services**. Most studies in ecology argue that the benefits of biodiversity are derived from the **portfolio effect**, where a few species with particular functional traits contribute disproportionately to ecosystem services. Diversity improves ecosystem services by increasing the chance of having one of these highly- functional species (Tilman 1999, Cardinale et al. 2006). Yet the purpose of UA varies from site to site, and food yields are not necessarily the most important service. Thus, understanding how species diversity contributes to many perhaps conflicting ecosystem services is an area of active research (Iverson et al. 2014).

Urban areas are comprised of diverse human populations that cultivate a diversity of crops for a variety of reasons (Guitart et al. 2012, Lin et al. 2017). Yet we know little about whether **species diversity** significantly overlaps from site to site and how this is related to human demographics, economics and the history of immigration in different cities. From an ecological perspective, adaptation of UA to particular climatic and social conditions is likely to impact the plant communities that can physically colonize the space, thus affecting how diversity influences ecosystem services in different UA environments. Some plants in UA appear to be distant cousins, but many can and do interbreed freely. Varieties develop when growers select individuals for specific traits and isolate populations for other members of the species. Though many growers plant previously purchased seeds or plants, some do save seed that may have interbred with other plants in the garden to create new landraces. Considering the diversity of crop varieties planted in UA, **genetic diversity and adaptation** to the novel urban environment is a fruitful avenue for future research.

Conventional agricultural production focuses on the reproduction of particular breeds, sometimes utilizing clonal propagation techniques to create a more consistent agricultural product (Bellon 1996). As a result, the diversity of human diets may be in decline (Clement 1999). Urban populations may, over time, loose cultural connections with food leading to a homogenization of diets (Goodman et al. 2002). However, UA systems have consistently shown high levels of crop diversity and may act as pockets of **social-ecological memory** (Barthel et al. 2014). Constant influx of people and seed exchanges with new crop varieties may create particularly high levels of genetic diversity in UA. Thus, there may be interesting relationships between food diversity in and surrounding cities. Understanding how the diversity of crops in cities arises and how it impacts and is impacted by crops in adjacent agricultural sites would be an interesting and exciting angle on the biodiversity and ecosystem services of UA.

12.4 Translating Methods from Ecology

While many methods used by ecologists working in other systems can be readily adapted to studies in UA, modifications must frequently be made to adapt to the distinctive social and environmental conditions in which UA systems are embedded. These include high degree of spatial heterogeneity, with complex patterns of land ownership or management, and high levels of human activity, among others (see Methodological issues and next steps, below). Here we detail some of the methods that have been successfully employed in studying the ecology of UA systems, while also highlighting areas where further methodological development and adaptation are needed. Table 12.1 summarizes much of this information, including the types of questions that may be addressed by different methods.

12.4.1 Sampling Biodiversity

Most biodiversity assays are easily adapted to UA systems, though they will likely require modification to deal with, for example, the small size of many UA parcels, the intensive human use these sites, and UA users' priorities and perceptions of science. Organisms of particular interest in UA include crop, ornamental, and weedy plants; pests; natural enemies; pollinators; and soil microbiota (Lin et al. 2015, 2017). These groups are not necessarily taxa specific, but sampling techniques usually are. Here, UA-applicable techniques are briefly summarized by taxonomic group:

12.4.1.1 Plants

Plants in UA can be broadly classified into crops, ornamental and weed species (Lin et al. 2017), though the distinction among these categories may not always be clear: a plant that some users classify as a weed may, to other users—both within and across UA sites—be seen as a crop or ornamental plant, and many plants in UA systems are multifunctional, serving as both food plant and ornamental (Lin et al. 2017). Techniques to assess plant diversity in UA include surveys of quadrats or other fixed areas (Ahrne et al. 2009, Glaum et al. 2017), or entire plots in community gardens (e.g. Egerer et al., 2017). Regardless of survey method, plants are generally identified by sight during the survey, though pictures and clippings may be necessary to identify less familiar species, cultivars or landraces.

The level of taxonomic resolution required will depend on the particular questions being asked; in some cases, coarse identification at the level of the genus may be appropriate, while in other cases identification to the subspecies or cultivar level may be required, particularly for crops and some ornamentals where multiple cultivars may be present in a single site. UA is unique for hosting particularly high levels of local cultivar diversity, making identification below the species level more important than in other systems. For example, *Brassica oleracea* has a variety of cultivars including broccoli, cauliflower, collard greens, and kohlrabi. This dramatic difference in form across cultivars may have implications for the associated biotic communities, and ecosystem function, that would be obscured by identification only to the species level.

Beyond the level of taxonomic resolution required, your research questions will determine other aspects of plant surveys as well. For example, research on pollinators in UA frequently seeks to characterize available floral resources, rather than plant diversity per se (e.g. Glaum et al., 2017; Ahrne et al., 2009). In this case, the researcher need not identify plants that are not in bloom, but will likely want to quantify the number of blooms (e.g. (Fitch 2017) or area covered by blooms (e.g. (Fitch et al. 2019)) rather than simply species richness or abundance.

Alternatively, if interest is primarily in herbivores with a narrow host range, plant surveys may simply focus on the abundance and spatial arrangement of particular host plants of interest. In some cases, it is sufficient to simply quantify the proportional coverage of plants or flowers (e.g. (Matteson and Langellotto 2010), which has the obvious advantage of requiring no knowledge of the particular plant species present.

Finally, given the high rates of temporal turnover in plant community composition in UA—particularly in terms of floral availability, but also plant abundance—consideration must be given to the need for temporal replication of plant surveys. Again, the answer to the question "How many times in a season do I need to do vegetation surveys" will depend on the nature of the research being conducted, but there is clear evidence that the choice of temporal resolution in vegetation surveys can influence inference (Fitch et al. 2019).

12.4.1.2 Invertebrates

Invertebrates play a variety of key ecological roles in UA. Among them are major pests, important natural enemies, vectors of disease, pollinators, and decomposers. Many invertebrates can be sampled efficiently with a sweep-net, but this process can easily damage vegetation. When research is conducted on garden property, care must be taken not to damage crops and ornamentals, which makes sweep-netting impractical in many cases. Alternatives to sweep-netting, including targeted netting, trapping, and direct observation, are discussed below. In general, sweep-netting should be reserved for tall, grassy plants that are not easily damaged, or surrounding non-garden vegetation. To standardize sweep-net samples, a fixed time window and space for observation should be used.

Another technique that readily samples a wide range of invertebrates—particularly flying arthropods—is the use of sticky traps. These are cards—available in various sizes—covered in a sticky substance, and often brightly colored to attract arthropods. Arthropods that are attracted to the trap get caught in the glue and are unable to escape. Sticky traps—generally mounted on a stake to avoid contact with the ground—can be placed in gardens for a set amount of time, then collected. This method is particularly useful for surveying for small insects that might otherwise go undetected, including winged herbivores (e.g. leafhoppers, thrips (Nicholls et al. 2000)) and parasitoids (Bennett and Gratton 2012). Note, however, that specimens caught on sticky traps are often difficult to identify, and frequently cannot be identified to species.

Fixed-area surveys, such as along transects or within randomly placed quadrats, can be useful for assessing the abundance of some invertebrates, particularly larger and less mobile species (e.g. beetles and caterpillars). Such surveys typically involve searching vegetation for the relevant species or collecting all species in a particular guild of interest (e.g. herbivores) from a subset of plants (e.g. individuals of a crop of interest). Note that this survey technique is particularly sensitive to observer bias; care should be taken to ensure that all observers are skilled at locating the species of interest.

Vacuum traps can be used on a wide range of arthropods, both flying and terrestrial (Burkman and Gardiner 2015, Sivakoff et al. 2018). They can be used either to collect individuals that have been noted in visual surveys but cannot be identified in the field (Sivakoff et al. 2018), or to conduct systematic surveys of a set area of vegetation or ground (Burkman and Gardiner 2015).

Other sampling techniques are most appropriate for particular kinds of invertebrates; these techniques are detailed below under the appropriate heading.

12.4.1.3 Herbivores

Arthropod herbivores can be very significant pests in UA systems. Many invertebrate herbivores are best sampled with sweep netting or surveys, as described above. An additional useful technique is the use of sentinel plants. Sentinel plants are generally placed in the garden by the researcher, and are selected because of their attractiveness to the organism(s) of interest or their own importance (Egerer, Li, et al. 2018). Once placed in a garden, sentinel plants can be monitored for colonization by herbivores as well as for herbivore damage.

More broadly, monitoring herbivory on a subset of plants in the study site can be an alternative to direct surveys of herbivores, one that requires considerably less skill and experience on the part of the researcher. Herbivory can be quantified on many scales, including the plot (how many plants show herbivore damage?), the plant (how many leaves show herbivore damage?), and the leaf (what proportion of the leaf has been removed?). Image processing software, such as the open-source platform ImageJ (available for free download at https://imagej.net/Welcome), can be very useful for quantifying herbivory, particularly at the scale of the leaf.

12.4.1.4 Pollinators

Pollinators, including bees, hoverflies, butterflies, and others, are important to ensure adequate pollination of many crop species. Moreover, urban areas have recently been identified as important sites for conservation of insect pollinators (Baldock et al. 2015, Hall et al. 2017). Targeted aerial netting is commonly used to survey flying insects, especially pollinators. As with sweep netting, effort should be standardized by time and/or area surveyed. Because traditional aerial netting, while less damaging than sweep netting, can still cause damage to vegetation, we have found the use of resealable, quart-sized plastic bags as nets to be a useful low-impact alternative. This method has the added benefit of making non-lethal identification easier, as the specimen can be easily observed through the plastic bag, and subsequently released. In surveys of pollinator abundance and diversity, aerial netting is typically paired with pan (also known as bowl) trapping, as the two methods effectively survey complementary subsets of the pollinator community (Roulston et al. 2011). Pan trapping involves setting out small bowls painted with UV-fluorescent paint (generally in yellow, blue, and white) filled with soapy water. The fluorescent pigment attracts pollinators, which are trapped in the water. Standardized protocols for pan-trapping involve moderate-to-large spatial extent, so must often be adapted to the confines of small UA sites (Quistberg et al. 2016, Glaum et al. 2017).

In addition to their utility for studying herbivory, sentinel plants can be very important in studies of pollinators and pollination, particularly regarding questions of pollinator effectiveness and impacts of pollination on crop yield. Sentinel plants may be monitored for pollinator visitation (Lowenstein et al. 2014, Fitch 2017) and/or yield (Potter and LeBuhn 2015, Bennett and Lovell 2019). For questions regarding the importance of pollination to fruit production, or the effectiveness of particular pollinator taxa, it will be necessary to additionally manipulate the ability of pollinators to visit particular flowers or inflorescences. This is most commonly achieved by covering flowers, inflorescences, or entire plants with mesh bags (Potter and LeBuhn 2015) for part or all of the flowering period.

Trap nests—also known as "bee hotels"—are artificial nest sites for cavity-nesting Hymenoptera that can readily be installed in UA sites. These generally consist of a series of tubes, made from e.g. bamboo stems or drilling into wooden blocks. The diameter of the tubes determines the range of species that will recruit to the trap nest. A wide range of questions regarding population and community ecology can be addressed using trap nests (reviewed in (Staab et al. 2018). Given current public interest in bees, and concern over their decline, installation and monitoring of trap nests may be an excellent way to engage UA users in citizen science (McNally et al. 2018).

12.4.1.5 Predators and Parasitoids

12.4.1.5.1 Ants

Ants may be important biocontrol agents in agricultural systems (Larsen and Philpott 2010, Edwards 2016). Protein-feeding ants can be readily sampled using tuna baits (e.g. (Edwards 2016). Baits, arrayed in a grid pattern or along a transect either on the ground or in vegetation, rapidly attract foraging ants. Baits of honey or jam may be used to attract ants that are less predaceous. Because many ant species do not recruit to baits, full characterization of the ant community requires additional sampling via e.g. pitfall traps (see above). As noted for pollinators, above, trap nests can be used to characterize the community of twig-nesting ants (using, of course, substantially smaller tubes; (Larsen and Philpott 2010)).

12.4.1.5.2 Parasitoids

Parasitoids—particularly parasitoid wasps—can be very important biological control agents in agriculture (Goulet and Huber 1993). Sticky traps (see above, under *General techniques*) are generally the most effective way to sample the parasitoid community, though note that the normal difficulty of identifying arthropods caught on sticky traps is amplified by the impressive, often cryptic diversity of parasitoids.

Parasitoids may also be sampled by rearing larvae from herbivore species of interest that have been previously collected or experimentally placed in the study site and collecting any parasitoids that emerge. This method, while time consuming, can give important information about the impact of parasitoids on their hosts and details of host-parasitoid relationships (Tylianakis et al. 2007).

12.4.1.5.3 Soil Arthropods

Soil-dwelling arthropods include significant predators, herbivores, and detritivores/decomposers. They can be sampled using passive trap methods. Pitfall traps are installed in the ground and collect organisms dwelling on the soil surface. They primarily catch fairly mobile and active arthropods. Mini-Winkler extraction involves removing and sifting a set volume of leaf litter and/or soil. This sample is then placed in the Winkler trap—essentially a mesh-bottomed bag suspended over ethanol, with a light source at the opposite end. Arthropods fall into the ethanol at the bottom of the trap and are preserved for future identification. These methods are complementary, in that they sample somewhat different components of the soil arthropod community (Silva et al. 2013). Baermann funnels are useful for sampling nematodes, which are both important consumers in soils and also indicators of soil health (Sharma et al. 2015). Note that all these methods involve disturbing the soil; care should be taken not to damage users' plants in the process. In community gardens in particular, where plots are often rented to users on an annual basis, one possibility is to rent a space in which to conduct these surveys.

Earthworms are often added to soils by growers to aid in decomposition, and are very significant drivers of decomposition (Milcu et al. 2008). They can be surveyed and collected by applying a mustard powder-water solution to the soil (Tresch et al. 2018); this causes the worms to move to the soil surface.

12.4.1.6 Vertebrates

While there is a sizeable literature on vertebrates in urban ecosystems generally, vertebrates in UA have received very little attention (but see (Sorace 2001). This is likely due in part to the small size, relative to territory size of many vertebrate species, of many UA sites. Nevertheless, many questions of practical and theoretical interest could be addressed through the study of vertebrates in UA. As we note below, barriers to the study of vertebrates in UA include concerns from site users about the impact of sampling techniques. In addition, research involving vertebrates generally requires approval from Institutional Animal Care and Use Committees (IACUCs) and potentially other bodies before research can begin.

Some mammalian vertebrates (e.g. rabbits, woodchucks, deer, squirrels, and rats) are significant pests of UA and are a cause of concern to UA stakeholders. Many sites attempt to exclude these species using fencing. At the same time, domestic cats and dogs may provide pest control services as predators (Mahlaba et al. 2017). Minimally invasive techniques for surveying mammals in urban areas include the use of camera traps or identifying animal scat and tracks. Live traps (e.g. Sherman or Longworth traps), baited with food to attract the species of interest, can be used to census populations, and also allow for the collection of tissue samples for DNA extraction (useful for population genetic studies; see below) or fecal matter (useful for diet analysis) (Munshi-South and Kharchenko 2010, Wilson et al. 2016). However, note that baiting and live-trapping activities are likely to raise concerns among UA users.

Other vertebrates of possible interest include birds, bats, and herps (i.e. reptiles and amphibians). Birds, bats, and herps are important biocontrol agents of arthropod pests in agricultural systems generally

(Jedlicka et al. 2011, Maas et al. 2016, Monagan et al. 2017), though their role in UA systems has been less well-studied. All three of these groups are also of conservation interest, so much of the existing research relating them to urbanization has been focused on the capacity of urban areas—including UA—to support robust populations and/or diverse communities (Germaine and Wakeling 2001, Liker et al. 2008, Aronson et al. 2014).

Birds can be surveyed using established point-count methodologies In cases where it is necessary to handle the birds—e.g. to take tissue samples to determine health and disease status, or fecal samples to determine diet—mist nets may be used, though this requires special permits and is likely to attract significant attention from UA users. Bats can also be studied using similar techniques.

Herps (reptiles and amphibians) can be censused by searching along transects, or with more thorough surveys of particular microhabitats (e.g. areas of standing water for frogs; minimally disturbed areas of wood or rock for snakes).

Finally, the large size of vertebrates makes it possible to use telemetry to study the movements of individual organisms. While radio telemetry tracking has been rarely used in urban landscapes, it has the potential to provide important insights on movement ecology and habitat use. For example, a study tracking raccoons in urban areas demonstrated strikingly small territories and intense intraspecific competition (Prange et al. 2004).

12.4.1.7 Microorganisms

Microorganisms play a variety of key roles in UA, though they have been relatively understudied. Mycorrhizae and endophytes play important roles in nutrient absorption by plants, while other free-living fungi and bacteria are central to decomposition and nutrient cycling. Fungal, bacterial or protist infections are common diseases of crops, while other microorganisms can provide beneficial pest and disease control services. Progress in technology has made sampling of microorganisms using molecular techniques very cost effective. At the same time, the rapid development of analytical methods in molecular ecology means that there is no stable consensus on best methodological practices. While research examining soil microbial communities in urban areas have shown that urbanization influences microbial communities (Yan et al. 2016, Wang et al. 2017), to date little research using modern molecular techniques to characterize microbial communities has been conducted in UA systems (Guilland 2018). Thus, the researcher interested in characterizing microbial communities in UA systems would be well advised to choose methods based on a review of the current literature in microbial molecular ecology more generally. Targeted assays involving key macroscopic effects on plants or pest species represent an alternative to the full characterization of microbial communities and may be most time efficient. For example, many entomopathogenic fungi infect pest arthropods. Collecting live insects or cadavers in tubes with sufficient humidity can trigger sporulation of the fungus, providing larger samples to aid in DNA-based identification (Sookar et al. 2008). Regardless of the methods used, more research on microorganisms in is needed. Our limited understanding of microbial communities in UA—particularly in urban soils—and their influence on ecosystem properties represents a major knowledge and research gap in the ecology of UA, one with major implications for management and use of these systems.

12.4.2 Population and Community Ecology

12.4.2.1 Population Dynamics

While surveys are able to provide a snapshot of populations or communities, often ecologists are more interested in the trajectories of population through time. Collecting such data is somewhat more complicated. In the simplest case, surveys can be repeated at multiple time points. For example, one might survey for a pest or natural enemy species several times over the course of the growing season to assess short-term dynamics and seasonal forcing, or over multiple seasons to gauge population trends.

Finer-grained analysis of population dynamics is also possible, at least for some organisms. Perhaps the best-studied example in agroecology is the use of sentinel colonies of social bees (especially of the European honeybee [*Apis mellifera*] and bumble bees [*Bombus* spp.]). The use of sentinel colonies allows for more precise measurement of short-term population change (Vaidya et al. 2018), and a more mechanistic understanding of the drivers of population growth (Williams et al. 2012). Note that some authors have raised concerns about the introduction of commercial or lab-reared bee colonies into the field, due to the potential for pathogen spillover. If managed bees are not already a part of the UA sites in which you are working, it may also be necessary to provide education on safe conduct in proximity to beehives.

12.4.2.2 Landscape Ecology and Movement Ecology

As noted above, the spatial arrangement of UA sites raises many interesting questions regarding the movement of organisms through urban landscapes. Population genetic approaches are very useful for understanding how organisms experience UA sites and the surrounding urban matrix. Such research involves extracting DNA from tissue samples of organisms caught across the landscape and using patterns of genetic differentiation to infer population history. In urban areas, such studies conducted on small mammals (Munshi-South and Kharchenko 2010, Wilson et al. 2016) and bees (Jha and Kremen 2013) have shown high levels of genetic structure, suggesting that urbanization strongly limits movement, though studies in plants find less effect of urban development on genetic structure (Culley et al. 2007, Dornier and Cheptou 2012).The influence of UA on this pattern has been scarcely evaluated, and many open questions remain.

Mark-recapture methods provide a less costly alternative to address similar questions of movement and dispersal. While less expensive, mark-recapture studies require considerable time, and limits inference to short-term and contemporary movements. Relatively few studies using mark-recapture methods have been conducted in UA; Egerer et al. (2018) provide a good example of the potential of such methods, evaluating site fidelity of ladybird beetles by releasing captive-reared individuals marked with fluorescent pigment into gardens in varying landscapes.

12.4.2.3 Predation Services and Community Ecology

Often, researchers in UA are interested not only in the diversity and abundance of biological control agents, but also in their effectiveness at controlling particular pest species of interest. The predation services provided by arthropod predators can be measured using sentinel pests. This involves placing individuals of one or more pest species (potentially including multiple life stages) in the field for a set time, and assessing pest removal as a proxy for predation rate (Yadav et al. 2012, Lowenstein et al. 2017, Philpott and Bichier 2017). Note, however, that users of UA sites may not look kindly upon the addition of pest species to a garden. For such experiments, it is particularly important to communicate with users about the goals and methods of the study, and secure approval. It may be helpful to install pests on potted plants brought to the site especially for that purpose, or in other ways, minimize the risk of the pests causing damage to non-target plants.

To date, while a number of studies have assessed the abundance and diversity of parasitoids in UA systems, there has been very little work explicitly evaluated rates of parasitism or the impacts of parasitoids specifically on host populations in UA.

Exclosures can also be used to evaluate the effect of potential biocontrol agents, particularly vertebrates. Exclosure studies generally involve placing replicated mesh-covered frames around one or more plants, with the size of the mesh determining the types of organisms that will be excluded. Arthropod populations, herbivory, etc. within the exclosures can then be compared to control plots to measure the effect of exclusion (see (Maas et al. 2019) for a recent review of methodology). Exclosure experiments are difficult and time-consuming to establish, are easily tampered with, and require substantial area. For all these reasons, they have been little used in UA systems, or urban areas more generally, despite their potential utility in addressing questions about biological control in UA.

12.4.3 Ecosystem Ecology

Many aspects of the abiotic environment may be of interest to ecologists working in UA; a partial list includes soil nutrient profile, heavy metal content, and moisture; precipitation and chemistry of falling or standing water; gas fluxes between air and soil; concentrations of airborne particulate matter; temperature; and insolation. The inquisitive researcher will surely expand on this list, but in most cases the types of data collected should be tailored to the specific question(s) of interest.

Data loggers, some relatively inexpensive, can easily monitor multiple aspects of the abiotic environment, including temperature, insolation, and soil moisture. Likewise, precipitation can be easily monitored with rain gauges deployed in the field or can be gathered from established monitoring station if coarser-grained data are sufficient. Water meters on irrigation pumps or hoses can reveal how much water users add to UA systems.

Analyses of soil constituents and water chemistry can be accomplished on-site, with a plethora of kits that provide rapid, if somewhat crude, results. For more precise and complete results, lab-based analyses will likely be necessary. For any of these analyses, it is important to have a standardized sampling technique across study sites, ideally including multiple subsamples from the same site (see e.g. Beniston et al., 2016; Harada et al., 2019 for examples). If the equipment to conduct these analyses are not available to you, many university extension services provide soil testing – for nutrient profiles and contaminants – at relatively low cost.

Some groups of organisms are valuable indicators of ecosystem health, due to their sensitivity to pollutants, nutrient deposition, and other features of degraded habitats. Soil and freshwater invertebrates are one such group. Sampling these organisms in and downstream of UA may help reveal levels nutrient retention or spillover. These observational approaches can help describe the current status of UA, but experimental approaches can reveal how UA may change under new management regimes or climatic conditions. Soils can be amended with fertilizers, mulch or artificial media, shade can be constructed, and irrigation can be altered, to provide a few examples.

12.4.4 Spatial Ecology

Spatial analyses are common and useful in UA, making the use of Geographic Information Systems (GIS) particularly important. Most remote sensing tools are not able to distinguish UA from other greenspace in cities, so sites are typically identified manually. Many satellite images of cities are publicly available, though the resolution may not be sufficient to identify all forms of UA. Typically, large community gardens, with characteristic raised box-shaped beds and rows of crops are readily identified. Smaller plots can also be identified if image resolution is high enough, but may need to be mapped on the ground.

Once UA sites are mapped, their spatial distribution and relationship to sampled species, or surrounding levels of impervious surface can be assessed. A variety of spatial analyses are available for determining the spatial distribution of UA or organisms inhabiting UA. These include Ripley's K and Moran's I (see Fortin and Dale, 2014 for an exhaustive discussion of spatial analysis tools for ecologists). There are also historical aerial images available for many cities at extremely high resolutions. Some of these maps are in print and can be retrieved at local libraries for digitizing. These maps provide time sequence data that can reveal changes in landscape characteristics and the gardens themselves over time.

12.4.5 Relevant Datasets

Many datasets covering large spatial and temporal scales are available to use for ecological studies of UA. Baltimore Ecosystem Study Long-term Ecological Research site (BES LTER) has compiled several datasets on ecological systems in Baltimore and other urban ecosystems since 1997 (available at https://baltimoreecosystemstudy.org/besdatav2/). These include biodiversity and ecosystem service surveys of urban greenspaces like urban gardens, though the studies do not tend to focus on UA. Urban gardens have been documented haphazardly in the U.K. and other European countries, in some cases since the 18th century. Many cities have GIS maps of UA submitted voluntarily by garden owners. In many cases, these datasets require extensive digitization and cleanup before analysis is possible, yet the spatio-temporal resolution that they offer is unparalleled.

Though this chapter focuses on natural science methods, studies of UA necessarily have a human component. Particularly useful human datasets for ecology include disease records from the U.S. Center for Disease Control (CDC) as well as demographic and economic information from the U.S. Census Bureau.

Some of this data has been compiled in searchable databases including Project Tycho and Integrated Public Use Microdata Series (IPUMS). IPUMS Terra integrates human demographics with environmental data on land cover, agricultural land use and climate, which can be useful for contextualizing the landscape and environment of UA. There is also data on food security collected by the U.S. Department of Agriculture (USDA) since 1992. Similar databases are available at a variety of municipal levels in other countries as well. Some of this data has yet to be digitized, but many can be accessed in person at local governmental offices and public libraries. Other potentially valuable resources include climate and environmental data from the National Oceanic and Atmospheric Administration (NOAA), historical records on diets, archaeological records on past human settlements, and geological surveys.

12.5 Methodological Issues and Next Steps

UA can exist on private or public land; receiving permission to conduct research at these sites is often a significant challenge. The first step to studying UA is to develop a strong working relationship with UA owners and managers. Developing projects with mutually beneficial outcomes is essential to a successful project. Moreover, creating projects in collaboration with UA stakeholders often strengthens research output and encourages long-term commitments by both parties. Participatory action research espouses these approaches, though the degree to which these collaborative techniques are applied in UA is demonstrably low (Table 12.1). Though natural science tends to focus on the most effective and efficient approaches to answering questions, it is important to remember that UA systems are managed by people, with rights to how data is collected and for what purpose. Privacy is a major concern and care should be taken to maintain anonymity of individuals and sites that participate in studies.

Successful UA sites are, by their nature, often areas of high human activity. This represents an exciting opportunity to study coupled human-natural systems, and to engage the public on research topics, but it also poses significant challenges to the researcher. Research sites are commonly disturbed, often inadvertently and with the best of intentions, by site users. Curious UA site users may want to chat about your research while you are conducting observations that require close attention. In addition, the priorities of site users may require adaptation of study design and methodology. For example, since UA is used for food production, destructive sampling of crops is highly discouraged. If crops do need to be harvested, efforts should be made to minimize impacts on yields, and site users may need to be compensated for the loss. Additionally, many UA stakeholders have animal welfare and conservation concerns. Thus, destructive sampling of other organisms, including some charismatic arthropods, may also be problematic.

In all these situations, we have found that open and clear communication with UA users is the best practice for ensuring a productive and, ideally, mutually beneficial research experience. This includes formal outreach events to explain the purpose of the research and communicate results, as well as openness to informal interactions while on site. Explaining your methods to site users can go a long way to avoiding mishaps and ensuring people are on board with the research. For example, most sampling of species in UA does not pose significant threats to populations in the long-term, so educational outreach events presenting that information to interested stakeholders can quell many concerns. Engaging site users in data collection, while logistically challenging, is also an exciting way to generate buy-in. See Chapters 13, 14, and 15 for more details on how to engage UA stakeholders through citizen science, participatory action research (PAR), and other methods.

The urban environment itself imposes significant challenges for ecological studies of UA. Many UA sites exist in neighborhoods with high crime rates. Thus, steps should be taken to ensure the safety of researchers. This can include traveling to sites in groups and restricting research to daytime hours. In addition, precautions should be taken to minimize the visibility of any equipment left onsite.

In terms of study design, replication is another challenging issue in ecological studies of UA. Most such studies focus on a single city. Specific case studies are important, but these results cannot be generalized. To move ecology of UA forward, comparative methods are needed, particularly to better understand how infrastructure and development patterns influence UA structure and function. For example, Egerer et al. (2018) found that ladybeetles found in UA are more diverse in urban environments in California, but less diverse in Michigan. While this is an intriguing pattern, the dearth of other comparative studies makes it difficult to draw inferences from it. More robust comparisons also have great potential in improving understanding of characteristics that maximize the persistence and dominance of UA or its ecosystem services in urban areas has great potential for development.

Our ability to generalize is further constrained by the limited geographic and temporal scope of existing ecological research in UA. For example, Guitart et al. (2012) found that 51 of 87 papers about community gardens were situated in the USA, mostly in low-income areas of industrial cities. Only 16 other countries were studied, mostly with one or two studies per country (Guitart et al. 2012). Additionally, despite the fact that UA exists in many forms, most research focuses on community gardens. Important microscale processes are likely obscured when we ignore smaller lots and transient guerrilla gardens. Finally, concentrating on the city itself may not be sufficient in understanding the role of UA in the overall food system. Pairing local, landscape, and regional scale analyses is currently at the forefront of much research in UA and stands to greatly improve our knowledge on how UA contributes to overall urban and broadly regional sustainability.

The lack of longitudinal data on UA sites also hobbles our progress in advancing UA ecology. Though land-use history is undeniably important for UA, few studies have sufficient data to understand the long-term consequences on UA of past land-use. Space-for- time substitutions are one potential solution – e.g. using an urbanization gradient as substitute for studying dynamic processes associated with urbanization-as-process (eg. Ahrne et al. 2009, Aronson et al. 2014, Yan et al. 2016). But, as efforts in other areas of ecology have demonstrated, the payoff for investment in long-term data collection is often extremely high. Future studies may do well to consider long-term surveys paired with well-documented methods that can be followed up and replicated in other cities. Some studies of UA have recently turned to the ancient past, where UA may have played important roles in urban resilience to social and environmental stress. Yet, ancient forms of cities are much less dense and have much less impervious surface than modern equivalents. Understanding the relationship between UA in ancient, recent past and contemporary times is another exciting avenue of UA research that will require the development of new methods capable of resolving disparate, fragmented data taken at different spatial and temporal scales.

Ultimately, UA is a socio-ecological system, and an interdisciplinary approach to research will likely prove most fruitful in advancing our ecological understanding of it. At the same time, engaging effectively in such interdisciplinary work is often challenging, as social and natural scientists often employ radically different methodologies and vocabularies (Table 12.1). Yet conceptual advances will often require bridging that gap. For example, if we are to understand how biological control systems composed of many interacting species function in UA, we may first need to understand what motivates people to construct UA. Collecting data that can address both ecological and sociological, anthropological, or economic questions appropriately is a challenge. Chapter 13 in this volume provides an in-depth look at social science methods for UA research, and there are clear opportunities for crossover between the methods presented in this chapter and those discussed there. For example, ethnographic studies may provide critical insight on the motivations of UA stakeholders, but quantitative translations of case studies may be necessary to answer questions beyond the scale of a single city or garden. Though graduate programs are beginning moving towards interdisciplinary studies, many scientists are still trained to specialize narrowly on natural or social science. Advance in studies of UA will require significant collaboration between social and natural scientists, and new methods to synthesize work from the diversity of disciplines represented.

12.6 Conclusions

Though gaps still exist in ecological studies of UA, ecological work in UA is growing rapidly. Research that continues to address ecological theories in UA's unique environment are sure to lead to novel insights. This chapter introduces some techniques that may aid in achieving that goal, with the expectation that new techniques will quickly emerge as the field advances.

Perhaps most significant of the many methodological challenges facing UA is the marrying of its social and ecological components. An examination of the literature cited in this chapter reveals a strong divide between natural science and participatory action research (PAR), with no studies strictly applying participatory methods in natural science (Table 12.2). These could theoretically include collaboration with citizen scientists in UA research of a strictly natural science nature. In fact, only one study in this chapter actually used PAR in tandem with natural science (Table 12.2). This study combined PAR, natural, and

TABLE 12.2

Incorporation of Social Science in UA Natural Science Research

Question Themes	Incorporation of Social Science	References
Succession- legacy effects, path dependence	NS	Adhikari et al. 2004, Bellemare et al. 2002, Beniston et al. 2016, Egerer, M. et al. 2018., Gardiner, M.M. et al. 2014., Katz, D.S.W. et al. 2014.
	NS+ PAR	N/A
	NS+ PAR+ SS	N/A
	NS+ SS	De Kimpe, C.R., Morel, J.-L., 2000.
Ecosystem ecology- soil, water and nutrients	NS	Beniston et al. 2016, Harada, Y. et al. 2019., Harada, Y. et al. 2018a., Harada, Y. et al. 2018b., Iram, S. et al. 2011., Pennisi, G. et al. 2016., Sharma K. et al. 2015, Small, G. et al. 2018., Tresch, S. et al. 2019.
	NS+ PAR	N/A
	NS+ PAR+ SS	Gregory et al. 2015.
	NS+ SS	De Kimpe and C.R., Morel, J.-L., 2000., Lin, B.B. et al. 2018., Minca, K.K. et al. 2013., Riskin, S.H. et al. 2013.
Population and community ecology	NS	Ahrné, K. et al. 2009, Edwards, N., 2016., Egerer, M. et al. 2018., Gardiner, M.M. et al. 2014., Gibb, H. and Hochuli, D.F., 2002., Matteson & Langelotto 2010, Pardee, G.L. and Philpott, S.M., 2014., Philpott, S.M. and Bichier, P., 2017., Sorace A. 2001.
	NS+ PAR	N/A
	NS+ PAR+ SS	N/A
	NS+ SS	Bernholt et al. 2009.
Habitat fragmentation and connectivity- what's in and outside of the garden?	NS	Bennett and Gratton 2012, Edwards, N., 2016., Egerer, M. et al. 2018., Egerer, Monika H. et al. 2017, Egerer, M. H. et al. 2017., Gibb, H. et al. 2015., Pardee, G.L. and Philpott, S.M., 2014. , Philpott, S.M. and Bichier, P., 2017., Quistberg, R.D. et al. 2016., Ricketts, T. and Imhoff, M., 2003., Rudd, H., Vala, J. and Schaefer, V., 2002.
	NS+ PAR	N/A
	NS+ PAR+ SS	N/A
	NS+ SS	Hunter, M.C.R. and Brown, D.G., 2012., Taylor, J.R. and Lovell, S.T., 2012.
Biodiversity and ecosystem services	NS	Akinnifesi et al. 2009, Bennett and Gratton 2012, Bennett and Lovell 2019, Colla, S.R. et al. 2009, Edwards, N., 2016., Egerer, M. et al. 2018., Egerer, Monika H. et al. 2017, Egerer, M. H. et al. 2017., Glaum, P. et al. 2017., Iram, S. et al. 2011., Otoshi, M.D. et al. 2015., Pardee, G.L. et al. 2014., Quistberg, R.D. et al. 2016., Ricketts and T., Imhoff, M., 2003., Sorace, A., 2001.
	NS+ PAR	N/A
	NS+ PAR+ SS	Gregory et al. 2015.
	NS+ SS	Daniels, G.D., Kirkpatrick, J.B., 2006b, Orsini, F., Kahane, R., Nono-Womdim, R., Gianquinto, G., 2013., Speak, A.F., Mizgajski, A., Borysiak, J., 2015., WinklerPrins, A.M.G.A., 2002.
Miscellaneous	NS	N/A
	NS+ PAR	N/A
	NS+ PAR+ SS	N/A
	NS+ SS	Buehler, D. et al. 2006., Martellozzo et al. 2014., Mougeot, L.J., 2010., Taylor, J.R. and Lovell, S.T., 2012.

NS = natural science, PAR = participatory action research, SS = social science social science (Gregory et al. 2016). A more thorough review is necessary to confirm this divide, but our initial findings suggest that natural science in UA could stand to work harder to include UA stakeholders in the design and implementation of future research projects. Some natural science questions, in particular, biodiversity and ecosystem services and ecosystem ecology more commonly combined natural and social science methods, at least for the UA studies reviewed in this chapter (Table 12.2). For other question themes, including succession, population and community ecology, and habitat fragmentation, significant gaps remain in our understanding of how social and ecological dimensions combine to influence UA ecosystem processes (Table 12.2). Though this chapter is meant as a primer on natural science approaches for UA, understanding how to collect data so that it is useful for both social and natural scientists is a key challenge for ecologists working in UA. It is our hope that this primer proves useful not only for the ecologist interested in conducting research in UA, but also can inspire other scholars of UA, from other disciplines, to consider the utility of ecological research methods and questions.

REFERENCES

Achlioptas, D., D'Souza, R.M., and Spencer, J., 2009. Explosive Percolation in Random Networks. *Science*, 323(5920), 1453–1455.

Adams, D. and Hardman, M., 2013. Observing Guerrillas in the Wild: Reinterpreting Practices of Urban Guerrilla Gardening. *Urban Studies*, 51(6), 1103–1119.

Adhikari, T.K., Manna, M., Singh, M.V., and Wanjari, R.H., 2004. Bioremediation measure to minimize heavy metals accumulation in soils and crops irrigated with city effluent.

Ahern, J., 2013. Urban Landscape Sustainability and Resilience: The Promise and Challenges of Integrating Ecology with Urban Planning and Design. *Landscape Ecology*, 28(6), 1203–1212.

Ahrne, K., Bengtsson, J., and Elmqvist, T., 2009. Bumble Bees (Bombus spp) Along a Gradient of Increasing Urbanization. *PLoS one*, 4(5).

Akinnifesi, F.K., Sileshi, G.W., Ajayi, O.C., Akinnifesi, A.I., Moura, E.G. de, Linhares, J.F.P., and Rodrigues, I., 2009. Biodiversity of the Urban Homegardens of São Luís city, Northeastern Brazil. *Urban Ecosystems*, 13(1), 129–146.

Aronson, M.F.J., Sorte, F. a L., Nilon, C.H., Katti, M., Goddard, M. a, Lepczyk, C. a, Warren, P.S., Williams, N.S.G., Cilliers, S., Clarkson, B., Dobbs, C., Dolan, R., Hedblom, M., Klotz, S., Kooijmans, J.L., Macgregor-fors, I., Mcdonnell, M., Mörtberg, U., Pysek, P., Siebert, S., Werner, P., Winter, M., Williams, S.G., Sushinsky, J., and Pys, P., 2014. A Global Analysis of the Impacts of Urbanization on Bird and Plant Diversity Reveals Key Anthropogenic Drivers. *Proceedings of the Royal Society B: Biological Sciences*, 281, 20133330.

Baker, W.L., 1989. A Review of Models of Landscape Change. *Landscape Ecology*, 2(2), 111–133.

Baldock, K.C.R., Goddard, M.A., Hicks, D.M., Kunin, W.E., Mitschunas, N., Osgathorpe, L.M., Potts, S.G., Robertson, K.M., Scott, A.V., Stone, G.N., Vaughan, I.P., and Memmott, J., 2015. Where is the UK's Pollinator Biodiversity? The Importance of Urban Areas for Flower-Visiting Insects. *Proceedings of the Royal Society B: Biological Sciences*, 282(1803), 20142849.

Barabási, A.-L. and Albert, R., 1999. Emergence of Scaling in Random Networks. *Science*, 286(5439), 509–512.

Barthel, S., Parker, J., Folke, C., and Colding, J., 2014. Urban Gardens: Pockets of Social-Ecological Memory. *In*: K.G. Tidball and M.E. Krasny, eds. *Greening in the Red Zone*. Springer Netherlands, 145–158.

Bazzaz, F.A., 1996. Plants in Changing Environments: Linking Physiological, *Population, and Community Ecology*. Cambridge University Press. UK.

Bellemare, J., Motzkin, G., and Foster, D.R., 2002. Legacies of the Agricultural Past in the Forested Present: An Assessment of Historical Land-use Effects on Rich Mesic Forests. *Journal of Biogeography*, 29(10–11), 1401–1420. https://doi.org/10.1046/j.1365-2699.2002.00762.x

Bellon, M.R., 1996. The Dynamics of Crop Infraspecific Diversity: A Conceptual Framework at the Farmer Level 1. *Economic Botany*, 50(1), 26–39.

Beninde, J., Veith, M., and Hochkirch, A., 2015. Biodiversity in Cities Needs Space: A Meta-Analysis of Factors Determining Intra-Urban Biodiversity Variation. *Ecology Letters*, 18(6), 581–592.

Beniston, J. and Lal, R., 2012. Improving Soil Quality for Urban Agriculture in the North Central U.S. *In*: R. Lal and B. Augustin, eds. *Carbon Sequestration in Urban Ecosystems*. Dordrecht: Springer Netherlands, 279–313.

Beniston, J.W., Lal, R., and Mercer, K.L., 2016. Assessing and Managing Soil Quality for Urban Agriculture in a Degraded Vacant Lot Soil. *Land Degradation & Development*, 27(4), 996–1006.
Bennett, A.B. and Gratton, C., 2012. Measuring Natural Pest Suppression at Different Spatial Scales Affects the Importance of Local Variables. *Environmental Entomology*, 41(5), 1077–1085.
Bennett, A.B. and Lovell, S., 2019. Landscape and Local Site Variables Differentially Influence Pollinators and Pollination Services in Urban Agricultural sites. *PLOS ONE*, 14(2), e0212034.
Bernholt, H., Kehlenbeck, K., Gebauer, J., and Buerkert, A., 2009. Plant species richness and diversity in urban and peri-urban gardens of Niamey, Niger. *Agroforestry Systems*, 77(3), 159.
Bowman, J., Cappuccino, N., and Fahrig, L., 2002. Patch Size and Population Density: The Effect of Immigration Behavior. *Conservation Ecology*, 6(1).
Brown, K.H. and Jameton, A.L., 2000. Public Health Implications of Urban Agriculture. *Journal of Public Health Policy*, 21(1), 20–39.
Brzuszek, R.F., Harkess, R.L., and Mulley, S.J., 2007. Landscape Architects' Use of Native Plants in the Southeastern United States. *HortTechnology*, 17(1), 78–81.
Buehler, D. and Junge, R., 2016. Global Trends and Current Status of Commercial Urban Rooftop Farming. *Sustainability*, 8(11), 1108.
Burkman, C.E. and Gardiner, M.M., 2015. Spider Assemblages within Greenspaces of A Deindustrialized Urban Landscape. *Urban Ecosystems*, 18(3), 793–818.
Callahan, J.T., 1984. Long-Term Ecological Research. *BioScience*, 34(6), 363–367.
Cardinale, B.J., Srivastava, D.S., Emmett Duffy, J., Wright, J.P., Downing, A.L., Sankaran, M., and Jouseau, C., 2006. Effects of Biodiversity on the Functioning of Trophic Groups and Ecosystems. *Nature*, 443(7114), 989–992.
Carpenter, S.R., Ludwig, D., and Brock, W.A., 1999. Management of Eutrophication for Lakes Subject to Potentially Irreversible Change. *Ecological Applications*, 9(3), 751–771.
Clement, C.R., 1999. 1492 and the Loss of Amazonian Crop Genetic Resources. I. The Relation between Domestication and Human Population Decline. *Economic Botany*, 53(2), 188.
Clements, F.E., 1916. *Plant Succession: An Analysis of the Development of Vegetation*. Washington DC: USA. Carnegie Institution of Washington.
Cofie, O.O., Bradford, A., Dreschel, P., and Veenhuizen, R. van, 2006. Recycling of urban organic waste for urban agriculture.
Colding, J., Lundberg, J., and Folke, C., 2006. Incorporating Green-area User Groups in Urban Ecosystem Management. *AMBIO: A Journal of the Human Environment*, 35(5), 237–244.
Colla, S.R., Willis, E., and Packer, L., 2009. Can Green Roofs Provide Habitat for Urban Beeds (Hymenoptera: Apidae). *Cities and the Environment*, 2(1), 1–12.
Connell, J.H. and Slatyer, R.O., 1977. Mechanisms of Succession in Natural Communities and Their Role in Community Stability and Organization. *The American Naturalist*, 111(982), 1119–1144.
Cowles, H.C., 1899. The Ecological Relations of the Vegetation on the Sand Dunes of Lake Michigan. Part I.-Geographical Relations of the Dune Floras. *Botanical Gazette*, 27(2), 95–117.
Craul, P.J., 1992. *Urban Soil in Landscape Design*. Hoboken, NJ, USA. John Wiley & Sons.
Craul, P.J., 1999. *Urban Soils: Applications and Practices*. Hoboken, NJ, USA. John Wiley & Sons.
Crumley, C.L., 2018. Historical Ecology. *In*: *The International Encyclopedia of Anthropology*. American Cancer Society, 1–5.
Culley, T.M., Sbita, S.J., and Wick, A., 2007. Population Genetic Effects of Urban Habitat Fragmentation in the Perennial Herb Viola pubescens (Violaceae) using ISSR MarkersCulley et al. — Genetic Effects of Urban Habitat Fragmentation in ViolaCulley et al. — Genetic Effects of Urban Habitat Fragmentation in Viola. *Annals of Botany*, 100(1), 91–100.
Daniels, G.D. and Kirkpatrick, J.B., 2006a. Does Variation in Garden Characteristics Influence the Conservation of Birds in Suburbia. *Biological Conservation*, 133(3), 326–335.
Daniels, G.D. and Kirkpatrick, J.B., 2006b. Comparing the characteristics of front and back domestic gardens in Hobart, Tasmania, Australia. *Landscape and Urban Planning*, 78(4), 344–352.
De Kimpe, C.R. and Morel, J.-L., 2000. Urban Soil Management: A Growing Concern. *Soil Science*, 165(1), 31–40.
Dornier, A. and Cheptou, P.-O., 2012. Determinants of Extinction in Fragmented Plant Populations: Crepis Sancta (asteraceae) in Urban Environments. *Oecologia*, 169(3), 703–712.

Edwards, N., 2016. Effects of Garden Attributes on Ant (Formicidae) Species Richness and Potential for Pest Control. *Urban Agriculture & Regional Food Systems*, 1(1).

Egerer, M.H., Bichier, P., and Philpott, S.M., 2017. Landscape and Local Habitat Correlates of Lady Beetle Abundance and Species Richness in Urban Agriculture. *Annals of the Entomological Society of America*, 110(1), 97–103.

Egerer, M.H., Li, K., and Ong, T.W., 2018. Context Matters: Contrasting Ladybird Beetle Responses to Urban Environments across Two US Regions. *Sustainability*, 10(6), 1829.

Egerer, M.H., Lin, B.B., and Philpott, S.M., 2018. Water Use Behavior, Learning, and Adaptation to Future Change in Urban Gardens. *Frontiers in Sustainable Food Systems*, 2, 71.

Epstein, G., Vogt, J.M., Mincey, S.K., Cox, M., and Fischer, B., 2013. Missing Ecology: Integrating Ecological Perspectives with the Social-Ecological System Framework. *International Journal of the Commons*, 7(2), 432–453.

Evans, K. L., Gaston, K. J., Sharp, S. P., McGowan, A., Simeoni, M. and Hatchwell, B. J. (2009), Effects of urbanisation on disease prevalence and age structure in blackbird Turdus merula populations. *Oikos*, 118, 774–782. doi:10.1111/j.1600-0706.2008.17226.x

Fitch, G., Glaum, P., Simao, M.-C., Vaidya, C., Matthijs, J., Iuliano, B., and Perfecto, I., 2019. Changes in Adult Sex Ratio in Wild Bee Communities are Linked to Urbanization. *Scientific Reports*, 9(1), 1–10.

Fitch, G.M., 2017. Urbanization-Mediated Context Dependence in the Effect of Floral Neighborhood on Pollinator Visitation. *Oecologia*, 185(4), 713–723.

Fortin, M.-J. and Dale, M.R.T., 2014. *Spatial Analysis: A Guide for Ecologists*,. 2nd edition. Cambridge, UK: Cambridge University Press.

Fukami, T., 2015. Historical Contingency in Community Assembly: Integrating Niches, Species Pools, and Priority Effects. *Annual Review of Ecology, Evolution, and Systematics*, 46(1), 1–23.

Gallo, T, Fidino, M, Lehrer, EW, Magle, S, Loison, A., (2019). Urbanization alters predator-avoidance behaviours. *J Anim Ecol*, 88: 793– 803. https://doi.org/10.1111/1365-2656.12967

Gardiner, M.M., Prajzner, S.P., Burkman, C.E., Albro, S., and Grewal, P.S., 2014. Vacant Land Conversion to Community Gardens: Influences on Generalist Arthropod Predators and Biocontrol Services in Urban Greenspaces. *Urban Ecosystems*, 17(1), 101–122.

Germaine, S.S. and Wakeling, B.F., 2001. Lizard Species Distributions and Habitat Occupation Along an Urban Gradient in Tucson, Arizona, USA. *Biological Conservation*, 97(2), 229–237.

Gibb, H. and Hochuli, D.F., 2002. Habitat Fragmentation in An Urban Environment: Large and Small Fragments Support Different Arthropod Assemblages. *Biological Conservation*, 106(1), 91–100.

Giller, K.E., Beare, M.H., Lavelle, P., Izac, A.-M.N., and Swift, M.J., 1997. Agricultural Intensification, Soil Biodiversity and Agroecosystem Function. *Applied Soil Ecology*, 6(1), 3–16.

Glaser, M., Krause, G., Ratter, B., and Welp, M., 2008. Human/Nature Interaction in the Anthropocene Potential of Social-Ecological Systems Analysis. *GAIA - Ecological Perspectives for Science and Society*, 17(1), 77–80.

Glaum, P., Simao, M.-C., Vaidya, C., Fitch, G., and Iulinao, B., 2017. Big City Bombus: Using Natural History and Land-Use History to Find Significant Environmental Drivers in Bumble-Bee Declines in Urban Development. *Royal Society Open Science*, 4(5), 170156.

Gleason, H.A., 1926. The Individualistic Concept of the Plant Association. *Bulletin of the Torrey Botanical Club*, 53(1), 7–26.

Gliessman, S.R., 2014. *Agroecology: The Ecology of Sustainable Food Systems, Third Edition.* 3 edition. Boca Raton, FL, USA. CRC Press.

Goddard, M.A., Dougill, A.J., and Benton, T.G., 2010. Scaling up from Gardens: Biodiversity Conservation in Urban Environments. *Trends in Ecology & Evolution*, 25(2), 90–98.

Goodman, D., Redclift, M., and Redclift, M., 2002. *Refashioning Nature : Food, Ecology and Culture.* Abingdon-on-Thames, UK. Routledge.

Gotelli, N.J., 1991. Metapopulation Models: The Rescue Effect, The Propagule Rain, and The Core-Satellite Hypothesis. *American Naturalist*, 138(3), 768–776.

Goulet, H. and Huber, J., 1993. *Hymneoptera of the World: An Identification Guide to Families.* Ottawa, Ontario Canada: Centre for Land and Biological Resources Research.

Gregory, M.M., Leslie, T.W., and Drinkwater, L.E., 2016. Agroecological and Social Characteristics of New York City Community Gardens: Contributions to Urban Food Security, Ecosystem Services, and Environmental Education. *Urban Ecosystems*, 19(2), 763–794. https://doi.org/10.1007/s11252-015-0505-1

Groffman, P.M. and Tiedje, J.M., 1988. Denitrification Hysteresis During Wetting and Drying Cycles in Soil. *Soil Science Society of America Journal*, 52(6), 1626–1629.

Guilland, C., 2018. Biodiversity of Urban Soils for Sustainable Cities. *Environmental Chemistry Letters*, 16.

Guitart, D., Pickering, C., and Byrne, J., 2012. Past Results and Future Directions in Urban Community Gardens Research. *Urban Forestry & Urban Greening*, 11(4), 364–373.

Hall, D.M., Camilo, G.R., Tonietto, R.K., Ollerton, J., Ahrné, K., Arduser, M., Ascher, J.S., Baldock, K.C.R., Fowler, R., Frankie, G., Goulson, D., Gunnarsson, B., Hanley, M.E., Jackson, J.I., Langellotto, G., Lowenstein, D., Minor, E.S., Philpott, S.M., Potts, S.G., Sirohi, M.H., Spevak, E.M., Stone, G.N., and Threlfall, C.G., 2017. The City as a Refuge for Insect Pollinators. *Conservation Biology*, 31(1), 24–29.

Hanski, I. and Gilpin, M., 1991. Metapopulation Dynamics: Brief History and Conceptual Domain. *Biological Journal of the Linnean Society*, 42(1–2), 3–16.

Harada, Y., Whitlow, T.H., Russell-Anelli, J., Walter, M.T., Bassuk, N.L., and Rutzke, M.A., 2019. The Heavy Metal Budget of An Urban Rooftop Farm. *Science of the Total Environment*, 660, 115–125.

Harada, Y., Whitlow, T.H., Templer, P.H., Howarth, R.W., Walter, M.T., Bassuk, N.L., and Russell-Anelli, J., 2018. Nitrogen Biogeochemistry of an Urban Rooftop Farm. *Frontiers in Ecology and Evolution*, 6.

Harada, Y., Whitlow, T.H., Todd Walter, M., Bassuk, N.L., Russell-Anelli, J., and Schindelbeck, R.R., 2018. Hydrology of the Brooklyn Grange, An Urban Rooftop Farm. *Urban Ecosystems*, 21(4), 673–689.

Höflich, G., Tauschke, M., Kühn, G., Werner, K., Frielinghaus, M., and Höhn, W., 1999. Influence of Long-Term Conservation Tillage on Soil and Rhizosphere Microorganisms. *Biology and Fertility of Soils*, 29(1), 81–86.

Huang, W., Yu, H., and Weber Jr., W.J., 1998. Hysteresis in the Sorption and Desorption of Hydrophobic Organic Contaminants by Soils and Sediments: 1. A Comparative Analysis of Experimental Protocols. *Journal of Contaminant Hydrology*, 31(1–2), 129–148.

Hunter, M.C.R. and Brown, D.G., 2012. Spatial Contagion: Gardening Along the Street in Residential Neighborhoods. *Landscape and Urban Planning*, 105(4), 407–416.

Ines M.G. Vollhardt, Teja Tscharntke, Felix L. Wäckers, Felix J.J.A. Bianchi, Carsten Thies, Diversity of cereal aphid parasitoids in simple and complex landscapes, Agriculture. *Ecosystems & Environment*, Volume 126, Issues 3–4, 2008, Pages 289–292.

Iram, S., Ahmad, I., Nasir, K., and Akhtar, S., 2011. Study of fungi from the contaminated soils of peri-urban agricultural areas.

Isendahl, C., 2018. Urban Ecology in the Ancient Tropics: Foodways and Urban Forms. *Handbook of Urban Ecology.*

Isendahl, C., Barthel, S., and Barthel, S., 2018. Archaeology, history, and urban food security : Integrating cross-cultural and long-term perspectives [online]. *Routledge Handbook of Landscape and Food.* Available from: https://www.taylorfrancis.com/ [Accessed 12 Nov 2018].

Iverson, A.L., Marín, L.E., Ennis, K.K., Gonthier, D.J., Connor-Barrie, B.T., Remfert, J.L., Cardinale, B.J., and Perfecto, I., 2014. REVIEW: Do Polycultures Promote Win-Wins or Trade-Offs in Agricultural Ecosystem Services? A meta-analysis. *Journal of Applied Ecology*, 51(6), 1593–1602.

Jedlicka, J.A., Greenberg, R., and Letourneau, D.K., 2011. Avian Conservation Practices Strengthen Ecosystem Services in California Vineyards. *PLoS One; San Francisco*, 6(11), e27347.

Jha, S. and Kremen, C., 2013. Urban Land use Limits Regional Bumble Bee Gene Flow. *Molecular Ecology*, 22(9), 2483–95.

Katz, D.S.W., Connor Barrie, B.T., and Carey, T.S., 2014. Urban Ragweed Populations in Vacant Lots: An Ecological Perspective on Management. *Urban Forestry & Urban Greening*, 13(4), 756–760.

Karp, D. S., Mendenhall, C. D., Sandí, R. F., Chaumont, N., Ehrlich, P. R., Hadly, E. A., & Daily, G. C. (2013). Forest bolsters bird abundance, pest control and coffee yield. *Ecology letters*, *16*(11), 1339–1347.

Kowarik, I., 2011. Novel Urban Ecosystems, Biodiversity, and Conservation. *Environmental Pollution*, 159(8), 1974–1983.

Lane, P., 2015. Just How Long does 'Long-Term' Have to Be? : Matters of Temporal Scale as Impediments to Interdisciplinary Understanding in Historical Ecology. *The Oxford Handbook of Historical Ecology and Applied Archaeology, Eds: Isendahl, C & Stump, D.* Oxford University Press. UK

Larsen, A. and Philpott, S.M., 2010a. Twig-Nesting Ants: The Hidden Predators of the Coffee Berry Borer in Chiapas, Mexico. *Biotropica*, 42(3), 342–347.

Larsen, A. and Philpott, S.M., 2010b. Twig-Nesting Ants: The Hidden Predators of the Coffee Berry Borer in Chiapas, Mexico. *Biotropica*, 42(3), 342–347.

Levins, R., 1969. Some Demographic and Genetic Consequences of Environmental Heterogeneity for Biological Control. *Bulletin of the Entomological Society of America*, 15(3), 237–240.

Li, X., Poon, C., and Liu, P.S., 2001. Heavy Metal Contamination of Urban Soils and Street Dusts in Hong Kong. *Applied Geochemistry*, 16(11–12), 1361–1368.

Liker, A., Papp, Z., Bókony, V., and Lendvai, Á.Z., 2008. Lean Birds in the City: Body Size and Condition of House Sparrows along the Urbanization Gradient. *Journal of Animal Ecology*, 77(4), 789–795.

Lin, B.B., Egerer, M.H., Liere, H., Jha, S., Bichier, P., and Philpott, S.M., 2018. Local- and Landscape-Scale Land Cover Affects Microclimate and Water use in Urban Gardens. *Science of the Total Environment*, 610–611, 570–575.

Lin, B.B., Philpott, S.M., and Jha, S., 2015. The Future of Urban Agriculture and Biodiversity-Ecosystem Services: Challenges and Next Steps. *Basic and Applied Ecology*, 16(3), 189–201.

Lin, B.B., Philpott, S.M., Jha, S., and Liere, H., 2017. Urban Agriculture as a Productive Green Infrastructure for Environmental and Social Well-Being. *In*: P.Y. Tan and C.Y. Jim, eds. *Greening Cities: Forms and Functions*. Singapore: Springer Singapore, 155–179.

Lowenstein, D.M., Gharehaghaji, M., and Wise, D.H., 2017. Substantial Mortality of Cabbage Looper (Lepidoptera: Noctuidae) From Predators in Urban Agriculture Is not Influenced by Scale of Production or Variation in Local and Landscape-Level Factors. *Environmental Entomology*, 46(1), 30–37.

Lowenstein, D.M., Matteson, K.C., Xiao, I., Silva, A.M., and Minor, E.S., 2014. Humans, Bees, and Pollination Services in the City: The Case of Chicago, IL (USA). *Biodiversity and Conservation*, 23(11), 2857–2874.

Lowenstein, D.M. & Minor, E.S. (2018). Herbivores and natural enemies of brassica crops in urban agriculture. *Urban Ecosyst*, 21, 519. https://doi.org/10.1007/s11252-018-0738-x

Maas, B., Heath, S., Grass, I., Cassano, C., Classen, A., Faria, D., Gras, P., Williams-Guillén, K., Johnson, M., Karp, D.S., Linden, V., Martínez-Salinas, A., Schmack, J.M., and Kross, S., 2019. Experimental Field Exclosure of Birds and Bats in Agricultural Systems — Methodological Insights, Potential Improvements, and Cost-Benefit Trade-Offs. *Basic and Applied Ecology*, 35, 1–12.

Maas, B., Karp, D.S., Bumrungsri, S., Darras, K., Gonthier, D., Huang, J.C.-C., Lindell, C.A., Maine, J.J., Mestre, L., Michel, N.L., Morrison, E.B., Perfecto, I., Philpott, S.M., Şekercioğlu, Ç.H., Silva, R.M., Taylor, P.J., Tscharntke, T., Bael, S.A.V., Whelan, C.J., and Williams-Guillén, K., 2016. Bird and Bat Predation Services in Tropical Forests and Agroforestry Landscapes. *Biological Reviews*, 91(4), 1081–1101.

MacIvor, J. S., Cabral, J. M., & Packer, L. (2014). Pollen specialization by solitary bees in an urban landscape. *Urban Ecosystems*, *17*(1), 139–147.

Mahlaba, T.A.M., Monadjem, A., McCleery, R., and Belmain, S.R., 2017. Domestic Cats and Dogs Create a Landscape of Fear for Pest Rodents around Rural Homesteads. *PLOS ONE*, 12(2), e0171593.

Malik, A., 2004. Metal Bioremediation through Growing Cells. *Environment International*, 30(2), 261–278.

Martellozzo, F., Landry, J.-S., Plouffe, D., Seufert, V., Rowhani, P., and Ramankutty, N., 2014. Urban Agriculture: A Global Analysis of the Space Constraint to Meet Urban Vegetable Demand. *Environmental Research Letters*, 9(6), 064025.

Matteson, K.C. and Langellotto, G.A., 2010. Determinates of Inner City Butterfly and Bee Species Richness. *Urban Ecosystems*, 13(3), 333–347.

Matteson KC, Ascher JS, Langellotto GA. (2008). Bee richness and abundance in New York City urban gardens. *Ann. Entomol. Soc. Am*, 101, 140–150.

Mayergoyz, I.D., 2012. *Mathematical Models of Hysteresis*. Springer Science & Business Media.

McNally, X., Goulson, D., and Fowler, R., 2018. Air Bee n'Bee: a citizen science study of man-made solitary bee hotels as a conservation approach. *In: ECCB2018: 5th European Congress of Conservation Biology. 12th–15th of June 2018, Jyväskylä, Finland*. Open Science Centre, University of Jyväskylä.

Milcu, A., Partsch, S., Scherber, C., Weisser, W.W., and Scheu, S., 2008. Earthworms and Legumes Control Litter Decomposition in a Plant Diversity Gradient. *Ecology*, 89(7), 1872–1882.

Minca, K.K., Basta, N.T., and Scheckel, K.G., 2013. Using the Mehlich-3 Soil Test as an Inexpensive Screening Tool to Estimate Total and Bioaccessible Lead in Urban Soils. *Journal of Environmental Quality*, 42(5), 1518–1526.

Monagan, I.V., Morris, J.R., Rabosky, A.R.D., Perfecto, I., and Vandermeer, J., 2017. Anolis Lizards as Biocontrol Agents in Mainland and Island Agroecosystems. *Ecology and Evolution*, 7(7), 2193–2203.

Mougeot, L.J., 2010. *Agropolis:" The Social, Political and Environmental Dimensions of Urban Agriculture"*. Routledge.

Mougeot, L.J.A., 1999. Urban agriculture: definition, presence, potentials and risks.

Morris, J. R., Vandermeer, J., & Perfecto, I. (2015). A keystone ant species provides robust biological control of the coffee berry borer under varying pest densities. *PloS one*, *10*(11), e0142850.

Munshi-South, J. and Kharchenko, K., 2010. Rapid, Pervasive Genetic Differentiation of Urban White-Footed Mouse (Peromyscus Leucopus) Populations in New York City. *Molecular Ecology*, 19(19), 4242–4254.

Neil J. Reimer, Mei-Li Cope, George Yasuda. (1993). Interference of Pheidole megacephala (Hymenoptera: Formicidae) with Biological Control of Coccus viridis (Homoptera: Coccidae) in Coffee, Environmental Entomology, 22 (2), 483–488, https://doi.org/10.1093/ee/22.2.483

Newman, M.E.J., 2001. Clustering and Preferential Attachment in Growing Networks. *Physical Review E*, 64(2), 025102.

Nicholls, C.I., Parrella, M.P., and Altieri, M.A., 2000. Reducing the Abundance of Leafhoppers and Thrips in a Northern California Organic Vineyard through Maintenance of Full Season Floral Diversity with Summer Cover Crops. *Agricultural and Forest Entomology*, 2(2), 107–113.

Ong, T.W.Y. and Vandermeer, J., 2018. Multiple Hysteretic Patterns from Elementary Population Models. *Theoretical Ecology*, 11(4), 433–439.

Orsini, F., Kahane, R., Nono-Womdim, R., and Gianquinto, G., 2013. Urban Agriculture in the Developing World: A Review. *Agronomy for Sustainable Development*, 33(4), 695–720.

Orwell, R.L., Wood, R.A., Burchett, M.D., Tarran, J., and Torpy, F., 2006. The Potted-Plant Microcosm Substantially Reduces Indoor Air VOC Pollution: II. Laboratory Study. *Water, Air, and Soil Pollution*, 177(1), 59–80.

Ostrom, E., 1990. *Governing the Commons: The Evolution of Institutions for Collective Action*. Cambridge University Press. Cambridge, UK

Otoshi, M.D., Bichier, P., and Philpott, S.M., 2015. Local and Landscape Correlates of Spider Activity Density and Species Richness in Urban Gardens. *Environmental Entomology*, 44(4), 1043–1051.

Pardee, G.L. and Philpott, S.M., 2014. Native Plants are the Bee's Knees: Local and Landscape Predictors of Bee Richness and Abundance in Backyard Gardens. *Urban Ecosystems*, 17(3), 641–659.

Pascual, M. and Dunne, J.A., 2005. *Ecological Networks : Linking Structure to Dynamics in Food Webs: Linking Structure to Dynamics in Food Webs*. Oxford, UK. Oxford University Press.

Pascual, M. and Guichard, F., 2005. Criticality and Disturbance in Spatial Ecological Systems. *Trends in Ecology & Evolution*, 20(2), 88–95.

Pavao-Zuckerman, M.A. & Coleman, D.C. 2005. Decomposition of chestnut oak (Quercus prinus) leaves and nitrogen mineralization in an urban environment. *Biol Fertil Soils*, 41, 343. https://doi.org/10.1007/s00374-005-0841-z

Pennisi, G., Orsini, F., Gasperi, D., Mancarella, S., Sanoubar, R., Vittori Antisari, L., Vianello, G., and Gianquinto, G., 2016. Soilless System on Peat Reduce Trace Metals in Urban-Grown Food: Unexpected Evidence for a Soil Origin of Plant Contamination. *Agronomy for Sustainable Development*, 36(4), 56.

Perini, K., Ottelé, M., Haas, E.M., and Raiteri, R., 2013. Vertical Greening Systems, A Process Tree for Green Façades and Living Walls. *Urban Ecosystems*, 16(2), 265–277.

Phillips, L., 2006. Food and Globalization. *Annual Review of Anthropology*, 35(1), 37–57.

Philpott, S.M. and Bichier, P., 2017. Local and Landscape Drivers of Predation Services in Urban Gardens. *Ecological Applications*, 27(3), 966–976.

Potter, A. and LeBuhn, G., 2015. Pollination Service to Urban Agriculture in San Francisco, CA. *Urban Ecosystems*, 18(3), 885–893.

Pouyat, R.V., Szlavecz, K., Yesilonis, I.D., Groffman, P.M., and Schwarz, K., 2010. Chemical, physical and biological characteristics of urban soils. Chapter 7. *In:* Aitkenhead-Peterson, Jacqueline; *Volder, Astrid*, eds. *Urban Ecosystem Ecology. Agronomy Monograph 55. Madison, WI: American Society of Agronomy, Crop Science Society of America, Soil Science Society of America: 119-152.*, 119–152.

Pouyat, R.V., Yesilonis, I.D., Dombos, M., Szlavecz, K., Setälä, H., Cilliers, S., Hornung, E., Kotze, D.J., and Yarwood, S., 2015. A Global Comparison of Surface Soil Characteristics Across Five Cities: A Test of the Urban Ecosystem Convergence Hypothesis. *Soil Science*, 180(4/5).

Prange, S., Gehrt, S.D., and Wiggers, E.P., 2004. Influences of Anthropogenic Resources on Raccoon (Procyon lotor) Movements and Spatial Distribution. *Journal of Mammalogy*, 85(3), 483–490.

Preston, F.W., 1962. The Canonical Distribution of Commonness and Rarity: Part I. *Ecology*, 43(2), 185–215.

Quistberg, R.D., Bichier, P., and Philpott, S.M., 2016. Landscape and Local Correlates of Bee Abundance and Species Richness in Urban Gardens. *Environmental Entomology*, 45(3), 592–601.

Raciti, S.M., Hutyra, L.R., Rao, P., and Finzi, A.C., 2012. Inconsistent Definitions of 'Urban' Result in Different Conclusions about the Size of Urban Carbon and Nitrogen Stocks. *Ecological Applications*, 22(3), 1015–1035.

Reddy Krishna R., Adams Jeffrey A., and Richardson Christina, 1999. Potential Technologies for Remediation of Brownfields. *Practice Periodical of Hazardous, Toxic, and Radioactive Waste Management*, 3(2), 61–68.

Ricketts, T. and Imhoff, M., 2003. Biodiversity, Urban Areas, and Agriculture: Locating Priority Ecoregions for Conservation. *Conservation Ecology*, 8(2).

Riskin, S.H., Small, G., Mikkelsen, R., Metson, G., Bateman, A., Cooper, J., Hanserud, O.S., Haygarth, P.M., Laspoumaderes, C., McCrackin, M., and Remington, S., 2013. Phosphorus in Urban and Agricultural Landscapes: P is for Preservation. Oxford University Press.

Rosset, P. and Altieri, M., 2017. *Agroecology: Science and Politics*. Warwickshire, UK: Practical Action Publishing.

Roulston, T.H., Smith, S.A., and Brewster, A.L., 2011. A Comparison of Pan Trap and Intensive Net Sampling Techniques for Documenting a Bee (Hymenoptera: Apiformes) Fauna. *Journal of the Kansas Entomological Society*, 80(2), 179–181.

Rudd, H., Vala, J., and Schaefer, V., 2002. Importance of Backyard Habitat in a Comprehensive Biodiversity Conservation Strategy: A Connectivity Analysis of Urban Green Spaces. *Restoration Ecology*, 10(2), 368–375.

Savage, A. M., Hackett, B., Guénard, B., Youngsteadt, E. K. and Dunn, R. R. 2015. Fine-scale heterogeneity across Manhattan's urban habitat mosaic is associated with variation in ant composition and richness. *Insect Conserv Divers*, 8, 216–228. doi:10.1111/icad.12098

Sharma, K., Cheng, Z., and Grewal, P.S., 2015. Relationship between Soil Heavy Metal Contamination and Soil Food Web Health in Vacant Lots Slated for Urban Agriculture in Two Post-Industrial Cities. *Urban Ecosystems*, 18(3), 835–855.

Silva, F., Delabie, J., Santos, G., Meurer, E., and Marques, M., 2013. Mini-Winkler Extractor and Pitfall Trap as Complementary Methods to Sample Formicidae. *Neotropical Entomology*, 42, 351–8.

Simon Tresch, David Frey, Renée-Claire Le Bayon, Andrea Zanetta, Frank Rasche, Andreas Fliessbach, Marco Moretti, 2019. Litter decomposition driven by soil fauna, plant diversity and soil management in urban gardens, *Science of The Total Environment*, 658, 1614–1629, https://doi.org/10.1016/j.scitotenv.2018.12.235.

Simberloff, D., 2003. Confronting Introduced Species: A Form of Xenophobia. *Biological Invasions*, 5(3), 179–192.

Sivakoff, F.S., Prajzner, S.P., and Gardiner, M.M., 2018. Unique Bee Communities within Vacant Lots and Urban Farms Result from Variation in Surrounding Urbanization Intensity. *Sustainability*, 10(6), 1926.

Small, G., Shrestha, P., and Kay, A., 2018. The Fate of Compost-Derived Phosphorus in Urban Gardens. *International Journal of Design & Nature and Ecodynamics*, 13(4), 415–422.

Smit, J. and Nasr, J., 1992. Urban Agriculture for Sustainable Cities: Using Wastes and Idle Land and Water Bodies as Resources. *Environment and Urbanization*, 4(2), 141–152.

Sookar, P., Bhagwant, S., and Ouna, E.A., 2008. Isolation of Entomopathogenic Fungi from the Soil and Their Pathogenicity to Two Fruit Fly Species (Diptera: Tephritidae). *Journal of Applied Entomology*, 132(9–10), 778–788.

Sorace, A., 2001. Value to Wildlife of Urban-Agricultural Parks: A Case Study from Rome Urban Area. *Environmental Management*, 28(4), 547–560.

Speak, A.F., Mizgajski, A., and Borysiak, J., 2015. Allotment Gardens and Parks: Provision of Ecosystem Services with An Emphasis on Biodiversity. *Urban Forestry & Urban Greening*, 14(4), 772–781.

Specht, K., Siebert, R., Hartmann, I., Freisinger, U.B., Sawicka, M., Werner, A., Thomaier, S., Henckel, D., Walk, H., and Dierich, A., 2014. Urban Agriculture of the Future: An Overview of Sustainability Aspects of Food Production in and on Buildings. *Agriculture and Human Values*, 31(1), 33–51.

Staab, M., Pufal, G., Tscharntke, T., and Klein, A.-M., 2018. Trap Nests for Bees and Wasps to Analyse Trophic Interactions in changing Environments—A Systematic Overview and User Guide. *Methods in Ecology and Evolution*, 9(11), 2226–2239.

Stephen C. Rottenborn, 1999. Predicting the impacts of urbanization on riparian bird communities, *Biological Conservation*, 88 (3), 289–299

Tansley, A.G., 1935. The Use and Abuse of Vegetational Concepts and Terms. *Ecology*, 16(3), 284–307.

Taylor, J.R. and Lovell, S.T., 2012. Mapping Public and Private Spaces of Urban Agriculture in Chicago through the Analysis of High-Resolution Aerial Images in Google Earth. *Landscape and Urban Planning*, 108(1), 57–70.

Tilman, D., 1999. The Ecological Consequences of Changes in Biodiversity: A Search for General Principles. *Ecology*, 80(5), 1455–1474.

Tresch, S., Moretti, M., Bayon, R.-C.L., Mader, P., Zanetta, A., Frey, D., and Fliessbach, A., 2018. A Gardener's Influence on Urban Soil Quality [online]. *Frontiers in Environmental Science*. Available from: https://link.galegroup.com/apps/doc/A537786535/AONE?sid=lms [Accessed 10 Oct 2019].

Tylianakis, J.M., Tscharntke, T., and Lewis, O.T., 2007. Habitat Modification Alters the Structure of Tropical Host-Parasitoid Food Webs. *Nature*, 445, 202–205.

Uno, S., Cotton, J., & Philpott, S. M. (2010). Diversity, abundance, and species composition of ants in urban green spaces. *Urban Ecosystems*, 13(4), 425–441.

Vaidya, C., Fisher, K., and Vandermeer, J., 2018. Colony Development and Reproductive Success of Bumblebees in an Urban Gradient. *Sustainability*, 10(6), 1936.

Vandermeer, J. and Perfecto, I., 2019. Hysteresis and Critical Transitions in a Coffee Agroecosystem. *Proceedings of the National Academy of Sciences*, 116(30), 15074–15079.

Vandermeer, J., Perfecto, I., and Philpott, S., 2010. Ecological Complexity and Pest Control in Organic Coffee Production: Uncovering an Autonomous Ecosystem Service. *BioScience*, 60(7), 527–537.

Wang, H., Marshall, C.W., Cheng, M., Xu, H., Li, H., Yang, X., and Zheng, T., 2017. Changes in Land use Driven by Urbanization Impact Nitrogen Cycling and the Microbial Community Composition in Soils. *Scientific Reports*, 7, 44049.

Widdows, C.D., Ramesh, T. & Downs, C.T. (2015). Factors affecting the distribution of large spotted genets (Genetta tigrina) in an urban environment in South Africa. *Urban Ecosyst* 18, 1401. https://doi.org/10.1007/s11252-015-0449-5

Williams, N.M., Regetz, J., and Kremen, C., 2012. Landscape-Scale Resources Promote Colony Growth but not Reproductive Performance of Bumble Bees. *Ecology*, 93(5), 1049–1058.

Wilson, A., Fenton, B., Malloch, G., Boag, B., Hubbard, S., and Begg, G., 2016. Urbanisation Versus Agriculture: A Comparison of Local Genetic Diversity and Gene Flow Between Wood Mouse Apodemus Sylvaticus Populations in Human-Modified Landscapes. *Ecography*, 39(1), 87–97.

WinklerPrins, A.M.G.A., 2002. House-Lot Gardens in Santarém, Pará, Brazil: Linking Rural with Urban. *Urban Ecosystems*, 6(1), 43–65.

Wortman, S.E. and Lovell, S.T., 2013. Environmental Challenges Threatening the Growth of Urban Agriculture in the United States. *Journal of Environmental Quality*, 42(5), 1283–1294.

Yadav, P., Duckworth, K., and Grewal, P.S., 2012. Habitat Structure Influences Below Ground Biocontrol Services: A Comparison Between Urban Gardens and Vacant Lots. *Landscape and Urban Planning*, 104(2), 238–244.

Yair Paker, Yoram Yom-Tov, Tal Alon-Mozes, Anat Barnea (2014). The effect of plant richness and urban garden structure on bird species richness, diversity and community structure. *Landscape and Urban Planning*, 122,186–195

Yan, B., Li, J., Xiao, N., Qi, Y., Fu, G., Liu, G., and Qiao, M., 2016. Urban-Development-Induced Changes in the Diversity and Composition of the Soil Bacterial Community in Beijing. *Scientific Reports*, 6, 38811.

Zezza, A. and Tasciotti, L., 2010. Urban Agriculture, Poverty, and Food Security: Empirical Evidence from a Sample of Developing Countries. *Food Policy*, 35(4), 265–273.

13

Navigating Urban Agroecological Research with the Social Sciences

Evan Bowness,[1] Jennifer A. Nicklay,[2] Alex Liebman,[3] Kirsten Valentine Cadieux,[4] and Renata Blumberg[5]

[1] Institute for Resources, Environment and Sustainability, University of British Columbia, Canada, Unceded Musqeam Territory
[2] Department of Soil, Water, and Climate, University of Minnesota, USA
[3] Department of Geography, Rutgers University, USA
[4] Environmental Studies and Anthropology Department, Hamline University, USA
[5] Department of Nutrition and Food Studies, Montclair State University, USA
[*] Corresponding Author: Email - HYPERLINK "mailto:evan.bowness@gmail.com" evan.bowness@gmail.com

CONTENTS

KEY WORDS: *Urban political agroecology; critical social science; reflexivity; praxis; urban agriculture; community food systems; research methods.*

13.1 Introduction: Urban Agroecology and Social Science

Who is involved in urban agroecology? Why are they involved? What are the challenges they face? Who is excluded?

What are social, political, historical, and economic conditions that enable or hinder urban agroecology? What concepts and perspectives help people understand those conditions?

What visions of the future do urban agroecology practitioners and researchers want to bring to life? Where are the points of leverage that can help support building toward such futures? What are the research outputs, outcomes, or byproducts that community partners may be interested in?

Urban agroecology, like agroecology more broadly, is a science, practice, and movement (Wezel et al. 2009). This chapter focuses on urban agroecology research that is informed by the *social sciences.*[1] In its broadest sense, social science contributes rich histories, insights, tools, and action strategies for exploring society–environment relationships within *urban agoroecologies*. Building on Arnold and Siegner's argument (Chapter 15) to use the language of urban agroecology over urban agriculture, we propose pluralization—urban agroecologies—to emphasize the diversity of urban agroecological practices and research communities. Regardless of discipline, researchers must all eventually grapple with the social context of that diversity and recognize that social actors, social relationships, and social institutions are central to urban agroecologies.

Informed by fields such as sociology, geography, philosophy, and history of science and scientific knowledge, social science research helps us understand these social phenomena. For example, ethnographic research can highlight how racialized exclusion from urban farm programming is connected to wider processes of racism and discrimination (Davenport and Mishtal 2019). Participatory Action Research can be used to develop indicators that assess the degree to which urban growers are democratically empowered to shape their food systems, exploring governance structures and distribution of decision-making power (Garcia-Sempere et al 2019). Visual methodologies (such as participatory video) can demonstrate how communication among community garden participants relates to broader power dynamics in self-organizing communities and social networks (Yap 2019). Evaluation research projects can contribute to community management, protection, and love of the food lands near where research participants live, such as in the Food Dignity project (Porter et al. 2018). Through these projects, we see examples of how social science methods can explore the conditions driving, enabling, or hindering urban agroecologies, as well as its effects or outcomes.

Social science *methods*—and the choice of when and how to use them—are grounded in broader *ways of thinking* about social relations. This distinction is explored by Opaskwayak Cree scholar Shawn Wilson in *Research Is Ceremony* (2008). *Methods*—such as qualitative interviewing, participatory modeling, discourse analysis, and surveys—are tools used to explore research questions, collect and analyze data, and communicate results. Specific methods are informed by *methodology, or a methodological approach*, which is a system that aligns methods with research topics and questions. Underpinning both are *axiology*, the "ethics or morals that guide the search for knowledge" (p. 34), or the value systems that guide the methodological choices we make as researchers (whether we are conscious of those value systems or not).[2] The ways of thinking encompassed by methodologies and axiologies points to a core acknowledgement in the *critical social sciences*: research is never value neutral.

Critical social science—an umbrella term used across multiple disciplines—investigates power and inequality in social relations (Box 13.1). It uses analytical concepts such as privilege, exclusion, and marginalization, usually with an intention to oppose domination and oppression by supporting more equitable social and political relationships. This is especially important for the relations between research participants and their social contexts, and also in light of a history of research (including social and natural science urban agroecology projects) that have been heedless of relationships and, as a result, have been extractive and unaccountable to research participants (Wilmsen 2008). Therefore, critical social science helps urban agroecology scholars across disciplines remember that even when their research or learning projects are not *primarily* focused on the social complexities of power and marginalization, these complexities are still present and *are always being navigated.*

Our goal in this chapter is to introduce social science concepts to help scholars navigate the complex social relations that are always present in urban agroecologies. We argue that these ways of thinking, grounded in the social sciences, are powerful tools for conceptualizing and critiquing power and inequality in urban agroecosystems—whether you are a social scientist, natural scientist, artist, organizer, urban planner, or anyone else involved.

---

[1] We use the terms social sciences and social science research synonymously, referring to the use of methodologies and methods designed to understand the way people relate to one another and the world.

[2] Consider an example of these distinctions: a project using a participatory action research *methodology* may choose to use a participatory evaluation survey as a specific *method*, both of which are grounded in an *axiology* that prioritizes collective governance.

BOX 13.1 CRITICAL SOCIAL SCIENCE

By analyzing social relations with an analytical focus on power, critical social science problematizes conditions where power is being concentrated or dominated by a particular social group. Terms such as "critical" or "problematize" are often viewed as "complaints" but here we use them with a more specific meaning; "problematizing," in this context, usually means challenging often taken-for-granted assumptions about how a given condition is seen as normal or natural, making it more possible to address how assumptions affect the dynamics being researched. Critical social science perspectives tend to focus on how groupings of people, designated by their gender, racialized identity, class position, or some other social positionality are either privileged (in that they have disproportionately greater access to power and resources) or marginalized (having less access to power and resources). Critical research identifies situated and context-specific causes of this social inequality and also explores mechanisms for redistributing power across lines of social difference—*addressing* inequitable access to power and resources. "Critical" approaches can be contrasted with positivism in social science; while both share an interest in causal mechanisms in society, critical social scientists engage with ethical questions that cannot be resolved through positivism's emphasis on scientific rationalism and objectivity.

In the second section, we discuss some concepts necessary to recognize, name, and reflect the value position implicit in this focus on inequality. We recognize that many of these concepts are often dismissed as "jargon," or the terminology and labels avoided because they seem inaccessible, and we have consequently used boxes throughout the chapter to define key concepts in order to make them clearer and more accessible, regardless of your disciplinary background. (We also use boxes to highlight the connections between the content of the chapter and the types of research that we, the authors of this chapter, conduct.) We make the case that social science concepts and terminology are valuable tools to organize observations and experiences and that competency in this area may be expected by the communities you will encounter if you are engaged in fieldwork. Thus, people conducting urban agroecological research need to build these competencies, or at least to start by building literacy and collaborating intentionally with people who are more familiar with critical social science.

In the third section, we consider five broad theoretical perspectives, or threads of scholarship (what we refer to as "lineages") from which many of our own social science research methods and methodologies have grown. We conclude with a discussion of how, informed by these lineages, urban agroecological research can be more attentive to its social context, engaged with and addressing inequalities in the literal and scholarly fields in which we work. While some scholars more readily identify as social scientists, we argue that everyone involved in urban agroecology can benefit from engaging with the ideas presented here. We hope that this chapter, which we treat as an evolving conversation, can help scholars, practitioners, and movement participants feel more comfortable framing, approaching, and answering questions about the social dimensions of urban agroecologies: as social science, as social practice, and as a diverse social movement.

13.2 From Value Neutrality to Reflexive Praxis

Adopting a critical approach in urban agroecological research is not a value-neutral decision because researchers are not objective instruments that discover scientific truths. Social science perspectives remind us that we are always humans, trying through research to understand the actions and reactions in social networks that extend far beyond our scope of experience. Remaining aware of our own position within these networks is a feature of research informed by the social sciences that we refer to as *reflexivity* (Box 13.2).

We can choose to be reflexive, to recognize and acknowledge the effects of our own social position on our work as researchers. We (the authors of this chapter) are just as enmeshed in our own biographies,

BOX 13.2 REFLEXIVITY

For social scientists, reflexivity often refers to *recognizing* our social locations in terms of class, race, gender and other forms of privilege and marginalization, also called "positionality," and *reflecting* on what that position means for the knowledge that we co-create and use when engaging in research. Unfortunately, this is different from the common use of the term, in which "reflexive" connotes a "knee-jerk," or reflex-like, reaction—the opposite of self-awareness of position and relationship processes. We use the term "reflexivity" cautiously, as the researcher is inseparable from the messy entanglement of socio-ecological relations in which they are embedded; consequently, there is no removed space from which to reflect.

histories, and values—and those of our disciplines—as you (the reader of this chapter) are. Practicing reflexivity about these positionalities is done not so much to focus on us as individual researchers but instead to focus on how, as Nagar and Geiger (2007) argue, researchers' "identities intersect with institutional, geopolitical, and material aspects of their positionality" (p. 356). This frame allows us, as researchers, to acknowledge the contingency of identity depending on our relationship to the land, communities, and institutions with which we are interacting—and the ethical implications involved (Wilson 2008); it also helps us understand how we as researchers can "produce knowledges across multiple divides (of power, geopolitical and institutional locations, axes of difference, etc.) without reinscribing the interests of the privileged" (Nagar and Geiger 2007, p. 267).

With this in mind, and without only *listing* potentially relevant social positions, we believe that the following identities are important to share within the context of this chapter. As researchers and practitioners talking to other researchers and practitioners, we seek to move with self- and context-aware reflexivity between positions where our identities (queer/straight, racialized/White with settler and mixed ancestry, gendered, cultured, institutionally-affiliated, etc.) confer more and less privilege and power. Collectively, we conduct community-based research: research with, for, and on behalf of communities of people who practice urban agroecology or are engaged with urban agroecology movements. We are geographers, political ecologists, sociologists, and soil scientists. We've conducted transdisciplinary natural and social science projects. We are urban farmers and gardeners, organizers, community land trust advocates, and peer-to-peer knowledge facilitators. Our work is critical (see Box 13.1). It is also reflexive (see Box 13.2), as we continually revisit, deconstruct, and rebuild the assumptions in ourselves and in the tools we use (also called "self-reflective" work) to identify the values present in our decisions, observing their impact on the social and ecological contexts within which our research is entangled, and making efforts to modify our actions and relations based on these reflections.

Reflexivity, informed by critical social science, directs our attention as researchers to identify power and inequalities across entire *fields* of agroecology. In using "fields" here, we refer to *literal fields* where agroecological practices and movement politics take place. Fields also refers to *fieldwork*, which describes the process of collecting data as well as other activities to build relationships with communities, practitioners, and activists with whom research is being conducted (while recognizing that the lines between these roles might be blurred). Finally, fields also refer to *fields of knowledge*, such as different disciplinary approaches to producing knowledge within agroecology as a science, including knowledge gained through research and also practitioner and traditional approaches to learning and answering research questions. In this chapter, we focus particularly on this last use of fields, as critical and reflexive social science illuminates how research is problematically enmeshed in processes of unequal racialization, gender marginalization, gentrification, and dispossession simply because agroecological research *exists in a world shaped by vast inequities across social difference*. In studying urban agroecologies, there is no outside, no perfectly objective space, no vantage point unaffected by these social dynamics.

We cannot emphasize this point enough: if you are doing urban agroecological research (from any disciplinary position), the research questions, the relationships, the results, and even you as a researcher are fully enmeshed in social relationships and institutions. This is why research is never value neutral, and reflexivity represents a powerful tool to navigate research—especially for those in natural science

BOX 13.3 PRAXIS

A rallying cry for transformative research is often traced to an oft-quoted statement from Karl Marx's *11 Theses on Feuerbach*, where he claimed that "the philosophers have only interpreted the world in various ways. The point, however, is to change it." The combination of research and effort to bring about change is often referred to as "praxis" (Lather, 1986). If a critical social science (operating within a social justice value frame, for example) identifies social inequities, then its extension to critical praxis aims to redistribute power and opportunity, and lessen, to the extent that it can, those inequities. Reflexive research praxis, therefore, also aims to identify inequities in the practice of science itself.

disciplines that value objectivity. Reflexivity in our own research means we attempt to learn about and to counter the role that research itself has played in perpetuating inequities, such as legitimating inequality by essentializing differences between "races" or "sexes" (Barkan, 1992). Reflexivity also allows us to challenge the inequalities pervasive in the history of mainstream sciences, and support change to social relations that bring about liberation, equity, and a redistribution of power and privilege in society. This approach is sometimes referred to as a research *praxis* (Box 13.3).

13.3 Lineages Through Which Social Science Research "Comes to Life" in Political Agroecology

As researchers we are all immersed in disciplinary training, and as a result, the ethical assumptions embedded in academic disciplines (axiologies) affect how we approach and conduct research (our methodology and the methods we use). Rather than listing and reviewing methodological approaches associated with specific social science disciplines and suggesting appropriate social research methods pairings, we take this as an opportunity to step back and trace some of the broad scholarly *lineages* that can inform a critical axiological stance. In the following sections we point to deeper roots growing beneath methodological decisions such as which research questions to ask, what methods to use, and why engage in research in the first place. We use the term "navigate" in this chapter's title to refer to a sense of orientation to a map drawn by our own intellectual, practical and movement ancestors—in other words, the ideas reviewed in this chapter guide our critical and reflexive research praxis.[3]

Agroecological research can therefore be thought of as a pathway through which diverse scholarly lineages "come to life." We have selected five lineages that are relevant in agroecology, and particularly urban agroecology. They are interdisciplinary in the sense of multiple academic disciplines interacting, or transdisciplinary in the sense of bridging academic and other forms of knowledge only possible through collaboration across sectors (Mendez et al. 2016). We imagine these lineages as clusters or a braided stream (rather than existing in isolation or within a single discipline) that together contribute to what is often referred to as "political agroecology." Political agroecology includes multiple ways of seeing agroecology in political terms—not only as a social movement, but as a political and economic project engaged with issues related to social inequality, the state, and power.[4] Political agroecologists (as

[3] We note that the language being used here, of navigation and lineage, are also subject to critique due to their connections with the colonialist tendencies of our inherited methodologies for studying social and environmental issues (see below on Development Studies, and see also common and unselfconscious invocation of "pioneering" in academic contexts). We consider the way that academics understand their lineages to be part of the ancestor wisdom we inherit to guide us—and seek to respect and also change in solidarity with those we work and live alongside.

[4] Such approaches are varied. For instance, Hernandez (2020) considers agroecology, adopted by the MST, the Brazilian Landless Rural Workers Movement (one of the largest agrarian social movements in the world), as a political tool for social peasant organizing for agrarian reform. In a similar vein, Dale (2019) examines how the National Farmers Union of Canada, a radical farming advocacy group, uses agroecology as a frame for farmer activists mobilizing against corporate-dominated and energy-intensive agrifood system actors and their effects on the climate crisis.

BOX 13.4 COLLECTIVE EXAMPLE OF PRAXIS AND REFLEXIVITY

Three of us (Nicklay, Liebman, and Cadieux) have been involved over the past decade in the development of a community agricultural land trust in the Twin Cities metro area in Minnesota. This represents one form of praxis for us, since we have used our research on urban agroecological land management and its potential benefits to cities and neighbors to advocate for structures, institutions, and other conditions that support urban agriculture—particularly through land access. Land in this area is Dakota and Ojibwe territory and exists in a social context distinguished by significant migration from the U.S. South and Central/South America, East Africa, and Southeast Asia. Because of these contexts, we enact critical reflexivity about our social positionality to enact research processes and praxes that are embedded in social networks that partner with Indigenous and immigrant communities and value the guidance of elders and growers. These are ways we can reflexively acknowledge and work to change the exploitative and dispossessing conditions set up as defaults in the Settler colonial definitions of land as "property" that exists with the goal of maximizing capital investment. Community land trusts work to decenter White and official expertise and charitable "helping" in favor of healing land relationships, supporting equitable access to land, and valuing relationships to land beyond monetary value.

scholars, advocates, and farmers) suggest that sustainability transitions require political change (Molina 2013). What makes political agroecology *political* is that it is informed by critical social science perspectives on intersectional issues, such as capitalist accumulation and domination of subordinate classes, patriarchy and gender inequality, racial injustice and racial privilege, and environmental degradation. It is also political in that it provides analytical tools and perspectives for understanding how to mobilize in response to these intersecting issues (see Box 13.4 for one example).

In the following sections, we discuss how development studies, agrarian and society–environment studies, feminist perspectives, environmental justice, and Indigenous food sovereignty align with the critical orientation of interdisciplinary political agroecology. This applies equally to political agroecology in urban contexts (Anderson 2017, Tornaghi and Dehaene 2019, Roman-Alcala and Glowa and Roman-Alcala, Chapter 8); placing political agroecology within broader disciplinary and intellectual fields bolsters our *call for urban agroecology as a critical and reflexive research praxis.*

13.3.1 Development Studies

The idea of "development" reflects the Eurocentric or "Western"-style trajectory toward competency, where non-western (or "Global South") societies are perceived as inferior, lacking, and needing intervention from more "developed" nations (Escobar 1995, Chakrabarty 2000, Power 2003, Lee and Ahtone 2020). Mainstream development studies influenced U.S. urban social science research through the proliferation and expansion of New Deal reforms (Gilbert 2015) and post-WWII development projects abroad (Escobar 1989), building on legacies of colonial science (Mitchell 2002). Against the backdrop of rural Communist and anti-colonial movements, the Vietnam War, and a global landscape of proxy battles between the United States and USSR, young, university-trained American technocrats attempted to forge rational, economistic futures for the global peasantry and new global urban classes (for a comprehensive history, see Offner 2019).

While the notion of "development" may have been reserved for the Global South, these researchers and development workers, upon returning to the U.S., could not help but compare the endemic poverty and racial differences in the U.S. to their experiences abroad. Black and brown urban dwellers would be studied and observed under the same expectant gaze, the "improving eye" (Pratt 2007). Echoing prior colonial removal of edible landscapes from cities in the Global South, which resulted in alienation from local food self-sufficiency (and the intended need to work for wages to meet basic needs), we see mid-century development ideals applied in urban areas of the Global North in "city beautiful" policing of "backwards," rural or lower-class land uses like gardens in cities (WinklerPrins 2017). Further, development ideologies and practices helped normalize a relationship between research and the "rule of experts" (Mitchell 2002, Gilbert 2015) that remains a source of distrust today, especially around relationships between communities and state-supported researchers, and even more so in the domain of allocation of land and resources.

In addition to the impact of development studies on urban planning, developmentalist paradigms around agriculture also impact urban agroecologies. Dominant explanations of food provision call urgently for the need to "feed the world" through more intensive, high-input, monocultural agriculture—what critics call "productivism." Norman Borlaug, hero of this narrative, explicitly connected the proliferation of productivist agriculture to the creation of logistical supports for capitalist democracy. When smallholders in hard-to-govern regions are transformed into farmers, commons are enclosed to become economically productive, farmers become dependent on global markets (e.g., buying inputs and equipment, selling products), which, in turn, supports taxation for road building to reach, and hence call, additional smallholders into the logic of the state and market (Cullather 2010, Scott 2010).

This cycle continues through proclamations that farms must "feed 9 billion by 2050" by expanding in size and adopting a suite of new technologies. Within this paradigm, urban agroecologies are either dismissed due to their seemingly inconsequential scale or justified using productivist logics of yield, employment, or tax revenue per area (WinklerPrins 2017). Often, this is further perpetuated by agronomy departments and Extension programs mobilizing the expert delivery of knowledge for "influencing nutrition and physical activity behaviors of low-income families" in ways that are shaped by Green Revolution-era programs[5] and in conceptions of land embedded in the very formation of U.S. Land Grant institutions.[6] Urban agriculture contests these traditions in urban planning, agriculture, and Extension because it recognizes that food production happens (and has always happened) in cities (WinklerPrins 2017) and often seeks to re-establish common land (Tornaghi 2017).

It is useful to understand some of underlying tensions between dominant models of economic "development" and what we refer to as the more critical fields studying societal well-being and quality of life (see Box 13.5). We note that many of these fields (e.g., geography, rural sociology, anthropology, and more participatory branches of Extension) have undergone reflexive turns to grapple with their history in development praxis as tools of imperialist expansion. Many large fields that focus on the ecological aspects of urban agroecologies, however, do not have reflexive or critical training. In other words, biology, ecology, agronomy, horticulture, and most Extension outreach programs have largely not problematized their roots in "feeding the world" within development initiatives. This tension is often visible in clashes over not only methodologies, but the very premises of urban agroecological projects—for example, in debates over whether agroecology is a science or movement, over whether agriculture is legitimate as a "highest and best use" of land in cities, or whether agroecologies can be normalized as part of urban form. "Developmentalist" ideologies govern against the emancipatory ideals identified with agroecology as a movement by dismissing traditional ecological knowledge in favor of expert "assistance" verging on saviorism. These strands of "development" research are important to consider in order to understand *what contemporary traditions of critical scholarship on food sovereignty, food justice, and community empowerment are reacting against*—traditions which we turn to in the following sections.

[5] The U.S Land Grant University system's "Cooperative Extension" arm provides technology transfer and domestic "development" units. One such program, for example, is the Expanded Food and Nutrition Education Program, which "addresses critical societal concerns by employing paraprofessional staff and *influencing nutrition and physical activity behaviors of low-income families*" (University of Minnesota Extension, n.d., emphasis ours). In this example, we see the continuation of developmentalist science in the way expert knowledge is portrayed as necessary, improving, and top-down.

[6] "Land-grant" universities are state-run universities that were given federal lands as endowments (for a robust investigation of land-grant university land, see Lee and Ahtone 2020). It is important to point out the history of land-grant universities is complex and deeply intertwined with the ongoing occupation of Indigenous lands, and as such agroecology practiced on university lands is an extension of colonial expropriation (see section below on Indigenous Food Sovereignty).

BOX 13.5 SITUATING LIEBMAN'S WORK IN RELATION TO DEVELOPMENT STUDIES

Alex studies the development of international agricultural science in Colombia and its relationship to historic and ongoing peasant struggles for land. He traces how a focus on production, attempting to resolve peasant problems through increased yield, has shaped development agronomy since its inception. Rather than addressing structural issues in Colombia's rural areas (such as land inequality), increased production of select, market-oriented crops has worked in conjunction with the motivations of agricultural industries. Under the guise of improving peasant livelihoods, development-oriented agriculture institutes have worked hand-in-hand with agribusiness and government agencies to consolidate land, increase agri-chemical use, and concentrate power over supply chains within which smallholders have little autonomy. Alex is a geographer who uses tools of environmental history, science and technology studies, and human-environment geography to understand how the scientific approaches of agricultural development—moving from genetic "improvement" to systems ecology to contemporary digitization—are interwoven with political and economic considerations of capitalist accumulation and the management of unruly natures and political subjects.

13.3.2 Agrarian, Peasant, and Society–Environment Studies

The lineages of critical scholarship that emerged in significant part as a reaction against development studies can be roughly grouped around society–environment studies and agrarian studies, or peasant studies. This work takes place in disciplinary contexts from rural and environmental sociology and political science to transdisciplinary programs in natural resource and environmental management. Given the history and prevalence of people growing food near where they live, it should not be surprising that there is no particular disciplinary "ownership" of studies of urban agriculture and agroecologies.

Current theoretical trends in agroecology draw heavily on critical agrarian studies, which, in turn, draws on Marxian and feminist scholarly traditions, as described further below. Critical agrarian studies explores critiques of the commoditization of land, labor relations, capital, and State processes of governance (Edelman and Wolford 2017). The conceptual frame of "corporate food regimes" focuses even more closely on these latter themes by investigating the strategic relationship between state power, commercial agriculture, and supply chains structuring the global economic system (Friedmann and McMichael 1989, McMichael 2009, Holt-Giménez and Shattuck 2011). Corporate food regimes can be traced to colonial food production that drove 19th century industrialization, which established the early political and economic infrastructure for contemporary multinational food corporations to amass power through strategic control of supply chains, trade, and financing opportunities (McMichael 2009). These changes have supported tendencies of food system commoditization, financialization, and consolidation with the attendant exploitation of agricultural laborers, dismantling of peasants' livelihoods, and ecological havoc—tendencies that often lead to the displacement of rural farmers to cities. The corporate food regime creates (and thrives on) divisions between the city and country and between nature and society—often described in terms of the "metabolic rift" (Moore 2016, Foster 1999). Using this analytical frame, researchers turn their attention to analyzing the displacement and massive ecological consequences caused by depeasantization and industrialization, as well as the movements of resistance against it.

Recurring themes in agrarian and peasant studies concern the ways that the countryside is emptied of low-input, subsistence smallholder farmers, or how agricultural regions shift away from markets towards subsistence food provisioning—processes referred to as "depeasantization" and "repeasantization." Agrarian studies scholars argue that these dynamics are important for understanding social organization around not only food and agriculture but also in labor and economic relations, global migration patterns, and broader social and political movements (Van der Ploeg 2011, Davis 2005). Research in these areas, consequently, while tending to be grounded in household-level economics and daily decision making, land-use patterns, and power imbalances, also ranges to global political economic analysis (Borras Jr. and Franco 2012, Schneider 2016), for example of the systemic "food regimes" described above. The role of smallholders in capitalist development and the degree to which small-scale agriculture is a relatively revolutionary or conservative influence upon political movements has also been a central theme in critical agrarian/peasant studies.

The focus of this lineage on smallholders, social relations, and global food regimes has significant application in urban agroecologies, where smallholders play a central role and urban disparities are highlighted. If "urban agroecologies" are defined by their proximity to where people live at urban densities, this includes village allotments, smallholder plots on fringes of collective farms, and informal settlement agriculture (Jacobs 2018). For example, Nathan McClintock applies the idea of a "metabolic rift" to urban agriculture to provide a theoretical framework "bridging political economy, urban geography, agroecology, and public health" (2010). McClintock, and others, describe how growing things amidst residential landscapes resists the consolidation of control implicit in corporate food regimes; the vast diversity of crops, cultivation and labor choices, land tenure arrangements, and scales of operations also may resist governmental surveillance and regulation, similar to rural agrarian resistance movements (Scott 2010). Because of this resistance (along with racialized urban disinvestments), inputs and outputs of urban agroecologies are systematically underestimated, and arguably marginalized (MacGaffey 1991, Taylor and Lovell 2014, Tornaghi 2017). Urban agroecologies are also systematically devalued in favor of privileging the rural as more "fundamentally" agricultural because urban agroecologies do not fit the developmentalist and productivist norms discussed in the previous section (WinklerPrins 2017). Paradoxically, it is this very opacity of urban agroecologies that has led to urban agriculture projects becoming sites of resistance and autonomy in the face of a litany of harms caused by industrial and globalized food supply chains. This focus on the political and social arrangement of smallholder agriculture is arguably one of the most relational contributions to agricultural research from the social sciences, because this view profoundly influences the question of what is similar and different between urban and rural places.

Society–environment studies build on this relational focus by including the agency of more-than-human participants. Fields such as landscape ecology have linked landscape patterns to ecological processes, acknowledging humans and human institutions as intrinsic to these understandings—even if the application of this acknowledgement in practice has lagged (Nassauer and Opdam 2008). In urban agroecology, for example, this includes studies exploring how the impacts of spatially distributed urban agricultural sites interact with habitat fragmentation implicit in current urban planning (Lovell and Johnston 2009, La Rosa et al. 2014). Critical physical geography also builds on linking more-than-human entities and ecologies with human institutions. One example of particular relevance to urban agroecology is soil quality, as we often see urban soils characterized as "bad," "nutritionally deficient," or "contaminated" (Beniston and Lal 2012). While soil quality frameworks are often proposed, critical physical geography critiques the development of such frameworks when they do not also examine how soil quality is defined, for what purposes, and for whom (Engel-Di Mauro 2014). As a result of unreflexive soil surveys, we see the repetition of a historical trend happening with the invention of the rural soil survey and differential valuation of White-owned (rural) and Black-owned (urban) farms (Van Sant 2018). In urban agroecologies, working at the society–environment nexus includes the importance of specific configurations and flows of power, and the persistence of commons as collaborative regimes of land stewardship, perhaps most actively visible in contemporary landscapes in urban agriculture (Cadieux and Slocum 2015).

Over the past fifty years, society–environment studies research on urban agroecosystems has focused on environmental change and ecological conditions increasingly in conversation with the political processes that have shaped them (Basset 1981, Dennery 1995, Germundsson 1996, Donahue 2001, Hovorka 2005, Van Veehuizen 2006, Pudup 2008). Cognate fields of environmental history, science and technology studies, and social movement studies have also contributed to the vocabulary and practice of society–environment studies. These efforts have often coalesced as the field of political ecology, which Robbins (2012: 20) notes "represents an explicit alternative to 'apolitical' ecology." For Robbins, apolitical ecology fields, such as biology, ecology, and land-use science, often rely on "objective" studies to eschew political commitments despite inadvertently lapsing into political narratives such as ecoscarcity and modernization. Political ecology suggests that the implications of these narratives should influence researchers to *reflect critically on how their studies contribute to* "equitable ways to access, manage, and control land and other resources," understanding "resources" as more-than-human (and as more than landownership and foodie land access) and build "diverse knowledge systems to grow food, make

BOX 13.6 SITUATING NICKLAY'S WORK IN ENVIRONMENT-SOCIETY STUDIES

Jennifer is a natural scientist; her main disciplinary backgrounds are in soil science and microbiology. In her research, Jennifer works with urban growers to expand our understanding of how their practices impact urban socio-ecological systems—and how they, in turn, are impacted by the land, institutions, and governance. Jennifer strongly believes the critical, reflexive praxis she learned from the social sciences is vital to her work—especially because her training in the natural sciences provided limited practical guidance about how to navigate social questions, relationships, and impacts in research.

Critical social science helps Jennifer continually revisit her research assumptions. For example, to explore socio-ecological impacts, Jennifer uses an ecosystem service framework (which includes things that urban agroecology provides humans, such as yield, biodiversity, connection to neighbors, and shared green space). She's able to critique the ways in which an ecosystem services framework perpetuates the capitalist valuation of nature (Schröter et al. 2014) while also acknowledging that it expands our understanding of benefits and challenges in a way that is legible to people in many roles, including researchers, growers, organizers, and policy makers.

Reflexivity helps Jennifer be cognizant that the data generated by her work is not neutral. For example, the growers Jennifer works with are very interested in how using compost impacts soil nutrients—including initial trends showing the accumulation of phosphorus (Nicklay et al. 2019, Small et al. 2019). Although growers use this information to choose practices that limit adding excess phosphorus, existing policies of displacement, racial policing, and de-valuing agrarian land uses in urban areas mean this data could very easily be used to limit urban food production, create additional barriers to land access, or impose fines. This is especially relevant because Jennifer works in neighborhoods whose residents and growers are primarily Black, Indigenous, and people of color.

It's not enough to just be aware of these social forces, though, and a critical praxis allows Jennifer to leverage her research to work with communities towards transformative change. In addition to practicing community-based participatory research, Jennifer relies on mentorship from elders and participation in community networks in her roles as a researcher, grower, and organizer to make sure her research grows out of community questions and provides tangible, relevant outputs to growers and communities. Furthermore, she is committed to supporting emerging agroecologists, regularly mentoring undergraduate students and collaborating with a cohort of faculty and graduate students to propose an agroecology pedagogy for graduate education that would provide tools for navigating social questions, relationships, and impacts in research (Nicklay et al. 2018).

change, and sustain societies" (Cadieux and Slocum 2015: 14). For example, if we look at land tenure/access, we may ask how has an urban agroecology site come to be seen as "vacant" and "abandoned" or as valuable and worthy of investment? What are the politics and institutional interests surrounding this change, and how will research conducted at that site potentially impact those processes? Even for a brief research engagement with a site, openness to considering these questions of the politics playing out, and the ethical implications of one's work there, can help reduce a number of problems that are likely to arise if a researcher blunders into situations that do not warrant the naive optimism of many possible urban agroecological research questions (see Box 13.6).

13.3.3 Feminisms

Women play a significant role in urban agriculture worldwide, and their ubiquitous presence in urban food provision is often assumed to signal improvements in social inclusion and gender inequality (Orsini et al. 2013). Although attention to the significance of women's agricultural labor is a welcome departure from decades of women's invisibility in development planning, feminist analyses in critical agrarian studies have long demonstrated that participation in agricultural activities alone does not ensure empowerment or gender equity (Razavi 2009). Indeed, gender shapes the meanings of places, divisions of labor, access to and control over resources, such as land and water, and livelihood opportunities (Jarosz 1999, Truelove 2011, Mollett and Faria 2013). Without an attention to these gendered power relations, agricultural development risks negatively impacting women's lives, such as by increasing demands on their labor or marginalizing their economic opportunities, as has often been documented by scholars deploying feminist methodologies to study agrarian change (Moore and Vaughan 1987, Alston et al. 2018). Overlooking how gender mediates relations with the environment risks marginalizing gendered ethnobotanical knowledge, local science, and ecological practices, all of which help to sustain communities (Rocheleau 1995a).

The inequalities produced through gendered power relations are also present in urban agroecologies (Hovorka et al. 2009), where they intersect with various dimensions of difference, including class, race, ethnicity, nationality, religion, caste, and age (Corcoran and Kettle 2015, Frazier 2018). In her study of urban farmers in Botswana, which combined quantitative and qualitative data analysis methods, Hovorka (2005) found that although women and men were engaged in urban agriculture in equal numbers, women farmers made lower incomes and faced greater barriers in accessing capital and land. In the United States, disparities exist in urban agriculture, which mirror broader place-based inequalities and are shaped by class, race, and ethnicity (Hoover 2013, Cohen and Reynolds 2014, Reynolds and Cohen 2016). Applying an intersectional lens, feminist geographers have demonstrated how these social differences produce urban gardens as places with meaning, often in ways which are exclusionary (Engel-Di Mauro 2018, Hoover 2013).

Methodologically, feminist scholars have focused on identifying intersectional oppressions and analyzing exclusionary practices, ultimately so that communities could be empowered to take transformative action (Kobayashi 2005). For women engaged in urban agroecologies, this involves recognizing that food work, from gardening to cooking, remains gendered (Little et al. 2009). As part of the broader processes that make up social reproduction and fall disproportionately on women's shoulders, engaging in urban agroecologies may allow women to improve their household's food security, but it does not necessarily challenge exploitative gender relations (Engel-Di Mauro 2018). Using qualitative methods, including interviews and thematic coding by two differently positioned authors, Parker and Morrow (2017) found that urban homesteaders, including avid gardeners, pursued ideologies of intensive mothering, which relied on a traditional gender division of labor, heightened expectations of normative motherhood, and relative privilege to allocate time and resources for homesteading.

While Parker and Morrow (2017) identify the limitations of intensive mothering, especially as it revolves around food work, they also detail how women engaged in these activities felt a sense of empowerment through their care work and celebrated their ability to reclaim food work, like gardening. These findings mirror other research that has found that urban agroecology provides a space where oppressive gender relations can be both reproduced and resisted (Parry et al. 2005, Hovorka 2006). Parker and Morrow (2017) also see possibilities in the ethics of care espoused by urban homesteaders, but they maintain that only institutional investments in care and healthy food production would really challenge power relations.

A commitment to an ethics of care, however undervalued and unrecognized, partially explains women's involvement in social reproduction, which encompasses all of the practices in everyday life needed to reproduce society, from subsistence food production to raising children (Bauhardt and Harcourt 2018). In previous sections, we noted how traditional development studies often contributed to dismantling social reproduction, and agrarian/peasant studies often focused on movements to reclaim social reproduction. Methodologically, feminist researchers have drawn attention to the importance of these practices by seeking to make them more visible through their research. Feminist geographers have documented how social reproduction is a necessity for the continuation of capitalism, even as capitalist development threatens to destroy the basis for social reproduction (Katz 2001a). Even with the withdrawal of the state, capital, and civil society, social reproduction will be accomplished despite the costs to families and the additional gendered labor required from households (Katz 2001a). However, it also can form the basis for oppositional and/or prefigurative politics, entailing a struggle to make institutional investments in care (Parker and Morrow 2017) by pushing capital and the state to take the responsibility for social reproduction (Katz 2001a) or by expanding shared spaces for collective practices, such as food commons (Tornaghi 2017).

As a site for social reproduction, urban agroecology can benefit from institutional investments in care. Without broader struggles to achieve such change, urban agroecology initiatives may remain disconnected and marginal within the broader food system (Tornaghi 2017). Feminist geographic scholarship provides the methodological tools to link struggles for social reproduction across space and scales, such as Katz's notion of countertopographies (Katz 2001b), as well as Pratt and Yeoh's (2003) transnational feminist countertopographies, "a metaphor that suggests tracing lines across places to show how they are connected by the same processes, and simultaneously embedding these processes within the specifics of fully contextualised, three-dimensional places" (2003, 163). While not providing an exhaustive overview, and recognizing that feminist geographers draw upon diverse and even conflicting epistemological traditions, Box 13.7 describes other possible methodologies used in feminist geography, which could generate insightful research on urban agroecologies.

BOX 13.7 SITUATING BLUMBERG'S WORK IN FEMINIST GEOGRAPHIES

Renata has used a feminist lens in her research on agrarian change in Eastern Europe, where she found that historical gender relations, combined with the uneven development and rural marginalization that have accelerated since the implementation of neoliberal reforms in the 1990s, have shaped women's empowerment through alternative food production and distribution in complex ways (Blumberg 2020). Women farmers also face the challenge of managing social reproduction and production for profit, which is especially difficult for those who remain the only workers on their farms. In the US, Renata has used feminist methodologies to challenge the academy as the privileged site of knowledge production by collaboratively generating scholarship with urban agroecology activists (Blumberg et al. 2018).

Additional feminist methodologies may involve:

- Using multiple techniques for data collection and analysis, including qualitative and quantitative methods, while recognizing that any perspective is partial (Rocheleau 1995b, Sharp 2005);
- Working as an ally with communities in struggle in a way that fosters accountability and reciprocity (Pulido 2007);
- Recognizing that marginalized communities have been exploited by researchers, and that research involves navigating power relations and straddling multiple divides rooted in geopolitical and institutional locations (Nagar 2002, Nagar and Geiger 2007);
- Developing analyses of the border crossings necessary to undertake research in a world created by the uneven development of neoliberal capitalism, (neo)imperialism, and scattered hegemonies (Grewal and Kaplan 1994, Nagar 2002, Nagar and Geiger 2007);
- Examining what is being researched or written about, for whom, how and why, to account for the tendencies of research projects and their products (scholarly papers) to further academic careers rather than benefit the communities being studied;
- Committing to slow scholarship to counter tendencies in the neoliberal university (Mountz et al. 2015);
- Deploying an intersectional lens (Crenshaw 1991) to understand how categories, such as gender and ethnicity, become fixed and operate in dynamic relationships with places and throughout space (Mollett and Faria 2018);
- Excavating and reevaluating subjugated knowledges and practices, including those that support social reproduction, and analyzing discursive formations and representations that influence power relations.

13.3.4 Environmental Justice and Movements

Environmental justice stems from activists working against environmental racism, the unequal distribution of toxic waste and ecological harm in communities of color and Indigenous communities (Bullard 1994, Wilson 2010). The issue of environmental racism first gained popular national attention in the United States in 1982 when civil rights activists in Warren County, North Carolina prevented the state from disposing of 120 million pounds of soil contaminated with polychlorinated biphenyl (PCB) in a predominantly African-American area (Mohai 2009). Five years later, the United Church of Christ published one of the first reports to systematically demonstrate the extent to which Black, Latinx,[7] Asian-American, and Indigenous Peoples were exposed to hazardous wastes in their communities (UCC 1987). Other examples abound[8]; activists and academics have collaborated to demonstrate how the environmental externalities of capitalist accumulation such as garbage dumps (Pellow 2002), pesticide residues (Arcury et al. 2014), and air pollution (UCS 2019) target communities unevenly according to race and class. Environmental justice movements, most often led by women marginalized by race, ethnicity, and class, have sought redress for the ecological harms perpetuated against communities of color alongside broader struggles against capitalism and patriarchy (Mann 2011).

In urban environments, for example, the remediation and redevelopment of brownfield sites has been linked to the displacement of local residents long subject to living in close proximity to toxicity (Lee and Mohai 2012, Dillon 2013). Urban agroecological research methodologies growing from environmental justice lineages might, therefore, not only ask whether the remediation and redevelopment reduces toxicity, but also study, for example, the economics of how surrounding land values change. Such research could also investigate the planning processes (from neighborhood and city level, to county, state, or federal) that governed decision making, and how it could alternatively support development without displacement. Such questions can help broaden what constitutes environmental harm. Recent work has highlighted that environmental justice, with its focus on harm and toxicity, can also reproduce narratives equating Blackness or being Black (as "Other") with death—narratives that foreclose possibility of escape or resistance (Wright 2018, Opperman 2019, McKittrick 2014). Therefore, it is equally important for urban agroecological research to explore spaces of resistance and liberation, even through forces such as displacement, toxicity, etc.[9]

While environmental justice scholars are primarily concerned with unequal distribution of environmental risks to people of color, they also study how communities organize and resist the powerful actors in society who create those risks (see Box 13.8). Toxicity *and* efforts to reverse the impacts, after all, are embedded within systems of unequal power. Therefore, environmental justice also sits in dialogue with

[7] We use "Latinx" here as an example of gender inclusive language, which opens space for those who identify on the gender spectrum beyond the male/female binary. Use of the term Latinx is not without debate; critics point to the ways in which changing language can perpetuate imperialism while supporters point to the ways in which it promotes "mixedness," malleability, and reflection (see Brammer 2019).

[8] It is important to note that we wrote this chapter in the midst of a global environmental justice crisis—the COVID-19 pandemic. This socio-ecological phenomenon demonstrates the extent to which structures of White supremacy shape ecologies, causing the premature and unequal deaths of people of color. Early reports indicate that, in the U.S., hospitalization and fatality rates for African-Americans are disproportionately high (Aubrey 2020); for example, in Chicago, African Americans represented approximately 72% of virus fatalities yet make up less than one-third of the city's population (Mays and Newman 2020). The inequitable impacts of COVID-19 across race, gender, and income is driven by the prevalence of pre-existing health conditions, stress and "weathering" due to racialized policing, housing, and economic policies, and greater viral exposure due to an over-representation as frontline workers; these drivers, in turn, are the result of long histories of dominant political alliances that created "land dispossession, exposure to toxic environments, failed and racist education systems, abandonment of social services, anti-worker and anti-union organizing, company towns, and a concentration of prisons, jails, and detention centers" (Woods 2017 cited in Freshour and Williams 2020). The racial pandemic is yet another chapter in a long history of unequal harm through the interconnected processes of public health, structural racism, and environmental change.

[9] For example, Tiffany Lethabo King (2016) narrates how toxic indigo processing at the edge of plantations both exposed slaves to increased harm *while simultaneously* providing a space of autonomy where resistance could be plotted. How might urban agriculture studies interrogate zones of toxicity and spaces of uneven development in a manner that both identifies increased risk and demonstrates how these spaces constitute particular forms of sociality made inaccessible and erased by dominant narratives of development and improvement?

BOX 13.8 SITUATING CADIEUX'S WORK IN ENVIRONMENTAL JUSTICE

As a participatory researcher and educator, Valentine primarily works with communities who, dispossessed from ancestral homes, are seeking food sovereignty through restored access to land. Valentine has found the frameworks of food justice and environmental justice important for orienting collaborators, policy makers, and students to fundamental questions of how to organize their work together. Raising up the work of Minnesota food justice practitioners in policy briefs, for example, has helped nudge many White-led organizations—such as the University of Minnesota Extension's Regional Sustainable Development Partnerships, the Common Table local foods exhibit at the State Fair, and gatherings convened by the Union of Concerned Scientists—to work more explicitly and successfully at being open to conversation about representation and participation in research and advocacy. In particular, communities want to narrate their own stories, identities, and futures (De Schutter 2012); they advocate for representation of their strengths and resilience in the face of structural inequality, not as deficient, in need of charity, or research "subjects."

This provides a good example of how research methods in urban agroecology may often not look like research, perhaps for years, as researchers build reciprocal relationships that allow their skills, interests, and resource networks to be of use to communities in real ways. Policy briefs, summaries of relevant research, and assistance with public education, along with basic tasks such as printing, laminating, making flyers, and organizing communications, can all be ways that scholars can participate in community life around urban agroecologies that may help offset tendencies to impose the felt urgency of the need to publish and apply for grants on tight deadlines onto communities already working on much thinner budgets than academics. This set of "meta methods" for urban agroecological field work, in Valentine's experience, also includes cross-subsidizing projects to allow for such an extended timeline, finding diverse funding supports to enable direct service along with evaluation and more theoretical research, along with recruiting student projects that align with community needs, and supporting workshops on participatory learning processes that bridge community and academic space (such as, for example, Anti-Oppression Resource and Training Alliance workshops on "how to make meetings awesome for everyone," to which all of the urban food systems community collaborators of her campus were invited along with the campus community, or a series of workshops with organizations like the Union of Concerned Scientists and Science for the People that helped re-orient academics to community-led research and learning processes).

The lineage of environmental justice is characterized by mutual aid, and "call in" rather than "call out" culture. Such supportive call-in culture fosters collaboration and mentoring for those of us coming up, by calling us in to assist on existing projects where we can learn from the elders in the movement, like Robert Bullard, Shirley Sherrod, Winona LaDuke, Sam Grant, Melvin Giles, Diane Dodge, Faith Spotted Eagle, and many others with connections here and across the larger environmental and food justice movement. Choosing to orient our institutions to follow the leadership of largely Black, Indigenous, and People of Color (BIPOC) community organizers—especially those focused on social and environmental repair and mitigation of the ongoing harms of the status quo—helps to decenter White academic perfectionism and urgency, instead choosing methods that can function as needed, "making the methods something to be tried out, practiced, changed, tried again, rather than something to be judged" only for end results (Cadieux 2016: 231). This approach to methods as a practical tool and platform for explicit communication about inclusive and meaningful process encourages steering community-engaged student research toward program evaluation tasks useful for community organizations strapped for resources to deliver evaluative results to funders, for example, and encourages intervention research and evaluation research methods in general, to keep research outputs—whether supportive or critical—aligned with movement goals, practical needs, and relational priorities.

social movement studies, a broad interdisciplinary literature engaged with questions around the composition and effects of social movements. A social movement is a difficult concept to define and the field contains multiple definitions; however, most tend to align in that a social movement consists of a group of associated people who share a common identity and engage in collective action to shape the direction of social change beyond the conventional channels of participation in established political systems or through market activity alone (Diani 1992). How those people are related to one another and how they pursue change can be analyzed through a range of analytical perspectives.

One influential approach in the social movement studies literature is the use of Collective Action Frames (Benford and Snow 2000). A collective action frame is a discursive articulation of collective demands that define the problem, define the solution, and incite action. In agroecological action frames, social movement participants define problems related to food and land, identify collective pathways to addressing those problems, usually via building alternative food economies or mandating that the state adopt agroecology-inspired policies, and issue calls to mobilize in activism through social movement campaigns. Much of this framing work is legitimated and sustained by a "collective identity" shared by the movement's participants (Polletta and Jasper 2001, Fominaya 2010). Urban farming groups can mobilize action frames that problematize industrial agriculture and connect participations through a shared commitment to creating more sustainable food systems (Jonason 2019, Lyson 2014). Urban agriculture, as a practice of growing food in cities, cannot usually be considered a social movement on its own; however, many engaged in it are also participants in broader social networks of activism against environmental injustice or in the social movement struggle food sovereignty.

13.3.5 Indigenous Food Sovereignty

The human relationship with land is arguably the central dynamic in any agroecological system. While design and policy proposals to support healthy human-land relationships abound, they often default to a conception of land as property, as a commodity to be owned publicly (e.g., by the state), institutionally (by a university) or privately (by an individual farmer). While environmental justice and food movement efforts often organize around inequitable access to food providing lands, rarely do these critiques go so far as to question *whose relationships to that land have been previously erased by turning that land to property* (Rotz 2017, Kepkiewicz & Dale 2018, Lee and Ahtone 2020). In settler-colonial contexts—where colonization of a territory brings settlers who displace Indigenous peoples, such as contemporary North and South America, Australia, and Aotearoa New Zealand—a historical lens points to the enclosure of Indigenous lands for purposes of "producing" food as an original source of environmental racism (see prior section on Environmental Justice and Movements).

Agroecology in settler-colonial contexts is therefore usually only possible through the ongoing preservation of property relations created through theft or acquisition through unethical or violent state-sponsored dispossession. For land-based communities living for millennia in traditional and ancestral territories, this dispossession is not just a separation from the land—it is a form of genocide (Wolfe 2006, National Inquiry into Missing and Murdered Indigenous Women and Girls 2019). In the context of urban agroecology specifically, the underlying tension with respect to land and land rights often remains unresolved—even in more radical proposals for food commoning (Vivero-Pol et al. 2019) and in alternative land access and tenure arrangements for urban agriculture in particular (Wekerle and Classens 2015, c.f., McClintock 2018).

While settler colonialism imposes generational cycles of violence and trauma, it also incites resurgence and resistance. With roots in decolonial thought, Indigenous food sovereignty advocates demand Indigenous reclamation of traditional food lands and territories as well as cultural ways of relating to land and food (Morrison 2011, Desmarais & Wittman, 2014, Daigle 2017). We can distinguish between two separate, yet often intertwined pathways that can lead to the reclamation of Indigenous sovereignty over foodways, foodlands and waters, and cultures: *Indigenization* and *decolonization*. Indigenization refers to the resurgence of Indigenous structures and ways of being and knowing, and decolonization refers to the dismantling of colonial structures and ways of being and knowing (Coté, 2016). On the one hand, settler efforts to support Indigenization within agroecological sites can be found in inviting Indigenous- led ceremonies, medicinal plantings, and traditional harvesting (Bowness and Wittman 2020). The settler positionalities in agroecological spaces at minimum require building relationships with Indigenous peoples when appropriate, learning cultural protocols tied to the lands when shared, and acting on responsibilities out of a commitment to reciprocity. However, given the harmful role agriculture plays in settler-colonialism, settler agroecological researchers, practitioners, and activists alike should also be aware of the limitations of farming, agroecological or otherwise, to Indigenization on colonized lands.

Agroecology, as part of the reclamation of democratic control of food systems under the evolving banner of "food sovereignty," also has a relationship with decolonization (see Grey and Patel 2014). In particular, rematriation/repatriation of agricultural land located on traditional territories, including urban agricultural spaces, is a pathway by which settlers can support Indigenous food sovereignty. However, *it matters who is employing the language of decoloniality and how it is mobilized,* as "decolonization is not a metaphor" (Tuck and Yang 2012). Decolonization means dismantling colonial structures, and property and its ownership by settlers is pivotal to this process. With this in mind, declarations to "decolonize the garden" in the context of small urban agroecology projects are open to critique as a non-sequitur and largely symbolic concern, overshadowed by other urban property issues such as extravagant rents, gentrification, and declining social services. Recognizing the settler-colonial context of land confers a responsibility upon agroecological practitioners and researchers to engage in decolonial land politics, meaning relinquishing control of land where appropriate, in particular, and critically interrogating their allyship efforts as part of an intersectional analysis of the equity issues present on agroecological lands (Bowness and Wittman 2020; see Box 13.9).

BOX 13.9 SITUATING BOWNESS' WORK IN DECOLONIAL LAND POLITICS

Urban agroecology includes not only growing practices in urban farms and gardens—it also relates to how urban people mobilize collectively to create a different food system beyond the city. Evan's research with Dr. Hannah Wittman at the University of British Columbia advances the idea of urban agrarianism to refer to urban people who defend, through activism, foodlands located both in the city but also in the countryside (Bowness and Wittman, 2020). Urban agrarians in the context of Metro Vancouver, Canada, acting on their sense of responsibility to create a sustainable food system, are involved in land struggles at different scales. These include struggles for Indigenous food sovereignty. As a settler-colonial state, the federal government has established treaties with many of the Indigenous Peoples living within the geopolitical borders of so-called Canada. There are several highly political energy development projects currently under construction in the province, one of which—the Site C hydroelectric dam—will flood thousands of acres of farmland, destroying traditional territories and affecting Indigenous hunting and fishing treaty rights, such as those of the West Moberly First Nation who have resisted the dam's construction. Even though the British Columbian government enacted the United Nations Declaration on the Rights of Indigenous Peoples (UNDRIP) in late 2019, projects such as these are still poised to move forward. This example is illustrative of the disillusionment experienced by Indigenous activists and their supporters of the "reconciliation" discourse that gained international attention through the final report of the national Truth and Reconciliation Commission of Canada (2015). Instead, the discourse has shifted to forms of *allyship* and *rising up* in solidarity with Indigenous-led struggles to defend against ongoing colonial violence and protect the land. As Evan and Hannah (both settlers) argue, there can be no food sovereignty without decolonization, and therefore urban agrarianism must be not only an agroecological practice but also a form of social mobilization in solidarity with Indigenous movements for self-determination, within and beyond the city.

13.4 Urban Agroecology as a Critical and Reflexive Social Science Praxis

In the literal fields that we study, power courses through relationships between research participants, their social networks and broader social contexts, and the land itself. Agrarian and society–environment studies direct our attention to the relationships between human and non-human nature, the corporate and political constraints on urban farmers, and the economic and cultural pressures influencing the use of chemicals for weed control and productivity. Feminist perspectives point to the intersection of social reproduction and urban agriculture, and to inequality between farmers and organizers of different genders. Environmental justice advocates show how brownfield sites may be sited unevenly based on race or how gentrification may seize upon beautification, displacing people under the guise of urban sustainability. Indigenous food sovereignty takes a step back from the struggles for access to land on which food is grown to recognize the history of dispossession of Indigenous peoples who in settler contexts have been removed from their territories through colonial urbanization and relocation programs.

In bringing together the threads in this chapter, we offer a re-visioning of the oft-quoted definition of agroecology as a science, movement, and practice (Wezel et al. 2009). The intersecting lineages we list in this chapter inform critical social science focused on dynamics of power and the potential of alternative, more equitable, power dynamics. They call for praxis or research that is engaged, action-oriented, and participatory, with the intention to address inequality and marginalization. They also invite reflexivity, critical not only of the world around us but also of the knowledge and tools we use to understand it. Urban agroecology can therefore be seen as an emerging branch of critical social science, an engaged research praxis embedded within scholarly traditions and part of wider agroecology movements.

What does a critical and reflexive urban agroecology praxis look like? As researchers, we should be aware of how the dynamics discussed above are present in our fieldwork. Our own social positions give us access to some field sites and not others. Our methodological choices can help us to recognize barriers to working with communities, including very reasonable historical legacies of mistrust in the professional class. Whenever we work with agroecological collaborators, part of the research praxis is to identify and counter those legacies whenever possible, but also accept that privilege is a feature of research itself and therefore will always be there in some form. For example, basic expectations to respond accountably through reflexivity call for research communication that uses language that is gender inclusive and accessible to communities for whom the research matters. Settler and White researchers should recognize their positions and refrain from imposing undue burdens on Indigenous and racialized people to educate and manage emotions associated with settler guilt and racial privilege—or to respond to researchers' felt urgency to "fix" situations. Researchers should also recognize their responsibilities to follow through on commitments made to community partners and research participants, especially in terms of mobilizing resources. These are a few praxis considerations within the challenging colonial, gendered, racialized, and class politics of participatory collaboration.

In writing this chapter, we also understand part of our task is to be knowledgeable and accountable to our own disciplinary lineages within the social sciences. In this way, we can both use the methodological tools we have been gifted to understand and challenge social and power dynamics, and also be aware of the baggage that they carry. As noted in the section on Development Studies, scholarly work can have harmful effects. Academic knowledge is hierarchical and structured to exclude particular ways of knowing: science is privileged over traditional ecological knowledge, folk understandings, and Indigenous ontologies. Further studies that report mainly on urban farm "productivity" risk reproducing the idea of land and nature as being separate from society, or of smallholder agrarian knowledge as outmoded, faddish, or merely inadequate for the developmentalist project of "feeding the world" through profitable export agriculture. Whether overtly political or not, urban agroecological research is embedded within the politics of knowledge and the power dynamics of institutional research. Urban agroecological research as a critical praxis can also challenge inequality among different knowledges.

We hope this orientation to social science perspectives helps bring social issues into the agroecological research frame—for social and natural scientists alike, as for all others learning through urban agroecologies. Agroecology is imbued with racialized, gendered, classed, and other differentials of power and marginalization. Highlighting some ways to engage political and social forces is our main intended contribution. But more radically, urban political agroecology research is also itself a social practice. In urban agroecological research, researchers should practice making choices about methodology with reflexivity and critical awareness of context—this is urban political agroecology in action. Whatever methods are appropriate to the questions negotiated in a research site, this stance may help researchers to adopt adequate humility about the scope of work possible, respect for relational values, and commitment to the challenging politics of truly participatory collaboration.

REFERENCES

Alston, M., Clarke, J. and Whittenbury, K. 2018. Contemporary feminist analysis of Australian farm women in the context of climate changes. *Social Sciences* 7(2), 16.

Anderson, C. 2017. Policy from below: Politicising urban agriculture for food sovereignty. *Urban Agriculture Magazine* 33, 72–74. https://ruaf.org/document/urban-agriculture-magazine-no-33-urban-agroecology/

Arcury, T.A. Nguyen, H.T., Summers, P., et al. 2014. Lifetime and current pesticide exposure among Latino farmworkers in comparison to other latino immigrants. *American Journal of Industrial Medicine* 57, 776–787.

Aubrey, A. CDC hospital data point to racial disparity in COVID cases. April 8, 2020. *National Public Radio.* https://www.npr.org/sections/coronavirus-live-updates/2020/04/08/830030932/cdc-hospital-data-point-to-racial-disparity-in-covid-19-cases

Barkan, E. 1992. *The retreat of scientific racism: Changing concepts of race in Britain and the United States between the world wars*, New York: Cambridge University Press.

Basset, T. 1981. Reaping on the margins: A century of community gardening in America. *Landscape* 25(2), 1–8.

Bauhardt, C. and Harcourt, W. (Eds.) 2018. *Feminist political ecology and the economics of care: In search of economic alternatives*, London: Routledge.

Benford, R. D. and Snow, D. A. 2000. Framing processes and social movements: An overview and assessment. *Annual Review of Sociology* 26, 611–639. DOI:10.2307/223459

Beniston, J., and Lal, R. 2012. Improving soil quality for urban agriculture in the north central U.S. In (Eds.) R. Lal & B. Augustin, *Carbon sequestration in urban ecosystems*, 279–313. New York: Springer Dordrecht Heidelberg. https://doi.org/10.1007/978-94-007-2366-5

Blumberg, R., Huitzitzilin, R., Urdanivia, C. and Lorio, B.C. 2018. Raíces del sur: Cultivating ecofeminist visions in urban New Jersey. *Capitalism Nature Socialism* 29(1), 58–68.

Blumberg, R. 2020. *Engendering alternative food networks in Europe through transnational feminist counter-topographies.* Unpublished manuscript.

Borras Jr., S.M. and J. C. Franco. 2012. Global land grabbing and trajectories of agrarian change: A preliminary analysis. *Journal of Agrarian Change* 12(1), 34–59.

Bowness, E. and Wittman, H. Forthcoming. 2020. Bringing the city to the country? Responsibility, privilege and urban agrarianism in Metro Vancouver. *The Journal of Peasant Studies* 1–26. DOI: 10.1080/03066150.2020.1803842

Brammer, J. P. 2019. Digging into the messy history of "Latinx" helped me embrace my complex identity: The term is gaining steam, but plenty of people still despise it. *Mother Jones. May/June.* https://www.motherjones.com/media/2019/06/digging-into-the-messy-history-of-latinx-helped-me-embrace-my-complex-identity/

Bullard, R. D. 1994. *Introduction in unequal protection: Environmental justice and communities of color.* (Ed.) R.D. Bullard. San Francisco: Sierra Club Books.

Cadieux, K. V. 2016. Methodologism: radical practice. In S. Matteson (Ed.) *Meandering methodologies, deviant disciplines: Four years of city art collaboratory*, 221–233., M St. Paul: Public Art St. Paul.

Cadieux, K. V. and Slocum, R. (2015). What does it mean to do food justice. *Journal of Political Ecology* 22(1), 1–26.

Chakrabarty, D. 2000. *Provincializing Europe: Postcolonial thought and historical difference*, Princeton: Princeton University Press.

Cohen, N. and Reynolds, K. 2014. Urban agriculture policy making in New York's "New Political Spaces" strategizing for a participatory and representative system. *Journal of Planning Education and Research* 34(2), 221–234.

Corcoran, M. P. and Kettle, P. C. 2015. Urban agriculture, civil interfaces and moving beyond difference: The experiences of plot holders in Dublin and Belfast. *Local Environment* 20(10), 1215–1230.

Coté, C. 2016. "Indigenizing" food sovereignty. Revitalizing Indigenous food practices and ecological knowledges in Canada and the United States. *Humanities* 5(3), 57. DOI:10.3390/h5030057

Crenshaw, K. 1991. Mapping the margins: Intersectionality, identity politics, and violence against women of color. *Stanford Law Review* 43(6), 1241–1299.

Cullather, N. 2010. *The hungry world: America's cold war battle against poverty in Asia*, Harvard University Press.

Daigle, M. 2017. Tracing the terrain of Indigenous food sovereignties. *The Journal of Peasant Studies* 18(2), 1–19. DOI:/10.1080/03066150.2017.1324423

Dale, B. 2019. Alliances for agroecology: From climate change to food system change. *Agroecology and Sustainable Food Systems* 44(5), 1–24. DOI:/10.1080/21683565.2019.1697787

Davenport, S. G. and Mishtal, J. 2019. Whose sustainability? an analysis of a community farming program's food justice and environmental sustainability agenda. *Culture, Agriculture, Food and Environment* 41(1), 56–65. DOI:10.1111/cuag.12227

Davis, M. 2005. *Planet of Slums*. New York: Verso.

De Schutter, O. 2012. The right to food: a weapon against global hunger. *Friedman seminar series*, Medford, MA: Tufts University

Dennery, P. 1995. Cities feeding people project fact sheets. *Cities feeding people series, Report 15*. Ottawa: International Development Research Centre (IDRC).

Desmarais, A. A. and Wittman, H. 2014. Farmers, foodies and first nations: Getting to food sovereignty in canada. *The Journal of Peasant Studies* 41(6), 1153–1173. DOI:10.1080/03066150.2013.876623

Diani, M. 1992. The concept of social movement. *The Sociological Review* 40(1), 1–25. DOI:10.1111/j.1467-954x.1992.tb02943.x

Dillon, L. 2013. Race, waste, and space: Brownfield redevelopment and environmental justice at the Hunters Point shipyard. *Antipode* 46(5), 1–17. DOI:10.1111/anti.12009

Donahue, B. 2001. *Reclaiming the commons: Community farms and forests in a New England town*, New Haven: Yale University Press.

Edelman, M. and Wolford, W. 2017. Introduction: Critical agrarian studies in theory and practice. *Antipode* 49(4), 959–976. DOI:10.1111/anti.12326

Engel-Di Mauro, S. 2014. *Ecology, soils, and the left*. Palgrave Macmillan US. DOI:/10.1057/9781137350138

Engel-Di Mauro, S. 2018. Urban community gardens, commons, and social reproduction: Revisiting Silvia Federici's Revolution at Point Zero. *Gender, Place and Culture* 25(9), 1379–1390.

Escobar, A. 1995. *Encountering development: The making and unmaking of the third world*, Princeton, NJ: Princeton University Press.

Escobar, A. 1989. The professionalization and institutionalization of 'development' in Colombia in the early post-world war II period. *International Journal of Educational Development* 9(2), 139–154.

Fominaya, C. F. 2010. Collective identity in social movements: Central concepts and debates: Collective identity in social movements. *Sociology Compass* 4(6), 393–404. DOI:10.1111/j.1751-9020.2010.00287.x

Foster, J. B. 1999. Marx's theory of metabolic rift: Classical foundations for environmental sociology. *American Journal of Sociology* 105(2), 366–405. DOI:10.1086/210315

Frazier, C. 2018. "Grow what you eat, eat what you grow": Urban agriculture as middle class intervention in India. *Journal of Political Ecology* 25(1), 221–238.

Freshour, C. and Williams, B. 2020. Abolition in the time of Covid-19. *Antipode Online*. Accessible at: https://antipodeonline.org/2020/04/09/abolition-in-the-time-of-covid-19/.

Friedman, H. and McMichael, P. 1989. Agriculture and the state system: The rise and decline of national agricultures, 1870 to the Present. *Sociologia Ruralis* 29(2), 1–25. DOI:10.1111/j.1467-9523.1989.tb00360.x

García-Sempere, A., Morales, H., Hidalgo, M., Ferguson, B. G., Rosset, P., and Nazar-Beutelspacher, A. 2019. Food Sovereignty in the city?: A methodological proposal for evaluating food sovereignty in urban settings. *Agroecology and Sustainable Food Systems* 43(10), 1145–1173. DOI:10.1080/21683565.2019.1578719

Germundsson, T. 1996. "Smallholdings and great expectations-state-funded owner-occupied farms in the Nordic countries," in *Nordic landscapes: Cultural studies of place*, J. O. Nilsson, A. L. Laursen (eds). 98–115, Nordic Council of Ministers. Accessible at: https://portal.research.lu.se/portal/sv/publications/smallholdings-and-great-expectations–statefunded-owneroccupied-farms-in-the-nordic-countries(ae6eef8d-eb4d-471c-b7a9-ec27e81c404c).html

Gilbert, J. 2015. *Planning democracy: Agrarian intellectuals and the intended New Deal*, New Haven: Yale University Press.

Grewal, I. and Kaplan, C. (Eds). 1994. *Scattered hegemonies: Postmodernity and transnational feminist practices*, Minneapolis, MN: University of Minnesota Press.

Grey, S. and Patel, R. 2014. Food sovereignty as decolonization: Some contributions from Indigenous movements to food system and development politics. *Agriculture and Human Values* 32(3), 431–444. DOI:10.1007/s10460-014-9548-9

Hernandez, A. 2020. The emergence of agroecology as a political tool in the Brazilian Landless Movement. *Local Environment* 25(3), 205–227. DOI:10.1080/13549839.2020.1722990

Holt-Giménez, E. and Shattuck, A. 2011. Food crises, food regimes and food movements: Rumblings of reform or tides of transformation. *The Journal of Peasant Studies* 38(1), 109–144. DOI:10.1080/03066150.2010.538578

Hoover, B. 2013. White spaces in Black and Latino places: Urban agriculture and food sovereignty. *Journal of Agriculture, Food Systems, and Community Development* 3(4), 109–115.

Hovorka, A., Zeeuw, H. D., and Njenga, M. 2009. *Women feeding cities: Mainstreaming gender in urban agriculture and food security*. Rugby, UK: Practical Action Publishing

Hovorka, A. J. 2005. The (re) production of gendered positionality in Botswana's commercial urban agriculture sector. *Annals of the Association of American Geographers* 95(2), 294–313.

Hovorka, A. J. 2006. The No. 1 Ladies' Poultry Farm: a feminist political ecology of urban agriculture in Botswana. *Gender, Place and Culture* 13(3), 207–225.

Jacobs, R. 2018. An urban proletariat with peasant characteristics: Land occupations and livestock raising in South Africa. *The Journal of Peasant Studies* 45(5-6), 884–903.

Jarosz, L. 1999. A feminist political ecology perspective. *Gender, Place and Culture: A Journal of Feminist Geography* 6(4), 390.

Jonason, A. 2019. Defining, aligning, and negotiating futures: New forms of identity work in an urban farming project. *Sociological Perspectives* 62(5), 691–708. DOI:10.1177/0731121419845894

Katz, C. 2001a. Vagabond capitalism and the necessity of social reproduction. *Antipode* 33(4), 709–28. DOI:/10.1111/1467-8330.00207.

Katz, C. 2001b. On the grounds of globalization: A topography for feminist political engagement. *Signs: Journal of Women in Culture and Society* 26(4), 1213–34. DOI:10.1086/495653.

Kepkiewicz, L., and Dale, B. 2018. Keeping 'our' land: Property, agriculture and tensions between Indigenous and settler visions of food sovereignty in Canada. *The Journal of Peasant Studies* 60(3), 1–20. DOI:10.1080/03066150.2018.1439929

King, TL. 2016. The Labor of (re)reading plantation landscapes fungible(ly). *Antipode* 48(4), 1022–1039.

Kobayashi, A., 2005. Anti-racist feminism in geography: an agenda for social action. In *A Companion to Feminist Geography*, (Eds.) L. Nelson and J. Seager. Malden, MA: Blackwell Publishing.

La Rosa, D., Barbarossa, L., Privitera, R., & Martinico, F. 2014. Agriculture and the city: A method for sustainable planning of new forms of agriculture in urban contexts. *Land Use Policy* 41, 290–303. DOI:10.1016/j.landusepol.2014.06.014

Lee, R. and Ahtone, T. 2020. Land grab universities: expropriated Indigenous land is the foundation of the land-grant university. *High Country News*. Available at: https://www.hcn.org/issues/52.4/indigenous-affairs-education-land-grab-universities.

Lee, S. and Mohai, P. 2012 Environmental Justice Implications of Brownfield Redevelopment in the United States. *Society & Natural Resources* 25(6), 602-609. DOI: 10.1080/08941920.2011.566600

Li, T. M. 2007. *The will to improve: Governmentality, development, and the practice of politics*. JN, Durham, N.C: Duke University Press.

Lather, P. 1986. Research as praxis. *Harvard Educational Review* 56(3), 257–278. DOI:10.17763/haer.56.3.bj2h231877069482

Little, J., Ilbery, B. and Watts, D. 2009. Gender, consumption and the relocalisation of food: A research agenda. *Sociologia Ruralis* 49(3), 201–217.

Lovell, S. T., & Johnston, D. M. 2009. Designing landscapes for performance based on emerging principles in landscape ecology. *Ecology and Society* 14(1), 44.

Lyson, H. C. 2014. Social structural location and vocabularies of participation: Fostering a collective identity in urban agriculture activism. *Rural Sociology* 79(3), 310–335. DOI:10.1111/ruso.12041

MacGaffey, J. 1991. *The real economy of Zaire: The contribution of smuggling and other unofficial activities to national wealth*, Philadelphia, PA: University of Pennsylvania Press.

Mann, S.A. 2011. Pioneers of U.S. ecofeminism and environmental justice. *Feminist Formations* 23(2), 1–25. DOI: 10.1353/ff.2011.0028

Mays, J. C. and A. Newman. Virus is twice as deadly for Black and Latino people than Whites in N.Y.C. The New York Times, April 8, 2020. Accessible at: https://www.nytimes.com/2020/04/08/nyregion/coronavirus-race-deaths.html

McClintock, N. 2010. Why farm the city? theorizing urban agriculture through a lens of metabolic rift. *Cambridge Journal of Regions, Economy and Society* 3(2), 1–17. DOI:10.1093/cjres/rsq005

McClintock, N. 2018. Urban agriculture, racial capitalism, and resistance in the settler-colonial city .*Geography Compass* 12(6), 1–16. DOI:10.1111/gec3.12373

McKittrick, K. 2014. Mathematics black life. *The Black Scholar: Journal of Black Studies and Research* 44(2), 16–28.

McMichael, P. 2009. A food regime genealogy. *The Journal of Peasant Studies* 36(1), 139–169. DOI:10.1080/03066150902820354

Méndez, V. E., Bacon, C. M., Cohen, R., and Gliessman, S. R. (Eds.). 2016. *Agroecology: A transdisciplinary, participatory and action-oriented approach*, Boca Raton: CRC Press.

Mitchell, T. 2002. *Rule of experts: Egypt, techno-politics, modernity*, University of California Press.

Mohai, P., Pellow, D., and Roberts, J. T. 2009. Environmental justice. *Annual Review of Environment and Resources* 34, 405–430.

Molina, M. G. de. 2013. Agroecology and politics: How to get sustainability? about the necessity for a political agroecology. *Agroecology and Sustainable Food Systems* 37(1), 45–59. https://www.tandfonline.com/doi/full/10.1080/10440046.2012.705810

Mollett, S. and Faria, C. 2013. Messing with gender in feminist political ecology. *Geoforum* 45, 116–125.

Mollett, S. and Faria, C., 2018. The spatialities of intersectional thinking: Fashioning feminist geographic futures. *Gender, Place & Culture* 25(4), 565–577.

Moore, H. and Vaughan, M. 1987. Cutting down trees: Women, nutrition and agricultural change in the Northern Province of Zambia, 1920-1986. *African Affairs*, 86(345), 523–540.

Moore, J. W. 2016. Metabolic rift or metabolic shift? dialectics, nature, and the world-historical method. *Theory and Society*, 46(4), 285–318. DOI:10.1007/s11186-017-9290-6

Morrison, D. 2011. Indigenous food sovereignty: a model for social learning. In H. Wittman, A. A. Desmarais and N. Wiebe, (Eds.) *Food Sovereignty in Canada: Creating Just and Sustainable Food Systems* (pp. 97–113). Halifax: Fernwood.

Mountz, A., Bonds, A., Mansfield, B., Loyd, J., Hyndman, J., Walton-Roberts, M., Basu, R., Whitson, R., Hawkins, R., Hamilton, T. and Curran, W., 2015. For slow scholarship: A feminist politics of resistance through collective action in the neoliberal university. *ACME: An international E-journal for critical geographies*, 14(4).

Nagar, R., and Geiger, S. 2007. Reflexivity, positionality and identity in feminist fieldwork revisited. In A. Tickell, E. Sheppard, J. Peck and T. Barnes (Eds.) *Politics and practice in economic geography*, London: Sage.

Nagar, R. 2002. Footloose researchers, 'traveling' theories, and the politics of transnational feminist praxis. *Gender, Place and Culture: A Journal of Feminist Geography* 9(2), 179–186.

National Inquiry into Missing and Murdered Indigenous Women and Girls. 2019. A legal analysis of genocide: A supplementary report of the National Inquiry. Accessible at: https://www.mmiwg-ffada.ca/wp-content/uploads/2019/06/Supplementary-Report_Genocide.pdf

Nassauer, J. I., and Opdam, P. (2008). Design in science: extending the landscape ecology paradigm. *Landscape Ecology*, *23*(6), 633–644. https://doi.org/10.1007/s10980-008-9226-7

Nicklay, J. A., Perrone, S., Wauters, V., Badger, S., Grossman, J., Hecht, N.D., Jordan, N. R., Lieman, A., Meyer, N., Runck, B., Sames, A., and Thurston, C. 2018. Moving agronomic pedagogy beyond disciplinary boundaries. *Oral session presented at the sustainable agriculture education association national conference*, Kapolei, HI, July 27.

Nicklay, J. A., Cadieux, K. V., Jelinski, N. A., LaBine, K., Rogers, M., and Small, G. 2019. Initial Trends in Ecosystem Service Metrics of Urban Agriculture in Minneapolis/St. Paul, MN. Oral Session presented at: Embracing the Digital Environment. ASA-CSSA-SSSA International Annual Meeting, San Antonio, TX. 10-13 Nov.

Offner, A. C. 2019. *Sorting out the mixed economy: The rise and fall of welfare and developmental states in the americas.* Princeton: Princeton University Press.

Opperman, R. 2019. A permanent struggle against and omnipresent death: Revisiting environmental racism with Frantz Fanon. *Critical Philosophy of Race* 7(1), 57–80.

Orsini, F., Kahane, R., Nono-Womdim, R. and Gianquinto, G. 2013. Urban agriculture in the developing world: A review. *Agronomy for Sustainable Development* 33(4), 695–720.

Parker, B. and Morrow, O. 2017. Urban homesteading and intensive mothering: (Re)Gendering care and environmental responsibility in Boston and Chicago. *Gender, Place and Culture* 24(2), 247–259.

Parry, D. C., Glover, T. D., and Shinew, K. J. 2005. 'Mary, Mary quite contrary, how does your garden grow?': Examining gender roles and relations in community gardens. *Leisure Studies* 24(2), 177–192.

Pellow, D. N. 2002. *Garbage Wars: The struggle for environmental justice in Chicago*, Cambridge, Mass.: MIT Press.

Polletta, F., and Jasper, J. M. 2001. Collective identity and social movements. *Annual Review of Sociology* 27(1), 283–305. DOI:10.1146/annurev.soc.27.1.283

Porter, C. M., Woodsum, G. M., and Hargraves, M. 2018. The food dignity issue. *Journal of Agriculture, Food Systems, and Community Development* 8, Special Issue 1.

Power, M., 2003. *Rethinking development geographies*, London: Routledge.

Pratt, G., and Yeoh, B. 2003. Transnational (counter) topographies. *Gender, Place and Culture* 10(2), 159–66. DOI:10.1080/0966369032000079541

Pratt, M. L. 2007. *Imperial eyes: Travel writing and transculturation.* 2nd Edition. London: Routledge.

Pudup, M. B. 2008. It takes a garden: Cultivating citizen-subjects in organized garden projects. *Geoforum* 39(3), 1228–1240.

Pulido, L. 2007. FAQS: Frequently (un)asked questions on being a scholar/activist. In C. Hale (Ed.) *Engaging contradictions: Theory, politics and methods of activist scholarship*, Berkeley, CA: UCIAS Press.

Razavi, S. 2009. Engendering the political economy of agrarian change. *The Journal of Peasant Studies* 36(1), 197–226.

Reynolds, K. and Cohen, N. 2016. *Beyond the kale: urban agriculture and social justice activism in New York City*, Athens, GA: University of Georgia Press.

Robbins, P. 2012. *Political ecology: A critical introduction*, Malden, MA: John Wiley & Sons.

Rocheleau, D., 1995a. Gender and biodiversity: A feminist political ecology perspective. *IDS Bulletin* 26(1), 9–16.

Rocheleau, D. 1995b. Maps, numbers, text, and context: mixing methods in feminist political ecology. *The Professional Geographer* 47(4):458–466.

Rotz, S. 2017. 'They took our beads, it was a fair trade, get over it': Settler colonial logics, racial hierarchies and material dominance in Canadian agriculture. *Geoforum* 82, 158–169. DOI:10.1016/j.geoforum.2017.04.010

Schneider, M. 2016. Dragon head enterprises and the state of agribusiness in China. *Journal of Agrarian Change* 17(1), 3–21.

Scott, J. C. 2010. *The art of not being governed: An anarchist history of upland Southeast Asia.* Nus Press.

Schröter, M., Van der Zanden, E. H., van Oudenhoven, A. P., Remme, R. P., Serna-Chavez, H. M., De Groot, R. S., & Opdam, P. 2014. Ecosystem services as a contested concept: A synthesis of critique and counter-arguments. *Conservation Letters* 7(6), 514–523.

Sharp, J. 2005. Geography and gender: Feminist methodologies in collaboration and in the field. *Progress in Human Geography* 29(3), 304–309.

Small, G. E., Osborne, S., Shrestha, P., & Kay, A. 2019. Measuring the fate of compost-derived phosphorus in native soil below urban gardens. *International Journal of Environmental Research and Public Health* 16(20), 1–9. DOI:10.3390/ijerph16203998

Taylor, J. R., & Lovell, S. T. 2014. Urban home food gardens in the Global North: Research traditions and future directions. *Agriculture and Human Values* 31(2), 285–305.

Tornaghi, C. 2017. Urban agriculture in the food-disabling city: (Re) defining urban food justice, reimagining a politics of empowerment. *Antipode* 49(3), 781–801.

Tornaghi, C., and Dehaene, M. 2019. The prefigurative power of urban political agroecology: Rethinking the urbanisms of agroecological transitions for food system transformation. *Agroecology and Sustainable Food Systems* 44(5), 1–17. DOI:10.1080/21683565.2019.1680593

Truelove, Y., 2011. (Re-)Conceptualizing water inequality in Delhi, India through a feminist political ecology framework. *Geoforum* 42(2), 143–152.

Truth and Reconciliation Commission. 2015. Truth and Reconciliation Committee report and recommendations. Accessible at: http://nctr.ca/reports.php

Tuck, E., and Yang, K. W. 2012. Decolonization is not a metaphor. *Decolonization: Indigeneity, Education and Society* 1(1), 1–40.

Union of Concerned Scientists. 2019. Inequitable exposure to air pollution from vehicles in the northeast and mid-atlantic. Union of Concerned Scientists Fact Sheet. Accessible at: https://www.ucsusa.org/resources/inequitable-exposure-air-pollution-vehicles

United Church of Christ. 1987. Toxic wastes and race in the united states: a national report on the racial and socio-economic characteristics of communities with hazardous waste sites. New York, New York. Accessible at: https://www.nrc.gov/docs/ML1310/ML13109A339.pdf

University of Minnesota Extension. Expanded food and nutrition education program (FENEP). Accessible at: https://extension.umn.edu/teaching-nutrition-education/expanded-food-and-nutrition-education-program-efnep

Van der Ploeg, J. D. 2011. The drivers of change: The role of peasants in the creation of an agro-ecological agriculture. *Agroecología* 6, 47–54.

Van Sant, L. 2018. "The long-time requirements of the nation". the US Cooperative Soil Survey and the political ecologies of improvement. *Antipode* DOI:10.1111/anti.12460

Van Veenhuizen, R. 2006. *Cities farming for future, urban agriculture for green and productive cities.* RUAF Foundation, IDRC and IIRP, ETC-Urban agriculture, Leusden, The Netherlands.

Vivero-Pol, J. L., Ferrando, T., De Schutter, O., and Mattei, U. (Eds.). 2019. *Routledge handbook of food as a commons.* Milton Park: Routledge.

Wekerle, G. R., and Classens, M. 2015. Food production in the city: (Re)negotiating land, food and property. *Local Environment* 20(10), 1175–1193. DOI:10.1080/13549839.2015.1007121

Wezel, A., Bellon, S., Doré, T., Francis, C., Vallod, D., and David, C. 2009. Agroecology as a science, a movement and a practice. A review. *Agronomy for Sustainable Development* 29(4), 503–515. DOI:10.1051/agro/2009004

Wilmsen, C. 2008. Extraction, empowerment, and relationships in the practice of participatory research. In B. Boog, M. Slager, J. Preece, & J. Zeelen (Eds.) *Towards Quality Improvement of Action Research: Developing Ethics and Standard*, 135–146, Brill | Sense.

Wilson, S. 2008. *Research is ceremony: Indigenous research methods*, Black Point, NS: Fernwood Publishing.

Wilson, S. M. 2010. Environmental justice movement: A review of history, research and public health issues. *Journal of Public Management and Social Policy.* Spring 2010, 19–50.

WinklerPrins, A. M. (Ed.). 2017. *Global urban agriculture* ,CABI.

Wolfe, P. 2006. Settler colonialism and the elimination of the native. *Journal of Genocide Research* 8(4), 387–409. DOI:10.1080/14623520601056240.

Wright, W. J. 2018. As above, so below: Anti-black violence as environmental racism. *Antipode* DOI:10.1111/anti.12425

Yap, C. 2019. Self-Organisation in urban community gardens: Autogestion, motivations, and the role of communication. *Sustainability* 11(9), 2659. DOI:10.3390/su11092659

Chapters in this Volume:

Glowa and Roman-Alcalá (Chapter 8, This Volume)
Arnold and Siegner (Chapter 15, This Volume)

14

Agroecological Transformations in Urban Contexts: Transdisciplinary Research Frameworks and Participatory Approaches in Burlington, Vermont

Martha Caswell,[1*] V. Ernesto Méndez[1], María A. Juncos-Gautier,[2] Stephanie E. Hurley,[3] Rachelle K. Gould,[4] Denyse Márquez Sánchez,[5] and Storm Lewis[6]

[1] *Agroecology and Livelihoods Collaborative (ALC), Department of Plant and Soil Science, University of Vermont, Burlington, Vermont, USA*

[2] *Faculty of Environmental and Urban Change, York University, Toronto, CA and Agroecology and Livelihoods Collaborative, University of Vermont, Burlington, Vermont, USA*

[3] *Department of Plant & Soil Science, University of Vermont, Burlington, Vermont, USA*

[4] *Rubenstein School of Environment and Natural Resources and Environmental Program University of Vermont, Burlington, Vermont, USA*

[5] *Department of Biology, Carlton College, Northfield, Minnesota, USA*

[6] *Department of Environmental Science and Policy, Smith College, Northampton, Masschusetts, USA*

[*] *Corresponding author: Email martha.caswell@uvm.edu*

CONTENTS

KEY WORDS: *Urban Agroecology, Participatory Action Research (PAR), agroecology principles, transformation*

14.1 Introduction

Historically, agrifood systems have been perceived mostly in the realm of activities and policies in rural areas (Wiskerke 2015). Consequently, the potential role of city-regions in developing innovative sustainable food systems has been largely ignored (Sonino 2009). Through integration into the urban fabric and urban ecosystems, urban and peri-urban agriculture (UPA) confront particular socioeconomic and environmental constraints and offer opportunities that set them apart from rural agriculture. This results in UPA filling a distinctive multifunctional[1] role. UPA intersects with food security, public health (physical and mental), community-building, redevelopment and restoration of distressed areas, economic opportunities for low-resource neighborhoods through microenterprises, and the broader spiritual and identity-related benefits of connecting with place in meaningful ways (Lovell 2010).

In this chapter, we assess the potential of agroecology to provide a series of benefits to human and environmental health in urban and peri-urban contexts. Specifically, we explore what differentiates urban agroecology from urban agriculture. According to Vaarst et al. (2017 p. 4), food systems that follow the principles of agroecology call for "resilience, multifunctionality, equity, and recycling of resources… (and offer) significant options for impacting sustainable development in city regions." Using commonly accepted principles to explore the multiple dimensions of agroecology (environmental, social, political and economic) we hope to provide "…a more holistic framework than urban agriculture to assess how well urban food initiatives produce food and promote environmental literacy, community engagement, and ecosystem services" (Siegner et al. 2019 p. 557). Grounding our research in agroecological principles, we explore the ways these spaces, actors and systems interact, in order to better understand how urban agroecosystems can be designed to support urban farms and gardens that achieve these diverse ends (Altieri et al. 2017). Parallel to this work, we evaluate the effectiveness of community-based participatory action research to assess the relative importance of, and interactions between, the environmental, social, political and economic elements of our partners' work within urban and peri-urban contexts.

Participatory Action Research (PAR), is an approach to collaborative inquiry in which researcher and non-researcher partners (e.g. farmers or community organizations) engage in an iterative process of research, reflection and action, with the goal of addressing mutually identified issues of interest (Méndez et al. 2017). The PAR process described in this chapter integrates agroecology (defined for this purpose as the combination of ecological science with other academic disciplines and knowledge systems to study food systems (Méndez et al. 2017)); eco-landscape design (which incorporates ecological principles and values to urban and landscape design (Lovell 2009)); and cultural ecosystem services (the non-material benefits that people derive from ecosystems, which can include spiritual importance, cultural heritage and psychological well-being (Milcu et al. 2013)). By drawing from these academic fields and integrating these perspectives with the knowledge and experience of our selected local partners, we set out to assess what differentiates urban agroecology from urban agriculture and how design choices can optimize these benefits. As a by-product of this process, we also hope to observe whether this type of PAR process serves as an effective mechanism for amplifying agroecology as a relevant option for food systems transformation.

[1] Agricultural Multifunctionality examines multiple functions and/or benefits that agriculture and agricultural land can provide, in addition to producing food, feed and fiber (e.g. ecosystem services, recreation, aesthetics, cultural value, etc.).

In summary, the goals of this project are to:

- assess the potential of agroecology to provide a series of benefits to human and environmental health in urban and peri-urban contexts,
- explore what differentiates urban agroecology from urban agriculture,
- understand how urban agroecosystems can be designed to support urban farms and gardens that achieve these diverse ends, and
- evaluate the effectiveness of community-based participatory action research to assess the relative importance of, and interactions between the environmental, social, political and economic elements of our partners' work within urban and peri-urban contexts.

In this chapter we will address the above-mentioned goals by defining our context and the framework that we are using to guide this research, providing brief descriptions of our community partners and their work, explaining our transdisciplinary PAR process, and then sharing an overview of the research we have conducted to date with one of our community partners. We will close with observations about what we hope to improve upon and pursue as this research continues to evolve.

14.1.1 Geographic and Research Context

We explore the above themes and topics using a case study of Burlington, Vermont, USA, where we have been engaged in agrifood systems work for over a decade, as scholars, researchers, and practitioners. Burlington is a small city located in the northwestern portion of the state of Vermont, on the shore of 150-mile-long Lake Champlain. Vermont is a predominantly rural state, where 78% of the land is forested (Lovell et. al 2010). Agriculture and tourism contribute significantly to both the character and economic viability of the state. Long, cold winters and a relatively short growing season impact the types of agricultural activities that are viable. Dairy continues to dominate the agricultural economy, but recent years of prices at or below the price of production are contributing to a decrease in the number of dairies across the state. With a population of around 43,000, Burlington is Vermont's largest city. Burlington is economically diverse, but relatively racially homogenous; 85% of residents self-identify as white (census.gov 2019). However, waves of refugee resettlement have contributed to increasing racial diversity within the city.

Some Vermonters joke that the best thing about Burlington is that it is so close to Vermont, hinting that its urban feel sets it apart from the traditional conceptualization of the state's culture and rural identity. However, Burlington's contributions to the state match recent trends where small- to mid-sized cities are rapidly growing, both in the US and in other countries (Forman 2008). These cities often have distinct and strong ties to their rural hinterlands (Arnosti and Liu 2018), and maintain more connected and cohesive local urban food movements, including diverse forms of urban agriculture (Bricas and Conaré 2019). Vermont is recognized as a leader within the farm-to-table movement (Benjamin and Virkler 2016), and boasts a strong tradition of successful food activism. Recent efforts include passing the 2016 GMO labeling law and Migrant Justice's "Milk with Dignity" campaigns, among others. In addition, a network of local organizations and universities support research and a growing movement to strengthen food systems and agroecology studies in the region. As actors within Burlington's foodshed, who also work internationally, this research team was drawn to exploring what urban agroecology looks like in our own context and what potential it offers for strengthening our local agrifood system. Our case study offers a framework and research process to guide future participatory and transdisciplinary work in urban agroecology.

14.2 An Urban and Peri-Urban Agroecological Framework

We sought to investigate the potential of urban and peri-urban agroecology by using established agroecological principles, combined with insight from ecological design and cultural ecosystem services. This approach offered a scaffold for identifying key factors to achieve more sustainable urban food systems.

Additional considerations included optimizing ecological agricultural production, urban sociopolitical dynamics, urban policies, planning and regulations, as well as a desire to deepen our understanding of the environmental implications of and for food production in urban settings (McClintock 2010; Newman and Jennings 2008; Peters et al. 2009). By structuring our examination of these diverse dimensions in ways that explicitly include stakeholders (farmers, gardeners, community stakeholders, decision-makers, etc.), we plan to facilitate the development of a set of scenarios that allow participants to envision "...a wide range of societal and environmental effects...for each scenario" (Nassauer et al. 2007: p.47). We hope that this allows our partners to see themselves, their work and their broader context in potentially new (and transformative) ways.

14.2.1 Agroecology – Definitions and Principles

As agroecology becomes more widely recognized across the globe, what it means[2] and what it represents has also become more contested. In 2018, the Food & Agriculture Organization of the United Nations (FAO) crafted 'the internationally agreed definition', which states that "...agroecology is an integrated approach that simultaneously applies ecological and social concepts and principles to the design and management of food systems" (FAO, 2018). A few years earlier, civil society participants in the 2015 International Forum on Agroecology produced the Nyéléni Declaration, in which they provided a clear outline of expectations for agroecology, specifically for its peasant constituency. This declaration is overtly political, and explicit in stating that one of the motivations for articulating their position is to prevent the cooptation of agroecology. In addition to emphasizing the central role of smallholder farmers and social movements within agroecology, the Nyéléni Declaration calls for solidarity between urban and rural populations as necessary for (re)establishing responsible production and consumption patterns, prioritizing shared benefits and exploring viable models for shared risks (Nyéléni Declaration 2015).

Drawing from its ecological roots, agroecology has been structured around core principles since its early beginnings. Initially, the principles were ecologically oriented and based on experiences in rural environments, as seen in the key earlier texts of the field (e.g. Altieri 1987; Gliessman 1998). As research and development have embraced the notion that there are no 'recipes', but rather that all studies and initiatives need to be adapted to specific contexts, interest in better understanding and applying agroecological principles has grown. According to Patton, "An evidence-based effective principles approach assumes that while the principles remain the same, there will necessarily and appropriately be adaptation within and across contexts in implementing them" (Patton 2017 p. 200). Victor Toledo describes agroecology as something that started as an alternative science, but has evolved into an emergent practice and innovative technology that includes social, cultural, and political movements (Toledo 2012). Other authors working with different agroecology frameworks go as far as to name agroecology as a new "knowledge paradigm" (Shiva 2016). Numerous agroecology frameworks proposed in the past decade have incorporated principles within both their conceptual and applied expositions. The FAO adapted the notion of agroecological principles by identifying 10 agroecology 'Elements' (FAO 2018), which align well with the five principles for its Sustainable Food and Agriculture approach (FAO 2014). The Nyéléni Declaration includes key principles and values (referred to as pillars) for agroecology; these are distinct from other lists of agroecological principles because they articulate who should be involved and mechanisms for how change needs to happen (Nyéléni Declaration 2015). In 2018, the CIDSE[3] network published a document that attempts to clarify and align previous interpretations of agroecology from a variety of actors within the movement (CIDSE 2018 p. 3). The result is a list of 15 agroecological principles that cross economic, political, social and environmental domains and aim to articulate "...what

[2] See the collection of definitions of agroecology on Biovision's Agroecology Info Pool, representing various organizations including perspectives from governmental entities, civil society and the research community: https://www.agroecology-pool.org/agroecology/definitions/

[3] The abbreviation CIDSE stands for the organization's historical name, originally in French: "Coopération Internationale pour le Développement et la Solidarité" which can be translated as International Cooperation for Development and Solidarity https://cidse.org/faq/ (Accessed 10/31/19)

agroecology is and what it is not in order to gather political support, for the discipline to flourish, to avoid co-optation, and fight against false solutions" (ibid.).

There is almost no published work that critically analyses the use of principles in agroecological research and applications. As the field evolves, agroecologists are starting to recognize the importance, potential, and need for critical examination of these principles (Bell and Bellon 2018). Given the increasing interest and popularity of applying principles frameworks in different contexts, this has become an important area of inquiry in agroecology work (see the forthcoming special feature on 'Principles-based approaches in agroecology', in the journal *Elementa*). One of the key questions to ask when using principles is: how can we best compare information collected across multiple contexts? In theory, principles allow us to use different, site-specific indicators, which would allow for comparison of the principles across contexts. However, this type of comparative analysis has been rarely done in practice, and it remains to be seen whether these comparisons are valid. Rural contexts remain the predominantly cited examples of agroecology principles in practice. However, recent work recognizes the relevance of agroecology beyond rural contexts, and a growing number of studies focus on the unique relevance and expression of these principles in urban areas (Tornaghi and Hoekstra 2017). Our understanding of the diversity of approaches in agroecology and the various interpretations of its principles has informed both our definition of agroecology and selection of a specific set of principles for this study.

14.2.2 How We Conceptualize Agroecology

The research team for this project represents a wide range of disciplines and life experiences. It includes scholars from the fields of agroecology, urban landscape design and cultural ecosystem services; multiple NGO partners; and both graduate and undergraduate students, affiliated with several programs within the University of Vermont. This question of what agroecology is, and what it looks like when expressed in urban contexts, is one of the foundational questions for this project and remains at the heart of our study. We appreciate that there are multiple interpretations of what agroecology is and how people engage with it; for us, it is inherently transdisciplinary and participatory (Méndez et al. 2017). When we refer to transdisciplinary, we mean moving beyond just the intentional valuing and integration of different academic disciplines, to also incorporate different forms of knowledge and knowledge systems (e.g. indigenous, local, practical/empirical). Agroecology provides us with a framework to better understand the complex systems and interactions that comprise our agrifood systems in order to work toward transforming them to be ecologically sound, economically viable, and socially just.

14.2.3 Participatory Action Research

Participatory Action Research (PAR) is "...an epistemological stance that values knowledge produced from lived experience as equal to that produced in the academy and, in so doing, expands traditional notions of expertise" (Torre 2014 p.1). This combination of perspectives and knowledge types creates a platform for stakeholders to feel joint ownership of the process and results (Hovmand 2014). It rejects historical power dynamics among 'researchers' and 'subjects.' In PAR, research and non-research partners investigate, reflect, act and/or investigate again through iterative cycles. Because PAR projects increase community members' access to, and ideally motivation for, taking part in research processes (including design, execution, analysis and results dissemination), the approach supports articulation of problems and potential solutions outside of 'hegemonic definitions' (Dlott, Altieri and Masumoto 1994).

This approach is especially appropriate for agricultural inquiries, and even more so for the context-specific inquiries that are central to agroecology. Drawing out the questions, perspectives and insight of multiple actors is central to PAR, considering the often tacit and obscured nature of critical agricultural knowledge and expertise (Milgroom et al. 2016). Researchers and farmers, along with other stakeholders, in diverse contexts, can benefit from processes that work to find ways to combine knowledge and articulate lessons. In this way, learning can be made explicit and accessible to wider audiences. This can serve both to foster connections among people who might not otherwise interact, and to reestablish awareness about food production and agricultural systems among urban populations. In Burlington, for example, urban and school gardens are a place where gathering and sharing across cultures is encouraged, and when those spaces can also include inquiry and dialogue around issues, the seeds of PAR can be sown.

14.2.4 Our Conceptualization of PAR

As a function of the relationships that develop, and the mutual curiosity and knowledge that emerge from PAR, the processes themselves are often difficult to describe in terms of discrete phases. The research cycle depicted in Figure 14.1 represents a generic PAR cycle. The notable change from our previous visualizations of PAR processes is that instead of depicting isolated moments for reflection, the reflection process is integrated with research and action throughout the cycle. The nodes then become moments to pause for making decisions around where effort and attention should be directed and how to move forward. The graphic applies for the project as a whole (as we aggregate findings about the expressions of agroecological principles across the sites, and engage in collective research, reflection and action), and also individually to each organization as they independently move through this process. An example of this process with one of our partner organizations is described later in the chapter.

14.3 A Transdisciplinary Research Approach

As previously mentioned, the objective of our work is to assess the current status of agroecological expression and its potential as a mechanism for growth and transformation in and around Burlington, Vermont, USA. In our research, we sought to identify and explore local expressions of agroecology, and how to strengthen/deepen urban and peri-urban agriculture utilizing agroecology principles. In line with our PAR commitment, we wanted to examine if this approach holds potential for mutual benefit to both scholars and community actors. The first step for this was to develop a general level of recognition/conceptual acceptance of agroecology, and then help our partner organizations to build confidence through using images to establish 'observational cues' whereby they were able to discern examples of agroecology in practice and opportunities for its future development. To achieve this, we integrated aspects of principles-based evaluation, eco-landscape design, and cultural ecosystems services/relational values frameworks. Each of these is described in greater detail below.

14.3.1 Principles-Based Evaluation and Agroecology

Principles-focused evaluation "...examines (1) whether principles are clear, meaningful and actionable, and if so, (2) whether they are actually being followed, and, if so, (3) whether they are leading to desired results." (Patton, 2018 p. viii) This matches our research objectives, and offers the potential for both quantitative and qualitative data to inform this agroecological assessment project. Given that principles can represent broad aspirations, principles eventually are broken down into sub-principles and/or practices. A scaffolding approach encourages on-the-ground exploration to identify points of strength and weakness that emerge from the data. In this case, and following the PAR cycle, through subsequent validation and reflection (involving researchers and partners) the team works together to identify potential actions to support agroecological transformations, refine the original research questions and/or follow emergent leads.

14.3.2 Agroecology Principles and Landscape Design

The fields of landscape architecture, design and planning offer a wide array of spatial and cultural analytical tools that are seldom explored in the contexts of agricultural systems or agroecology. Combining these fields is one of the innovations of this project, as we incorporate agroecological design processes at multiple socio-spatial scales. The *community scale* includes initial mapping of partner sites and their surrounding land uses, in order to examine relative influence of surroundings upon sites, and sites upon surroundings. We are using a combination of larger-scale geographic information system (GIS) imagery and localized site-scale concept plans and landscape visualizations, which illustrate existing conditions and

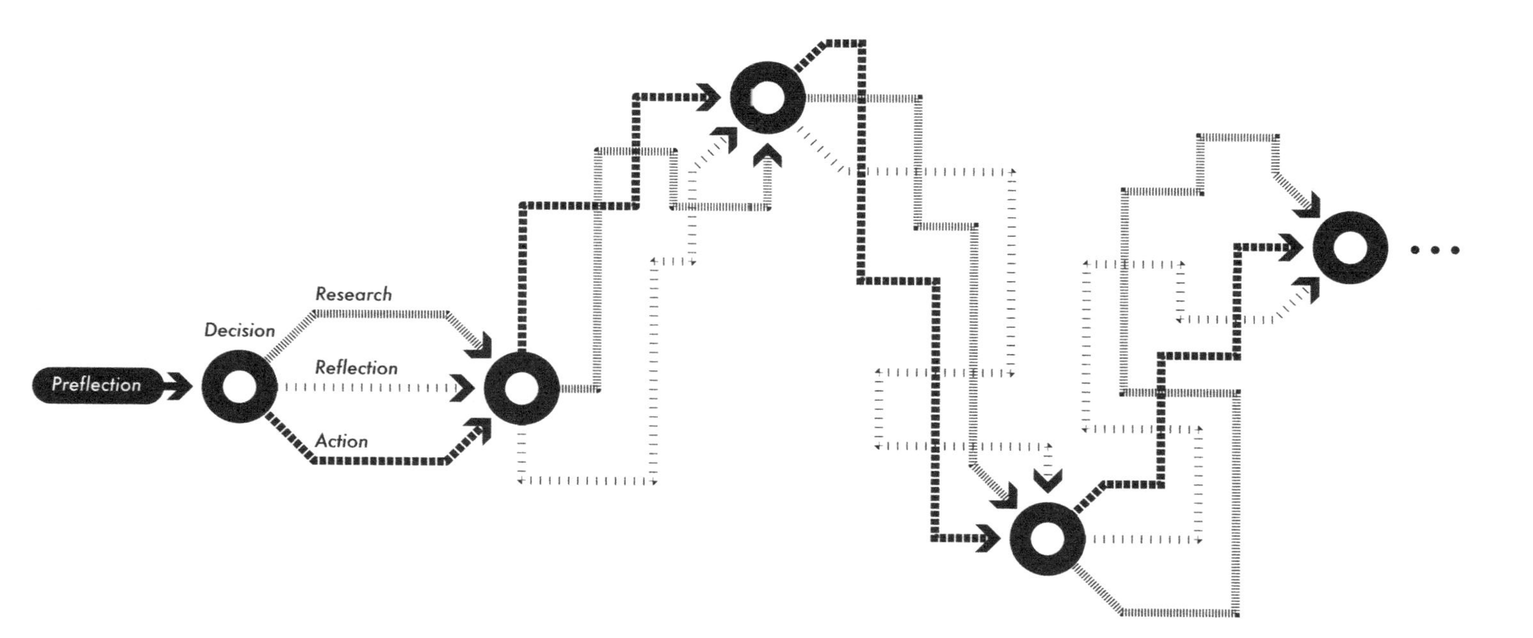

FIGURE 14.1 PAR research cycle with decision points represents the unpredictable path and necessary components that accompany a successful PAR process (some combination of research, reflection, action and moments for consensus around decisions), including steps outlined in Participatory Action Research Toolkit (Pain et al. 2011), integrated with the phases of the PAR cycle as described in Méndez et al. 2017.

proposed alternative future design scenarios (Nassauer et al. 2007; Santelmann et al. 2004; Werthmann and Steinitz 2008). For this, we are considering development density, transportation systems, infrastructure, water, vegetation types, etc., as well as locations of cultural importance. At the *site scale*, design opportunities and interventions are being identified in light of the agroecology principles our partners prioritize. Site plans and landscape visualizations will be developed to receive partner feedback and serve as examples of designs that could be implemented in order to enhance agroecological performance at the site. This represents a move toward 'democratization of design, on a small scale, where community members are encouraged to consider their own spaces and be active participants in envisioning potential changes (Angheloiu et al. 2017).

14.3.3 Cultural Ecosystem Services and Relational Values

As the field of agroecology has expanded, consideration of socio-political domains has increased. These domains are especially important for urban agroecology, because in urban areas social factors are arguably more ever-present than biophysical ones. Indeed, in some urban areas, "…food may be the only remaining connection to nature" (Francis et al. 2003 p. 102). Questions about how agroecological activity may differ in urban versus rural contexts engage heavily with the social side of agroecology. One lens on these social dimensions is found in the fields of cultural ecosystem services and relational values. These fields explore non-material dimensions associated with human-ecosystem interaction. Cultural ecosystem services describe "ecosystems' contribution to the nonmaterial benefits (e.g., experiences, capabilities) that people derive from human–ecological relations" (Chan et al. 2011 p. 206). Relational values encapsulate "…elements of human-nature relationships that do not fit into the provider-receiver or stock-flow metaphor of ecosystem services" (Chan et al. 2018 p. A5). Both concepts relate closely to many of the socially focused principles of agroecology, and can provide helpful tools for investigation related to those principles.

14.4 The Sites and the Partners

14.4.1 Why Urban Agroecology in Vermont?

In a predominately rural state like Vermont, it may seem incongruous that we are exploring examples of urban or peri-urban agroecology, instead of conducting a similar effort in our abundant rural landscape. However, we see Burlington's position as a hub of both food distribution and consumption as an important factor that facilitates the expression of agroecological principles. It therefore lends itself well to research on urban agroecology. We chose to focus on sites involved in food production for this first phase, but are open to expanding to additional realms in the future. As Vaarst articulates, "The deep mutual embeddedness of farming and food systems emphasizes that 'agroecological food' is not only food which is produced using agroecological agricultural methods, but also food going into a system which is built on the basis of agroecological principles, and where resources are part of full cycles, that is, also going from where food is eaten to where food is grown" (Vaarst et al. 2017 p. 704).

Given Burlington's reputation as a 'food city', and the strong connections we observe between farmers and city dwellers, we are keen to learn from our microcosm in ways that could both help to conceptually advance urban agroecology, as well as apply to other contexts. Those who live in larger cities know all too well, that in urban areas the human-nature interaction can feel out of balance (too much human, not enough nature). One opportunity for responding to this gap is through gardens and other forms of urban food production that, in Vandana Shiva's words, maximize 'health per acre' and 'nutrition per acre' (Shiva 2016). Because we use agroecological principles as our lens, we also carefully consider social and cultural factors within our agricultural system. "In the agroecological paradigm of knowledge…food is the web of life. Humans are part of this web as co-creators and co-producers, as well as eaters" (Shiva 2016 p. 13). The combination of cultures and dietary preferences that are present in urban areas reinforces the benefit of sovereign agricultural models. Burlington's history as a site for refugee resettlement has influenced its food culture, which includes new hyper-local pockets of production, such as a small goat farm and production of culturally preferred corn varieties, African eggplant, bitter melon, and okra (Bose and Laramee 2010).

In 2012, an Urban Agriculture Task Force (UAFT) coalesced in response to a perceived lack of specific policies related to the agricultural activities in Burlington. During a year-long process, the group collected data and facilitated conversations with the goal of providing recommendations to city officials for how urban agriculture could be encouraged and supported in Burlington. When they surveyed residents on their motivations for participating in and/or caring about urban agriculture, responses reflected values focusing "...on the importance (of) place-based food production with the goal of building an environmentally sustainable, resilient, socially just, and secure food supply" (Nihart et al. 2012 p. 22). Additional values of social capital, environmental stewardship, recreational outlets and cross-cultural interactions were also highlighted. A year earlier, the city's own planning document 'Plan BTV' highlighted the importance of Burlington's local food system; it asserted that "...(a) local food supply accelerates economic development, fosters a stronger and more sustainable community, improves the health of those who live and work in Burlington, and supports a system that regenerates and protects our natural resources and the environment" (Plan BTV 2012).

Though Burlington is a city, it has a small-town feel, and there is general familiarity around 'who does what' within certain circles. This is especially true in the urban/peri-urban agricultural sector, as there is overlap not only in the land that is used, but also in professional and social circles. While the UATF is no longer active, key actors from that effort were included in the first conversations around our explorations of urban and peri-urban agroecology in Burlington. In early spring, 2016, we gathered four organizations working on urban/peri-urban agriculture to explore potential areas of synergy. For the purposes of this chapter we have chosen to highlight the experience with one of the partners for the results section, but descriptions of progress with each of the partners is forthcoming. To provide a sense of the full group, brief profiles of each of the core partners follow:

New Farms for New Americans (NFNA - https://www.aalv-vt.org/farms) is an agriculture program for refugees and immigrants based in Burlington, VT. NFNA provides plots and support for over 275 farmers and gardeners, primarily refugees from various countries in Africa and Asia, to grow culturally significant crops, increase access to food, land and agricultural resources, and learn about growing food in Vermont. NFNA originally supported their participants in growing food for sale in local markets, but the participants were more interested in cultivating for their own consumption. Now the program focuses on family production, with exception of a few farmers who have found markets and have interest in producing particular crops on a larger scale.

The Vermont Community Garden Network (VCGN - https://vcgn.org/) leads the state's community garden movement by educating, supporting and connecting garden leaders. Established in 2001, as a non-profit organization, VCGN has helped initiate and sustain hundreds of gardens all over Vermont, and has connected thousands of children, teens, and adults to fresh, healthy food and sustainable food production practices. VCGN gardens are located in both urban and peri-urban settings, and the network serves as an organizing/support body for community gardens across the state of Vermont. One of the unique partnerships that VCGN has established is with the **Champlain Housing Trust (CHT -** https://www.getahome.org/). VCGN provides technical support for establishing and maintaining gardents at CHT-managed apartments/coops within Chittenden County. The Champlain Housing Trust is the largest community land trust in the United States. Their mission is to 'support the people of Northwest Vermont and strengthen their communities through the development and stewardship of permanently affordable homes.'

The University of Vermont's Horticultural Research Center (HREC - http://www.uvm.edu/~hortfarm/) is a 97-acre university farm, in a peri-urban/suburban setting, within the limits of the city of South Burlington, and about four miles from downtown Burlington. Set behind car dealerships and housing developments, HREC is an oasis of green in aerial maps. **Catamount Farm** is a 12-acres, organically certified area embedded within the HREC. The mission of Catamount Farm is to model sustainable farming practices through a working vegetable and fruit farm that provides educational and research opportunities for the UVM community, including a six-month Farmer Training Program (FTP). The FTP draws participants from all over the country each summer.

The **Intervale Center (IC -** https://www.intervale.org/) is a farm and food non-profit organization in Burlington, founded in 1988 (originally as the Intervale Foundation), with a mission to strengthen community food systems. The IC manages a 340-acre mixed-use campus that includes organic farms, wildlife areas, a native tree nursery, recreational paths, community gardens, a food distribution center (hub), and a suite of other programs and enterprises. They also provide business planning to farms across Vermont, restore riparian buffers in all of Vermont's watersheds and network with groups from across the country and around the world. The Intervale Center operates in a peri-urban context, literally in the backyard of the city of Burlington (about 2 miles from downtown). Sited along the Winooski River, the Intervale is located in a floodplain that was originally an Abenaki sacred site. This land was being used as the city dump before being purchased and rezoned in the mid-1980s. Now the Intervale is an agricultural and recreational destination year-round thanks in part to a calendar of events hosted by the IC, including music and cultural gatherings, educational and volunteer opportunities, a food hub and harvesting/gleaning activities.

The group of partners that we have assembled represent different models of organized urban/peri-urban agroecology that we see around us. NFNA is a critical component of refugee resettlement in our community, which offers a connection to agriculture for many people who come from displaced traditions of agriculture; VCGN is a neighborhood resource with statewide reach; Catamount represents the land grant mission of agricultural education and outreach; and the Intervale is a multi-faceted model of what integrated urban agricultural initiatives can include. While in some ways they are more different than similar, they all operate within a 20-mile radius and are familiar with and appreciative of each other's work.

14.5 The Process

14.5.1 Preflection

All PAR processes have to begin somewhere. When we are entering either a new geographic or content area, we like to start slow to make sure we have our bearings and are clear that our questions and contributions will be relevant. We call this phase 'preflection', and it essentially amounts to a period of confirming compatibility between research and non-research partners (both in interests and ability to work together) (Méndez et al. 2017). As mentioned above, this project was born from general curiosity, within our research group, related to expressions of urban agroecology in our local context. We perceived there to be ample evidence of urban agroecology in practice, but wondered whether there was familiarity with the term/concept of agroecology, and, if local actors self-identified with, or were using the term to describe their work.

We received a seed grant to explore the differences and similarities between urban agroecology in Havana, Cuba and Burlington, VT, which provided a jump-start to this research initiative. Parallel to international exchanges that accompanied that work, we conducted interviews with representatives of our partner organizations in Burlington, to solicit general opinions and determine topics for further study. The most salient issues that emerged across all of the organizations included: access to land (leasing vs. ownership); soil (erosion and quality); understanding the landscape in order to optimize land use and implement best practices; infrastructure challenges; equipment/appropriate technologies; a lack of economic/financial resources (problems accessing these resources); support from local authorities/government; and a desire for more information on agroecological management practices.

Analysis of these interviews indicated that NFNA and VCGN wanted to better understand and describe the value they provide to their participants and the community at large, whereas Catamount Farm wanted to have a clearer idea of whether and how they exemplify agroecology. The Intervale Center (IC) wanted to test whether they qualify as an 'agroecological organization' and further, if/how this might connect with whether their model could work in other contexts. Confusion persisted about the term 'agroecology.' The Executive Director of VCGN confirmed our perception that agroecology was not a familiar concept, saying "…I'm willing to bet money that most of the gardeners we work with will not recognize the word agroecological." The NFNA program coordinator was skeptical, "Agroecology? The whole concept, is something totally foreign to me…I'm not sure what I'm supposed to be learning from other people or what I might offer other people under this lens of agroecology." The IC Executive Director hoped that this project might help them to "…communicate, show, demonstrate…the power of peri-urban agriculture in our community. And by power, I mean…the value of it."

Through our repeated interactions, we started to see interest among the group to further understand agroecology, to be able to identify its expressions, and explore if greater alignment with agroecology (theoretically, by changing practices, and/or by linking up with the movement) could create positive change. Recognizing that education around agroecology, especially active or engaged learning can promote transformation (Francis et al. 2003, David and Bell 2018), we responded to a request from the partners to organize a workshop to explore the question, 'What is agroecology?' In April of 2018, the partner organizations all gathered at UVM, where we shared several agroecological principles frameworks, including the FAO Elements, the CIDSE principles and a set of principles ALC had used for previous projects. After a brief theoretical presentation, we divided into small groups where each organization selected one set of principles (we shared the FAO Elements, the CIDSE principles and a set of principles ALC had used for previous projects). The assignment was to see if and where the organizations could identify themselves in what was articulated as agroecology. In the discussion that followed, there was a tangible shift as each group began to identify how and why we saw their work as embodying agroecology.

14.5.2 Decision Point

From this workshop, we interpreted clear commitment for moving forward from each of the partners. We saw articulation of interest not just in the question of 'where/how is agroecology expressed in your context', but also a desire to either incorporate more agroecology or do agroecology 'better' if this project revealed that there were gaps/missed opportunities. In other words, we heard requests for working explicitly to identify potential areas for these partners to enhance agroecology within their current contexts. This articulation of preferences by the partners (for valuation, and then designing future plans/actions) led us to realize that the initial ALC team needed to expand to include other disciplines– including faculty and students with expertise/interest in CES, landscape design, and GIS.

To formalize and fund the subsequent phases of this PAR process, a small group of faculty and staff, representing four disciplines within UVM, applied and received a USDA Hatch award in the fall of 2018. We proposed an exploration of if and how conceptualizing urban/peri-urban agricultural initiatives as 'urban/peri-urban agroecology' (specifically attaching this name to things that might have previously been referred to in other ways), would alter the way stakeholders perceived their actions and impact. In addition, we set out to explore how intentional design, based on agroecological principles, might also contribute to a more ecological, vibrant and just food system in and around Burlington. The proposal centered around the following hypotheses: 1) agroecological principles might be unfamiliar concepts, but associated practices would be familiar, and recognized as beneficial; 2) participatory agroecological design processes could facilitate adoption of additional practices or improvement of existing ones; and 3) principles-based assessments could serve as a useful framework for making strategic choices. Figure 14.2 represents the steps we outlined for the project.

14.6 First Research Cycle: Selected Results

14.6.1 Identification of AE Principles of Interest

Once we had funding, we worked with the organizations to identify what specific agroecological principles were of greatest interest to each of them. Based on our partners' feedback and our own deliberation, we decided to use the CIDSE principles framework, which consists of 15 principles that are divided into four dimensions—economic, political, environmental and socio-cultural. Using a common framework helped to support a shared definition of what agroecology looks like and means (CIDSE, 2018). We noted that partners were drawn to the clarity and simplicity of the CIDSE framework, especially as outlined in

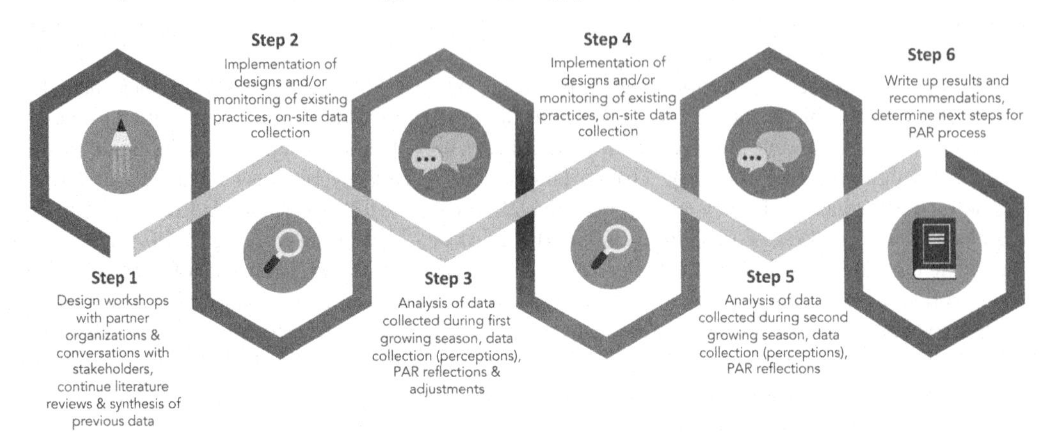

FIGURE 14.2 Project timeline.

their summary infographic[4], and subsequently decided to use these principles to orient our work. Each organization determined their particular research priorities (represented in Table 14.1 below) and we began to map our individualized plans with each site.

14.6.2 Assessing Agroecology at the Intervale Center

We have chosen to highlight our work with the Intervale Center as representative of our project to date, for the purposes of this chapter. The presentation of results is followed by a short discussion section, lessons to date and the opportunities we see as this PAR process continues to unfold. Major project activities with the Intervale to date include: 1) cataloguing previous research, 2) assessing the presence of agroecological principles at the Intervale, and 3) early developments for a participatory design process. We describe these steps in more detail below, categorizing them by the PAR component that they most represent (i.e. research, reflection or action).

14.6.2.1 Action – Cataloguing Previous Research to Avoid Replication

As mentioned above, the ALC's partnership with the Intervale Center goes back many years and includes an earlier study of landscape multifunctionality (Lovell et al., 2010). Because of its proximity to the university, and its unique organizational model and landscape features, the Intervale Center (IC) is a frequent subject of UVM-related research and educational activities. Through reflective conversations with Intervale staff, concerns around duplication in research emerged, due to the lack of a central system for organizing results/publications from previous work. We saw this as an opportunity to complement our first round of research with action. Two members of the research team (graduate students focusing on IC for their dissertations) worked with UVM library staff to create a research repository. The repository's objective was to provide interested researchers and IC staff with an organized resource for exploring past research, and better direct future efforts towards remaining gaps instead of replicating previous work.

This work reinforced the value of building from our earlier study of multifunctionality to identify the expressions of agroecological principles in practice and opportunities for land use optimization. With this foundation, we would then facilitate a participatory design process to enhance current agroecological practices and/or implement new ones. We anticipated that after identifying the presence of agroecological principles, we could then turn to cultural ecosystem services tools to layer on additional assessments of their meaning and relative importance.

[4] https://www.cidse.org/wp-content/uploads/2018/04/CIDSE_AE_Infographic_EN.pdf

TABLE 14.1

Selected AE Principles of Focus and Aligning Activities (by organization)

	Selected AE Principles of Focus and Aligning Activities (by organization)	Initial activities	Next steps
	  Strengthens food producers, local communities, culture, knowledge, spirituality Promotes healthy diets and livelihoods	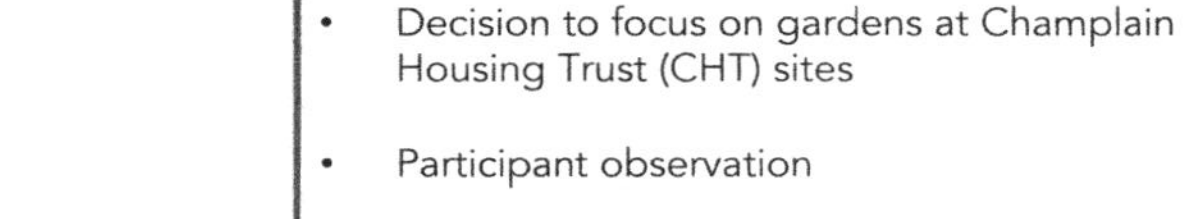 • Decision to focus on gardens at Champlain Housing Trust (CHT) sites • Participant observation • AE design prioritization game	• Plan for assessing gardener motivation and satisfaction • Continuing participatory design processes
The University of Vermont CATAMOUNT FARM	Enhances integration of various elements of agro-ecosystems Nourishes biodiversity and soils	 • Permaculture Course – student designs for pollinator habitat • GIS mapping • Data collection for pollinator and biodiversity inventories	• Land use/AE principles mapping • Continuing participatory design processes (implement/monitor)
New Farms for New Americans	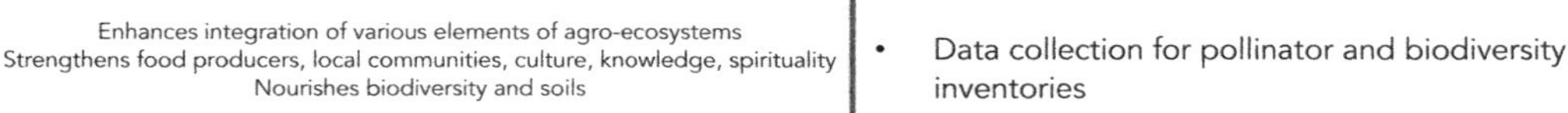 Enhances integration of various elements of agro-ecosystems Strengthens food producers, local communities, culture, knowledge, spirituality Nourishes biodiversity and soils	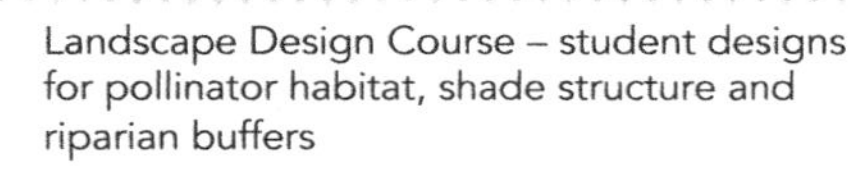 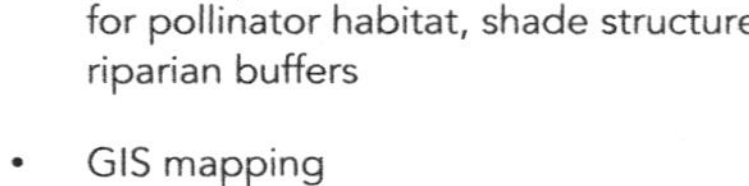 • Landscape Design Course – student designs for pollinator habitat, shade structure and riparian buffers • GIS mapping • Data collection for pollinator and biodiversity inventories	• Continuing participatory design processes (implement/monitor) • Plan for assessing gardener motivation and satisfaction
 	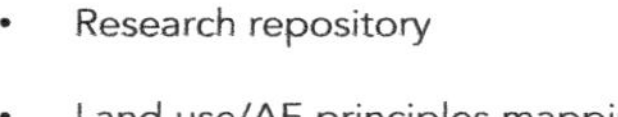  Enhances integration of various elements of agro-ecosystems Strengthens food producers, local communities, culture, knowledge, spirituality	• Research repository • Land use/AE principles mapping • Participant observation, photo documentation • Educational interventions and CES	• First round of participatory designs • Network analysis

14.6.2.2 Research – Expression of Agroecology (AE) Principles by Land Use

The idea to start by focusing on links between agroecological principles and land-use optimization, was inspired by the recommendation of the optimized performance you can achieve by designing a quilt-like pattern of agroecosystems within particular landscapes (Altieri, 2016). In order to study the various expressions of AE principles by land use, the research team designed an interactive methodology with multiple visual surveying tools to use during semi-structured interviews. The interviews included a short conversation to collect basic demographic information and gauge the participants' general understanding of agroecology. Participants then completed a brief exercise where they reviewed the 15 agroecological principles (numbered for ease of identification) represented in the CIDSE infographic. Participants were then asked to mark the principles they recognized as being expressed within the different Intervale land uses and operational areas, providing at least one example to support each selection Participants to date (n = 30) represent a diverse group of actors associated with the Intervale Center (including Intervale Center staff in the different areas of operation, farmers and community gardeners, among others).

Subsequently, a subsample of key participants (n = 15) were asked to mark the locations they associated with the expressions of the principles they had identified in a map of the property delineating various land uses and landscape features (see Figure 14.3).

FIGURE 14.3 Map-based data collection exercise with IC stakeholders. Participants (n = 30) were asked to identify agroecological principles they perceived to be present within the Intervale, using the CIDSE infographic. A subsample of participants (n = 15) then located the principles they had identified on a map of the property and surrounding area, categorized by land-use and area of operation. (Photo credits – María A. Juncos-Gautier)

As a follow up to this interview, the lead researcher for this site (María A. Juncos-Gautier), used a modified photovoice exercise (Wang, 1997), where the subsample of key participants were asked to take photos that represented their interpretation of the agroecological principles they had identified. This activity was designed to elicit additional dialogue around each participant's understanding of the principles, and contribute toward a set of images or 'observational cues' that could be used as future examples to help others to 'see' agroecological principles expressed in their own contexts. In a second semi-structured interview, this subsample of participants described their photos and the links with the AE principles to the researcher, writing a short description of their photos and including the number(s) of the principle(s) depicted in a caption area provided under each photo. During this time, Juncos participated in various volunteer opportunities at the IC and engaged in participant observation and photo-documentation so that she could triangulate the perceptions of the actors with her own observations and experiences.

14.6.2.3 Reflection – Preliminary Analysis of Results and the Process to Date with the Intervale

In early 2020, the research team facilitated a workshop with the subsample of Intervale study participants, in order to validate the preliminary analysis from this phase. In addition, the workshop served to assess whether the exploration of agroecological principles and their expression at the Intervale had shifted perceptions, understanding and/or motivations around agroecology. As demonstrated in Figure 14.4, the Intervale Center contains expressions of each of the 15 CIDSE principles, albeit with both strengths and opportunities for improvement. Initial analysis shows strengths in the categories of socio-cultural and environmental-related principles (e.g., the principles—"strengthen food producers, local communities, culture, knowledge and spirituality" and "nourishes biodiversity and soils"). The observations provided by the actors (practical perspective) with those of the researcher (academic/theoretical perspective), and the multi-methods approach (semi-structured interviews with visual tools, the photovoice exercise, participant observation with field notes and photo-documentation), provided a robust triangulation to validate common concepts and themes around the principles. The Intervale has also mentioned their

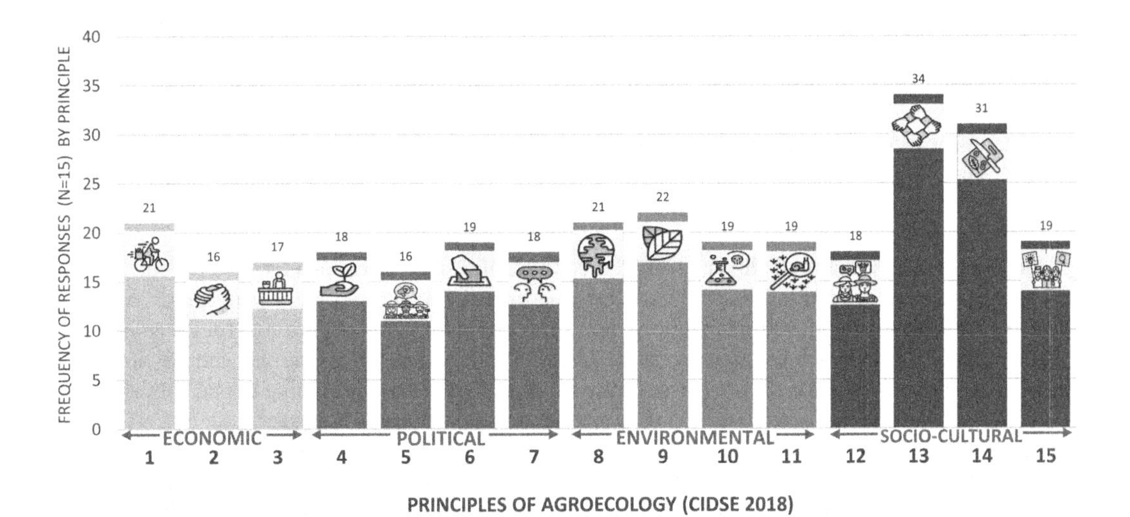

FIGURE 14.4 Frequency of responses related to the presence of the 15 CIDSE agroecological principles. After sorting the responses by CIDSE dimension from the initial mapping exercise, the frequency of responses (x-axis) represents the total number of times each principle was noted by the subsample of actors (n = 15) across the categories of land-use and landscape features in the Intervale map. Single principles may have been signaled more than one time by the same person (i.e., occurrences in multiple locations within the Intervale map or property based on the actors' perceptions).

appreciation for the PAR model, noting that Juncos' contributions of volunteer hours and the team's effort to make sure they were clear on the purpose and potential benefits of this research, makes them feel like this is 'their research project' in a more tangible way than they have felt with previous studies.

During the workshop, the subsample of participants identified areas of strength, challenges and opportunities for advancement. Two areas that were identified for additional focus emerged from principles within the political and economic domains. The Intervale Center Land Manager is excited to review the results to assess participant perceptions, both to consider opportunities for land use optimization that are not yet being realized, and also to further their desire for integrating additional practices to advance agroecology. An interesting byproduct of this process has been the increased affinity with agroecology as a concept, exemplified by comments such as:

> I now have a really good understanding of this framework and I can see it in the work that we do, and before I didn't...When I was out walking around and I would think ... here's actually an example of how we promote farmer-to-farmer exchanges and it's happening right here in front of me....Before I didn't know the agroecology principles well enough to be able to go out and identify that. (IC staff member)

14.7 Discussion

Our work in and around Burlington, demonstrates the potential of using a PAR approach – combined with theory and principles from agroecology, ecological design, and cultural ecosystem services – in an urban agroecology research framework to better understand and support the work of organizations involved in urban and peri-urban agricultural/food contexts. Montenegro (2014) discusses the importance of the transdisciplinary research approach and transformative role that urban agroecology can play in achieving more sustainable urban food systems. Citing an urban agroecology course as an example, she highlights the importance of facilitating agroecological knowledge co-creation, among a diversity of relevant actors, to advance initiatives in the urban context. In this phase of the study, we attempted to provide a holistic framework, through multiple exercises with the CIDSE principles, so that the Intervale actors could 1) better understand where they are currently positioned in terms of agroecological expression; and 2) have a new tool for helping them envision where they might want to focus their attention and efforts as part of a future transformation. The Intervale Center's Executive Director noted that the process of everyone becoming familiarized with the CIDSE agroecological principles contributed to 'different conversations' because now they share a 'common language' and a collective perspective for interpreting their context. Participants also mentioned that once they better understood agroecology, they were interested in using it to guide their work. The next step is using the principles not only as a tool for understanding the current situation, but also as a frame for focusing on how to improve and assess future performance in factors such as land use and opportunities for social engagement.

Holt Giménez and Shattuck (2011) describe a continuum from neoliberal/enterprise to radical/food sovereignty models, and suggest that we must navigate toward the latter for agroecological transformations to occur. Working with multiple groups across Burlington, we see an opportunity for using the principles to create this 'common language' around agroecology in pockets across the city, and are curious to see what happens to the dialogue on a larger scale. Agroecology is only beginning to make its way into research, practice and movement within urban and peri-urban settings (Fernandez et al. 2012; Renting 2017). Hence, we must test and develop promising frameworks to continue to advance agroecological work in cities and their periphery. Instead of only considering production and consumption interactions between urban and rural actors in the agrifood system, reworking these dynamics "...will require processes of negotiation, adjustments and development of common understandings, shared knowledge, and collective action to ensure that everybody at all times will have access to healthy nutritious food" (Vaarst et al. 2017 p. 699). The Intervale Center's tagline is 'Farms, Land, People.' Their educational efforts, multiple food production, distribution and rescue (gleaning) programs, make them in many ways a one-stop shop with the potential to respond to the call made by Vaarst and co-authors. However, there is still much ground to cover to fully realize an agroecological transformation of the Intervale Center or the other organizations participating in this project.

Parallel to our deep dive with the CIDSE principles, and to contextualize our choice of this particular framework, we conducted a literature review of ten sources that explicitly refer to agroecological principles. We originally planned to use the four CIDSE dimensions (economic, political, environmental, and social-cultural) to sort the themes in the other versions of agroecological principles, but it quickly became clear that additional dimensions were necessary to capture the full range of principles described in the other texts. As a result of this exercise, we plan to explore additional themes in our future work with the partner organizations, to more fully capture urban agroecological expressions and potential as described in other contexts. One area of additional focus will be 'resource management,' and considering what actions are needed to help the environment; for example, lowering/eliminating the use of fossil fuels and looking for other ways to be more efficient in resource use (Anderson et al. 2019). We also want to specifically consider examples of 'localization,' which can be summarized as promoting local ideas, strategies, and resources (Bell and Bellon 2018; Gliessman 2016). This will push us to find examples related to the use of traditional ecological knowledge, efforts to shorten supply and distribution chains, and modifying practices based on the local context. Finally, we will seek out cases where human/nature interactions demonstrate a prioritization of the mental and physical wellbeing of people, and not just a general concern for the environment. We believe modifying CIDSE's categories to consider these three additional dimensions with our partners will result in a more comprehensive assessment of viability and desirability for agroecological transformation.

14.7.1 Next Steps in the PAR Process

As we continue to explore the CIDSE principles and these additional dimensions of resource management, localization and ecological health, we look forward to using tools and concepts from relational values and cultural ecosystem services to help incorporate some of the intangible aspects of these principles into our assessment metrics. An important contribution of these methods is that they offer new ways to determine value. Increasing levels of urbanization around the world have motivated a growing interest in analyzing ecosystem services dynamics in urban contexts (Gómez-Baggethun et al. 2013). Research also demonstrates that the cultural ecosystems services lens can help to inform conversations about what is meaningful in urban agricultural settings (Riechers et. al. 2016). As our experience with the Intervale Center demonstrates, we see potential for inquiry that uses cultural ecosystem services and relational values approaches to contribute to new "....ways of 'knowing landscapes'" (Chan et al. 2016 p. 1464). This expanded suite of valuation tools may increase "local appreciation for systematic, science-based approaches, and vice versa" (ibid. p. 1464). In our next phase we will move beyond the circle of actors with direct responsibilities (jobs, farms or garden) within the Intervale, to the wider public who access this space. Especially within that broader population, we anticipate that discussions of cultural ecosystem services may provide a potential 'foot in the door' that awakens interest in the environment, since, particularly in urban areas, the nonmaterial and social benefits of agriculture are often more obvious and sometimes appear more meaningful, than the biophysical benefits of the food produced (Camps-Calvet et al. 2016).

One of the innovations of our project was to explore the potential of ecological design in agroecological participatory action research, both as critical for conceptualization, and as a tool for action/application. In all of the sites, the need for shared understanding of space has emerged as a critical theme. This surfaced early on, when we requested maps from our partners and found that many of them were relying on coarse or outdated versions of maps to visually represent the land they manage. We have since worked to update satellite imagery maps for each of the partner organizations. By encouraging further precision in the maps, we have been affirmed in the power of looking at something together and the utility of visual forms to enrich conversations (Schattman et al. 2019). Our examination of the Intervale's land, and the activities it supports, was complemented by internal Intervale conversations interrogating the differences between restoration and rehabilitation, and what it implies for design and ecosystem benefits. As we discussed how this related to our work with the agroecology principles, we were reminded of the value of holistic and creative exercises to teach ourselves to think differently (Orion 2015). During the recent results validation workshop with the subsample of interviewees, we used examples of the maps and photos from the exercise to convert their observations into tools for identifying needs and actions. Continuing to use the agroecological principles as a frame, our next step is to use these maps as a basis for designing interventions that either reinforce current expressions of agroecology, or support new efforts. We will then track these developments and their impact over time.

14.7.2 Lessons/Conclusions

The integration of agroecology, cultural ecosystem services and ecological design proved to be an effective approach from a conceptual perspective. Each one of these fields provided unique contributions to make our approach truly transdisciplinary, and pertinent, for urban and peri-urban contexts. As expected, none of the organizations we are working with knew about agroecology, and aligning concepts and definitions around the topic proved to be a critical first step. The principles framework has facilitated a shift from the initial focus on characterizing urban/peri-urban landscapes from an agroecological lens, to catalyzing stakeholder interest and action in strengthening agroecological dimensions. The principles also provide an opportunity to monitor and assess performance of agroecology over time, which we will be analyzing as the process evolves.

In addition, the use of participatory action research (PAR) proved pivotal to our process. PAR principles provide a guide on how to establish successful relationships between research and non-research actors (Méndez et al. 2017), which have resulted in strong partnerships, yielding both important academic and actionable results. Successful PAR processes move at the speed of trust, which can cause impatience in a world accustomed to immediate results. PAR requires time, and can generate a high level of complexity that requires skillful management from the research team. Besides learning with and from each other, this PAR process has offered a regular forum for these partners to gather—something that they mentioned as a strength, and which was not happening prior to this project. In our experience, PAR also often leads to the need for inter- or transdisciplinary work, since most people outside of academia do not limit their thoughts or work to particular disciplinary silos.

Through building a shared vocabulary and developing observational cues that help people connect with the principles, we see exciting potential for urban agroecology to become more recognized and recognizable in and around Burlington, Vermont, and beyond. This is related in part to the enthusiastic community engagement we have seen so far in the early stages of this work. In addition, we also see explorations around agroecology helping to connect dots among people and ideas in our small city. We continue to observe new work emerging that ties in directly with food sovereignty, land reform, and protection of native species. Through our PAR process, we hope to support and amplify these efforts. Ideally, by reinforcing its presence and relevance, agroecology can become a unifying catalyst for broad food system transformation in Burlington.

Acknowledgments

Our community partners are co-creators of our questions and accompany us in our learning. This chapter represents an investment of time and generous sharing of expertise from many individuals. We especially want to thank the following: Patrick Dunseith and Travis Marcotte - The Intervale Center, Alisha Laramee - New Farms for New Americans, Rachel Stievater and S'ra DeSantis - Catamount Farm, Carolina Lukac and Michelle Gates - Vermont Community Garden Network, Meghan Tedder - Champlain Housing Trust, Dan Ma and Phoebe Little - 2019 UVM Doris Duke Fellows, mapmaker Emily Brodsky, and other members of the ALC community of practice including Annie White, Gabriela Bucini, Tatiana Gladkikh and Josh Taylor.

REFERENCES

Altieri, M.A. (1987) Agroecology: the scientific basis of alternative agriculture. Westview Press: Boulder, CO, U.S.A.

Altieri, M.A., C.I. Nicholls, P. Rogé and J. Arnold (2017) Urban Agroecology: principles and potential. 33:18–20.

Altieri, M. A. 2016. Agroecology: Principles and strategies for designing sustainable farming systems. www.agroeco.org/doc/news_docs/Agroeco_principles.pdf (accessed August 2, 2019).

Anderson C.R., J. Bruil, M.J. Chappell, C. Kiss and M.P. Pimbert. 2019. From Transition to Domains of Transformation: Getting to Sustainable and Just Food Systems through Agroecology. *Sustainability* 11 (19):5272.

Angheloiu C., G. Chaudhuri and L. Sheldrick. 2017. Future Tense: Alternative Futures as a Design Method for Sustainability Transitions. *The Design Journal* 20:sup1, S3213–S3225, DOI: 10.1080/14606925.2017.1352827

Arnosti N. and A. Liu. 2018. Why rural America needs cities. Brookings Institute Report. https://www.brookings.edu/research/why-rural-america-needs-cities/ (accessed January 23, 2020).

Bell M.M. and S. Bellon. 2018. Generalization without universalization: Towards an agroecology theory. *Agroecology and Sustainable Food Systems* 1–7. DOI: 10.1080/21683565.2018.1432003.

Benjamin, D. and L. Virkler. 2016. *Farm to Table: The Essential Guide to Sustainable Food Systems for Students, Professionals, and Consumers.* White River Junction: Chelsea Green Publishing.

Bose P. and A. Laramee. 2010. Taste of home: migration, food and belonging in a changing Vermont. Center for Rural Studies. Burlington, VT.

Nicolas B. and D. Conaré. 2019. Historical perspectives on the ties between cities and food. *Field Actions Science Reports,* Special Issue 20, 6–11.

Camps-Calvet, M., J. Langemeyer, L. Calvet-Mir and E. Gómez-Baggethun. 2016. Ecosystem Services Provided by Urban Gardens in Barcelona, Spain: Insights for Policy and Planning. *Environmental Science & Policy*, 62: 14–23.

U.S Census Bureau. 2019. Burlington Quick Facts Data Table, accessed on December 17, 2019. https://www.census.gov/quickfacts/burlingtoncityvermont

Chan, K.M.A., J. Goldstein, T. Satterfield, et al. 2011. Cultural services and non-use values. In *Natural Capital: Theory and Practice of Mapping Ecosystem Services*, eds. P. Kareiva, H. Tallis, T.H. Ricketts, G.C. Daily and S. Polasky, 206–228. Oxford: Oxford University Press.

Chan, K.M.A., P. Balvanera, K. Benessaiah, et al. 2016. Why protect nature? Rethinking values and the environment. *Proceedings of the National Academy of Sciences of the United States of America, 113*(6), 1462–1465. https://doi.org/10.1073/pnas.1525002113

Chan, K.M.A., R.K. Gould and U. Pascual. 2018. Editorial overview: Relational values: what are they, and what's the fuss about? *Current Opinion in Environmental Sustainability, 35,* A1–A7. https://doi.org/10.1016/j.cosust.2018.11.003

CIDSE. 2018. The principles of agroecology: towards just, resilient and sustainable food systems. https://www.cidse.org/2018/04/03/the-principles-of-agroecology/ (accessed April 25, 2018).

David, C. and M.M. Bell. 2018. New challenges for education in agroecology. *Agroecology and Sustainable Food Systems*, 612–619. DOI: 10.1080/21683565.2018.1426670

Dlott, J., M. Altieri and M. Masumoto. 1994. Exploring the theory and practice of participatory research in US sustainable agriculture: A case study in insect pest management. *Agriculture and Human Values, 126–139*(2–3), 12. http://www.springerlink.com/index/w4138r8625551758.pdf

FAO. 2014. Building a common vision for sustainable food and agriculture: principles and approaches. Food and Agriculture Organization of the United Nations (FAO). http://www.fao.org/3/a-i3940e.pdf

FAO. 2018. The 10 elements of agroecology: guiding the transition to sustainable food and agricultural systems. Food and Agriculture Organization of the United Nations (FAO). http://www.fao.org/3/i9037en/I9037EN.pdf

Fernandez, M., V.E. Méndez, T.M. Mares and R.E. Schattman. 2016. Agroecology, food sovereignty and urban agriculture in the United States. In *Agroecology: a transdisciplinary, participatory and action-oriented approach*, eds. V.E. Méndez, C.M. Bacon, R. Cohen and S.R. Gliessman, 161–175. Boca Raton: CRC Press.

Forman, R.T.T. 2008. *Urban Regions: Ecology and Planning Beyond the City.* Cambridge: Cambridge University Press. 10.1017/CBO9780511754982.

Francis, C., G. Lieblein, S. Gliessman, et al. 2003. Agroecology: The ecology of food systems. *Journal of Sustainable Agriculture* 22 (3):99–118. doi:10.1300/J064v22n03_10.

Gliessman, S.R. 1998. *Agroecology: ecological processes in sustainable agriculture.* Ann Arbor: Ann Arbor Press.

Gliessman, S.R. 2015. *Agroecology: the ecology of sustainable food systems, 3rd Edition.* CRC Press/Taylor & Francis: Boca Raton, FL.

Gliessman, S. 2016. Can food production be re-localized? *Agroecology and Sustainable Food Systems* 40 (2):115–115. DOI: 10.1080/21683565.2015.1118422.

Gómez-Baggethun E., A. Gren, D.N. Barton, et al. 2013. Urban Ecosystem Services. In *Urbanization, Biodiversity and Ecosystem Services: Challenges and Opportunities*, eds. Elmqvist T., M. Fragkias, J. Goodness et al. 175–251. Dordrecht: Springer.

Holt Giménez, E. and A. Shattuck. 2011. Food crises, food regimes and food movements: Rumblings of reform or tides of transformation? *Journal of Peasant Studies, 38*(1), 109–144. https://doi.org/10.1080/03066150.2010.538578

Hovmand, P.S. 2014. *Community based system dynamics.* New York: Springer.

Lovell S.T. and D.M. Johnston. 2009. Creating multifunctional landscapes: how can the field of ecology inform the design of the landscape? *Frontiers in Ecology and the Environment*, 7:212–220. DOI: doi:10.1890/070178.

Lovell S.T. 2010. Multifunctional Urban Agriculture for Sustainable Land Use Planning in the United States. *Sustainability*, 2:2499.

Lovell S.T., V.E. Méndez, D.L. Erickson, C. Nathan, S. DeSantis. 2010. Extent, pattern, and multifunctionality of agroforestry systems in Vermont, USA. *Agroforestry Systems*, 80:153–171

McClintock, N. 2010. Why farm the city? Theorizing urban agriculture through a lens of metabolic rift. *Cambridge Journal of Regions, Economy and Society* 3(2): 191–207. 10.1093/cjres/rsq005

Méndez, V.E., M. Caswell, S.R. Gliessman and R. Cohen. 2017. Integrating Agroecology and Participatory Action Research (PAR): Lessons from Central America. *Sustainability* 9(5): 705. 10.3390/su9050705

Milcu, A.I., J. Hanspach, D. Abson, and J. Fischer. 2013. Cultural Ecosystem Services: A Literature Review and Prospects for Future Research. *Ecology and Society* 18 (3). https://doi.org/10.5751/es-05790-180344.

Milgroom, J., J. Bruil and C. Leeuwis. 2016. Co-creation in the practice, science and movement of agroecology. *Farming Matters* 03/2016: 6–8. https://www.ileia.org/2016/03/23/editorial-co-creation-practice-science-movement-agroecology/

Montenegro de Wit, M. 2014. A Lighthouse for Urban Agriculture: University, Community, and Redefining Expertise in the Food System. *Gastronomica* 14, no. 1 (2014): 9–22. https://doi.org/10.1525/gfc.2014.14.1.9.

Nassauer, J.I., M.V. Santelmann and S. Donald. 2007. From The Corn Belt to the Gulf: Societal and Environmental Implications of Alternative Agricultural Futures. New York: Routledge. https://doi.org/10.4324/9781936331406

Newman, P. and I. Jennings. 2008. *Cities as sustainable ecosystems: principles and practices.* Island Press: Washington, D.C.

Nihart A., W. Robb, and J. Hyman. 2012. Burlington Urban Agriculture Task Force Report. Burlington City Council: Burlington, VT.

Nyéléni Declaration. 2015. Declaration of the International Forum for Agroecology, Nyéléni, Mali: 27 February 2015. *Development* 58 (2): 163–68. https://doi.org/10.1057/s41301-016-0014-4.

Orion, Tao. 2015. *Beyond the War on Invasive Species.* White River Junction: Chelsea Green Publishing.

Pain, R., G. Whitman, D. Milledge, and Lune Rivers Trust. 2011. *Participatory action research toolkit: An introduction to using PAR as an approach to learning, research and action.* Durham: Durham University.

Patton, M.Q. 2017. *Principles-focused evaluation: the guide.* Guilford Press: New York, NY.

Peters, C.J., N.L. Bills, J.L. Wilkins and G.W. Fick. 2009. Foodshed analysis and its relevance to sustainability. *Renewable Agriculture and Food Systems* 24(1): 1–7. https://doi.org/10.1017/S1742170508002433.

Plan BTV. 2012. Burlington Vermont Downtown Plan. produced by Town Planning & Urban Design Collaborative, LLC (www.tpudc.com) in partnership with the City of Burlington, Vermont. https://www.burlingtonvt.gov/sites/default/files/PZ/planBTV/Downtown_Plan/planBTV_MasterPlan_APPROVED_061013_LowRes.pdf

Renting, H. 2017. Exploring urban agroecology as a framework for transitions to sustainable and equitable regional food systems. *Urban Agriculture Magazine* no. 33–11–12.

Riechers, M., J. Barkmann and T. Tscharntke. 2016. Perceptions of cultural ecosystem services from urban green. *Ecosystem Services* 17: 33–39.

Santelmann, M.V., D. White, K. Freemark, et al. 2004. Assessing alternative futures for agriculture in Iowa, U.S.A. *Landscape Ecology* 19(4): 357–374.

Schattman R.E., S. Hurley and M. Caswell. 2019. Now I See: photovisualization to support agricultural climate adaptation. *Society & Natural Resources* https://doi.org/10.1080/08941920.2018.1530819.

Shiva, V. 2016. *Who Really Feeds the World? The failures of agribusiness and the promise of agroecology.* Berkeley: North Atlantic Books.

Siegner, A.B., C. Acey and J. Sowerwine. 2019. Producing urban agroecology in the East Bay: from soil health to community empowerment. *Agroecology and Sustainable Food Systems* 556–593. https://doi.org/10.1080/21683565.2019.1690615

Sonino, R. 2009. *Feeding the world: A challenge for the 21st century.* Cambridge, MA: MIT Press.

Toledo, V. 2012. La agroecología en Latinoamerica: tres revoluciones, una misma transformación. *Agroecología.* 6. 37–46.

Tornaghi C. and Hoekstra F. 2017. Editorial: Special Issue on Urban Agroecology. *Urban Agriculture* 33:3–4.

Torre, M.E. 2014. Participatory Action Research. In *Encyclopedia of Critical Psychology*, ed. T. Teo. New York: Springer.

Vaarst, M., A.G. Escudero, M.J. Chappell, et al. 2017. Exploring the concept of agroecological food systems in a city-region context. *Agroecology and Sustainable Food Systems.* https://doi.org/10.1080/21683565.2017.1365321.

Wang, C. and M.A. Burris. 1997. Photovoice: Concept, Methodology, and Use for Participatory Needs Assessment. *Health Education & Behavior* 24(3): 369–387. https://doi.org/10.1177/109019819702400309.

Werthmann C. and C. Steinitz. 2008. The Rebirth of the Tajo River (Spain), Assistant Editor Stephanie Hurley. Funded by Foro Civitas Nova, Fundacion+SUMA, Castilla-La Mancha, and Harvard Graduate School of Design.

Wiskerke, J.S.C. 2015. Urban food systems. In *Cities and Agriculture: Developing resilient urban food systems*:1–25, eds. H. de Zeeuw & P. Drechesel. London & New York: Earthscan, Routledge Taylor & Francis.

15

Multidimensional Challenges in Urban Agricultural Research

Joshua Earl Arnold[1*] **and Alana Bowen Siegner**[2]
[1] *Department of Environmental Science, Policy and Management, University of California, Berkeley, CA USA*
[2] *Energy and Resources Group, University of California, Berkeley, CA USA*
[*] *Corresponding author: Email j.earl.arnold@berkeley.edu*

CONTENTS

KEY WORDS: *Urban Agroecology (UAE), Urban Agriculture (UA), Interdisciplinary (ID), Transdisciplinary (TD), Extension, Participatory Action Research (PAR)*

15.1 Introduction

Urban agricultural (UA) systems vary in size, mission, and practice, but often share a common goal: increasing food security and sovereignty of urban residents (Haletky and Taylor 2006; Santos, Palmer, and Kim 2016). In 2012 there were more than 16,000 UA sites identified in the United States with an estimated increase of 10% per decade since the 90s (Lawson and Drake 2013). Research on UA and related topics has mirrored the increase of UA sites. In 2008, the database Web of Science recorded 258 publications associated with UA, and in 2018, over 1000 publications, a four-fold increase within the decade. Published research on UA generally highlights its virtues, is supportive of UA efforts, and spans a spectrum of investigations into ecological, social-ecological, economic, and political phenomena (Horst, McClintock, and Hoey 2017; Golden 2013; Lin, Philpott, and Jha 2015; McIvor and Hale 2015; Colasanti, Hamm, and Litjens 2012).

While some critical perspectives regarding the ability of UA to transform food systems exist (McClintock 2014; Sbicca 2012; Sbicca et al. 2019; Horst, McClintock, and Hoey 2017; Van Dyck, Tornaghi, and Halder 2018; Tornaghi 2014, 2017), few studies critically assess or discuss the processes or associated challenges of UA research. In this chapter, we discuss these challenges and argue that they manifest in each stakeholder's expectation of benefitting from UA research. Without a critical analysis of the processes and results of UA research, researchers can fail to acknowledge the complex social and economic factors inherent to UA. In doing so, UA researchers can valorize UA research and community-academia partnerships—failing to account for negative externalities in the form of extracted knowledge and labor from at-risk communities (McClintock 2018). Exploring the community-academia partnerships created during UA research can help reveal these multidimensional challenges and connect them to substantive actions to be implemented during research development.

Urban agroecology (UAE) provides a holistic process for assessing the multifaceted social, political, economic, and ecological implications of urban food production sites, practices, practitioners, and products (Hoekstra and Tornaghi 2017). This emerging distinction between the act of urban agriculture (UA) and a more holistic concept of UAE encourage stakeholders to move past a productivist focus to one of a full value stream analysis that includes education, community building, and urban greening impacts (Hoekstra and Tornaghi 2017; Siegner, Acey, and Sowerwine 2019). A UAE research perspective incorporates interdisciplinary (ID) and transdisciplinary (TD)[1] research with input from those trained in social science research methods, political economy, geography, and sociological elements of urban planning, and questions the ability of research efforts to be transformational, and genuinely supportive of practitioner-collaborator on-ground improvements.

In the sections that follow, we summarize the multidimensional challenges that are faced by all stakeholders in UA research efforts and explain how those challenges manifest into disparate outcomes for researchers and the UA community. We elaborate on the community-academia dichotomy, exploring what community practitioners are asking for from researchers and whether researchers are delivering desired results. We report on the state of current UAE ID and TD research through both primary and secondary source data, including literature review, surveys, and our experiences in participatory action research (PAR) in UA. Finally, we present a framework for conducting collaborative UAE research and concluding remarks. Through our methodological contributions, we aim to improve reciprocal researcher-practitioner collaborations and thus advance the successful practice of agroecology in urban areas across the United States and the globe. This chapter contributes most directly to the scientific method of conducting transdisciplinary UAE research and the movement towards strengthening and facilitating agroecological food system transitions.

It is not our intent to call into question established participatory research methodologies often used in UAE research but rather question their efficacy to translate into improvements for urban farmers in the context of the multidimensional challenges encountered in UAE research and community-academia partnerships. In doing so, we hope to strengthen current methods and translate critical inquiry into praxis. For those who approach UAE research from a purely observational perspective without accounting for their impact in a greater social context, we emphasize that in engaging with the UA community, our presence and actions have long-term effects on stakeholders and can reproduce existing inequalities.

A note on terminology: In this chapter, we will discuss the roles of stakeholders in community-academia partnerships. We understand that classification of stakeholders as either practitioners, farmers, or

[1] Note on terms "interdisciplinary" and "transdisciplinary": interdisciplinarity refers to integrating knowledge and methods from different disciplines, using a real synthesis of approaches, and transdisciplinary refers to creating a unity of intellectual frameworks beyond the disciplinary perspectives. Transdisciplinary can also be thought of as "transcending" the boundaries of academic disciplinary perspective and integrating interdisciplinarity with a participatory approach.

researchers and their corresponding positionality can be problematic, complex, and can at times work to reinforce existing inequalities (Janes 2016). We ask the reader to allow us some leeway during this process and to acknowledge that all stakeholders can fall within a spectrum of roles in these community-academia partnerships. Most importantly, we want to make sure to delineate positionality in the context of roles and power. Explicitly, that community, community partners, and/or urban agriculture practitioners can lack economic and political power in a greater societal context, and that researchers/academia by proxy of institutional affiliation have greater actual or perceived power.

15.2 Methods

Throughout this chapter we incorporate primary source data (personal communications) and responses from several published and unpublished surveys conducted with urban farmers between 2014 and 2019 in the San Francisco Bay Area. Each section utilizes data from these sources to ask critical questions regarding UA stakeholder research goals, the challenges encountered by stakeholders, and the condition of ID/TD research in UA.

Unpublished survey responses from urban farm managers collected from 2014 to 2015 during the UC Berkeley Urban Agroecology survey informed the concept of this chapter and identified challenges faced by UA practitioners (Altieri et al. 2016). Published survey results (A. B. Siegner, Acey, and Sowerwine 2019), and unpublished survey results from a 2017 policy brief on urban agriculture identified urban farmer needs (Driscoll 2017).

In addition to previous surveys that explicitly engaged with UA practitioners, from Fall 2019 to Spring 2020, we surveyed UA researchers identified in two published literature reviews spanning both social and ecological research in UA systems (Arnold, Egerer, and Daane 2019; A. Siegner, Sowerwine, and Acey 2018). Twelve researchers responded and reported information about the research processes that informed 34 UA specific publications. This survey explored the process before, during, and after UA research efforts, the prevalence of ID/TD research efforts, how UA communities are impacted through UA research, and the extent to which community members were involved with research. While we cannot infer that respondent sample size is reflective of the UA research community, these responses can give us some insight into the challenges faced by well-established UAE researchers that work in community-academia partnerships. Further data collection will be of great value to future UAE research, and the survey will remain open for submission.[2] All data were collected using Qualtrics survey software and will be anonymized and made available upon request (*Qualtrics* 2019).

15.3 The Community/Academia Dichotomy: Conflicting or Complementary Goals?

Urban agroecology research is inherently reliant on community-academia partnerships. Researchers must build trust with community members to gain access to research sites and information. Community member trust is often earned through the promise of research results that will benefit the community. Stakeholder expectations and goals can be complementary or in conflict creating a tension between expectations and outcomes. The tension between conflicting or complementary goals forms the foundation of the challenges set forth in the rest of this chapter. A failure of academia to account for community needs can limit community involvement and commitment to research goals and exacerbate logistical challenges for ongoing and future research efforts.

[2] TD/ID Urban Ag Research Survey: https://berkeley.qualtrics.com/jfe/form/SV_cUbYM6GkEE16jDT.

15.3.1 Extractive or Additive?

When critically assessing community-academia relationships in the context of UA research efforts, a simple heuristic can conceptualize the nature of the relationship: are research goals and deliverables meeting the needs of UA practitioners (e.g., creating value) or failing to address UA practitioner needs? If the latter, researchers risk the establishment of research goals that are extractive rather than additive. Even under the best circumstances, wherein a researcher acknowledges power differential and community needs, a rift can develop between what UA practitioners need and what researchers seek; academic research goals and advancement processes can be innately incongruous with community needs. Academic goals such as publishing or submitting a project for academic credit might conflict with protecting and increasing financial viability and food production on an urban farm.

Whether UA research is extractive or additive, academia must realize that in the absence of robust collaboration, consensus building, time, and resources, these research-community relationships can ultimately fail to acknowledge and advance community needs, communicate findings in a way that benefits community partners, or add value to an already tenuous social-ecological system. Exemplifying UA practitioner awareness of academia's failure to deliver value to a UA operation through research, one urban farm leader required that researchers perform several hours of volunteer labor in exchange for participating in an interview. This practitioner was seeking to ensure some "value generation" for their operation in advance of a promised "research product" benefiting their UA site. Academia must realize that UA research itself is realized only through the time and labor of urban residents. Urban agriculturalists labor to make urban spaces available not only for UA but also for UA research, by cleaning and clearing land, building soil, organizing to maintain access to land, and cultivating crops.

15.3.2 Community Research Needs: What are UA Practitioners Asking for?

One metric for understanding whether UA research is impacting community needs would be to compare stated community needs and UA research goals. Between 2014 and 2018, researchers at the University of California, Berkeley distributed several surveys to UA practitioners in the San Francisco Bay Area to inquire about their needs, greatest challenges, and suggested policy interventions. Results of this survey, which included responses from 50 urban farm managers (Table 15.1), indicated that land access, security of tenure issues, financial support, lack of skilled and affordable labor, lack of infrastructure for farm operations (especially storage and distribution of food), education, training, and technical support were the most pressing needs for urban farmers within the San Francisco Bay Area. These responses indicate that urban farmer needs cannot be remedied solely by filling in knowledge gaps related to UA practices, but reflect that social and economic conditions are the most challenging hurdles to effectively farming the city.

TABLE 15.1

Bay Area, CA: Urban Farmer Needs Assessment

Urban Farmer Needs	Num/n=58	Percent
Public land access/land banks/protect land	37	64%
Financial support	25	43%
Labor/skilled and able to pay/volunteers	21	36%
Infrastructure	16	28%
Education/training/technical support	12	20%
Access to markets/help w/ access/marketing	11	19%
Network of resources	10	17%
Local government support/acknowledgement	8	14%
Community involvement	4	7%
Water	4	7%
New technology	2	3%

The identified needs from urban farmers in the San Francisco Bay Area reflect an underlying lack of support for local food production and the underinvestment in social safety programs, rather than an inability of urban agriculturalists to conduct successful agricultural operations (McClintock 2018; McClintock, Miewald, and McCann 2018; Sbicca et al. 2019; Tornaghi 2017; Van Dyck, Tornaghi, and Halder 2018). Urban farmers are incredibly adept at utilizing agroecological practices in the built environment. Surveys of urban farms in the San Francisco Bay Area region recorded high rates of adoption of agroecological practices, including crop rotation, cover crops, inter-cropping, use of non-crop habitat for pollinators and natural enemies, soil building practices such as no-till, and incorporation of perennials and high habitat diversification (Siegner, Acey, and Sowerwine 2019). Measured yields were consistently high, and most crop losses occurred in the field and were the result of poor crop planning, lack of labor availability, and lack of infrastructure to process, store, transport and distribute crops (Driscoll 2017; Altieri et al. 2016). In other words, political-economic and structural challenges underpin the needs expressed by farmers, and not necessarily a need for agricultural training.

Urban agroecology researchers are uniquely positioned to engage community members in mutually beneficial participatory research. Conversely, they must realize that when academic research goals fail to address UA community needs, little value is generated for the community. When research efforts fail to address community member problems, research activities become extractive, generating knowledge and academic "goods" from the labor of urban agriculture practitioners.

15.3.3 How are Researchers Responding?

As researchers, we are often poorly suited to address many of the identified urban farming community needs (Table 15.1). In many cases, researchers are constrained by similar economic forces that affect the UA community, especially when it comes to direct funding and the ability to divert resources to address problems identified by UA farmers. Research funding is often limited in focus to measured outcomes, failing to incorporate deliverables for community members or policy change. For example, in the grant funds that support the authors' current UAE research, no funds are allocated to pay community partners or key stakeholders for their time, or for specific technical or social improvements that grant-funded research identifies as important for urban farmers (e.g., shared refrigerated truck for improved distribution of urban produced food).

While some UA research may lack the time and resources to address problems in UAE systems directly, they often try to contribute to community efforts through diverse strategies, including volunteer labor, direct payments, and donations for access to UA sites for research purposes. One approach we have used in our own research efforts is to sponsor and/or co-author grant applications for our community partners. Many UA operations in the San Francisco Bay Area are non-profits that are grant-dependent, so incorporating research findings into collaborative grant writing efforts is a potentially promising and synergistic form of practitioner-research collaboration. Another example from the authors' work is a community-based soil testing project that acquired funding to test for common contaminants in backyard gardens and community farms. While the ultimate results of that work did not contribute to future research goals, it helped develop knowledge within the community and directly addressed a community need.

Stakeholder involvement during the development phase of UAE research may be one mechanism for strengthening the community-academia partnerships. Our survey inquired about stakeholder involvement during development of research questions, if community members were involved in the research process, and whether a community agreement was signed with community members. Community members were rarely involved with directly developing research questions (8%) and only participated in a formalized agreement with researchers in less than 10% of the reported publications. During research efforts, stakeholders only participated in research activities in 33% of the reported publications.

Post-research efforts in UAE research appear to be more oriented to community needs. Our survey indicated that research results were shared back to the community about 80% of the time through two formats: 1) online/remote and 2) workshops. Results were most often shared directly with community partners through email or blog posts (38%). Education efforts through a workshop series that directly targets UA communities were the second most prevalent format at 36%. One survey respondent commented that "managers are not interested in working on research publications, but we interact in several ways to get the information out from our projects. My students provide programs as part of the curriculum of the learning farms programming; we also write outreach bulletins and hold public workshops" (UAE Researcher 2020). Another researcher has completed their research with their community partner but is "still helping facilitate an urban farmer dinner series and online networking platform to allow urban farmers to connect with each other more easily without having to go through a university research intermediary as was happening during parts of our project work" (UAE Researcher 2020).

Presumably, if research is developed with community needs in mind, some substantive results at the community level should occur. However, post-research interventions in community-academia partnerships are rarely reported. Surveyed researchers indicated they were aware of very few instances (16%) where actionable results were implemented after research results were shared. Many of the responses to this portion of our survey indicated research results may be less tangible to on-the-ground practices, such as influencing policy at the local level or promoting internal discourse among community organizations. Capturing post-research changes in UA is a compelling research topic for future study.

Assessing and mending the rifts that occur between research goals and tangible results for community partners is a great challenge for future UAE researchers, complicated by institutional forces that expect research results with minimal funding and support. While researchers are making great efforts to ensure that research results are being shared, future researchers in this field may be able to greatly impact on the ground results by including stakeholders in the development and execution of on the ground UAE research efforts. Moreover, pre-research organizing seems to have the greatest capacity for improvement, especially in the context of constraints on the researcher/academia portion of the relationship. Specific requirements and budgeting for community stakeholders should emphasize needs assessments and implementation of collaboratively identified "solutions." Researchers should strive to be more holistic and inclusive from the outset.

15.4 Multidimensional Challenges in UA/UAE Research

In previous sections, we introduced the concepts of UAE research in the context of complementary or conflicting goals based on identified community needs. Here we conceptualize and introduce a nonhierarchical structure of challenges encountered during UA research that inherently affects our ability to form equitable and reciprocal research-community partnerships (Figure 15.1). These challenges are multidimensional and act on both academia and community partners. These social, cultural, economic, and logistical challenges can complicate and disincentivize long-term, community-oriented efforts in UAE research.

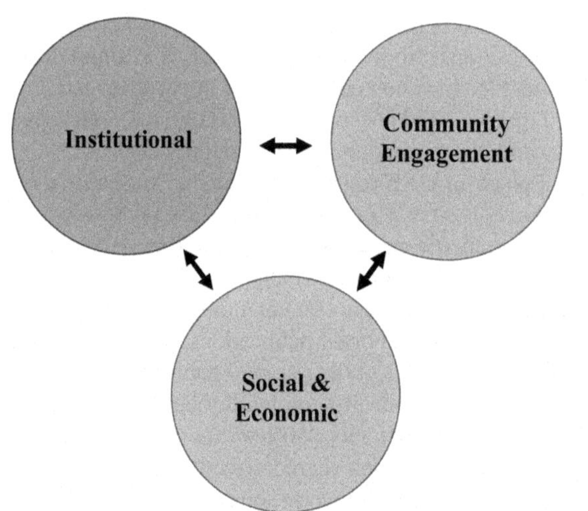

Institutional: Logistical and financial support for research, deadlines, other academic commitments, lack of ID/TD or community engagement training

Community Engagement: Power dynamics, history of community engagement, privilege and race dynamics, lack of trust of institutions

Social and Economic: Economic status of community and community orgs., security of tenure of research sites, economic status of urban farmers

FIGURE 15.1 Multidimensional challenges of urban agriculture research.

15.4.1 Research Challenges from an Academic Perspective

From an academic perspective, challenges are often encountered regarding structural and institutional-based limitations. Time-based requirements limit long-term research efforts and resource-based complications may limit the scope of the research itself or reduce the resources available to complete projects (e.g., lab space, soil tests, research assistants, transportation resources). Ideally, UAE researchers would commit to community-academia partnerships based on reciprocity and grounded in principles of PAR, but the reality of these relationships reveals a tension between academic deliverables, timelines, and the resources available for long-term projects. These limitations can constrain the ability to carry out meaningful community-based PAR that adequately addresses priority issues for the UA community.

Funding can be a restrictive issue for researchers in both agroecology and sustainable agriculture research broadly. Both the duration and extent of funding has been identified as a significant challenge (DeLonge et al. 2020; DeLonge, Miles, and Carlisle 2016; Pimbert and Moeller 2018). Ostensibly, an even smaller fraction of available agroecology research funding goes to UAE research on urban farms, which are often considered an insignificant impact on overall agricultural production despite their documented impacts on local food security in urban areas (A. Siegner, Sowerwine, and Acey 2018). Moreover, scarcity of funding and requirements associated with specific awards can shape research foci and may work to constrain research goals and limit the terms and extent of research projects for both the community and academia. This dynamic gives funding organizations an indirect but substantial say in what types of UA research take place, where, and whom they serve (Horst, McClintock, and Hoey 2017; Reynolds and Cohen 2016).

Labor retention of qualified research assistants can also be a limiting factor. Surveying organisms or gathering data on landscape-scale effects of UA systems may require extensive field survey work and can be arduous and time-consuming, necessitating capable and skilled research assistants. Because research assistants in an academic setting are often unpaid volunteer labor, a significant amount of time can be spent training research interns but may not result in high rates of retained skilled labor from season to season. In response, many researchers seek funding to pay undergraduate assistants, but funding is often limited, and rates of pay are often low.

Academic requirements, including deliverables and timelines, can also work to constrain time available to complete longer-term research projects. Many phenomena in UA that attract attention from researchers would lead to more valuable results if they were able to capture temporal changes more effectively. For example, a recent review on urban biological control shows that biological control research in urban farms and gardens rarely extends further than one growing season, failing to capture changes in insect populations and their relevant impacts over time (J. E. Arnold, Egerer, and Daane 2019). Social and economic studies can also be impacted by shorter observation periods. For example, security of tenure of UA sites is very low, but no research has tracked the loss of urban farms and gardens over any period of time (J. Arnold and Rogé 2018).

Both temporal and funding/resource limitations can be exacerbated by recent trends in the demographics of UAE researchers. Increased interest in UA research seems to be heavily invested in by junior researchers who are in graduate programs or have recently accepted academic positions. Positionality within academia can frequently lead to reduced access to resources and increased pressure to complete research on timelines out of sync with UA phenomena. The research community often fails to recognize that positionality within academia and relative power dynamics are determinants of access to resources, networks, funding, trans- or interdisciplinary educational background. These more junior scholars may or may not be sufficiently trained or experienced in PAR, navigating race/class/power dynamics, and conducting multi-year trans- and interdisciplinary UAE research. These academic challenges (lack of funding, lack of time, ability to capture shifting temporal and spatial dynamics of urban farms) point toward a need for improved, targeted training opportunities for aspiring UA researchers.

15.4.2 Research Challenges from a Community Partnership Perspective

Like other community-oriented research, UA research projects rely on community relationship building, participation, and approval. That being said, UA research is often complicated by a variety of unique social, economic, and logistical challenges, including UA security of tenure issues, high turnover for farm managers and farmers, and intermittent site accessibility.

Included in these challenges are concerns from urban farmers over power imbalances created by institutional affiliation and the historical relationships between the community and academic actors who may have previously interacted with urban farmers and/or UA sites. Many have concerns about being taken advantage of or having sensitive information revealed during research, including their site location, which may or may not be legally sanctioned for food production. Many UA practitioners may have insecure legal status or may be reluctant to discuss topics that reveal funding or financial information relevant to garden or farm operations. Many UA organizations have deep roots in social movements that started to protect their community from racism and aggression. Community member apprehension of working with institutions is the result of a long history of exploitation and institutional racism in and out of research settings.

Urban agroecology researchers take on the responsibility of not only developing experiments in dynamic ecosystems but end up acting as representatives for larger institutions. Historical context, nuance, and awareness of local race/class/power dynamics are essential during UAE research. Exacerbating problems in community-academia research collaborations is a lack of diversity and representation in higher education. More often than not, urban farming initiatives are undertaken by people of color in underserved neighborhoods - unpublished mapping data for Contra Costa and Alameda counties in the San Francisco Bay region show urban farms occurring in census tracts averaging 49% people of color (Siegner and Arnold unpublished data). However, academic partners they are most likely to interface with are most often White. Representation of faculty of color in the University of California is ~8% and, more relevant to urban farmers in most metropolitan cities, 2.6% Black and 4.6% Latino. STEM disciplines remain the least diverse among UC's disciplines. This lack of representation is also mirrored by graduate student admission, and graduate students, more often than not, would be urban farmers' point of contact for on the ground research initiatives. Importantly, these differences of ethnicity can also be connected to issues of class, where White middle-class researchers are less likely to understand through first-hand experience the realities of living in underserved neighborhoods.

Economic issues and urban land use regimes also create challenges by reducing access and time available to study UA systems, as is further discussed in Chapter 8 by Glowa and Roman-Alcala. Insecurity of tenure, which is ubiquitous among urban farms and gardens, can create issues with long-term monitoring of ecological processes. Research sites can be developed or may be impacted by significant ecological disturbances nearby that are out of control of the researcher or community members (neighboring construction projects, waste disposal activities, roads, and highways). Unpredictable funding issues can impact farm operations, especially with farm management turnover. For example, one of our own research partners has had four farm managers in three years. Building rapport, trust and social capital with new managers required to implement experiments can significantly slow research efforts or even render individual sites inaccessible.

The multi-dimensional challenges that arise from UA research, including a lack of diversity in academia, economic issues that define UA farm function, and urban land-use regimes that temporally challenge UA research efforts, create unique challenges for both the community and academia. A nuanced perspective can help acknowledge and develop strategies that take account of issues that arise from all stakeholders. Trans- and interdisciplinary capacity at the individual, as well as the research group level, are crucial in building capacity to connect with the community in a non-extractive way and to address issues from diverse perspectives (environmental history, social/racial theory, and ecological history). The extent to which ID/TD research methods and training are structured elements of undergraduate, graduate, and UAE research team training is an important topic that needs further investigation. While this training will not ameliorate the lack of racial diversity in higher education, fostering increased awareness of race/class dynamics in community-based PAR as a precursor to UA research efforts could have a significant impact on future research efforts.

15.5 Thematic Review of ID/TD Research in UA

Urban farming operations come into being through a variety of social, economic, and ecological processes that ultimately determine their positionality in the urban landscape and determine function, both spatially and temporally. Given the diversity of actors and processes that enable UA site existence (e.g., schools, churches, community centers, public parks, private lots), research efforts are inherently ID/TD. However, the process by which researchers engage with UA communities is often prefaced with highly specialized educations that have a narrow social or ecological focus, with little training in community or participatory based research efforts. Current paradigms in higher education have failed to formalize TD or ID training, favoring disciplinary specialization instead as a signal of "expertise." A recent publication on the state of sustainable agriculture and agroecology research states, "agroecology research requires training in interdisciplinary, systems-science approaches, which are relatively rare and difficult to pursue at U.S. research institutions" (DeLonge et al. 2020; DeLonge, Miles, and Carlisle 2016). Seventy-six percent of UA researchers surveyed for this chapter reported that they had not received ID/TD training at their institution but had sought it out on their own.

There is an implicit assumption that community relationships are naturally occurring, intuitive, and simple enough not to require formalized or required instruction and guidance from skilled researchers. Among sustainable agriculture and agroecology researchers, many express satisfaction over the relationships that arise out of PAR at the same time as they express dissatisfaction with the "lesser amount of interdisciplinary, farmer-driven, and community-based research" that they can conduct relative to other research demands (DeLonge et al. 2020).

15.5.1 TD/ID Literature Review

To further explore the question of how much existing UA research is ID/TD, we evaluated two UA-specific literature reviews (one focused on biological control of pests in urban gardens and the other focused on food security related to UA activities) to identify the occurrence of ID/TD. In both literature reviews we analyze the referenced articles (representing both peer-reviewed journal articles and grey literature) and determine how many articles represent interdisciplinary author teams (at least one author from a different discipline/department) or transdisciplinary author teams (which we operationalize as at least one author from outside of academia, broadening the focus of the article to include community outcomes or specific policy proposals).

In our first literature review regarding food security impacts and UA, we find that 23 of the articles (18%) represent interdisciplinary author teams, and 23 (18%) represent transdisciplinary scholarship (in this case, community-academia co-authorship). Our second literature review from 2018 focused on biological control of common garden pests, and the abundance and richness of insects relevant to agriculture in UA. This literature review was a much smaller sample size of only 13 publications. This particular research topic is unique in the sense that results should benefit urban farmers by helping them understand how to manage landscape features to promote beneficial insect populations on farms. While this field is heavily populated by entomologists, many researchers who study biological control have varied research interests and are hosted by a variety of different departments at universities around the world. That being said, researchers working on urban biological control rely heavily on community partnerships to gain access to urban farms for extended periods while conducting research and interface closely with urban farm managers. Working closely with UA practitioners on a topic of great importance epitomizes the community-academia partnership: where the researcher has the potential to create actionable research, the farmers allow access for research, and then researchers finish the project and provide some information to UA farmers. Because of this whole spectrum community-academia experience, it is important for researchers on the ground to have ID/TD training in working with and then educating the community. Our literature review reveals that teams working on urban biocontrol projects are often interdisciplinary, coming from a variety of different departments. But, more often are neither ID/TD (30%), and in only one case, we considered a research team transdisciplinary.

15.5.2 UA Researcher Survey

Our survey of UA researchers revealed that over half (69%) of respondents did not receive mandatory ID/TD training as part of their academic development, and 84% of respondents sought out training they would consider ID/TD on their own. Similarly, less than half (53%) did not receive training on working with vulnerable populations, defined by Institutional Review Board (IRB) guidelines as ethnic minorities and economically disadvantaged populations, which researchers are more likely to encounter during UA research. When not trained by their institution, over half (61%) pursued training on their own to assist them in working with vulnerable communities. These results indicate that academic programs often hosting UA research are less likely to mandate education that can help strengthen community-academia partnerships.

A vast majority of our respondents' publication results were shared back to the community in the form of workshops. However, 76% did not participate in training during their academic development that focused on community outreach and education. While many UAE researchers are educators and are likely to teach at the undergraduate level, communicating results to community members can be especially challenging. Community workshops often necessitate reframing research results in a way that laypeople can understand and implement. Ostensibly, extension agencies should be of great utility in this pursuit. Engaging local extension agents during UAE research and incorporating their networks into helping disseminate research results should be considered a best practice.

These results point toward the need for appropriate training and preparation to support the development of positive, non-extractive relationships between researchers and practitioners as well as proactive communication of research findings with media and policymakers.

15.6 PARticipating in Urban Agroecology: A Research–Practice Ethical Framework

15.6.1 Struggling with PAR

The gold standard in community-academia research partnerships is participatory action research (PAR), where authority over and execution of research is a collaborative process between academia and community members (Fortmann 2008; Pretty 1995). In the case of UAE research, "community members" are stakeholders in community farms and gardens. Partnerships occur on a spectrum of "participatory," and can be formalized by co-creating community agreements ahead of embarking on research activities. The spectrum of "participation" ranges from "manipulative participation" to "self-mobilization," where manipulative participation invokes the term participation without actual participation (Pretty 1995).

The multidimensional challenges presented in this paper represent the difficulties of executing "participatory" research in an environment that is often ephemeral and economically impacted. While these challenges are similar to many other types of participatory research, we believe that PAR is especially challenging to execute in UA spaces due to the myriad of factors that work to constrain the breadth and depth of UA. Moreover, these challenges also put UA practitioners at risk - their time and labor being realized in academic deliverables, but lacking on the ground benefits to UA farmers. That being said, we propose a framework for those struggling or aspiring to implement successful, interdisciplinary PAR on urban farms that is beyond participatory. UAE research should help operationalize communities' existing needs and emphasize "self-mobilization," where communities take the initiative independent of external institutions, and UAE researchers provide an enabling framework of support. UAE research that is self-mobilized would challenge existing distributions of academic power (Pretty 1995). An example of self-mobilization from our work comes from an urban farmer taking on a leadership role in developing an urban farmer online communication platform and dinner series, after discussing this need and receiving support from an academic graduate student researcher.

15.6.2 A Framework within a Framework

We propose a framework for training future food system researchers at multiple levels (undergraduate, graduate, and faculty) for proactively engaging in PAR/UA interdisciplinary research projects. Our framework (within the PAR framework) employs agroecology as a scientific method of inquiry, inherently recognizing and bringing in social, ecological, economic, and cultural dimensions into the collection of

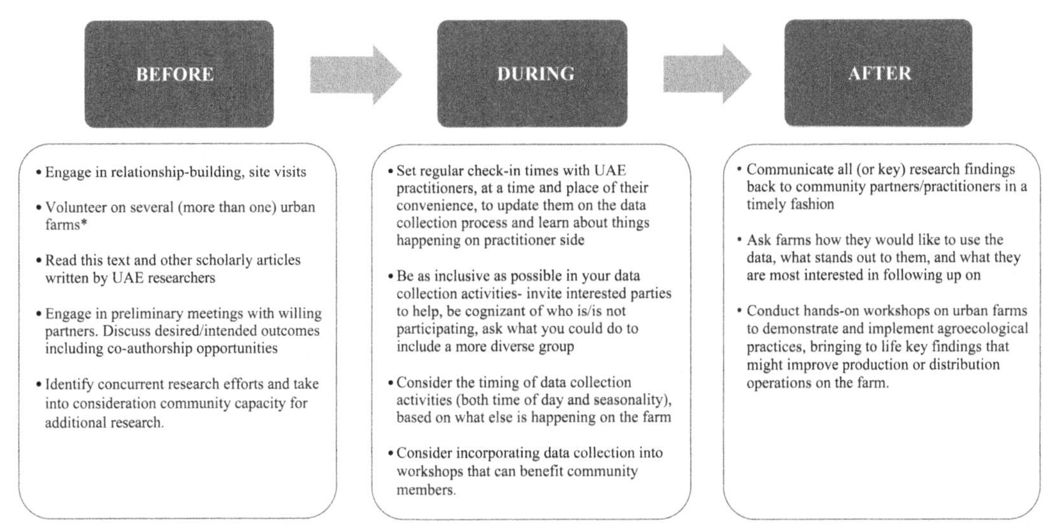

FIGURE 15.2 PARticipating in urban agroecology: Framework at-a-glance.

data, communication of results, and implementation of collaboratively identified action steps. It is broken up into three sections: Before, During, and After research activities/data collection on urban farms and summarized in Figure 15.2, followed by elaboration below. This framework is not meant to be a prescriptive, comprehensive, or list of rules, but rather a foundation for dialogue and education, upon which other UAE researchers and practitioners can add and build. We would also like to highlight the role that graduate schools and universities can play in more proactively requiring courses and training around conducting ID/TD research, working with vulnerable communities, conducting community outreach/education, and facilitating workshops. Workshops were the most common form of follow-through among researchers from our survey, and thus is a particular leverage point where universities could invest in providing training opportunities for students and faculty.

BEFORE:

1. Engage in relationship-building, site visits, and conversations with/about urban farms of interest.
2. Volunteer on urban farms. Most websites have posted open volunteer hours or special farm events seeking community volunteers.
3. Read text compiled by current UAE researchers (and other scholarly articles, especially those taking place in your city of interest).
4. Participate in training related to conducting ID/TD research, working with vulnerable communities, conducting community education and outreach, and facilitating workshops. If these are not offered by your academic institution, bring this to the attention of your colleagues and seek out training opportunities elsewhere (or online).
5. Seek IRB approval for UAE research requiring human subjects (e.g., interview, survey or participant observation methods).
6. Engage in preliminary meetings with willing research partners to co-produce research questions, define the research process, and determine preferred strategies for communicating results. Discuss desired/intended "outcomes," including co-authorship opportunities and on-ground outcomes.
7. Identify concurrent research efforts and take into consideration community capacity for additional research. Accept that your positionality as a researcher may make community members feel pressured into hosting research whether or not it directly addresses their needs.

DURING:

1. Set regular check-in times/meetings with UAE practitioners, at a time/place of their convenience, to update them on the data collection process, material needs, and learn about things happening on the practitioner side (events, land access threats, needs, concerns, and changes in circumstances). These can be "informal" in that they can happen over a meal or while working but should happen at regularly scheduled intervals (e.g., once/week, once/month, whatever is mutually desired by you and your research partners).
2. Be as inclusive as possible in your data collection activities. Invite interested parties to help. Be cognizant of who is/is not participating, ask what you could do to include a more diverse group of partners.
3. Consider the timing of data collection activities (both time of day and seasonality), based on what else is happening on the farm, and how seasonal activities might influence data.
4. Consider incorporating data collection into workshops that can benefit community members. Your methodologies may not be of use to the community but can act as a mechanism to invest labor/time/materials into acts that will be beneficial.

AFTER:

1. Communicate all (or key) research findings back to community partners/practitioners in a timely fashion (e.g., within a month of analyzed data becoming available), in a format that is desirable and understandable to partners (ask what format is preferred in BEFORE phase).
2. Ask farms how they would like to use the data, what stands out to them, and what they are most interested in following up on. What is still missing for them after seeing results?
3. Conduct hands-on workshops on urban farms to demonstrate and bring to life key findings that might improve production or distribution operations on the farm.

15.6.3 Ethics within UA Research

Moving toward an ethical research framework for conducting collaborative, mutually beneficial urban agroecological research requires asking a set of questions throughout the research planning, execution, and reporting process. These questions include:

- Who is initiating the proposed project? Who stands to benefit, and how?
- Who is getting paid (and for what), and who is contributing unpaid labor to the collaboration?
- What are the race and class dynamics of those who are initiating, benefiting, and getting paid? Who cannot afford to participate (in terms of time or financial considerations) but stands to benefit from successful UA operations?
- Who are urban farms seeking to serve, who are urban farmers, and are the producers and intended consumers from the same communities?
- How can this research dismantle systems of oppression and/or provide opportunities for employment and empowerment among those underserved in the community?
- What defines "success" for the researcher and community partners, and how can a collaborative arrangement be created to ensure success in both domains?

Honesty, transparency, and up-front communication are required to work through these questions before moving too quickly into data collection and analysis. Training from skilled and experienced urban agroecologists and PAR researchers can help develop the mental models for engaging in these types of questions and communication patterns.

15.7 Conclusions/Future Directions

Current academic institutional arrangements and performance expectations challenge collaboration opportunities in the realm of UAE PAR. Through pointing out common tensions between research and practitioner communities, based on a combination of personal experience, literature review, and primary data collection, we strive to create more space in academia for ID/TD UAE research training. We aspire to spark more collaborative research, outreach, and process-based education focused on outcomes for communities and UAE sites researched rather than outcomes in the form of research publications, presentations, and other academic materials.

5.7.1 Concluding and Moving Forward with UAE Research

Urban farming systems operate in the context of the broader food system and its inherent inequities and environmental injustices. It is not a panacea for solving all food access, security, sovereignty, and justice issues, but it is a promising avenue through which to address many of these concerns. It, therefore, merits rigorous and multidisciplinary research training, funding, and institutional as well as policy support in order to investigate, implement, and disseminate "best practices" in terms of urban ecology, urban food security, and social justice. Our analysis of existing published research, primary source data, and surveys of UA authors reveals a lag in proactively addressing and reversing food system inequities and injustices through PAR. Interdisciplinary, collaborative research-into-action must start from an understanding of the overarching food system, and work to dismantle oppressions without putting the onus for action and advocacy on already disadvantaged individuals or communities. Support of on-the-ground improvements in UA practice is a primary goal of UA research, and must not be subordinate to publication and academic advancement goals. Our proposed framework, "PARticipating in Urban Agroecology: A Research–Practice Ethical Framework" seeks to advance synergistic goals of urban food system sustainability, equity, and resilience in the face of climate change, through creating a skilled cohort of systems-thinking urban agroecological researchers.

REFERENCES

Altieri, Miguel A., Joshua Arnold, Celine Pallud, Courtney Glettner, and Sarick Matzen. 2016. "An Agroecological Survey of Urban Farms in the Eastern Bay Area." Berkeley Food Institute. https://food.berkeley.edu/priorities/agroecology/urban-agriculture/.

Arnold, Joshua E., Monika Egerer, and Kent M. Daane. 2019. "Local and Landscape Effects to Biological Controls in Urban Agriculture—A Review." *Insects* 10 (7): 215. doi:10.3390/insects10070215.

Arnold, Joshua, and Paul Rogé. 2018. "Indicators of Land Insecurity for Urban Farms: Institutional Affiliation, Investment, and Location." *Sustainability* 10 (6): 1963. doi:10.3390/su10061963.

Colasanti, Kathryn J. A., Michael W. Hamm, and Charlotte M. Litjens. 2012. "The City as an 'Agricultural Powerhouse'? Perspectives on Expanding Urban Agriculture from Detroit, Michigan." *Urban Geography* 33 (3): 348–69. doi:10.2747/0272-3638.33.3.348.

DeLonge, Marcia, Tali Robbins, Andrea Basche, and Lindsey Haynes-Mawlow. 2020. "The State of Sustainable Agriculture and Agroecology Research and Impacts: A Survey of U.S. Scientists." *Journal of Agriculture, Food Systems, and Community Development*, February, 1–26. doi:10.5304/jafscd.2020.092.009.

DeLonge, Marcia, Albie Miles, and Liz Carlisle. 2016. "Investing in the Transition to Sustainable Agriculture." *Environmental Science & Policy* 55 (January): 266–73. doi:10.1016/j.envsci.2015.09.013.

Driscoll, Laura. 2017. "Urban Farms: Bringing Innovations in Agriculture and Food Security to the City." Policy Brief. Berkeley, CA: Berkeley Food Institute. https://food.berkeley.edu/programs/policy/policy-briefs/.

Fortmann, Louise, ed. 2008. Participatory Research in Conservation and Rural Livelihoods: Doing Science Together. Conservation Science and Practice, no. 3. Chichester, UK ; Hoboken, NJ: Wiley-Blackwell.

Golden, Sheila. 2013. "Urban Agriculture Impacts: Social, Health, and Economic: A Literature Review." UC Sustainable Agriculture Research and Education Program. http://ucanr.edu/sites/CEprogramevaluation/files/215003.pdf.

Haletky, N, and O Taylor. 2006. "Urban Agriculture as a Solution to Food Insecurity: West Oakland and People's Grocery." *URB. ACTION.*

Horst, Megan, Nathan McClintock, and Lesli Hoey. 2017. "The Intersection of Planning, Urban Agriculture, and Food Justice: A Review of the Literature." *Journal of the American Planning Association* 83 (3): 277–95. doi:10.1080/01944363.2017.1322914.

Janes, Julia E. 2016. "Democratic Encounters? Epistemic Privilege, Power, and Community-Based Participatory Action Research." *Action Research* 14 (1): 72–87. doi:10.1177/1476750315579129.

Lawson, Laura J., and Luke Drake. 2013. "2012 Community Gardening Organization Survey." Community Greening Review. American Community Gardening Association. www.communitygarden.org.

Lin, Brenda B., Stacy M. Philpott, and Shalene Jha. 2015. "The Future of Urban Agriculture and Biodiversity-Ecosystem Services: Challenges and next Steps." *Basic and Applied Ecology* 16 (3): 189–201. doi:10.1016/j.baae.2015.01.005.

McClintock, Nathan. 2018. "Cultivating (a) Sustainability Capital: Urban Agriculture, Ecogentrification, and the Uneven Valorization of Social Reproduction." *Annals of the American Association of Geographers* 108 (2): 579–90. doi:10.1080/24694452.2017.1365582.

McClintock, Nathan, Chistiana Miewald, and Eugene McCann. 2018. "The Politics of Urban Agriculture: Sustainability, Governance, and Contestation." In *The Routledge Handbook on Spaces of Urban Politics*, 361–364. Routledge.

McIvor, David W., and James Hale. 2015. "Urban Agriculture and the Prospects for Deep Democracy." *Agriculture and Human Values* 32 (4): 727–41. doi:10.1007/s10460-015-9588-9.

Pimbert, Michel, and Nina Moeller. 2018. "Absent Agroecology Aid: On UK Agricultural Development Assistance Since 2010." *Sustainability* 10 (2): 505. doi:10.3390/su10020505.

Pretty, Jules N. 1995. "Participatory Learning for Sustainable Agriculture." *World Development* 23 (8): 1247–63. doi:10.1016/0305-750X(95)00046-F.

Qualtrics. 2020. *Qualtrics XM - Experience Management Software.* [online] Available at: <https://www.qualtrics.com/> [Accessed 22 May 2019].

Reynolds, Kristin, and Nevin Cohen. *Beyond the Kale: Urban Agriculture and Social Justice Activism in New York City.* Athens: The University of Georgia Press, 2016.

Santos, Raychel, Anne Palmer, and Brent Kim. 2016. "Vacant Lots to Vibrant Plots: A Review of the Benefits and Limitations of Urban Agriculture." Baltimore, MD : Johns Hopkins Center for a Livable Future.

Sbicca, Joshua, India Luxton, James Hale, and Kassandra Roeser. 2019. "Collaborative Concession in Food Movement Networks: The Uneven Relations of Resource Mobilization." *Sustainability* 11 (10): 2881. doi:10.3390/su11102881.

Siegner, Alana Bowen, Charisma Acey, and Jennifer Sowerwine. 2019. "Producing Urban Agroecology in the East Bay: From Soil Health to Community Empowerment." *Agroecology and Sustainable Food Systems* 0 (0): 1–28. doi:10.1080/21683565.2019.1690615.

Siegner, Alana, Jennifer Sowerwine, and Charisma Acey. 2018. "Does Urban Agriculture Improve Food Security? Examining the Nexus of Food Access and Distribution of Urban Produced Foods in the United States: A Systematic Review." *Sustainability* 10 (9): 2988. doi:10.3390/su10092988.

Tornaghi, Chiara. 2017. "Urban Agriculture in the Food-Disabling City: (Re)Defining Urban Food Justice, Reimagining a Politics of Empowerment." *Antipode* 49 (3): 781–801. doi:10.1111/anti.12291.

UAE Researcher, Anonymous. 2020. Email interview.

Van Dyck, B., Chiara Tornaghi, and S Halder. "The Making of a Strategizing Platform: From Politicizing the Food Movement in Urban Contexts to Political Urban Agroecology." In *Urban Gardening as Politics*, edited by Chiara Tornaghi, and Chiara Certomà, 183–201. Routledge, 2018.

Conclusion: Future Directions in Urban Agroecology

Hamutahl Cohen[1] and Monika Egerer[2]
[1] *Department of Entomology. University of California, Riverside. Riverside, California. 92507.*
[2] *Department of Ecology and Ecosystem Management. School of Life Sciences - Weihenstephan. Technische Universität München. 85354 Freising, Germany.*

Introduction

Urban agroecology is rooted in the desire to grow our own food, reconnect with nature, and to right social injustices. We must understand urban agroecosystems as *social systems*, where people's values and management take center stage, and as *ecological systems*, where biotic and abiotic interactions emerge to shape ecosystem processes and human wellbeing. Multiple facets of urban agroecology exemplify the duality of the social-ecological in cities. Take soil: the bed and basis of the garden, soil is a textured entity composed of living organisms, organic matter, residue and detritus. Soil also chronicles the consequences of history and land-based policy and warfare; the pollutants and heavy metals harbored deep in the soil will determine what crops a gardener can or cannot grow. Finally, soil can be amended—restored with cover cropping and compost. Soil is a subject matter for physicists, chemists, ecologists, political ecologists, sociologists, and for community activists. The multifaceted nature of soils, and the dialectic among scientific disciplines is illustrated by Raimsey et al. in Chapter 4, who sketch a biophysical, social, and cultural framework for soil regeneration and land-use transformations in urban systems.

The body of work in this book furthers our understanding urban agroecosystems as ecological and social systems. Philpott et al. in Chapter 2, describe how ecological networks, the mapping of complex interactions between birds, bees, and other bugs, not only influence ecosystem processing, but provide services important to human wellbeing. Pauline Marsh (Chapter 6) reveals how public health approaches can promote the importance of urban agriculture as therapeutic landscapes, and Fischer et al. (Chapter 5) describe the intersection of cultural knowledge and biodiversity in urban foraging practices. As editors, we had to decide how to present these and other contributions to our understanding of the socio-ecological. We clustered chapters together based on whether they contribute primarily to the ecological or social literature, and divided the book into these two components, beginning with a focus on ecological and biophysical research to understand biodiversity and associated ecosystem processing (Chapters 1–4), then moving into social understandings of the human-dimensions of urban agriculture (Chapters 5–11). While the rationale behind this organizational logic was to divide the content into digestible information for the reader, we acknowledge that our decision blatantly points to what is possibly the largest challenge facing urban agroecology: integrating research from multiple fields towards a transdisciplinary approach of understanding urban agroecology. Such an approach moves beyond acknowledging and utilizing knowledge from different disciplines, but utilizes an intellectual framework that unifies these perspectives (see Introduction and Chapters 12–15 for a discussion around urban agroecology methodology and challenges).

Caswell et al. offer a perspective on how to develop transdisciplinary approaches to explore local expressions of agroecology in heterogeneous city landscapes (Chapter 14). They argue that a transdisciplinary approach in urban agroecology must be grounded in participatory action research, a methodology bringing participants and researchers together to set, execute, and refine research objectives (a fantastic deep delve into PAR is also featured in Chapter 13 by Bowness et al.). Such an approach is distinctive in that it has the potential to benefit both academic scholars and community members, creating partnerships that strengthen the field of urban agroecology. Together with with the book contributors, Caswell et al. point to the importance of working both across disciplines *and* outside of academic walls—with urban farmers and city planners and policymakers. Developing these partnerships, for example through cooperative extension efforts (Diekmann & Ostrom, Chapter 11), supports urban agroecological transitions through the act of research transfer. Another way, described by López-García and de Molina (Chapter 9), requires researchers to directly engage with city policy structures to co-develop policies that will secure the future of agroecology in cities.

In this conclusion, we discuss steps to forward urban agroecology as an interdisciplinary and transdisciplinary field. We observe how the chapters herein contribute to our understanding of urban agroecology as more than "urban agriculture," but as a science, practice, and social movement. We summarize and synthesize the unique contributions of each author to highlight key priorities the field must address.

Future Directions for Urban Agroecology

Address Economic and Social Dimensions of Human Well-Being and Health

Much work has been conducted on the social benefits and political currency of urban gardening, but few studies have addressed how to design urban agricultural systems to sustain individual food security. This will involve quantifying gardener yields, identifying markets around urban agriculture, and analyzing the costs incurred to gardeners for supplies and equipment. It will also involve a political economic approach to develop solutions for a just economy around urban agriculture, where there is equal access to important materials such as seed and soil and water. Economic perspectives have a long history in rural agriculture under the field of agricultural economics, but we are still scratching at the surface of what we know about the role of economy (and how it relates to society, politics, and ecology across local and global scales) in urban agricultural spaces. To realize the potential of urban agriculture, we need to work towards equitable ecosystem service provision, land access for all, and social-environmental justice.

To improve human well-being, our goals must also interface with ecological research that identifies practices that increase ecosystem services important for people: we need more work understanding how horticultural practices improve soil health, promote pest control, and protect water resources, while maintaining productive urban garden spaces. Urban gardeners are not *inherently* growing sustainability just because they work on a small scale, though this is often assumed. Indeed, many gardeners rely on heavy usage of water, agrochemicals, conventionally produced seed and starters, and imported soil substrates from extractive industries (with peat moss, perlite, vermiculite, coco coir and wood substrates, gardeners have more additives available to them than ever before). In rural agroecology, we know a lot about how strategies like cover cropping, crop rotation, IPM, organic production, and irrigation strategies affect system sustainability, but have made little strides to translate this knowledge into on-the-ground practices that make sense in an urban sphere. Philpott et al. (Chapter 2) offer an in-depth discussion on how the state of urban agroecology differs from rural agroecology, and suggest ways we can frame ecological research questions towards comprehensive, comparable studies. Let us continue to refine urban horticultural practices, and to disseminate this information to urban growers.

Integrate Participants in Research, From Design Through Dissemination

Urban agriculture offers a perfect system to involve diverse participants in research because practitioners often represent the human diversity of urban areas, and involving participants in the design and undertaking of research can promote more sustainable urban agriculture management through experience and learning. Research approaches shaped by civic ecology frameworks can better incorporate science, stakeholder experience, and policy for "action-based" research. Urban gardeners and urban agroecology stakeholders often include community members, local Cooperative Extension specialists and agents, and organizations that work to promote food justice or monitor biodiversity. There are several forms of participant research, with varying levels of engagement and configurations of both community participation and professional direction and guidance. One such form is participant action research (PAR), discussed in Chapter 13, 14, and 15. Another form, often referred to as "citizen science," entails recruiting members of the general public to conduct scientific work in collaboration with (or under the direction of) scientists and scientific institutions (mentioned in Chapter 12). While PAR often requires long-term research development and high investment from both researcher and community partners in a scaffolded knowledge co-production framework, citizen science projects are often shorter-term projects, already organized by researchers and guided explicitly by the interests of academia. Both such participatory research approaches in 'civic science' integrate the realms of science, engagement and education for promoting (ideally) scientific discovery and social change.

Participatory approaches face several challenges. For citizen science research, data accuracy has often been called into question. However data produced by citizen science can be cost-effective and high quality when paired with proper training, and can help garner community support for science. In PAR, practitioners need increased academic recognition and research funding. This is because the iterative nature of PAR necessitates funding timelines that span longer durations needed for participatory research across communities. Currently, there are few academic and government incentives for long-term, collaborative research between practitioners and researchers (Arnold et al. Chapter 15). But integrating participants in research, using PAR in particular, has the capacity to democratize science through knowledge co-production by scientists and stakeholders. Doing science for stakeholders, by stakeholders, means that the knowledge generated can be directly applied by the practitioners involved. This is illustrated by Diekmann and Ostrom (Chapter 11), who introduce a US-centered approach in which government-supported Extension Services use dissemination strategies to bridge academic research with local communities towards developing urban food production networks. Both Diekmann and Ostrom (Chapter 11) and Galvis Martinez et al. (Chapter 10) point to the importance of engaging the participation of low-income and minority groups that are disproportionately represented in urban areas in urban agroecology programs. They describe educational frameworks rooted in social justice, humanistic values, and indigenous knowledge that promote a more equitable and sustainable urban agroecology. Together, these chapters point to the necessity and nuances of inclusive, deliberate, and meaningful community engagement.

Address Global Environmental Change

Under climate change, variability and extremes in temperature and precipitation in cities will have a range of effects on plants, animals, and related ecosystem services. Egerer (Chapter 3) describes future directions for climate change research and suggests we are only beginning to understand the role of urban agroecology in adapting to climate change impacts. For example, future research could address the effects of urban heat on community assembly of organisms important for ecosystem function and urban food security, such as pollinators. As urban heat extremes will heighten in some cities, the availability of urban water will also likely change, requiring investigations on soil and adaptive management to conserve water. Ultimately, an important focus should be on how to design urban gardens and farms to act as a source of resilience within changing city climates. This means understanding climate change, landscape ecology, and geophysical sciences, but also examining the role of social behaviors, politics, and geography in shaping city climates and agroecosystems therein.

Bridge the Urban-Rural Divide Around Public Interest in Agriculture

Urban agriculture has boomed in popularity in recent years in response to a variety of stressors. These include shortages of arable land, dissatisfaction with the diversity of commercially available food, and disconnection with our food. Urban agriculture has also been offered as an antidote to the dangers and corruption of industrial agriculture, including agrochemical exposure, pollution, and capitalism. While Morales et al. (Chapter 7) describe how the act of urban gardening is political expression in the face of industrial agriculture, one challenge still facing urban agroecologists is using urban agriculture to bridge the disconnect between the urban and rural production spheres. Beyond individual practice, we must scale political action to legitimize agroecological approaches to both urban and rural production. As urban agriculture grows and develops, urban agroecologists should leverage this momentum towards system-wide food change. For Glowa and Roman-Alcala (Chapter 8), land access is a key component of broader food systems change. They map differing articulations of land politics around urban agriculture in the U.S., demonstrating how urban agriculture projects must subvert, rather than reproduce, existing norms around property relations in urban areas. We need to extend these analyses to rural spaces to understand the legacy of land-use politics and to restore indigenous and popular access to the earth. A different, comprehensive and multi-pronged approach is described by López-García and de Molina in Chapter 9 to harness transformative movements in urban spaces towards sustainable transitions outside of the city. This approach entails the mobilization of participants from multiple backgrounds, including women, new farmers, and underrepresented groups towards both a) grassroots movements to spread the use of agroecological practices and b) placing pressure on policies, institutions, and markets through public policy approaches.

Unify Multiple Fields of Study and New Methodological Approaches

Urban agroecology must integrate multiple fields of study and harness diverse methodological approaches to elucidate agroecosystem complexity. Indeed, many chapters in this book argue or demonstrate that urban agroecology requires inter- or transdisciplinary research. Yet, while it is easy to identify this gap, this type of research is difficult to conduct in practice. These challenges are described by Arnold et al. (Chapter 15). Nevertheless, strides should be made to integrate approaches. Transdisciplinary research can explicitly recognize and incorporate complex problems within the research and develop knowledge and practices for on-the-ground change. While the complexity and context-dependence of trans- and interdisciplinary research seems to defy a single definition or template for the research, there are some long to short term goals that researchers can aim towards. The long-term goals that researchers should take include: reconfiguring academic institutions to allow student training in multiple fields, incentive faculty through hiring incentives and tenure, broker collaborations between different scholars (often housed in different parts of campuses), and develop a communal lexicon to synthesize mutual understanding of theoretical concepts, models and products from across disciplines. Smaller, short-term goals towards achieving transdisciplinary research include: creating working groups that can meet often and virtually and identifying sources of funding for collaborative work.

We need new methodological approaches in urban agroecology that integrate, for example molecular methods, public health frameworks, and new conceptualizations of land and cultivation. For example, consider Heath et al.'s (Chapter 1) contribution to our understanding of how urban agroecology supports biodiversity. Heath et al. summarize studies around the world that document urban biodiversity and identity key habitat features for biodiversity maintenance. Next, we should aim as a field to complement this work with gene and micro-level approaches to examine largely "invisible" biodiversity this is critical for ecosystem processes. This includes soil microbes, plant pathogens, and insect symbionts that structure the functioning of agricultural systems. Public health approaches, as described by Marsh (Chapter 6), could be harnessed to complement social and ecological studies of urban agriculture design and benefits. Marsh does this well by showing how small-scale urban agroecological communities can regenerate human health and remedy cultural health inequality in the Tasmanian food system. Finally, we need to extend the scope of what urban agroecology entails. Fischer et al. (Chapter 5) reconceptualizes urban agroecology by examining the drivers and impacts of urban foraging, a practice not traditionally associated with agroecosystems but that plays a key role in urban food systems. In the methodological sections of this edition, Ong and Fitch (Chapter 12) provide novel recommendations on methodological tools that can guide ecological investigations in urban agroecology. In Chapter 13, Bowness et al. highlight how diverse methodological toolkits can be assembled to change social and power dynamics.

Final Thoughts

These research goals will forward urban agroecology and sustainable urban agri-food systems in cities across the world. New research priorities will also certainly evolve over time with changing city landscapes, social needs, policy, and governance. We hope that the collection of work within this book will inspire research in urban agroecology that will provide solutions to the challenges and opportunities of growing food in our future cities.

Index

Note: Locators in *italics* represent figures and **bold** indicate tables in the text.

D

E

J

K

L

M

W

X

Y

Z

Printed in the United States
By Bookmasters